PARADIGMS
LOST

PARADIGMS LOST

Tackling the Unanswered Mysteries of Modern Science

JOHN L. CASTI

AVON BOOKS ◆ NEW YORK

Permissions on page 567

AVON BOOKS
A division of
The Hearst Corporation
105 Madison Avenue
New York, New York 10016

The William Morrow and Company edition contains the following Library of
Congress Cataloging in Publication Data:

Casti, J. L.
  Paradigms lost : images of man in the mirror of science / John L. Casti.

  p. cm.
  Includes index.
1. Science—Miscellanea. I. Title.
Q173.C35 1989 500—dc19   88-37232
                                                                  CIP

First Avon Books Trade Printing: November 1990

To anyone who's ever wondered "Why?"
and especially to those visionaries whose work described here
takes us well down the road to "Because!"

PREFACE

Quo Vadimus? The eternal question: "Where are we going?" In more colloquial terms, we might ask, "Where are you coming from?" And if we're in a contemplative mood, we might even vary the basic theme and extend our inquiry to arrive at that deepest question of all speculative thought: "What is the true nature of mankind?" Fundamentally, this is a book putting forth science's best guesses regarding the ways to assemble the pieces of this ultimate, eternally difficult, and ever-tantalizing puzzle. More precisely, each of the scientific stories I've chosen to tell here addresses in its own characteristic way the issue of the uniqueness of mankind, in our lives here on Earth, our place in the galaxy, and even our role in the universe at large. In short, our most basic concern here is to explore what science has to say about the perpetually elusive question "Is there anything special—or unique—about human beings?"

Eternal questions have a nasty habit of remaining eternally impenetrable when left on the lofty plane of philosophical discourse. Consequently, I've tried to decompose the "uniqueness of mankind" question into a set of bite-sized, individually digestible pieces involving our human (1) physical and biochemical structure, (2) social behavioral patterns, (3) linguistic communication capabilities, (4) cognitive thought processes, (5) presence in the galaxy, (6) role as observers in the universe. Each of these aspects of our lives and activities is paired with what I think of as one of the Great Problems of modern science: the origin of life, sociobiology, language acquisition, thinking machines, the search for extraterrestrial intelligence, and quantum reality. As Francis Crick once remarked in a similar context, "To show no interest in these topics is to be truly uneducated," a good example of the well-known Crick irony. Personally, I would vary Crick's thesis a bit by saying that to show no interest in these topics is to be *uninformed* about the true nature and beauty of the problems. My hope is that by telling the story of where science stands today on each of these problems, it will be possible to shed light on the more general question of where *Homo sapiens* as a species fits into the cosmic scheme of things. I would also hope to make a small contribution to education by displaying a few of the fascinating interconnections between these seemingly diverse pebbles strewn about on the seashore of science.

One of the more deceptive aspects of a scientific research article is that the path of development from hypothesis to conclusion as traced in the paper is almost never a faithful account of how the results were really obtained. So it is with this book as well. The noble theme of our special position in the cosmos never entered my mind when I set forth on this project. My original aims were far more modest, involving little more than trying to trace for myself and my students the multiple threads of a number of interesting questions scattered across the landscape of modern intellectual thought. It was only after I began to gather together these individual strands that it dawned on me that the work's overall theme is really what I've grandly termed the uniqueness of Mankind.

Originally this book came about as the outgrowth of another work of mine, *Alternate Realities: Mathematical Models of Nature and Man* (New York: Wiley, 1989), which is a semitechnical textbook on the modeling of natural and human systems. While preparing that volume, I had occasion to wander over a pretty diverse mindscape of topical areas ranging from chaos, game theory, catastrophe theory, and cellular automata to their applications in physics, engineering, evolutionary biology, and cognitive psychology, as well as ecological and economic cycles and beyond. More than ever before, this venture brought home to me the heretofore unappreciated fact that virtually all knowledge is intertwined at some level in nontrivial ways, and that these interrelationships are important enough in their own right that they should be included in the course programs of all serious students and aspiring researchers. This conviction led to my resolve to put together a course of lectures for the general science student. These lectures focused on several topical areas in modern intellectual thought in which the central problems lie in that no-man's-land between the boundaries of the classical disciplines. It is these Great Problems—the origin of life, sociobiology, quantum reality, and all the rest—that form the heart of this book. As a result of the overwhelmingly encouraging response to the lectures, I felt emboldened enough to undertake the task of trying to put the competing ideas, approaches, and personalities down on paper in a form accessible not just to university students, but to the proverbial educated layman as well. The result is the book you now hold in your hands.

When reflecting on the volume's overall structure, I'm continually reminded of one of those frosted layer cakes that certain Viennese pastry shops of my acquaintance specialize in serving up to their gluttonous clientele. The deepest layer running through every chapter of the book is the eternal question discussed above: How special are humans, either here on Earth or in the universe? Set on top of this delectable base is a second layer consisting of the individual stories taken on their own merits: How does science come to its conclusions about what is "true"? How did life get its start here on Earth? Are our social behavioral patterns "programmed" into our genetic makeup? What is the mechanism by which we learn to speak? Can machines think? Are there other intelligent beings in the Milky Way? Does reality itself require our presence as observer/participants? Even without the underlying philosophical layer, the individual stories are sweet and juicy enough in their own right to provide anyone with a tasty, intellectually fattening treat. Finally we have the frosting on the cake: the scientists themselves seen in all their glory (and some of their frailty, too) as they act to close the circle of self-reference in their singular role as humans investigating humanity. Taken as a whole, this particular layer cake is, I think, one that any of my favorite *Konditoreien* would proudly feature on its menu of irresistible attractions.

For the sake of exposition, I use the format of a jury trial to present the competing positions on each of the topical issues of the book. Consistent with this courtroom motif, each chapter begins with a "claim" phrased to represent the Prosecution's charge. The negation of that basic claim constitutes the position of the Defense. Following the customary sequence for jury trials, each chapter proceeds through opening statements, witnesses for the Prosecution and Defense, testimony from expert witnesses, summary arguments, and finally the verdict. In this last connection, I step out of my role as court reporter, don the hat of a typical member of the jury, and try to assess the merits of the competing arguments from the position of an uninvolved, but keenly interested, neutral observer. It's my hope and expectation that each reader will also serve as a member of this jury, coming to his or her own conclusions at the end of the competing arguments.

In attempting to address such a wide array of topics within

the confines of a few hundred pages, compromises necessarily had to be made. On the one side, to do justice to the ideas, arguments, and genius of the various scholars, I have perhaps described some of the material in a bit more detail than the average reader might care to confront head on. But if you find yourself starting to lose sight of the forest for the trees, don't despair. To help you stay in the game, I have employed several types of attention-focusing devices. First of all, in each chapter where the sheer weight of terminology begins to become burdensome, I have inserted a terminological table at a strategic location early in the proceedings. This table can be used as a convenient point of reference for the nomenclature as you wend your way through the arguments that follow. But the arguments themselves are not of uniform difficulty, either in the concepts they expose or in the twists and turns of their logic. Consequently, I have provided a variety of amplifying remarks in the notes and references for each chapter, material to which the reader can turn for further elaboration of some of the trickier-than-average passages of the chapter itself. Finally, each chapter is liberally sprinkled with a number of figures and diagrams that I trust will illustrate the main points far more efficiently and clearly than any amount of prose ever could. It is hoped the combination of these various devices will enable the general reader to stay afloat while navigating through the more dangerous rapids of our fast-flowing stream of knowledge.

At the other end of my potential reader spectrum are professional researchers and students. To these experts I offer my sympathies for what must appear at times to be gross caricatures of their beloved disciplines. My only defense is that such an approach is necessary in a broad, general treatment of this sort. As partial compensation, I trust that the admittedly incomplete treatment of the expert's territory given here will at least bring that territory to the attention of a wider audience, thus focusing a few rays of the public spotlight where it might do some good. Finally, I should recall here the fact that the book originally arose out of a course of lectures for both university students *and* faculty. These lectures were somewhat more technical and detailed than the treatment in the book, containing far more material from the professional literature, more mathematical pyrotechnics, more finely detailed arguments, and so forth. For those readers who want to examine this additional material or, perhaps, use this book as the basis for a lecture course of their own, I will be happy to provide my raw research notes, containing many additional references and sidelights that for various reasons didn't find their way into the book itself. Readers wishing to obtain this material should contact the author c/o Institute for Econometrics, OR, and System Theory, Technical University of Vienna, Argentinierstrasse 8, A-1040 Vienna, Austria.

So on balance, in attempting to navigate the fine line between boring the professional and overwhelming the layman, I tried to follow here what I think of as the Three-*E*'s-Minus-One Rule: make the book *e*ducational, *e*nlightening and *e*ntertaining without making it *e*ncyclopedic. As Anatole France once remarked, "I prefer the errors of enthusiasm to the indifference of wisdom." But as always, I'll let the reader be the final judge of the degree to which I have succeeded in walking this tightrope between triviality and impossibility.

A quick peek at the Contents will probably generate the impression that each of the book's chapters is an independent module that can be read without reference to any of the others. To confirm your suspicions, this is indeed the case. I had two considerations in mind when structuring the book in this way. The first was to reshuffle the deck every now and then, so that if you run hopelessly aground somewhere along the line, salvation is no more than a few pages away. And second, while both Francis Crick and I would find it hard to understand how anyone could fail to be interested in every one of the topics dealt with here, empirical observation forces me to the unhappy conclusion that this really could happen—such people do indeed exist! So if your tastes run toward extraterrestrial life and you couldn't care less about the genetic determination of human behavior, you may with confidence proceed directly to Chapter Six. Or if you're worried about a thinking machine's taking over your job (or your life), you may safely skip our deliberations on the origin of life and move with all due dispatch to Chapter Five. Without exception, each chapter is totally independent of the others, and you won't be hindered in the slightest if you just open the book at random and start reading.

But while I'm dispensing this largesse, let me introduce a precautionary note as well. If you want to get right down to the fundamental layer of the cake constituting the entire book and learn about the uniqueness of mankind, then the more chapters you read, the better position you'll be in to understand the many facets of the problem and the truly staggering magnitude of the task involved in providing even a partial answer. Consequently, if it's the essence of humanity you're after, my recommendation is at least to skim *all* the chapters. Some of them, like the chapter on thinking machines or the one on life's origins, involve slightly more abstract notions and hence are probably a bit tougher sledding than the average reader might want to enter into immediately. Nevertheless, each chapter is a piece in the mosaic of mankind, and to see the Big Picture you need to know at least something about the Great Problems—all of them. So skim if you must, but do so at your own peril.

A last bit of advice on reading the book: Don't start with Chapter One! I'm sure this admonition would strike my old high-school English teacher as nothing short of sheer heresy. Nevertheless, there is at least a little method in this seeming madness. I have chosen the ordering of the book's chapters to reflect a certain progression from life to behavior to mind, from Earth to galaxy and beyond. The opening chapter is designed to provide the philosophical and sociological underpinnings to the scientific doings recounted in following this progression. So from a logical standpoint, the chapter ordering is airtight and almost foreordained. However, experience shows that most people are like me when they get a new toy (or computer program): They want to start playing with it right away. And the last thing they want to do is read the instruction manual cover to cover before they begin having some fun. So think of the first chapter as constituting the book's instruction manual. But since we all know that you can have lots of fun without knowing the rules (or, at least, without following them), my advice is initially to pick out one or two of the topical chapters that capture your fancy. After digesting this material and getting a feel for how science operates in practice, you can then go back and compare how things *really* work with the way theory and armchair speculation say they should.

Several months ago during the course of discussing this project with a colleague, I made the offhand remark that I certainly hoped that the book would turn out to be a success. Unfortunately, he isn't the type of friend to let me get away with any such throwaway remark. "So what is your personal criterion for success?" he asked. Resisting the natural impulse to say sales of a hundred thousand copies (or more) on day one, together with glowing reviews in all the right places, I finally replied that I would consider the whole effort to have been worthwhile if I sat next to someone on a long flight who was reading the book, and at the end of the flight this nameless companion turned to me and asked, "Have you read this book?" At this moment, disavowing any knowledge of the book, I would hope to hear the magical words "Well, I recommend it highly. Not only did I learn something I didn't even know I was interested in, but I had fun doing it." Happily, this is still my principal criterion. So if I chance to drop into the seat beside you on my next flight, and you enjoyed reading the book as much as I enjoyed writing it, then perhaps . . .

JLC
Vienna, 1989

ACKNOWLEDGMENTS

There are two characteristics that every inhabitant of that vast universe of books seems to share. The first is the appearance of embarrassing typos, literary gaffes, and conceptual errors that no author's or editor's brand of "weedkiller" ever seems able to eradicate completely. The second is the presence in the book of the hearts, hands, minds, and souls of others. Like all authors, I hope that this book will be the exception that proves the rule for the first universal property, but I'm not placing any bets on it. As to the second general feature, it pleases me greatly to announce that this book is no exception. I have been luckier than most in having had the benefit of the support, encouragement, opinions, advice, and even services of a large number of people without whom this project would still be languishing in that shadowy world of ideas that almost were but aren't. So it's both a pleasure and a privilege for me to bring these unsung heros to the reading public's attention here.

Beginning this roll call of honor, the following hardy souls have through the years acted as sympathetic ears, as well as intellectual inspirations, in conversations ranging over the topics of this book and much, much more. In addition, many of them served as willing guinea pigs for a critical reading of one or more preliminary versions of the chapters of the book. So, in no particular order, I thank Karl Sigmund, Clint Perkins, Amy Okuma, Manfred Deistler, Gustav Feichtinger, Lucien Duckstein, Mel Shakun, Jesse Ausubel, Mary McCusker, David Berlinski, Hugh Miser, Nebojsa Nakicenovic, and Peter Schwed.

In a book of this sort, keeping the technical details straight is a job for three men and a boy, not to mention a massive computer database. For their valiant efforts to keep me on the straight and narrow, technically speaking, I am indebted to Professors John Bell, Michael Hart, David Lightfoot, Robert K. Merton, Michael Ruse, Abdus Salam, John Searle, Robert Shapiro, John Maynard Smith, and John A. Wheeler. Of course, these thanks are accompanied by the customary absolutions for whatever errors of fact and/or interpretation that remain.

For specific help far beyond the call of duty, let me also bow deeply and tip my hat to:

Eddy Löser, librarian *extraordinaire,* whose genius in tracking down important, but seemingly inaccessible, references accounts for the unseemly length of the book's bibliography;

Paul Makin, maestro of the computer terminal, who taught me what little I know about the ways of computers and their virtues (and vices) for writing a book;

John Ware, the kind of literary agent every author dreams of: one who believes in, continually encourages, and works tirelessly for his clients;

Bruce Giffords, a copy editor with an eye like a hawk, a mind like an encyclopedia, and a heart like a lion. If you actually *understand* this book, he's the reason; if not, blame the author;

Alex Grey and Dolores Santoliquido, artists of rare perception and talent, whose creations grace the dust jackets (AG) and pages (DS) of the book, illuminating that which was only darkness before;

Maria Guarnaschelli, the kind of editor every author dreams of: one who not only protects authors from themselves, but does it with such grace, humor, talent, artistry, and skill that the author can still write *his* book;

Peter de Janosi, my most faithful reader and perceptive critic, as well as a quintessential role model for that hardy but vanishing breed, the proverbial educated layman;

Joe Tabacco, Peggy Schmidt, and Teddy Tabacco, friends who not only provide the most congenial of environments for an itinerant visitor and the expression of his outlandish opinions on science and life, but who sometimes even agree with him.

To all these long-suffering friends, my thanks and appreciation for their many contributions reflected in almost every page of this work.

Finally, and most important, heartfelt thanks to my wife, Vivien, not only for her constant encouragement and support in all the usual ways too numerous to list, but especially for not asking to look at the manuscript of this book until it was too late to do anything about it.

CONTENTS

PREFACE   vii

ACKNOWLEDGMENTS   xv

1 / FAITH, HOPE, AND ASPERITY   1
BELIEF SYSTEMS, SCIENCE, AND THE
INVENTION OF REALITY   1
WORLD VIEWS IN COLLISION   1
DID YOU SAY SCIENCE?   10
THE NATURAL PHILOSOPHER'S STONES   15
RATIONALITY FOR REALISTS   26
BUDDY, CAN YOU PARADIGM?   38
PHILOSOPHICALLY SPEAKING   46
A TALE OF TWO SUICIDES   48
ON THE FRINGE OR AT THE CUTTING EDGE?   56
THE PULPIT AND THE LAB   62
INTO THE COURTROOM OF BELIEFS   66

2 / A WARM LITTLE POND   68
CLAIM: LIFE AROSE OUT OF NATURAL
PHYSICAL PROCESSES TAKING
PLACE HERE ON EARTH   68
OUT OF THE FIRE AND INTO THE SOUP   68
A CRASH COURSE ON HOW LIFE LIVES   74
POTHOLES ON THE ROAD TO LIFE   84
MONSTERS, HYPERCYCLES, AND NAKED GENIES   88
THE CHICKEN'S STORY   95
LIFE: A TWICE-TOLD TALE   100
ASHES TO ASHES, LIFE FROM DUST   108
IT CAME FROM OUTER SPACE   115
AND GOD CREATED...FROM FISH TO GISH   121
THE LOGIC OF LIFE   127
SUMMARY ARGUMENTS   139
BRINGING IN THE VERDICT   140

3 / IT'S IN THE GENES 143
CLAIM: HUMAN BEHAVIOR PATTERNS ARE DICTATED PRIMARILY BY THE GENES

NATURE/NURTURE: SENSE OR NONSENSE? 143
NEO-NEO-DARWINISM AND SOCIOBIOLOGY 147
ANIMAL ANTICS 155
THE STRANGE CASE OF ALTRUISM 171
THE GENETIC IMPERATIVE 173
GETTING INTO HER GENES: SEXISM AND SOCIOBIOLOGY 182
CANT VS. KANT 186
SO-SO BIOLOGY 192
CONFLICTING RATIONALITIES AND THE
  DILEMMA OF COOPERATION 198
SUMMARY ARGUMENTS 203
BRINGING IN THE VERDICT 205

4 / SPEAKING FOR MYSELF 209
CLAIM: HUMAN LANGUAGE CAPACITY STEMS FROM A UNIQUE, INNATE PROPERTY OF THE BRAIN

DUMB DOGS AND CLEVER HANS 209
VERBAL BOTANY AND UNIVERSAL GRAMMAR 213
THE NOAM OF CAMBRIDGE 218
POSITIVELY REINFORCING 232
OUT OF THE MOUTHS OF BABES 237
IT'S ALL A QUESTION OF SEMANTICS 241
SHOOT-OUT AT THE ROYAUMONT CORRAL 249
RULES AND REPRESENTATIONS 253
SUMMARY ARGUMENTS 257
BRINGING IN THE VERDICT 258

5 / THE COGNITIVE ENGINE 261
CLAIM: DIGITAL COMPUTERS CAN, IN PRINCIPLE, LITERALLY THINK

THE TURING TEST AND THE CHINESE ROOM 261
FORMAL SYSTEMS, MACHINES, AND TRUTHS 268
"STRONG" VS. "WEAK" AI, BRAINS, AND MINDS 285
TOP-DOWN SYMBOL CRUNCHING 290
BOTTOM-UP EMERGENCE 299
PHILOSOPHERS AGAINST: THEY'LL NEVER THINK! 314
THE MORALIST AND THE MYSTIC 324
SUMMARY ARGUMENTS 330
BRINGING IN THE VERDICT 332

6 / WHERE ARE THEY?   340
CLAIM: THERE EXIST INTELLIGENT BEINGS
IN OUR GALAXY WITH WHOM
WE CAN COMMUNICATE

THE FERMI PARADOX AND PROJECT OZMA   340
THEORETICAL ETI: THE DRAKE EQUATION   343
SLICES OF THE ETI PIE   345
ANTHROPOMORPHISMS, CHAUVINISMS, AND
 ETI NUMEROLOGY   362
EXPERIMENTAL SETI: HOW SHOULD WE LISTEN?   368
WHAT ARE WE LISTENING FOR?—THE SYNTAX AND
 SEMANTICS OF SETI   373
$N > 1$: ETI EXISTS!   387
THE SHAPE OF ETIS TO COME   391
ETI? THERE'S NO SUCH THING: $N = 1$ 397
SUMMARY ARGUMENTS   409
BRINGING IN THE VERDICT   411

7 / HOW REAL IS THE
''REAL WORLD''?   414
CLAIM: THERE EXISTS NO OBJECTIVE REALITY
INDEPENDENT OF AN OBSERVER

BUILDING THE STAGE   414
GHOSTS IN THE ATOM   417
MEASUREMENT TO MEANING   429
THE ROMANTIC REALITIES   441
THE DOGWORK REALITIES   456
THE BELL TOLLS FOR LOCALITY   467
IN THE BEGINNING, THE VERY BEGINNING   476
SUMMARY ARGUMENTS   488
BRINGING IN THE VERDICT   489

CONCLUSION /
THE BALANCE SHEET   492
ARE HUMANS REALLY SOMETHING SPECIAL?

WHERE DO WE STAND?   492

TO DIG DEEPER   500

INDEX   555

1

FAITH, HOPE, AND ASPERITY

===

BELIEF SYSTEMS, SCIENCE, AND THE INVENTION OF REALITY

WORLD VIEWS IN COLLISION

On the night of February 24, 1987, Canadian astronomer Ian Shelton was looking through the telescope at the Las Campanas Observatory in Chile; what he saw became *the* scientific event of the decade in the astronomical world. On that night, Shelton became the first to see the star Sanduleak −69° 202 come to the end of its cosmic tether in that most spectacular of celestial fireworks displays, a supernova. According to current astrophysical wisdom, such events occur when the hydrogen that fuels the thermonuclear furnaces of stars a little bigger than our sun runs out, allowing the contracting force of gravity to gain the upper hand over the expanding forces of thermal radiation. The star's mass then collapses in on itself until the pressures build to the point where the star literally blows its top, scattering most of its mass into the interstellar void, leaving behind a small, rapidly spinning ball consisting solely of neutrons at an incredibly high density. In fact, so dense is the material of such a "neutron star" that one cubic inch of it would weigh more than a billion tons, and a pinhead's worth several million. Although many supernovas have been seen in distant galaxies, the importance of supernova 1987A was twofold: It was the first time that astronomers had extensive observations of a star before it became suicidal, and it happened in the Large Magellanic Cloud, a galaxy "only" 170,000 light-years distant—essentially next door on the astronomical scale of things. While supernovas have been observed from Earth for centuries, going back at least as far as the Chinese accounts of what is now the Crab Nebula in A.D. 1054, observation of their neutron star residue dates back only a few years and constitutes one of the major science stories of the 1960s. Since the discovery of these neutron stars or, as they as more colloquially termed, *pulsars* (for "pulsating radio sources") serves as an admirable case study of the ways of science in the late twentieth century, let's climb into a time machine and go back to those exciting days to retrace the steps leading to this momentous discovery.

The story begins in 1965 with the decision by Jocelyn Bell, a young woman from Northern Ireland, to seek a doctorate at Cambridge University in the then-new field of radioastronomy. As Bell (now Jocelyn Bell Burnell) tells it, she had become fascinated with astronomy as a young girl when her architect father was hired to design the observatory in the small Irish town of Armagh. Unfortunately, even then she saw that a necessary condition for successful pursuit of the astronomer's nocturnal art is to have a night owl's constitution, easily being able to interchange the normal hours for sleeping and working. Despite her passion for the stars, in the 1950s her constitutional need for a good night's sleep at the normal hours looked like a fatal obstacle to any budding astronomical aspirations. But as luck would have it, this was the time when Martin Ryle of Cambridge was developing one of the first telescopes devoted to searching the skies in the radio rather than visible light part of the electromagnetic spectrum. Since the best time for "seeing" at these frequencies is during the daylight hours, Cambridge was the place for her, and off she went armed with an undergraduate degree in physics to work for her Ph.D. in a group led by Anthony Hewish.

One of the most sacred rules of academic institutions everywhere is that the graduate students perform the slave labor, the Cambridge Institute of Theoretical Astronomy being a staunch upholder of this venerable principle. Consequently, Bell spent her first two years as a graduate student wielding a 20-pound sledgehammer, helping to construct the radiotelescope that she would later use to gather the material for her doctoral dissertation. Following completion of the telescope in 1967, team leader Hewish assigned Bell the thesis topic of measuring the angular diameter of radio galaxies (*quasars*) from the way their signals "twinkled" when seen from Earth due to the solar wind of material emitted from the Sun. Her job was to operate the telescope singlehanded and analyze the output until she accumulated enough data for a respectable thesis. Since the telescope spewed out 96 feet of three-track paper each day and covered the entire sky in four days, Bell's data analysis activity was hardly less energy-intensive than building the telescope itself, involving as it did eyeballing the telescope record and separating the wheat of true twinkling signals from the chaff of French television, military radar, aircraft altimeters, and other Earth-based sources of interference. The telescope was turned on in July 1967 and, not surprisingly, by October she was already 1,000 feet of chart paper behind. It was at this point that the fun, both galactic and earthly, began.

In one of the 400 feet of chart readings produced with each scan of the sky, Bell noticed that there was about half an inch of what she termed "scruff" that resisted classification. She saw that the scruff was neither twinkling or man-made interference, and then recalled having seen similar patterns before on another record from the same part of the sky. Furthermore, she noticed that the mysterious signals seemed to be appearing periodically on sidereal time of twenty-three hours, fifty-six minutes, i.e., the time needed for a given location on Earth to return to the same position relative to the fixed stars (the sidereal day is four minutes shorter than the terrestrial day due to the Earth's orbital motion about the Sun).

At this juncture Bell discussed the signals with Hewish, and they decided to look at them again on a faster recorder that would allow them to pick out more detail. This recorder was occupied at the moment, so they had to wait until mid-November to make the new reading. As so often happens in life, just when you want a taxi (or a cop) there's not one to be found anywhere; astronomical anomalies are similar, and Bell had to wait several weeks before she could reacquire the odd signal. Imagine her surprise when she finally found it again and discovered that it was pulsating at the metronomic rate of almost *exactly* 1⅓ seconds. She immediately phoned Hewish, who promptly dismissed the signals as man-made in light of their extreme regularity. However, an Earth-based source would keep terrestrial time, not sidereal, casting a very dark shadow over Hewish's offhand conclusion. But the fastest variable star then known had a period of one third of a *day,* and it was difficult to conceive of what kind of star would rotate in little more than a second.

The first attempt to reconcile these conflicting facts was to conjecture that the observations were radar signals bouncing off the Moon, or a satellite in an odd orbit. But such an explanation didn't wash, and since only astronomers and the stars keep sidereal time, Hewish thought that perhaps some other observatory had a program under way that would account for the unusual signals. His queries to other radioastronomers turned up no such program. The next trial explanation was the LGM Hypothesis, postulating that the signals were intelligent communications from "little green men." As a test of this conjecture, Hewish calculated the Doppler shift of the pulses assuming that the LGM would be on a planet, and that the planet's orbital movement around its star would create a clustering of the pulses as the planet moved toward Earth and a spacing-out of the signals as it moved away. This explanation also came a cropper when the only Doppler shift noted was that due to the Earth's motion around our sun. At this point, theory gave way to another observation, which definitively settled the matter.

Just before leaving for her Christmas holiday in December 1967, Bell was working late one night analyzing a record from a different part of the sky. She noticed some more scruff that looked remarkably similar to that of the LGM signal. As serendipity would have it, the telescope was due to scan that part of the sky again that very night, and she luckily got a strong reading showing an extremely regular train of pulses coming in at the rate of about 1¼ seconds per pulse. Since another rule of graduate student life is that you don't telephone your professor at 3 A.M. (at least you don't if you value finishing your degree program), Bell just dropped the recording on Hewish's desk with a note asking him to keep the recorder going over the vacation period, and left for her holiday. Hewish himself then made a recording in mid-January confirming the second source, thereby removing the LGM hypothesis from further consideration on the grounds that it was extremely unlikely that there could be two groups of LGM trying to signal us on different frequencies at the same time. So when Bell returned from her Christmas break, she had two important problems to deal with: (1) there was more than one pulsar, and (2) it was time to start writing up a thesis describing her original work on the angular diameter of quasars (although it ultimately contained an appendix describing the pulsar observations, too).

Forced into accepting that the sources of these pulses were some sort of stellar phenomena, Hewish, Bell, and three others from the Cambridge team coauthored the first paper on the subject, which was published in February 1968, and which vacillated between identifying the sources as neutron stars and as white dwarfs, the kind of object our own sun will contract into a few billion years from now. Six months later, the astrophysical community accepted Thomas Gold's interpretation that they were neutron stars as being the only plausible explanation fitting all the observations. This proposal followed up a theoretical suggestion that Fritz Zwicky and Walter Baade made in 1934. The general picture of how a neutron star acts to produce the observations seen by Bell and Hewish is shown in Figure 1.1. While the scientific excitement ended here, the story was still far from over.

In 1974 the Nobel Committee awarded its prize in physics for the first time to astronomers, citing Martin Ryle and Anthony Hewish for their "decisive work in the discovery of pulsars." Not a word was said about the actual discoverer of pulsars, Jocelyn Bell! Shortly after the award ceremony in December, another member of the Cambridge astronomical group, Fred Hoyle, said in a speech in Montreal that Bell's findings had been kept secret for six months while her supervisors "were busily pinching the discovery from the girl, or that was what it

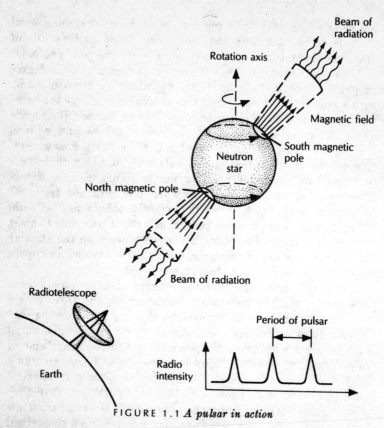

FIGURE 1.1 *A pulsar in action*

amounted to," Hewish admitted that he was "angry" over Hoyle's allegation, calling it "untrue," and noting that "Jocelyn was a jolly good girl but she was just doing her job. . . . If she hadn't noticed it, it would have been negligent." He went on to state that she had made the discovery using his telescope, under his instructions, making a sky survey that he had initiated. Other astronomers were less certain. The historical fact remained that Bell was the first person who had recognized the pulsar signals, and in fact she and Hewish, presumed to have shared equally in the work by the exacting standards of the Franklin Institute's awards committee, were jointly awarded the institute's prestigious Michelson Medal in 1973 for the discovery.

Personally, I've always felt that Hollywood missed a good bet by not putting this story on film, showing an upset, slightly bookish Jane Fonda or Meryl Streep look-alike publicly denouncing a suave, but faintly sinister, James Mason–ish professor on the steps of the Stockholm City Hall for casting her and her contribution aside in pursuit of personal fame and glory. Unfortunately for Hollywood, real life as usual had quite a different ending in mind. In response to the various claims and counterclaims, Jocelyn Bell had the last word when she stated that Hoyle "has overstated the case so as to be incorrect." But still, given the proclivity of the film industry for warping and distorting reality in pursuit of art and entertainment, not to mention hard cash, maybe there's hope yet for realization of my vision. In any case, the entire pulsar episode serves as a sterling example of the bright side of the folkways, mores, and byways of contemporary scientific life. For a look at the dark side, let's return to our time machine and go back a few more years to examine another tempest in the astrophysical teapot.

In the writings of Plato and Herodotus we find the assertion that the Sun now rises where it once set. How could they make such a bizarre claim? And why do so many cultures have legends of global floods, manna from heaven, darkness on the Earth, and other such strange phenomena? In 1950 the Macmillan Publishing Company put out the volume *Worlds in Collision* by a Russian-born psychoanalyst, Immanuel Velikovsky, who purported to explain these and many other phenomena as the result of a series of celestial cataclysms taking place during historical times. This book so enraged the scientific community that Macmillan, under pressure of a boycott of its textbook division, handed the best-selling project over to Doubleday and fired the editor responsible for dealing with the manuscript. It's instructive to examine Velikovsky's claims and methods as an example of the sort of thing that sends the scientific establishment into apoplectic fits.

The gist of Velikovsky's argument is that a large comet was expelled from Jupiter sometime around the year 1500 B.C. This comet passed very close to us, with its tail touching the Earth and causing a rain of petroleum, as well as darkening the sky for several days with its dust and debris. In addition, the Earth's rotation rate was slowed down by the comet, resulting in earthquakes, hurricanes, tidal waves, and a variety of other dra-

matic environmental shenanigans. Electrical discharges between
the Earth and the comet caused a reversal of the Earth's magnetic field, the polar regions shifted, and the Earth's axis of rotation was altered, resulting in a change in the order of the
seasons. Furthermore, the Earth was pushed into a larger orbit,
lengthening the year to 360 days.

Velikovsky correlates this first pass of the comet with the Ex-
odus of the Israelites from Egypt, claiming that the plagues of
blood, vermin, and hail noted in the Bible were the result of the
Earth's contact with the comet's tail. He also explains the part-
ing of the waters of the Red Sea as being due to the stopping of
the Earth's rotation, and that the manna from heaven sustain-
ing the Israelites in the desert was composed of carbohydrates
from the comet. *Worlds in Collision* then asserts a second passage
of the comet fifty-two years later, this time interfering with the
Earth's rotation just at the time when Joshua commanded the
Sun to stand still. And what does Velikovsky say about the identity of this celestial molester? He claims that the comet is now
what we call the planet Venus! But the story doesn't end there.

In Velikovsky's scenario there was another close cometary encounter around the year 800 B.C., this time with the planet Mars.
This near collision knocked Mars out of its orbit, bringing it
close to the Earth on at least three occasions. These near misses
shifted the Earth's orbit even further away from the Sun,
bringing about the current year of 365¼ days. At this point, all
three planets settled into their current positions, thus folding up
the tent on Velikovsky's celestial circus.

One might well inquire as to what kinds of arguments and
methods Velikovsky employed to explain these catastrophic goings-on. Fundamentally, *Worlds in Collision* is based upon ancient manuscripts, legends, and traditions. In a later volume,
Earth in Upheaval, he cites evidence such as the existence of coal
beds in Antarctica, rock formations with reversed magnetic po-
larity, fossil beds containing animals from both desert and forest, as well as other geological and paleontological facts. The
cometary origin of Venus also gave rise to Velikovsky's speculations that Venus was hot and that the material for the comet
had originally been expelled from Jupiter, leaving behind what
we now know as the giant Red Spot.

It probably goes without saying that mainline astronomers,
geologists, astrophysicists, and paleontologists speak with one

loud voice in their condemnation of both Velikovsky's methods and his conclusions. While his work represents an imposing piece of sustained scholarship, there are just too many inconsistencies in far too much of his historical, archaeological, astronomical, and physical data to take the arguments seriously. For instance, while it did turn out that Venus was scorchingly hot, just as Velikovsky had predicted, this is almost certainly due to an atmospheric "greenhouse effect" and not to any kind of cometary origin. Furthermore, the atmosphere of Venus is almost totally devoid of the hydrocarbons that Velikovsky claimed would be found as its main constituents. Moreover, the surface of Venus appears to be over 1 billion years old, instead of just a few thousand years as predicted by Velikovsky. For these reasons and many more, Velikovsky's vision of the solar system has now been relegated to that corner of the scientific attic where sit ancient astronauts, the Piltdown man, phrenologists, astrologers, and all the other playmates of the pseudoscientist.

Despite the truly devastating holes in his theory, Velikovsky died in November 1979 convinced that he had been the victor in his war against the Brahmins of science. And, in fact, his ideas live on to this day in some circles. In our quest here to uncover the essence of what constitutes "scientific" knowledge, it's worth taking a moment to examine the pulsar and *Worlds in Collision* theories as antipodes of the spectrum of what is commonly termed scientific research.

At first glance, there appear to be a number of similarities between the work of Bell and Hewish on pulsars and that of Velikovsky: unexplained astronomical phenomena, conjectures and refutations of various theoretical explanations, a physically unobservable explanation interpreted to fit the observations— even a public controversy over some sociological aspects of the way the world of science goes about distributing its accolades. With these points of contact, why is it that the scientific community chose to reward Hewish with its highest honor, the Nobel Prize, while at the same time vilifying Velikovsky and dismissing him as what could charitably be termed a misguided crank? Just what was it *exactly* about the pulsar work that made it the height of respectability and was so obviously lacking in the efforts of Velikovsky?

The long and proper answer to the question will occupy us for much of the remainder of this chapter; the short answer is that, by common consensus in the scientific community, certain standards have been set for what constitutes acceptable evidence and methods, with the pulsar work adhering to them while Velikovsky's did not. The central point for us in this volume is the degree to which those commonly accepted standards generate real rather than virtual knowledge of the universe *in itself.* Put another way, do the methods and standards of science produce a brand of knowledge that is somehow more certain or of higher intrinsic pedigree than the methods and standards of other seekers after truth like Velikovsky? The first step toward a resolution of this overarching question is to address a different question: Just what does constitute the practice of "science" as that term is commonly used in today's world?

DID YOU SAY SCIENCE?

Back in the days when I still attended cocktail parties, the most awkward situations always arose at those odd moments when the music stopped and social convention dictated that I make some feeble effort to "mix." Generally at these times, life conspired to place me next to some slightly frenetic, upwardly mobile yuppie type suffering from an overdose of adolescent enthusiasm for drinking deeply from the brackish waters of life, not to mention our host's bar. Inevitably such encounters began with the question "What do you do?" Resisting the temptation to reply, "Ah, yes, the eternal question," or give some other equally sophomoric response, in the early going I used to answer honestly that "I'm a mathematician." The reactions to this bit of ill-advised candor fell into one of two categories: a petulant pout followed by the curious compliment that "I was always terrible in math," or what was even worse, a bright smile and the remark "Oh, you'd love my uncle. He's an accountant." Being a slow learner, I needed some time to realize that such frank confessions of professional perversion were not the road to success on the cocktail-and-corn-chip circuit. So I began experimenting with other, less esoteric replies: "I'm an electrical engineer, a chemist, an agronomist ["What's *that?*"], a scientist." The results could hardly have been worse if I'd claimed to have been a psychiatrist, an undertaker, or, heaven forbid, some back-slapping politico type. Finally, I hit upon the winning solution of just

saying that I was an unemployed tennis coach, at which point
my Social Interaction Index shot up like a Minuteman missile.
But the sad conclusion to be reached from this very statistically
insignificant sample is that there is a wide variety of gross mis-
conceptions and nontrivial misunderstandings floating around,
even among the educated public, as to the nature of both scientists and the ways in which they spend their days (and nights!).

Trying to distill the essence out of the aforementioned encounters, I eventually came to the surprising realization that the
term *science* seems to be used interchangeably in general conversation in at least three quite distinct and inequivalent ways:

Science =
- a set of *facts* and a set of *theories* that explain the facts
- a particular *approach,* the scientific method
- whatever's being done by *institutions* carrying on "scientific" activity

As a general rule, the nonscientific public usually opts for the
third interpretation, occasionally the first, but virtually never
the second—just the opposite ordering from that given by the
scientific community itself. It's no wonder C. P. Snow could develop a lengthy essay on the "two cultures."

The fundamental misunderstanding on the public's part of
what constitutes a "scientific" activity gives rise to an array of
subsidiary misperceptions about the goals of science and the way
scientists go about their business of trying to achieve them. Let
me list just a few of the more important popular fictions:

- *The primary goal of science is the accumulation of facts.* Unfortunately, the mere cataloguing of data is not enough; we also
  require some overall organizing principles and a relationship
  between these principles and the data. Actually, for scientists
  the more reliable a fact is, the more trivial and unimportant it
  becomes. For instance, the atomic weight of carbon can confidently be given as 12.011 atomic units. Yet this fact is basically just a curiosity until it's correlated with similar facts
  about the other chemical elements, using the laws and theories
  of chemistry and physics.
- *Science distorts reality and can't do justice to the fullness of human
  experience.* Every human undertaking must somehow pick and
  choose as to what aspects of reality to omit in order to probe
  other aspects of the world. In this regard science is no differ-

ent from religion, art, literature, mysticism, or any of its
other competitors in the reality-generation business.

* *Scientific knowledge is truth.* Science is not in the business of
providing ultimate explanations. Every scientific law or theory is subject to modification; there are no universal, absolute,
unchangeable "truths" in science.

* *Science is concerned primarily with solving practical and social problems.* I can't think of a single statement about science that could
be further from the actual case. For most scientists, science is
a game played for understanding, not for obtaining practical
information about how to build a better radio, mix more nutri-
tious dog food or iron out the wrinkles of middle-aged dowag-
ers. In fact, this "science = technology" misperception is so
pervasive that it merits a few additional words all its own.

Some time back, I had the enervating experience of working
for a man who suffered from the delusion that doing science
meant finding answers to practical problems posed by industrial-
ists, government policymakers, and other dreamers, schemers,
and so-called men of affairs. One conversation that I ruefully
recall involved my temerarious claim that if you focus attention
on finding well-defined answers, then you're not doing research,
at least not scientific research. Research involves ideas, not answers. In my view, what counted was developing a deep understanding of the question itself; whatever "answers" there might
be would then follow as corollaries of this insight into the real
nature of the question. A solution itself is not the ultimate goal;
what's important is understanding why an answer is possible at
all, and why it takes the form that it does. The point I was making was that technological advancement and the acquisition of
scientific knowledge have only the feeblest points of contact with
each other. Technology is primarily engineering, and new technologies come more from fighting with physical reality than
from scientific theories. Besides, it's not clear that new technologies give us a better *understanding* of nature anyway, e.g., modern medicine vis-à-vis Chinese acupuncture.

The moral of the foregoing little tale is that even many people
who practice under the rubric of what in the vernacular is called
a scientist hold to a view of science and scientific work that at
best falls into the third category noted earlier, which we might
compactly describe as "the General Electric Syndrome." That is,
if GE is doing it, it must be science. Well if GE is doing it, it

probably *isn't* science, at least not the kind of science that most members of the global scientific community would recognize. It may be high-grade technology or world-class engineering or even pathbreaking developmental research, but definitely not science. I hasten to point out here that this observation is in no way intended to minimize the truly outstanding and genuine scientific work that *is* carried out at places like GE, IBM, Bell Labs, Exxon, and so on. But it's not the real science going on in these corporate research labs that members of the public have in mind when they think of, say, IBM. What comes to mind is computers, typewriters, and all the other office paraphernalia that carries the IBM logo and that people use in their day-to-day affairs. The development of these gadgets is the main business of such an institution, and that development is definitely not science; it's technology. Now let's get back on course and examine just what it is that *does* constitute science as it's seen by the scientists themselves.

Paradoxically, scientists usually think of science as one area of life in which ideologies play no role. Nevertheless, there is a collection of beliefs and ideals about the practice of science that the scientific community clings to with such universal tenacity that it's difficult to describe it as anything other than an ideology—the ideology of science. The scientific ideology is a mixture of logical, historical, and sociological ideals about how science should operate in a Panglossian world, and rests upon the following pillars:

• *The logical structure of science:* This pillar represents what many of us learn in our early schooling about the procedures followed in science. Here we find the sequence:

$$\text{Observations/Facts}$$
$$\downarrow$$
$$\text{Hypothesis}$$
$$\downarrow$$
$$\text{Experiment}$$
$$\downarrow$$
$$\text{Laws}$$
$$\downarrow$$
$$\text{Theory}$$

To many, this diagram represents the essence of what we think of as *the scientific method.* Observations give rise to conjectures and hypotheses, which in turn are checked out by performing experiments. If the experiments don't confirm the hypothesis, then new hypotheses are formed, just as in the pulsar work described earlier. Those hypotheses that survive are encapsulated into empirical relationships, or laws, which in turn are embedded in larger explanatory theories. It is this sequence of steps that's been the focus of most of the philosophical analyses of the process of science, as we shall discuss in detail later. However, to the practicing scientist there is much more to the scientific enterprise than mere philosophy.

• *Verifiability of claims:* Science is a public undertaking with many filters that a claim must pass through before it's accepted as part of the current conventional wisdom. Two of the most important are the refereeing process for scientific articles and the repeatability of experimental results. Before a reputable scientific journal will publish a research announcement, it's sent out for review to other workers in the field, not only as insurance that the results are correct, but also to substantiate their significance within the framework of current knowledge in the area. In a similar manner, published work is supposed to report all the details of the investigator's experimental setup so that any interested party can, in principle, repeat the experiment and try to replicate the reported results. Thus, in the utopian world where the scientific ideology reigns, refereeing and repeatability keep the scientific process (and the scientist) honest.

• *Peer review:* The modern scientist is in much the same situation as the artisan of the Renaissance, at least when it comes to needing a patron to finance pursuit of the muse. The only difference is that nowadays everyone has the same patron—the federal government. As a result, most funds are allocated by federal agencies, making liberal use of the so-called peer review process. This involves committees of experts from the various fields getting together and recommending to the funding agencies those projects and those scholars whose work they feel merits support. According to the ideology, this process ensures that money is channeled to those ideas, institutions, and individuals showing the clearest evidence of being able to do something productive with it.

Given the highly egalitarian, logical, meritocratic nature of the scientific ideology, it comes as no surprise that many scientists accept it as at least a very close approximation to the way science really is. I'll defer detailed consideration of this point to a later section. At the moment let me just remark that a neutral skeptic would almost certainly raise an eyebrow or two over the rather obvious fact that the conventional ideology focuses entirely upon the *process* of science, leaving aside all considerations of the motives and needs of the scientists themselves. The degree to which this omission casts a cloud over the rosy picture painted above will occupy our attention throughout the book. For now, let's stick to the scheme above and turn the spotlight on the cognitive structure of science, in an attempt to get back to the questions of just what kind of knowledge the process of science is able to offer us about the nature of the world as it is, and whether that kind of knowledge is in some way superior to any other kind.

THE NATURAL PHILOSOPHER'S STONES

The issue before the house for the next couple of sections is consideration of the dual questions:

Do scientific theories in any sense tell us about the way the world *is?*

Does science have anything like a *method* for creating and/or evaluating theories?

Since all theories must necessarily be expressed in some kind of language (natural, symbolic, mathematical), the first question takes us into the province of the philosophy of language as a tool for representing reality. The second question deals more with science per se, forcing us to confront the natural query "What's so special about science?" In other words, why should we believe that scientific knowledge is any more correct or reliable than any other sort? So our short-term objectives are to explore the question marks in the following diagram:

Scientific theory ⟶ Objective reality

↑ ?

Scientific methods

To address these two foundational question marks, it will be necessary for us to dip briefly into the work of several twentieth-century philosophers of language and science. But before delving into the ideas of these thinkers, let's first go back a couple of millennia and fix our attention on some of the pivotal ideas of the ancient Greeks that ultimately led to the confused state we find ourselves in today.

In his last will and testament, Aristotle offers the following logical sequence of steps—i.e., an algorithm—for disposition of his estate. Until his chosen son-in-law, Nicanor, came of age, the estate was to be managed by three executors. If Nicanor died prior to the time when Aristotle's daughter, Pythias, would be old enough to marry him, then Theophrastus was to step in and fill Nicanor's designated role. But if Pythias married someone else who, in the opinion of the executors, didn't disgrace Aristotle's name, then she was given permission to use the family ancestral home at Stagira, which was then to be furnished to her satisfaction by the executors. Even after death, Aristotle leaves no stone unturned and no possibility unaccounted for—just the kind of detailed, step-by-step prescription that we might have expected from the man who invented the idea of formal logical *deduction.*

For Aristotle, the procedure for uncovering the truth of things was to postulate premises, then use the now-familiar rules of logical deduction to derive the consequences implicit in the premises. The classical example of this procedure, which we're all familiar with from Philosophy 101, is:

> Premise I: All men are mortal.
> Premise II: Socrates is a man.
> $\Downarrow$
> Conclusion: Socrates is mortal.

Note that nothing is said here about the actual truth or falsity of the premises. Maybe some men are not mortal or maybe Socrates is really a woman or a hermaphrodite or whatever. Physical reality and truth play no role in the deductive method; the premises are *assumed* to be true, with the conclusion following from this assumption.

Prior to Aristotle the traditional means for structuring experience was the *myth,* a term deriving from the Greek *mythos,* meaning "word," in the sense that it is the definitive statement on the subject. A myth presents itself as an authoritative account of the facts that is not to be questioned, however strange it may seem. According to the famous mythologist Joseph Campbell, myths serve several functions:

- *Metaphysical:* Myths awaken and maintain an "experience of awe, humility and respect" in recognition of the ultimate mysteries of life and the universe.
- *Cosmological:* Myths provide an image of the universe and explanations for how it works.
- *Social:* Myths validate and help maintain an established social order.
- *Psychological:* Myths support the "centering and harmonization of the individual."

Myths need be neither true nor false, just useful fictions; however, they are not the kind of fiction that has entertainment value alone, and makes no pretensions to truth. Religion, as we shall see later, goes one step further than the useful fiction of a myth by making assertions about what is indeed the case. It is at this point that the age-old conflict between science and religion starts to take off.

To illustrate the use of myths, imagine a band of prehistoric hunters who have spent several days stalking a herd of mammoths. Just at the moment of truth when they've laid their ambush and are about to attack, a thunderbolt from the sky comes flashing down, scattering the herd and undoing all the hunters' carefully laid plans. Somehow it's comforting at such times for the hunters to have a belief system that provides some explanation for what would otherwise seem a capricious whim of the cosmos. A myth provides such a system of beliefs by offering a scheme by which to order and explain the thunderbolt. Perhaps the gods were angry because they had not been properly honored, or maybe the spirits of dead mammoths from the past warned their living brethren, or it might have been that the hunters hadn't approached from the right direction. Whatever, the important point is that the myth serves as a schemata whereby the events of daily life can be given an interpretation in terms of mysterious forces and beings whose powers transcend lowly human concerns. Aristotle began the process of replacing myth with what has now come to be termed science.

The opposite side of the reality coin from *mythos* is *logos,* the Greek term for an account whose truth can be demonstrated and debated. It is this kind of truth that Aristotle was trying to grasp when he developed *logos* into "logic" by use of the process of deduction. One of the main uses of myths as outlined above is to provide an explanation of how real-world events work. In everyday speech, an "explanation" is usually taken to be the answer to a question that begins "Why?" Such answers inevitably begin with "Because," and the question and answer together constitute what we generally call a statement of *cause and effect.* Thus, "Why is the sky blue?" is answered with "Because the air molecules absorb all frequencies of visible light except those in the blue part of the spectrum." And "Why does water boil at 100°C (at sea level)?" is answered by "Because at that temperature the thermal motion of the water molecules is able to overcome the external atmospheric pressure"—cause and effect, stimulus-response. The method of logical deduction is Aristotle's theoretical, or some might say mathematical, counterpart to the explanation of physical happenings by cause and effect.

In his *Physics,* Aristotle attempted to combine the purely logical method of deduction with his ideas about the nature of physical reality in order to draw conclusions about the way the world really works. In Aristotle's view physical matter was composed of three things: qualities, form, and spirit. He felt that there was only one kind of matter, which could take many forms, the fundamental forms being air, earth, fire, and water. Because these four fundamental forms were not elements in any sense in which we might understand that term, they could be transformed into each other. To illustrate, this scheme gave rise to what today we might term Aristotle's version of the hydrologic cycle: The Sun's heat changes water into air; heat rises, so the heat in this air pulls the rest of it up to the skies; the heat then leaves the vapor, which becomes progressively more watery again, and this process results in cloud formation. There ensues a positive feedback effect in which the more watery the cloud, the more the water drives away its opposite, the heat. Thus, the cloud gets colder and contracts. The contraction then restores true wateriness to the water, which falls as rain or, if the cloud's heat has now fallen below the freezing point, hail or snow. So we see here the relentless chain of cause and effect being employed to "explain" the observed behavior of water, air, heat, rain, and snow. What's amazing about the whole setup is how all the wrong reasons somehow combine to produce something remarkably close to the way things really do work!

For almost two thousand years Aristotelian logic and physics served as the "science" of the time, explaining various aspects of nature, body, and mind by logical consequences of assumptions of the foregoing type about the nature of matter. Oddly enough, despite Aristotle's main occupation as an observational biologist, the biggest flaw in his entire world picture was that he advocated no experiments or even use of observations to serve as a check on the validity of his underlying premises. Basically, his was an epistemology in which one inferred specific instances (conclusions) from general observations (premises). It was not until the work of Francis Bacon in the seventeenth century that someone had the courage to challenge the authority of Aristotle and suggest turning the situation around, i.e., trying to infer general instances from specific observations.

Bacon's argument was that if one wants to come to grips with the way the world really is, it's necessary to begin the investigation with the facts of life rather than prejudices about what those facts might be. Thus followed the principle of *induction,* whereby conclusions about future events are drawn on the basis of repeated past observations. Such an approach is just what we might come to expect from a man who was not only a philosopher, but also a lawyer who rose to the post of lord chancellor of England before being dismissed for taking a bribe (an indication, perhaps, that the current dubious ethical state of the legal, financial, and political professions are not late-twentieth-century aberrations, after all). In Bacon's view of things, if we observe the Sun rising in the east for fifty consecutive days, then we can predict that it will rise in the east on day 51. And the longer we observe such regular behavior, the more confidently we can speak about its continuation. In a nutshell, this is the method of induction—lots of individual observations eventually resulting in the inductive leap to a general conclusion.

On the one hand, it's satisfying to have a method that takes into account what Nature is actually doing; on the other hand, why should such a procedure provide reliable information about the way things work? On what grounds can I be certain that every time I put water into my ice-cube trays and leave them in the freezer for a few hours I'll soon have ice for my scotch on the rocks? Just because it's always happened this way before, does that give me any assurance that today's drink will have the customary satisfying "clink"? The short answer is that there's

absolutely no justification at all for my concluding that I'll soon
be enjoying a scotch on the rocks and not a scotch and water.
This is the Problem of Induction: Why should induction work?
Why is it a reliable guide to the future?

To illustrate the Problem of Induction, consider the following
exchange:

WOMAN: Professor, professor. You must help me. My husband
uses an inductive argument to justify the use of inductive arguments.

PROFESSOR HUME: That's terrible. How long has he acted this
way?

WOMAN: As long as I can remember.

HUME: Then why didn't you see me sooner?

WOMAN: I would have, but we needed (the conclusions of) the
inductive arguments.

HUME: I'm afraid I need them too.

Philosophers beginning with Hume have grappled with this
problem, and I'll consider some of their conclusions in the next
section. For now we leave it as a gaping hole in the attempt to
repair the difficulties in Aristotle by introducing actual observations into the creation of a world view.

Galileo and Newton are the last two supporting actors in our
cursory sketch of developments leading up to the modern era of
scientific "truth." Galileo was a contemporary of Francis
Bacon, and although there appears to be no record of direct contact between the two, there is a clear connection between the idea
of Nature as the arbiter of what's what as advocated by Bacon,
and Galileo's refinement of the idea by instituting the notion of
a *controlled experiment*. In effect, Galileo said that if you have a
theory about how some phenomenon works, you must construct
an experiment in which all the variables except the one you're
interested in are controllable. Then, by fixing the controlled variables, you can measure the variable of interest, thereby checking
your theoretical hypothesis against the supreme court of observation. Thus follows the oft-recounted legend (for which there's
not a shred of documentary evidence) of his experiment of dropping two different weights from the Leaning Tower of Pisa, and
measuring their respective rates of fall as a "laboratory test" of
the hypothesis that objects fall at a uniform rate in the absence
of air resistance, irrespective of their mass.

Newton added the idea of the description of nature in *mathematical* terms—the keystone in the arch of scientific knowledge whose foundations were laid by Aristotle. More than his remarkable experimental results in optics, mechanics, and chemistry, Newton's legacy as writ large in his *Principia* is the idea of what we would today call the *mathematical model.* Newton showed not only how to "encode" Bacon and Galileo's world of observation into mathematical form, but also invented the method (calculus) for using the mathematical machinery to grind out theorems that could be "decoded" into new implied statements about Nature. The essence of this procedure is depicted in Figure 1.2, where the physical system to be modeled (e.g., the solar system, an electrical circuit, or whatever) is on the left, while the formal mathematical system that represents it appears on the right. Also on the left is our earlier notion of causality, represented as a property of the physical system in which certain parts of the system exert influences "causing" things to happen elsewhere in the system. The term *implication* is used on the right to represent either the process of Aristotelian deduction or that of Baconian induction as the means of proving mathematical statements to be logically correct. These statements are usually called *theorems* and follow from axioms and the above logical rules of inference. The set of implications is the logical counterpart of the physical causality noted on the left side of the diagram. These implied statements are then *interpreted*—i.e., decoded—into assertions about the way the material system really *is.*

With the ideas of deduction, induction, observation, and experiment welded together by the symbolic formalism of mathematics, the stage is now set for a brief account of the alphabet by which modern science tries to inscribe the secrets of nature. The main letters in this alphabet are facts/observations, laws, theories, and models. Let's take a look at what each of these concepts means in the context of modern science.

In Dickens's tale *Hard Times,* the schoolmaster Thomas Gradgrind opens the story with the statement "Now, what I want is, Facts. Teach these boys and girls nothing but Facts. Facts alone are wanted in life. Plant nothing else, and root out everything else. You can only form the minds of reasoning animals upon Facts: nothing else will ever be of any service to them. . . . Stick to Facts, Sir!" While Gradgrind is hardly a role model of the

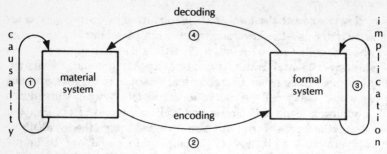

FIGURE 1.2 *Newton's scheme for mathematical modeling*

kindly, scholarly schoolmaster, his view forms the starting point of what many think of as constituting "reality": the world we can see, touch, smell, and hear; the world of Facts. But this commonsense view is only the starting point for a scientific investigation of Nature's scheme of things. As noted earlier, isolated facts are useless curiosities until they are put together with other facts into some kind of pattern. This requires the development of laws.

Suppose we do the following experiment: Take a long cylinder with a movable piston and fill it with gas (e.g., one of the cylinders in the motor of your car). Imagine now that we move the piston to various positions, and for each position measure the pressure that the enclosed gas exerts upon the walls of the cylinder. Further, suppose that after performing many such measurements, we note that whenever the volume of the cylinder is decreased by a certain fraction, the pressure increases by the same fraction; similarly, if we increase the volume by a fraction A by letting the piston rise, we find that the pressure decreases by the same amount A. By an inductive argument, after many repetitions of this experiment we would eventually conjecture (hypothesize) that there is a direct relationship between the pressure and the volume of the gas in the cylinder. Specifically, we would probably assert that the pressure P is inversely proportional to the volume V. And if we were mathematically inclined, we would compactly write this relationship as $PV = k$, where k is a constant determined by the nature of the particular gas and the units of measurement being used. This relationship is an example of what is called an *empirical law*. The law enables us to summarize a large number of individual facts (the results of the individual experiments) in one general statement.

The characteristic properties of laws of the foregoing type are that they:

1. are about *kinds* of events (experiments involving the pressures and volumes of gases in cylinders), not about any singular event (a particular experiment with a particular cylinder using a particular gas);
2. show a *functional* relationship between two or more kinds of events;
3. are supported by a large amount of *experimental data* containing little or no disconfirming evidence;
4. are applicable to *different events* (other types of gases and/or cylinders).

It's important to observe here that there are many different types of laws, not all of which are scientific. The reader might like to try to distinguish among the following in regard to their scientific character: parking regulations, the Ten Commandments, the Law of Conservation of Energy, the Law of the Excluded Middle.

Useful as it is, the above pressure-volume relationship (Boyle's Law) still doesn't tell us *why* an increase in pressure is linked with a decrease in volume. For this we need a *theory* of gases. An explanation for Boyle's Law can be obtained only if we invoke the atomic nature of the gas, and think of it as being composed of a large number of little "billiard balls" randomly moving about, occasionally colliding with each other and with the walls of the cylinder. Newtonian mechanics describes the motion of each such ball, and by combining their individual motions we can in principle calculate the pressure on the container walls by determining how many balls are colliding with the walls at each instant, and the strength of each such collision. With this picture in mind, it's easy to see why when the volume of the cylinder is halved, the pressure doubles. Since the cylinder's surface area has been cut in half, the likelihood that a randomly moving ball will collide with the wall doubles. Newton's laws of mechanical motion in the context of this gas situation form the basis for what is termed the Kinetic Theory of Gases, a framework that enables us to *explain* Boyle's Law.

The characteristic feature of a theory is that it offers a means of relating the laws describing a class of events to a framework and a set of principles described in terms differing from those used for the laws. Thus, the Kinetic Theory of Gases doesn't make use of the idea of pressure or volume at all, but only the notion of a particle, together with its associated mass and velocity. We obtain an explanation of Boyle's Law by deriving the law from the principles (Newton's laws of motion).

The idea of the gas molecules as little billiard balls flying about inside the cylinder also illustrates the notion of a *model* of a physical situation or, more precisely, a *physical* model as contrasted with a *formal,* or *mathematical,* model. No one takes seriously the idea that the gas molecules really are hard little inelastic spheres, but this turns out to be a very useful picture upon which to let common sense feed in order to generate intuitions about how the physical system will act under various circumstances. The same technique is employed in other types of physical models, as, for instance, in the use of scale models of cars and aircraft in wind tunnels to test for various sorts of aerodynamic properties. In these situations, many aspects of the real car or plane are neglected so that attention can be paid solely to the aerodynamic properties. Similarly, in the gas example many real properties of the gas, like its reactivity, color, temperature, and so forth, are neglected to study its pressure-volume relationship. Facts, laws, models, and theories—such are the tools that the scientist uses to prospect for the gold of reality in the mountainous doings of Nature. Figure 1.3 depicts the interconnections between these landmarks on the terrain of science.

Depending upon your inclination, there are several different philosophical positions that can be taken as to whether the nuggets of reality that turn up in the scientist's prospecting pan are fool's gold or the mother lode. In the philosopher's game, each of these positions is associated with a particular philosophical point of view, or "-ism," the most important for our purposes being:

• *Realism:* Realists believe that there is an objective reality "out there" independent of ourselves. This reality exists solely by virtue of how the world is, and it is in principle discoverable by application of the methods of science. I think it's fair to say that this is the position to which most working scientists subscribe. They believe in the possibility of determining whether or not a theory is indeed *really* true or false. Indica-

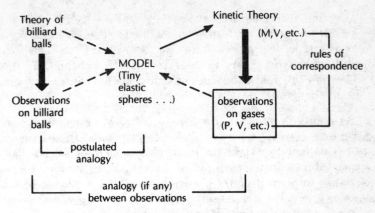

FIGURE 1.3 *Observations, laws, theories, and models*

tive of this position is the outcome of a straw poll taken recently in a small university department of physics consisting of eleven faculty members, ten of whom claimed that what they were describing with their symbols and equations was objective reality. As one of them remarked, "Otherwise, what's the use?"

- *Instrumentalism:* This school clings to the belief that theories are neither true nor false, but have the status only of instruments or calculating devices for predicting the results of measurements. Basically, this amounts to the belief that the only things that are genuinely real are the results of observations, i.e., Gradgrindian Facts. A typical statement along these lines comes from the engineer Rudolf Kalman, who remarks in the context of mathematical model building: "[Prejudice] means assumptions unrelated to data, independent of data; assumptions which cannot be (or simply are not) checked against the data." In light of the engineer's hunger for any solution that "works," perhaps such an extreme position is acceptable in engineering, but it's hard to see how it can be defended on any other than pragmatic grounds. As we'll see later in the book, the same problem arises at a much deeper level than mere practical engineering when one passes to foundational questions of epistemology in quantum mechanics. There, too, the principal defense of instrumentalism is that "it works."

- *Relativism:* In this increasingly popular position, truth is no longer a relationship between a theory and an independent reality, but rather depends at least in part on something like the social perspective of the person holding the theory. Thus, for a relativist as one passes from age to age, or from society to society, or from theory to theory, what's true changes. In this view it's not what is taken to be true that changes; *au contraire*, what changes is literally truth itself.

So reality is out there, in here, or what your measuring instruments (senses) tell you it is—take your pick! In an attempt to tell us how to weight the odds, philosophers of science have expended inordinate amounts of energy, thought, and heated verbiage in pursuit of the elusive essence of the process of science as a vehicle for unmasking the imposters on the "-ism" list. We can summarize their Herculean task as:

THE FUNDAMENTAL QUESTION OF THE PHILOSOPHY OF SCIENCE

Do scientists proceed as they do because there are objective reasons for doing so, or do we call those procedures *reasonable* merely because a certain group sanctions them?

To dig deeper into the ways science *might* be able to vindicate the creed of the realists and gain a glimpse of their nirvana of objective reality, there's no choice but to step into the twentieth century and look a little harder at the logical structure of science as seen by the philosophers. While most practicing scientists, not to mention laymen, find the discussion of such matters irksome, they are inescapable and cannot be ignored in a work such as this. Besides, as David Hawkins wisely noted, "Those who most ignore, least escape." So with this credo as our battle cry, let's briefly consider what the philosophers have to say about the correlation between the *praxis* and the *theoria* of science and their connection with any kind of objective reality.

RATIONALITY FOR REALISTS

If Plato's Academy in Athens served as the geographical focal point for Greek philosophy and its view of the world, then its twentieth-century counterpart can only be a small seminar room in the Mathematics Department of the University of Vienna, where a group of physicists, mathematicians, and philosophers met every Thursday evening for several years in the 1920s and 1930s to debate the relationship between the theories of science and objective reality. This group, christened the Vienna Circle in 1929, eventually came to what amounts to the instrumentalist position that the only meaningful statements that can be made are those for which we can give a definite prescription (method, algorithm) for their verification. Thus, use of a word like "yellow" would be equivalent to specifying a procedure for verifying that any particular object possessed the property of being yellow. In this way, the *meaning* or *reality* of "yellow" became equivalent to the statement of the procedure for its verification. This, in essence, forms the basis for the notorious Verification Principle, which lay at the heart of the school of *logical positivism,* the term later given to the philosophy expounded by the Vienna Circle. But to understand this blend of empiricism and logic, it's necessary to go back a few years and look at the work of another Viennese philosopher of the time, Ludwig Wittgenstein.

WITTGENSTEIN, LOGIC, AND LANGUAGE

For ordinary men, the middle of a battlefield with bullets flying and bombs bursting amid cries of human pain and agony is hardly the kind of place in which to engage in contemplative philosophical speculation. But Ludwig Wittgenstein was no ordinary man, and during the course of his valiant service with the Austrian Army during World War I, he developed ideas about the relationship of thoughts expressed in language to the actual state of affairs in the world, ideas that were later enshrined in the pages of his classic work *Tractatus Logico-philosophicus.* The basic tenet of this seminal volume, containing the only ideas of Wittgenstein's published during his lifetime, is that there must be something in common between the structure of a sentence and the structure of the fact that the sentence asserts. In this view, representation of the world in thought is made possible by logic, but the propositions of logic do not in and of themselves represent any actual state of the world. Thus, logic was necessary but not sufficient to describe any kind of objective reality. However, for Wittgenstein logic did reveal which states were theoretically possible, reflecting his underlying belief that reality was at least consistent—e.g., if the statement "Water boils at 100°C at sea level" is true, then the statement "Water does not boil at 100°C at sea level" cannot also be true.

Wittgenstein illustrated these ideas by what he called a "picture theory" of language, in which he compared logical propositions to pictures. A picture can represent some physical state using certain types of symbols; language can do likewise but with a different set of symbols. The picture bears some relationship to the physical reality that it represents. So, for example, if we see a human face in a photograph, the nose may appear in the center of the face both in physical reality and in the picture. However, if the picture is by Salvador Dali we might find the nose appearing in some quite different location, or not at all. Of course we might try to clarify the relationship between the picture and the object—for example, by introducing color or perspective—but such an attempt at clarification only gives rise to another picture, which itself will require additional analysis. At some stage the essence of the picture has to be understood directly, or we fall into an infinite regress.

In the picture theory of language, propositions making up the language are thought of as analogous to a series of pictures. Furthermore, since Wittgenstein assumes that the logical structure of language mirrors the logical structure of reality, the language "pictures" represent *possible* states of the world. It follows that linguistic statements are meaningful when they can, in principle, be correlated with the world. Actual observation of the world will then tell if they are true or false. To illustrate, we can meaningfully say that "the United Nations is in New York," but it is meaningless to state that "is United the New in York Nations." Of course, different logical rules (grammars) could be developed in which the latter statement is meaningful, but within the context of conventional English grammar it has no logical structure at all. So the main claim of the picture theory—namely, that there must be something in common between the logical structure of the language and the structure of the fact that it asserts—cannot really be "said" in terms of the language being used to make the statement; it can only be "shown." This conclusion gave rise to Wittgenstein's famous metaphor in the penultimate section of the *Tractatus:*

My propositions serve as elucidations in the following way: anyone who understands me eventually recognizes them as nonsensi-

cal, when he has used them—as steps—to climb up beyond them.
(He must, so to speak, throw away the ladder after he has climbed
up it.) He must transcend these propositions, and then he will see
the world aright.

So Wittgenstein's punch line is that the sense of the relationship
between reality and its description in language cannot be expressed in language.

Thus ended Wittgenstein's "early period" studies on the interplay of logic, language, and reality. The essence of his ideas
can be summarized in the following steps:

1. There is a world that we want to describe.
2. We try to describe it in some language, scientific, mathematical, or otherwise.
3. There is a problem about whether what we say about the
   world corresponds to the way the world really is.
4. We want to know the true nature of the correspondence between what we say and the way things are, but we can only
   use language itself to describe that correspondence.
5. Words of a language can never express the desired correspondence, and we must take recourse merely to *showing* it,
   i.e., using the picture theory, since otherwise we would fall
   into the infinite regress of descriptions of descriptions of
   descriptions . . .

At Step 5 we come to one of the most famous statements in all of
philosophy, with which Wittgenstein concluded the *Tractatus:*
"What we cannot speak about we must pass over in silence."

It's easy to see how Wittgenstein's exploration of the interplay of language, logic, and observation of the world would appeal to the members of the Vienna Circle, with their concerns
about constructing a coherent philosophy of science from an
amalgamation of logic and empirical epistemology. And indeed
the *Tractatus* did serve as a point of departure for many of their
deliberations, with several members of the circle in regular contact with Wittgenstein in Vienna, although Wittgenstein himself seems never to have participated in the Thursday night
discussions. As an ironic twist, while the Vienna Circle was busy
putting together the tenets of logical positivism using Wittgenstein's work as a basis, Wittgenstein himself was in the process
of undermining the entire effort by the development of his ideas
on the rules of language.

* * *

Remember those old IQ tests where some sequence of numbers is given and you're supposed to pick the "right" continuation of the sequence as a demonstration of your smarts? This kind of problem lies at the heart of what started to bother Wittgenstein about his picture theory of language, ultimately resulting in his repudiation of the entire idea. Consider the following simple example. Suppose the initial sequence is {1, 2, 4, 8} and you're asked, what's the "natural" or "right" continuation. On those absurd high-school IQ and College Board tests, the examiners would probably give full credit only if you answered with the sequence {16, 32, 64, 128}. Presumably this is the "correct" answer because you're supposed to recognize that each term in the original sequence is twice as large as its predecessor. Now there's no doubt that this is one logically defensible reason for guessing that the right continuation is one that extends this pattern. But there can be other continuations that, depending upon the context, would be equally logical and correct. For instance, if the context were the high-school football stadium rather than the examination room, then the most logical continuation might be {1, 2, 4, 8} → {"Who do we appreciate?"}. Or even in the examination room you might think of continuing with {9, 11, 15}, a pattern that reflects the jumps in the original sequence. The point is that in the absence of context, i.e., additional information, there's just no such thing as a "natural" continuation of the sequence. The reader will recognize this situation as just another illustration of the Problem of Induction stated earlier, and it's just this kind of difficulty that began to trouble Wittgenstein after the *Tractatus*.

Following the First World War, Wittgenstein spent time as a high-school teacher in village schools in Austria, where it is rumored he taught some of his pupils about the Liar Paradox ("This sentence is false"). By all accounts he was very popular with the students, but was eventually run out of the village by their parents, most likely on account of his homosexuality and inability to relate to the concerns of the peasant families in the regions where he worked. In any case, during this time he began to become dissatisfied with his picture theory of language, since it gave no clear-cut answer to questions like "Why should we see the principles of logic to be true, even though it's not possible to express the reasons in words?" (Because we can only "show" their truth, not "say" it.) Or "Is there some kind of underlying logical structure either to the world or to our thought systems that somehow can be held responsible for the apparent self-evidence of the propositions of logic?" In other words, is there a set of rules for organizing sense experiences that is fixed within our brains, but that we cannot articulate even though we all follow these rules automatically when we "see" in the same way and when we talk to each other?

In his later work Wittgenstein considered this kind of question, coming to the unhappy conclusion that there could be no underlying logical structure to the world to which our minds must adhere, or vice versa. In the final analysis, he claimed that the propositions of logic reflect the rules of language, and these are known to us by our use of language in everyday life and by linguistic experience. Consequently, Wittgenstein's solution as to why the right continuation of the sequence {1, 2, 4, 8} is {16, 32, 64, 128} and not {"Who do we appreciate?"} is that we know how to go on "in the same way" because we share a form of life. Thus the continuation is dictated by sociological considerations, and bears no contact with any kind of objective reality for number sequences. He then concluded that there are no private rules; rules are the property of a social group. Hence, Wittgenstein gave a "sociological" solution to the Problem of Induction by concerning himself not with how we could be certain *in principle* about the continuation, but rather with how we come to be certain about it *in practice*. The implication of all this for science is that science rests upon a foundation of taken-for-granted reality, a crucial aspect of the relativist school of scientific thought. We'll come back to this relativistic notion of scientific reality later, but for now let's return briefly to the Vienna Circle and its attempts to use the early Wittgenstein to clarify the meaning of language, thereby trying to uncover the "realness" of scientific propositions about the world.

THE LOGICAL POSITIVISTS AND VERIFICATION

In his account of the evolution of knowledge, Auguste Comte identified three stages of development: (1) the *theological,* in which reality is comprehended in terms of the conflicts and creations of gods and spirits; (2) the *metaphysical,* in which there is the use of abstractions and generalities; (3) the *positivistic,* which relies upon the quantitative description of sensory phenomena. The Vienna Circle was interested in formalizing the last stage by marrying Comte's quantification of empirical observations and data with the logical structure of language and its relationship to the physical world as outlined by Wittgenstein. The result was the philosophy of logical positivism, whose core element was the Verification Principle discussed earlier. For the logical positivists, there were only two sorts of statements or propositions: analytic statements and those that could be empirically verified. Only the latter had meaning, with analytic statements being either tautologies or literally meaningless.

The basic difficulty with the positivist approach is the Problem of Induction: General empirical statements just cannot be verified. For example, if I make the empirical claim that the Sun will rise in the east tomorrow on the grounds that it always has risen there up to now, the Problem of Induction prevents me from offering an empirical procedure for verifying this claim. Consequently, according to the positivist's creed my statement is meaningless, and certainly not scientific. Also, the Verification Principle had difficulties in verifying things like the wave function in quantum mechanics and, in general, failed to make a clear-cut distinction between meaningfulness and meaninglessness, thus coming up empty as a criterion for meaning or reality. As the source of this difficulty is the Problem of Induction, what could be more natural than to try to get around it by the simple expedient of rejecting the use of induction altogether? Enter Karl Popper and the idea of falsification.

POPPER, CONJECTURES, AND REFUTATIONS

Popper, the son of a Viennese lawyer, was originally interested in developing methods for separating scientific statements from pseudoscience. He also took an active part in the discussions of the Vienna Circle, whose members at first thought Popper shared their interest in meaning, a misunderstanding that was soon cleared up. While still a teen-ager, Popper recognized that no amount of supporting data will ever be sufficient to confirm a hypothesis, but all it takes to refute it is one piece of negative evidence. So, for instance, if I hypothesize that all Ferraris are red, no matter how many red Ferraris I see, the Problem of Induction will still prevent me from stating with certainty that this is the color of *all* Ferraris. However, all I need do is go to the Ferrari factory in Maranello and see that there is even one white car being built, and I can then confidently assert that my original hypothesis is false. This chain of argument constitutes what Popperians call the method of *falsification,* and forms the heart of Popper's view as to how science, as opposed to pseudo-science, is to be carried out. In his own words, "The criterion of the scientific status of a theory is its falsifiability, or refutability, or testability."

Popper is a realist and believes that there is an objective reality out there that science can acquire increasingly accurate information about. His method is conjecture and refutation: We make a hypothesis and then look for evidence to falsify it. For Popper, one theory of a given situation is to be preferred to another if there are more potential observations that can refute the theory than can refute its competitor. In other words, the more statements that could be refuted by direct observation a theory makes, the better the theory is. The classic example is the hypothesis that the Earth's orbit around the Sun is circular, as compared to the hypothesis (theory) that it is an ellipse with the circular orbit as just a special case. Since there are more potential observations that will falsify, or refute, the circular hypothesis, the theory that the orbit is circular would have more empirical content for Popper. To understand clearly the distinction between Popper's views and those of the logical positivists, it's instructive to examine the comparison given in Table 1.1.

While Popper seems to have banished the Problem of Induction from the philosophical banquet table, his conjectures-and-refutations methodology is not without a few flaws of its own. The most difficult obstacle is what is known as the Problem of Auxiliary Hypotheses. To illustrate, let's go back to the red Ferrari problem. If I happen to see a white Ferrari on the road, thereby refuting my original contention, the "red Ferrari" hypothesis can always be resurrected by adding some new background condition to the situation, such as "It wasn't really a Ferrari, but a Lamborghini," or "It was a red car that had only been painted white," and so on. Following this line of attack, any theory in trouble can always be saved by the introduction of suitable auxiliary hypotheses, since it may then be claimed that the original assertion wasn't wrong; the error was in one of the background assumptions.

POSITIVISTS		POPPER
	IDEAS	
	that is	
DESIGNATIONS		STATEMENTS
or TERMS		*or* PROPOSITIONS
or CONCEPTS		*or* THEORIES
	may be formulated in	
WORDS		ASSERTIONS
	which may be	
MEANINGFUL		TRUE
	and their	
MEANING		TRUTH
	may be reduced by way of	
DEFINITIONS		DERIVATIONS
	to that of	
UNDEFINED		PRIMITIVE
CONCEPTS		PROPOSITIONS

The attempt to establish (rather than reduce) by these means their

MEANING   TRUTH

leads to an infinite regress

TABLE 1.1 *Logical positivism versus Popper*

Popper's ideas place great emphasis upon scientific method. He is telling scientists about how they *ought* to behave, neglecting entirely how they actually do behave in practice. The hard facts are that very few scientists, if any, spend much time looking for data or trying to develop experiments that would falsify their hypotheses—just the opposite, in fact. This commonplace observation leads us into consideration of the way social conventions and ideas determine what we take to be scientific truth, a position that Popper himself ultimately came around to acknowledging in connection with his original problem of distinguishing science from pseudoscience. He finally concluded that if we want to know whether or not a theory is scientific, we should look and see how it is handled by people, rather than consider its logical structure—a position remarkably similar to that arrived at by Wittgenstein in his deliberations on many of the same issues.

LAKATOS AND SCIENTIFIC RESEARCH PROGRAMS

An important way station on the road from the purely realist position of the positivists and early Popper to the completely relativistic stance of today's Kuhnians, as discussed in the next section, is the work of the Hungarian educator and philosopher Imré Lakatos. After serving in the anti-Nazi resistance during World War II, Lakatos became a high-ranking official in the Ministry of Education, later fleeing to the West during the Hungarian uprising of 1956. At this time Lakatos went to England, where he began work on his Ph.D. thesis at Cambridge on the theme of mathematical discovery. This novel work, presented in the form of a dialogue centering on the proof of Leonhard Euler's famous formula relating the number of faces, vertices, and edges of a polyhedron, led Lakatos to a deeper interest in the question of the "dynamics" of theories. Thus he went one step further than Popper and the positivists by centering attention not just on the structure of scientific theories, but also upon how they change. The vehicle for this study was what Lakatos termed a *scientific research program* (SRP).

For Lakatos, an SRP is a *sequence* of theories in which certain methodological rules are followed. The primary components of an SRP are:

- *The hard core*—an inviolate cluster of hypotheses at the center of the program
- *The protective belt*—a set of auxiliary hypotheses
- *The negative heuristic*—assumptions underlying the hard core that are not to be questioned
- *The positive heuristic*—a set of suggestions or hints saying how the SRP is to be altered

A good example of the kind of SRP that Lakatos had in mind is the Ptolemaic view of the solar system, in which the Earth sits at the center with the various planets moving about on orbits that are described as complicated epicycles. These curves are just the path traced out by a fixed point on, say, the rim of a coin as you roll it along the top of a flat table. Coins of different sizes give rise to different epicycles, and Ptolemaic theory used combinations of these curves to describe the planetary orbits. The hard core of the Ptolemaic program is the geocentric hypothesis, together with the necessity of the planetary orbits being given by epicycles. The protective belt consists of the details of the various types of epicycles, while the positive heuristic would consist of a plan for developing increasingly sophisticated models of the planetary system. Note that this positive heuristic is not a vague, general set of principles, but a quite specific set of procedures giving definite advice on how to proceed, including instructions on how to handle anomalies.

On the positive side of the ledger, Lakatos's ideas were an improvement over Popper's since they acknowledged the social dimensions of science. In this sense they served as forerunners to the ideas of Kuhn. Furthermore, the Lakatos vision of what constitutes scientific truth had the virtue of showing that no particular research program is unambiguously to be preferred to any other. In this way, the SRPs opened the door for the anarchical views of Paul Feyerabend, which we'll look at in a moment. Also to his credit, Lakatos discerned two important facts about scientific procedure: (1) scientists have sufficient faith in the hard core that anomalies are explained away, and (2) scientists have general ideas about how one should try to cope with anomalies (the positive heuristic).

As to liabilities of SRPs, there are many, not the least of which is that the choice between two SRPs for Lakatos is no easier than the choice between two theories for Popper. The assessment of which of two programs to prefer eventually comes down to a situation analogous to having Donald Trump and Harry Helmsley tossing pennies off the top of the World Trade Center, the title Grand Real Estate Baron of Manhattan being awarded to the one whose penny lands first; it's a meaningless game without a criterion that they can employ to see who will reign as king of the towers. But there is no operational way for them to decide whose penny lands first without invoking outside agents, i.e., additional information outside the two "programs." Lakatos's SRPs had other drawbacks as well.

There were great difficulties in coming to agreement as to just what constitutes the hard core of an SRP in any specific situation. For instance, Newton's view of planetary motion used the inverse-square law of gravitational attraction as an inviolate hypothesis, i.e., as part of the hard core of Newtonian mechanics.

Yet in considering the motion of the planet Uranus, both George Airy and Friedrich Bessel suggested modifying the inverse-square law to account for the observations, while Urbain Jean-Joseph Leverrier and John Adams suggested keeping the law and explaining the motion by the presence of a hitherto-unobserved celestial body (which turned out to be the planet Neptune). Similarly, before the Theory of Relativity was promulgated in 1905, some suggested modifying the inverse-square law to account for aberrations in the perihelion of the planet Mercury. In fact, the *Encyclopaedia Britannica* (1910 edition) stated that the gravitational law should have the exponent 2.0000001612 instead of 2 to make things come out right! So even in that most solid of scientific bastions, Newtonian mechanics, there were heated disagreements as to what should and should not be in the hard core. A final difficulty for Lakatos is that the idea of the positive heuristic is hopelessly vague. This part of the program is supposed to tell us what to do to modify the program but, in fact, emerges during the course of the research. As a result, it says nothing about what one is supposed to do to carry out an investigation successfully.

Lakatos's vision of the scientific enterprise is far richer than Popper's in that his notion of heuristics directs attention to important aspects of scientific practice not stressed by Popper at all. Nevertheless, the difficulties with his SRPs cast aspersions on the kinds of views of scientific "reality" that can be expected from any such program.

So we see the various attempts by Wittgenstein & Co. to provide a solid, logical foundation, or *method,* for the scientific pursuit of knowledge all come to one bad end or another. Dare we entertain the idea that perhaps there is no method? Well, Paul Feyerabend not only entertains the notion, he insists upon it.

FEYERABEND: THERE AIN'T NO METHOD

In studies of scientific method, there are two principal branches:

A. Rules or techniques to use in the discovery of theories
B. Rules for the objective evaluation of rival theories

The Vienna Circle claimed that only B was the legitimate province of the philosophy of science; Paul Feyerabend denies that there is any valid distinction between the two.

In *Against Method,* his famous manifesto for scientific anarchy, Feyerabend states his basic theme in the following way:

"No set of rules can ever be found to guide the scientist in his choice of theories, and to imagine there is such a set is to impede progress. The only principle that does not impede progress is *anything goes* [italics added]." Feyerabend is claiming that there is no such thing as a scientific method. His argument is that science is just one tradition among many, and is privileged neither in terms of methods nor in terms of results. He goes on to advocate removing science from its pedestal and trying to create a society in which all traditions have equal access to power and education. Among the traditions he suggests giving equal weight with science are astrology, witchcraft, mysticism, and folk medicine! If this all sounds like the grumblings of a failed scientist to you, it's perhaps worth noting that Feyerabend did at one time study physics and astronomy.

Feyerabend was also active in the Berkeley Free Speech Movement, and became interested in the so-called alternative society ideas bandied about in the 1960s. But he eventually redeems himself by confessing that he doesn't have the seriousness of purpose of a true anarchist and would like to be remembered as a "flippant Dadaist."

The central thesis of incommensurability of theories brought out in such stark fashion by Feyerabend takes us from the ideas of realism and the work of Wittgenstein and the Vienna Circle clear across town to relativism and the offbeat ideas of Feyerabend. Despite their shade of lunacy, the visions of Feyerabend contain just enough good sense to suggest there's something worthwhile lurking at their core. This kernel of sense hiding in the flamboyant noise is the notion that there are many methods and ways of coming to scientific truth, and what is taken to be true at any moment is more a matter of social convention in the scientific community than it is a product of logical methods and procedures. Recognition of this startling fact constitutes the theme song for Thomas Kuhn, whose ideas about paradigms in science lie at the heart of what is by far the most talked-about view of the scientific enterprise in the second half of this century.

BUDDY, CAN YOU PARADIGM?

Julian Bigelow, an electrical engineer who helped John von Neumann build the Johnniac computer at the Institute for Advanced Study in Princeton in the early 1950s, tells a story about how when he drove down from Cambridge, Massachusetts, to be interviewed by von Neumann for the job, he met with the great man at his home in Princeton. As the story goes, there was a large dog romping on the lawn, and as von Neumann opened the door to let Bigelow in, the dog ran into the house and started running from room to room, sniffing everything in sight in the manner commonly practiced by dogs everywhere. Busy in their discussion, neither von Neumann nor Bigelow paid much attention to these canine antics for quite awhile, but finally von Neumann's curiosity overcame his courtly Central European manners and he asked Bigelow if he always traveled with his dog. Bigelow replied, "It's not my dog. I thought it was yours." Such are the presuppositions that pervade every aspect of human activity, science (and scientists) being no exception. And it's exactly these kinds of presuppositions that constitute the nucleus of the idea underpinning Thomas Kuhn's notion of a scientific *paradigm.*

In 1947 Kuhn, a young professor at Harvard, was asked to organize a set of lectures on the origins of seventeenth-century mechanics. As preparation, he began tracing the subject back to its roots in Aristotle's *Physics,* being struck time and again by the total and complete wrongheadedness of Aristotle's ideas. As noted earlier, Aristotle held that all matter was composed of spirit, form, and qualities, the qualities being air, earth, fire, and water. Kuhn wondered how such a brilliant and deep thinker, a man who had single-handedly invented the deductive method, could have been so flatly wrong about so many things involving the nature of the physical world. Then, as Kuhn recounts it, one hot summer day the answer came to him in a flash while he was poring over ancient texts in the library: Look at the universe through Aristotle's eyes! Instead of trying to squeeze Aristotle's view of things into a modern framework of atoms, molecules, quantum levels, and so forth, put yourself in Aristotle's position, give yourself the prevailing world view of Aristotle's time, and lo and behold, all will be light. For instance, if you adopt Aristotle's world view, one of the presuppositions is that every body seeks the location where by its nature it belongs. With this presumption, what could be more natural than to think of material bodies as having spirits, so that "heavenly" bodies of airlike quality rise, while the spirit of "earthly" bodies causes them to fall?

FIGURE 1.4 *Two visual gestalts or "paradigms"*

This stroke of inspiration resulted in Kuhn's developing the idea that every scientist works within a distinctive paradigm, a kind of intellectual gestalt that colors the way Nature is perceived. The situation is vaguely analogous to the picture in Figure 1.4, where one way of looking shows what appears to be two men face to face in profile, while another way shows a flower vase.

According to Kuhn's thesis as presented in his enormously influential 1962 book *The Structure of Scientific Revolutions*, scientists, just like the rest of humanity, carry out their day-to-day affairs within a framework of presuppositions about what constitutes a problem, a solution, and a method. Such a background of shared assumptions makes up a paradigm, and at any given time a particular scientific community will have a prevailing paradigm that shapes and directs work in the field. Since people become so attached to their paradigms, Kuhn claims that scientific revolutions involve bloodshed on the same order of magnitude as that commonly seen in political revolutions, the only difference being that the blood is now intellectual rather than liquid—but no less real! In both cases the argument is that the underlying issues are not rational but emotional, and are settled not by logic, syllogisms, and appeals to reason, but by irrational factors like group affiliation and majority or "mob" rule. As Kuhn states it: "There is no standard higher than the assent of the relevant community. The transfer of allegiance from one paradigm to another is a conversion experience that cannot be forced." With these ideas in mind, just what constitutes a paradigm anyway, at least as that term is used by Kuhn? The answer is not easy, and Kuhn has come in for plenty of criticism for the vagueness of the notion. But the basic concept can be made clear by the following map-making analogy.

Let's imagine scientific knowledge of the world as being the *terra incognita* of the ancient geographers and map makers. In this context, a paradigm can be thought of as a crude sort of map in which territories are outlined, but not too accurately, with only major landmarks like large rivers, prominent mountains, and the like appearing. From time to time, explorers venture into this ill-defined territory and come back with accounts of native villages, desert regions, minor rivers, and so on, which are then dutifully entered on the map. Often such new information is inconsistent with what was reported from earlier expeditions, so it's periodically necessary to redraw the map totally in accordance with the current best estimate of how things stand in the unknown territory. Furthermore, there is not just one map maker but many, each with a different set of sources and data on the lie of the land. As a result there are a number of competing maps of the same region, and the adventurous explorer has to make a choice of which map he will believe before embarking upon an expedition to the "New World." Generally, the explorer will choose the old, reliable firm of map makers, at least until gossip and reports from the Explorers Society show too many discrepancies between the standard maps and what has actually been observed. As these discrepancies accumulate, eventually the explorers shift their allegiance to a new firm of map makers whose pictures of the territory seem more in line with the reports of the returning adventurers.

This exploration fable gives a fair picture of the birth and death of a scientific paradigm. Kuhn realized that revolutionary changes in science overturning old theories are not in fact the normal process of science, nor do theories start small and grow more and more general as claimed by Bacon, nor can they ever be axiomatized as asserted by Newton. Rather, for most scientists major paradigms are like a pair of spectacles that they put on in order to solve puzzles. Occasionally a paradigm shift takes place when the spectacles get smashed, and they then put on a new pair that transforms everything into new shapes, sizes, and colors. Once this shift takes place, a new generation of scientists is brought up wearing the new glasses and accepting the new vision of "truth." Through these new glasses, scientists see a whole new set of puzzles to be solved in the process of carrying out what Kuhn called *normal science.*

The paradigms have great practical value for the scientist just as maps have value for the explorer: Without them no one would know where to look or how to plan an experiment (expedition) and collect data. This observation brings out the crucial point that there is no such thing as an "empirical" observation or fact; we always see by interpretation, and the interpretation we use is given by the prevailing paradigm of the moment. In other words, the observations and experiments of science are made on the basis of theories and hypotheses contained within the prevailing paradigm. As Einstein put it, "The theory [read paradigm] tells you what you can observe." According to Kuhn's paradigmatic view of scientific activity, the job of normal science is to fill in the gaps in the map given by the current paradigm, and it's only seldom, and with great difficulty, that the map gets redrawn when the normal scientists (explorers) turn up so much data not fitting into the old map that the map begins to collapse into a morass of inconsistencies. But what happens during these times of paradigm crisis?

Imagine we are at the initial stages of such a crisis, where the old paradigm can't account for certain anomalies, strange observations, and the like. Two new theories emerge, which offer different explanations for these aberrations. These theories represent different maps or sets of spectacles, i.e., different realities. After a period of competition, one of these theories begins to gain the acceptance of the scientific community. The reasons may not be objective at all, but may revolve about matters like simplicity, elegance, the social position of the theory's adherents, government science policies, and so forth. This support leads to experiments that then "corroborate" the theory, and the more evidence that accumulates, the more supporters the theory gathers, especially among the young Turks in the scientific community. Soon "reality" begins to take on the look of the new theory, and scientists universally begin to see and test for certain features of this reality and ignore others.

But what if the community had given its initial support to the other, competing theory? According to Kuhn, in that event "reality" would have taken a quite different turn, and the scientific view of the world would have been seen through that pair of spectacles rather than the first. This means that there is no such thing as scientific "progress," at least not in the sense that one paradigm builds upon its predecessor. Rather, the new paradigm turns in an entirely different direction, and as much knowledge is lost with the abandonment of the old paradigm as is gained from the new. Now we "know" a *different* universe.

If Kuhn's thesis is true, then it also destroys one of the main pillars of the scientific method, since the whole idea of a scientific experiment rests upon the assumption that the observer can be essentially separate from the experimental apparatus that tests the theory. Kuhn contends that the observer, his theory, and his equipment are all essentially an expression of a point of view, and the results of the experimental test must be an expression of that point of view as well. This position effectively asserts that science is not objective, but at the same time we know that science is not totally subjective either, since paradigms are eventually overthrown. So we're back to consideration of the central question: What is the relationship of the scientist to the universe he observes?

The most revolutionary aspect of Kuhn's claims is that they entirely omit things like knowledge, truth, and external reality. In fact, Kuhn states that in science truth is an entirely optional and gratuitous concept. As he puts it, "Does it really help to imagine that there is some one full, objective, true account of nature and that the proper measure of scientific achievement is the extent to which it brings us closer to that ultimate goal?" I think most practicing scientists would say that yes, such a belief helps a hell of a lot! But apparently Kuhn doesn't think so, since he says that there's no way for science to get hold of the "truth" anyway, so you can't measure scientific progress as getting closer to the way things are in themselves. Returning to the map-making analogy, Kuhn's claim is tantamount to the belief that not only are there many map makers, each emphasizing different aspects of the territory, but that it is in principle impossible ever to produce a complete map of the entire region. So you can't judge a map by how close it comes to this ideal Platonic map, since such a map is literally undrawable. In some ways this line of argument is reminiscent of Wittgenstein's claim that language cannot describe the intrinsic logical structure of the world.

Just like the revolutions they describe, Kuhn's arguments were met with fierce opposition from the philosophical community, although he was a minor saint to humanists since he seemed to be putting the human being back into the scientific enterprise. One of Kuhn's sharpest critics has been the philosopher Dudley Shapere, who complained that Kuhn was a relativist denying the objectivity and rationality of science. Shapere felt that science according to Kuhn is nothing more than a series of fads dressed up to look presentable, and offered the counterargument that even though we may be wearing rose-colored glasses, there's still a lot that shines through unaffected. The colors may be skewed, but other qualities like shape, size, and texture come through loud and clear. In short, the glasses may distort our view of reality but they don't create it—a staunch realist position.

Another criticism of Kuhn's ideas is that he places too little emphasis upon the social determinants of scientific revolutions. On the one hand, Kuhn argues that a paradigm shift takes place when there's an accumulation of anomalies; on the other hand, he says an anomaly can be ignored to preserve the paradigm. Question: At what point does a mass of discrepancies become irritating enough to bring about a paradigm shift? Kuhn offers little help in addressing this dilemma.

While Kuhn denies the label of an "irrationalist," he does assert that there are no methods or methodological rules for creating or evaluating scientific theories. His argument is that only propagandizing plays a role in changing allegiances from one paradigm to another. What makes reasons for theory change "good" is that they are generally accepted by the community, and if you want to be a member of that community it behooves you to operate within the framework of this system of reasons. As an immediate consequence, we find Kuhn's statement that rival paradigms cannot really be compared, although he does offer what we might term a Fivefold Way for characterizing the features of a good theory. Kuhn's way consists of the following points stating that a good theory must be

• *Accurate:* Consequences of the theory should be in agreement with experiment.

- *Consistent:* The theory should contain no internal contradictions and, moreover, it should be consistent with currently accepted theories applicable to related aspects of Nature.
- *Broad:* The scope of the theory's consequences should extend beyond the particular observations, laws, or subtheories that it was created to explain.
- *Simple:* It should bring order to phenomena that without it would be individually isolated.
- *Fruitful:* The theory should disclose new phenomena or previously unobserved relationships.

Kuhn's claim is that these criteria offer the shared basis for theory choice, but that there is no possible way of giving a justification for this selection of criteria.

To compare Kuhn with Feyerabend, Kuhn says there are rules (the Fivefold Way) for theory choice, but their application may be problematic and they cannot be given objective justification. Feyerabend says there are no rules whatsoever but, like Kuhn, rests much of his case on the existence of incommensurable theories.

We can also compare Kuhn with Popper and Lakatos by noting that, roughly speaking,

$$\text{Paradigm} = \text{Hard core} + \text{Positive heuristic}$$

enabling us to connect Lakatos's SRPs to the notion of a paradigm. As far as Popper is concerned, his central themes of conjecture, test, refutation, are also present in Kuhn's world, but only during the course of practicing normal science. Popper's contention that there is no rationale for the introduction of new conjectures in science, but only for the exposure of such conjectures to falsifying tests, is basically similar to Kuhn's claim that there is no rationale for the introduction of a new paradigm, but only for the attempt to "articulate" the paradigm and make it deal successfully with anomalies. The point of divergence between Kuhn and Popper arises when it comes time to shift from one paradigm to another. Popper believes this can and should (and is) done rationally, logically, and with little fuss; Kuhn says this method may be fine in the abstract, but *real* science just doesn't work that way.

With Kuhn we have come to the end of the line as far as contemporary views on the ways science operates both to form and

to validate its view of the world. Since the path from Wittgenstein to Kuhn has been a complicated one filled with lots of
switchbacks and strange meanderings, in the next section I'll try
to summarize the competing positions as well as briefly reexamine our original question: How real is scientific reality?

PHILOSOPHICALLY SPEAKING

When embarking upon this whirlwind tour of twentieth-century
philosophy of science, our point of departure was to explore the
two basic issues: What is the connection between scientific theories (language) and objective reality, and does science have any
special sort of procedures or methods for either generating new
theories or evaluating competing ones? Note again here the important point that when we use the term *method* in this setting,
we're referring to a method for generating theories and not to
the more common concept of the "scientific method" as constituting the potentially infinite sequence hypothesis → experiment → hypothesis . . . These questions led us to divide beliefs
on the nature of reality into three categories:

- *Realism* = Objective reality exists.
- *Instrumentalism* = Reality is the readings noted on measuring
  instruments.
- *Relativism* = Reality is what the community says it is.

We also saw that beliefs as to whether or not there's method
in the madness of science determine one's position as a rationalist or an irrationalist, with rationalists believing in method, ir-
rationalists not. The various philosophers and philosophical
schools took differing views on these matters, and to expound
them occupied a lot more time and space than I'd intended, but
necessarily so. Consequently, before going on to consider what
the practicing scientists themselves, as well as competing ideologies, have to say about these matters, I have tried to summarize
the story so far in Table 1.2. As the table shows, the overwhelming conclusion of the philosophers is that, as Einstein said, "it's
all relative." But we saw earlier that ten out of eleven everyday
physicists supported the idea of an objective reality "out there"
that their equations were describing. To address this paradox,
let's quickly hear from the laboratory instead of the ivory tower

SCHOOL	REALITY BELIEF	METHOD	ARGUMENT
Wittgenstein I	realism	rationalist	picture language
Wittgenstein II	relativism	irrationalist	language rules
logical positivists	instrumentalism	rationalist	verification principle
Popper	realism	rationalist	falsification
Lakatos	relativism	rationalist	SRPs
Feyerabend	relativism	irrationalist	"anything goes"
Kuhn	relativism	rationalist	paradigms

TABLE 1.2 *The battle of the philosopher kings*

and listen to what the players rather than the Monday morning quarterbacks have to say about the whole business.

In 1979 the Institute for Advanced Study in Princeton held a celebration to honor the one hundredth anniversary of the birth of Einstein, the institute's first and most celebrated resident genius. To plan for this celebration, a committee was formed at the institute to arrange a program and invite scholars from around the world to participate. Just as Caesar divided all Gaul into three parts, the IAS committee decided to organize the Einstein centennial similarly, focusing on Einstein's science, the historical genesis of his ideas, and, finally, the philosophical impact of his work. As Freeman Dyson tells it, the committee solicited names and put together lists of scholars who could be invited in each of the three areas. The committee was personally acquainted with almost everyone on the list of scientists. As to the historians, the committee didn't know them personally but at least had heard of most of them and knew of their work. But when it came to the philosophers of science, Dyson remarks that the committee was not only unfamiliar with them personally, but had never even heard the names of most of them! More than any abstract argument could ever hope to show, this little episode conveys the level of contact between the activities of the working scientist and the arguments of the philosopher: It is exactly zero! In Dyson's words, "There's a whole culture of philosophy out there somewhere with which we have no contacts at all. . . . there's really little contact between what we call science and what these philosophers of science are doing—whatever that is."

Dyson's observation serves to unravel the contradiction noted a moment ago between the beliefs of scientists and those of philosophers. As far as most practicing scientists are concerned, there's nothing more dangerous than a philosopher in the grip of a theory. In fact, there appears to be something of an unrequited love affair between the scientists and philosophers, in which the scientists by and large spend their days ignoring the attempts by the philosophers to press their attentions upon them. As an indicator of the state of affairs, the physicist Murray Gell-Mann at all times carries with him a doctor's prescription forbidding him to argue with philosophers on the grounds that it could be dangerous to his health!

So we come to the perhaps not so surprising conclusion that if you want to know about how scientists really think and work, you'll get no help from a philosopher of science. However, if your concerns go beyond what scientists do and encompass the broader issues of the *significance* of what they do and its relationship to other knowledge-generating mechanisms, then, as noted before, a consideration of matters philosophical is unavoidable. Most of our stories in this volume center upon what scientists are really doing, but in each one of them there is a strong undercurrent of philosophical presupposition conditioning the interpretation of the results. The reader should try to keep these deeper issues in mind as we go along, as a guide to evaluating the myriad competing arguments.

While philosophical factors probably are honored more in the breach than in the practice of science, sociological pressures are another matter. Science is not yet done by impersonal, uninvolved machines, but by real, live, thinking and feeling human beings, and it's impossible for this fact not to have some impact upon the way science proceeds to its conclusions about the way the universe functions. Let's take a few pages to consider the sociology of science rather than its philosophy, as another avenue to walk down on our way to learning about the way science comes to what it sees as "truth."

A TALE OF TWO SUICIDES

Ludwig Boltzmann and Paul Kammerer were both professors at the University of Vienna in the early part of this century; they were both popular with their students and held in great esteem by their colleagues; and they both committed suicide. While perhaps extreme in the outcome, these two cases serve as examples of one aspect of the way scientific truth is determined at least as much by the social climate of the times as by the dictates of reason and logic alone.

Boltzmann, a physicist, is perhaps best remembered for his work in thermodynamics and the connections he discovered between the theory of heat and the more general issues of randomness and order. He is today credited with having introduced the notion of *entropy* as a measure of the disorder present in a collection of objects of any sort, an idea that later served as the basis for the theory of information, which turned out to be so crucial to the development of modern communications technology. In fact the formula $S = k \log W$, expressing the entropy S as being proportional to the logarithm of W, the number of possible states that a system can assume, is engraved on Boltzmann's tombstone in Vienna's Zentralfriedhof, a fitting memorial to the importance of this fundamental idea. In this expression, the constant of proportionality k is even today termed *Boltzmann's constant* in recognition of this magnificent achievement. But at the time he was carrying out this pioneering work, the achievement was anything but magnificent, at least if one was listening to the leading scientists of the day.

Boltzmann's problem was that his theory of heat involved an assemblage of atoms moving according to the usual rules of Newtonian mechanics. He used this concept of an atom as a particle of matter to construct his theory of heat as a statistical property emerging out of the overall motion of these atoms. Note that this idea was put forth around the turn of the century, several years before the work of Ernest Rutherford, J. J. Thomson, and Niels Bohr gave the concept of an atom its modern birth. As a result of his atomistic speculations, Boltzmann came into heated conflict with several of the giants of the scientific community, most notably his Viennese colleague Ernst Mach and the German physical chemist Wilhelm Ostwald, who argued forcefully against the idea of the atom. Ostwald, in particular, preferred a theory of heat based upon the notion of energy rather than matter. Depressed by the acrimony of this opposition, as well as his failing eyesight and what he thought of as the decline of his mental faculties, Boltzmann took his life in Duino, Italy, on September 5, 1906.

Tragically, Boltzmann's suicide took place almost cotermi-
nously with the work by Thomson and Rutherford in Britain
that would lead to a complete vindication of his ideas. So here
we have a textbook illustration of how the social climate of the
scientific community, as well as the influence of two great men,
acted to delay introduction of what ended up being a major contribution to our way of thinking about the way the world works.
Now let's move the clock forward almost exactly twenty years
and examine the case of another Viennese professor as illustration of how these same social forces can work to rid science of
equally controversial, but this time erroneous, ideas.

Paul Kammerer was a professor of biology at the University
of Vienna in the 1920s. Accounts credit him with an almost magical skill at breeding amphibians and other types of animals.
They also note that he was an ardent socialist and crusader for
the political causes of what today we would term the liberal left.
Given this combination of scientific and political leanings, it's
perhaps not surprising that Kammerer supported the idea that
acquired characteristics can be pass on to offspring, i.e., Lamarckian inheritance. For ideologues bent upon improving the
human race, the idea that behavioral traits like learning, altruism, and the like can be acquired holds great appeal. So it was
for Kammerer, too, and he set out to prove the idea with his now
infamous experiments on the midwife toads.

Generally these toads breed on land, with the male lacking the
so-called nuptial pads of the male members of other species of
toads that breed in the water. These pads are rough patches on
the hands of the male that he uses to grab on to the back of the
slippery female during the course of mating in water. Kammerer's experiment involved forcing the midwife toad to breed
in water for several generations, his claimed results being that
such toads then developed the nuptial pads characteristic of
their naturally water-breeding cousins. The supporters of Kammerer focused upon this experiment as clear-cut evidence for
Lamarckism; opponents remained highly doubtful and requested
a closer look at the evidence.

These experiments with the midwife toad came under heavy
attack from naturalists in both Europe and America, especially
William Bateson in England and Kingsley Noble in New York.
On a visit to Vienna in 1923, Bateson saw Kammerer's last re-

maining specimen of a midwife toad with nuptial pads and later asked to reexamine it in his own lab. Kammerer replied that it could not be sent from Vienna. At the same time, Noble was having doubts about some of the particulars of the physical structure of Kammerer's claimed nuptial pads, and visited Vienna in 1926 to examine the last specimen personally. His results, published later that year in *Nature,* claimed that the so-called pads were nothing more than black markings made with India ink.

At the time of Noble's report, Kammerer was preparing to leave Vienna for a position at Moscow University as head of a new laboratory in Lamarckian biology. Noble's *Nature* article appeared on August 7, 1926. In a letter of September 22 to the Soviet Academy of Sciences, Kammerer wrote that he had examined Noble's claims and found them to be totally accurate. He went on to protest his ignorance of how the inking had been done, but acknowledged that his experimental conclusions about Lamarckism were baseless. After withdrawing from the post in Moscow, the letter concluded with the poignant statement "I am not in a position to endure this wrecking of my life's work, and I hope that I shall gather together enough courage and strength to put an end of my wrecked life tomorrow." And, in fact, during a walk in the Wienerwald the next day, Kammerer shot himself in the head. This was another extreme example of scientific peer-group pressure and its sometimes tragic effect upon the lives of scientists deviating from the group norms. Only this time the pressure acted to discredit wrong results rather than to suppress correct ones.

The tales of these two Viennese professors serve to underscore the sometimes dramatic influence that the social component of science plays in establishing what we take to be the scientific "truth" of the moment. These social factors operate within the scientific community itself as well as in the outside world, shaping not only the way scientific activity is carried out but also the manner in which certain ideas, like Boltzmann's, are buried while others thrive. One of the pioneers in studying these social determinants, at least inside science itself, is the sociologist of science Robert K. Merton, who in 1942 identified a small set of what he termed *norms* characterizing the scientific enterprise. Roughly speaking, in modern terms we can give Merton's norms as:

- *Originality:* Scientific results should always be original, i.e., novel. Studies that add nothing new to what is already known are not part of science.
- *Detachment:* Scientists undertake their work with no motive other than the advancement of knowledge. They should have no personal axes to grind insofar as the results of their work go, and they should have no psychological commitment to a particular point of view. The impersonal style of most scientific communications is a direct consequence of this norm.
- *Universality:* Claims and arguments should be given weight according to their intrinsic merits alone, and should not depend upon religious, social, ethnic, or personal factors surrounding the individuals who make them. In short, there are no privileged sources of scientific knowledge.
- *Skepticism:* No scientific statements of fact should be taken on faith. All claims should be carefully scrutinized for invalid arguments and errors of fact, and any such mistakes should be made public immediately. To put it simply, scientists should trust no one, at least not when it comes to claims of scientific truth.
- *Public accessibility:* All scientific knowledge should be freely available to anyone. Thus, results of research are not the private property of the scientist, but are public goods that should be transmitted immediately to the community of science. This norm lies at the heart of debates as to whether or not engaging in classified military research is scientifically ethical.

Anyone involved with the way scientific practice actually works will immediately recognize that these prescriptions are violated every day of the week in both trivial and not so trivial ways, serving the same role in science that general laws serve for society at large. There's nothing particularly disturbing about this gap between theory and practice, just as the fact that human beings jaywalk, rob banks, and drive their cars too fast is not really news either. What is disturbing, to some anyway, is what appears to be an increasing incidence of such violations of the spirit of science, at least as it's embodied in these norms. Such an increased pace of corner cutting in science seems especially evident in the last decade or so, certainly aided and abetted by science's Faustian bargain with government funding agencies. Nevertheless, the Mertonian norms are still the ethos to which the community of scientists subscribes, and form the heart of the code by which the behavior of most scientists is judged by their peers. And in exactly this way the norms make their contribution to the way scientists think, hence to what they ultimately come to accept as the way things are. But these factors working inside the scientific community are not the only social components influencing the work of science. Of equal importance are the forces affecting science from the outside, especially in today's mass-media-saturated and cash-hungry world.

In his 1971 State of the Union address, President Richard M. Nixon declared that the time had come for the country to wage war on cancer, with the "same kind of concentrated effort that split the atom and took man to the Moon. . . ." This pronouncement led to an avalanche of money pouring into the nation's cancer research laboratories, and resulted not only in a war on cancer but also in a war among the various research establishments for a generous hunk of the federal government's cancer war chest. One of the foot soldiers in both of these conflicts was William T. Summerlin, a young skin specialist at the prestigious Sloan-Kettering Institute for Cancer Research in New York City.

Amid the high-pressure political climate surrounding cancer research and the feverish hustling and grantsmanship, in March 1973 Summerlin applied for a five-year federal research grant from the American Cancer Society to pursue his special interest in skin grafts and immunology. In particular, Summerlin felt that he was on the track of developing procedures whereby skin treated by his technique could be transplanted without rejection. Thinking that a little favorable publicity never hurt the case of a relatively obscure, but ambitious, young researcher, Summerlin presented an outline of his work in progress at a science writers' convention. The results were predictable: a three-column headline the next day in *The New York Times* declaring LAB DISCOVERY MAY AID TRANSPLANTS. Summerlin was on his way, or so it seemed.

During the course of the next year, while Summerlin traveled the country presenting seminars and lectures on his work, colleagues were finding it increasingly difficult to confirm his results by independent experiments. In fact, even workers in

Summerlin's own laboratory at Sloan-Kettering were unable to reproduce the claimed properties of the specially treated "Summerlin skin," leading to a showdown between Summerlin and Sloan-Kettering Director Dr. Robert A. Good in March 1974. On his way to this fateful meeting, Summerlin pulled out a black felt-tip pen and hurriedly inked in some dark patches on the white mice he was bringing as evidence for his claims. At the time Good didn't notice the Summerlin embellishments, and it was only when the mice were returned to the lab assistant that Summerlin's "help" was discovered. The assistant immediately reported the matter to his boss, at which point Summerlin was instantly suspended. While he denied any wrongdoing, asserting that he had inked in the skin grafts on the mice only to make them more easily identifiable, Summerlin's credibility was shattered by the incident, along with the credibility of his supposed technique for skin grafts.

Interestingly enough, the Summerlin episode bears some strange similarities to that of Kammerer and the midwife toads, although without the same tragic suicidal ending. The point in raising these cases here is not so much the issue of whether or not Kammerer or Summerlin was really guilty of fraud, but rather to illustrate the degree to which forces outside the world of science, in this case the federal research-funding establishment and the public at large, contribute to creating a climate that can drive scientists to manufacture and/or artificially enhance what they claim are "the facts." And money is not the only such pressure. Political considerations, especially those involving what is often termed "human nature," can and do play a dramatic role in influencing what's scientifically "right." A good illustration of this kind of effect was the controversy over social Darwinism in the first half of the century, a debate about which we shall have much more to say later when we consider its modern incarnation: the Sociobiology Problem. In this context, it may even be safe to say that the real issue is the conflict between the norms of science, as exemplified by Merton's list, and the "norms" of politics as encoded in the ideologies of certain political movements (in the case of sociobiology, Marxism).

The foregoing stories barely scratch the surface of the many ways in which sociological considerations shape what science thinks of as being true, with many far more detailed accounts noted under "To Dig Deeper" in this volume. For our purposes here, the main consideration is the manner in which these social factors influence the way science validates its claims and comes to a consensus on a given issue. The heart of the difficulty is that knowledge is underdetermined. Thus, there are always many different theories, each of which can give a plausible account of the available facts. So how are we to choose one and let the others go? The basic problem is encapsulated in the remark of the philosopher Willard Van Orman Quine, who noted that "any statement can be held true, come what may, if we make drastic enough adjustments elsewhere in the system." One natural place to make these drastic adjustments is in the cultural background to the problem, thereby creating a climate in which only one or at most a few of the contending theories can survive. Again, we will see ample evidence of this kind of "cultural imperialism" in the raging sociobiology debate covered in Chapter Three.

As to arrival at a consensus, the key factor is the Mertonian norm relating to the public character of scientific knowledge. The rule that scientific information is communicated explicitly and unambiguously influences both the form and the content of knowledge that is labeled "scientific." For example, this norm goes a long way toward accounting for why experimental verification involving neutral instrumentation occupies such a hallowed position in science, as well as the great value attached to quantitative observation and expression of results in mathematical form. All of these features contribute to the public accessibility of the information and the reproducibility of the results, at least in principle. One need only consider other fields like literature or the arts, where such a norm is not the norm, to see some of the ways in which scientific knowledge differs in significant ways from these other forms of reality representation.

Since we'll see many concrete instances of these sociological factors entering into the stories that follow, there's no need to belabor the point here in the abstract. For now, it's of somewhat more interest to look at some of the knowledge-generating devices that make some pretense to a degree of scientific character, in their goals if not their methods. With the above ideas as prelude, the reader should be in a better position to distinguish those groups doing what we would now term science from those practicing at the fringe.

We began this chapter with the dual stories of Jocelyn Bell and Immanuel Velikovsky, noting their positions at opposite ends of the spectrum of what's currently held to be "good science." We are finally in a position to give the long answer to the question posed earlier about why Velikovsky's work has been relegated to the dustbin of pseudoscience, while Bell's was rewarded with the Nobel Prize for physics (although not to her).

ON THE FRINGE OR AT THE CUTTING EDGE?

As editor of a scientific journal, I'm regularly faced with the unpleasant task of telling potential contributors that their papers are not suitable for publication. Generally the reasons are the usual ones: trivial or nonexistent results, poor writing, work outside the scope of the journal, and so on. However, occasionally I get a paper that I don't even bother to send out for the customary refereeing process, rejecting it out of hand. Such papers are the bane of the editor of almost every scientific publication, and every editor soon becomes sensitized to their telltale aroma of nonsense masquerading as science. Since my own journal is devoted to mathematics, papers of this sort tend to involve such well-known impossibilities as squaring the circle, trisecting an angle, and doubling the cube, although they occasionally address famous outstanding problems like Fermat's Last Theorem or the Riemann Hypothesis (in which case I'm compelled to look at them seriously, even though there's not yet been one that was correct). Luckily for me, mathematics is an area where it's difficult to try to dress up such pseudoscience in respectable clothes and not have it show. Certainly my colleagues in biology, medicine, and the social sciences must have it much worse in this regard. But just what is it about this kind of paper that immediately stamps it as pseudoscience to the trained (and jaundiced) scientific eye? To answer this puzzling query, let's briefly recall what's been learned so far about the actual practice of science in today's world.

Our deliberations up to now allow us to summarize compactly the *practice* as opposed to the *philosophy* of science in the following two principles:

A. There is an ideology of science consisting of a *cognitive structure* (facts → hypothesis → experiment → laws → theory), together with the processes of *verification* and *peer review*.

B. Science is a *social activity,* with the standards for what constitutes good science determined by the norms of a particular community.

With these facts of modern scientific life in mind, let me now offer a short checklist of "sights and sounds" (and smells) for detecting pseudoscience. If you're reading a paper and catch the whiff of even one of the items on this list, be assured that the author is dealing in pseudoscience, at least by the standards prevailing in today's world of science. For the following list I am indebted to the outstanding work *Science and Unreason* by Michael and Daisie Radner, to which I direct the reader's attention for a far more extensive account of the whole culture of pseudoscience and pseudoscientists.

HALLMARKS OF PSEUDOSCIENCE

* *Anachronistic thinking:* Cranks and pseudoscientists often revert to outmoded theories that were discarded by the scientific community years, or even centuries, ago as being inadequate. This is in contrast to the usual notion of crackpot theories as being novel, original, offbeat, daring, and imaginative. Good examples of this kind of crankishness are the creationists, who link their objections to evolution to catastrophism, claiming that geological evidence supports the catastrophic rather than uniformitarian view of the kind of geological activity they associate with evolution. The argument is anachronistic insofar as it presents the uniformitarianism-catastrophism dichotomy as if it were still a live debate.

* *Seeking mysteries:* Scientists do not set out in their work to look for anomalies. Max Planck wasn't looking for trouble when he carried out his radiation emission experiments and Michelson and Morley certainly were not expecting problems when they devised their experiment to test for the luminiferous ether. Furthermore, scientists do not reject one theory in favor of another solely because the new theory explains the anomalous event. On the other hand, there's an entire school of pseudoscience devoted to enigmas and mysteries, be they the Bermuda Triangle, UFOs, yetis, spontaneous combustion, or other even more offbeat phenomena. The basic principle underlying such searches seems to be that "there are more things in heaven and earth than are dreamt of in your philosophy," cou-

pled with the methodological principle that anything that *can* be seen as a mystery *ought* to be seen as one.

- *Appeals to myths:* Cranks often use the following pattern of reasoning: Start with a myth from ancient times and take it as an account of actual occurrences; devise a hypothesis that explains the events by postulating conditions that obtained at that time but that no longer hold; consider the myth as providing evidence for support of the hypothesis; argue that the hypothesis is *confirmed* by the myth as well as by geological, paleontological, or archaeological evidence. This is a pattern of circular reasoning that is absent from the blackboards and laboratories of science.

- *A Casual approach to evidence:* Pseudoscientists often have the attitude that sheer quantity of evidence makes up for any deficiency in the quality of the individual pieces. Further, pseudoscientists are loath ever to weed out their evidence, and even when an experiment or study has been shown to be questionable, it is never dropped from the list of confirming evidence.

- *Irrefutable hypotheses:* Given any hypothesis, we can always ask what it would take to produce evidence against it. If nothing conceivable could speak against the hypothesis, then it has no claim to be labeled scientific. Pseudoscience is riddled with hypotheses of this sort. The prime example of such a hypothesis is creationism; it's just plain not possible to falsify the creationist model of the world, as we'll see in the next chapter.

- *Spurious similarities:* Cranks often argue that the principles that underlie their theories are already part of legitimate science, and see themselves not so much as revolutionaries but more as the poor cousins of science. For example, the study of biorhythms tries to piggyback upon legitimate studies carried out on circadian rhythms and other chemical and electrical oscillators known to be present in the human body. The basic pseudoscience claim in this area is that there is a similarity between the views of the biorhythm theorists and those of the biological researchers, and therefore biorhythms are consistent with current biological thought.

- *Explanation by scenario:* It's commonplace in science to offer scenarios for explanation of certain phenomena, such as the origin of life or the extinction of the dinosaurs, when we don't have a enough data to reconstruct the exact circumstances of the process. However, in science such scenarios must be consistent with known laws and principles, at least implicitly. Pseudoscience engages in explanation by scenario *alone,* i.e., by mere scenario without proper backing from known laws and theories. A prime offender in this regard is the work of Velikovsky, who states that Venus's near collision with the Earth caused the Earth to flip over and reverse its magnetic poles. Velikovsky offers no mechanism by which this cosmic event could have taken place, and the basic principle of deducing consequences from general principles is totally ignored in his "explanation" of such phenomena.

- *Research by literary interpretation:* Pseudoscientists frequently reveal themselves by their handling of the scientific literature. They regard any statement by any scientist as being open to interpretation, just as in literature and the arts, and such statements can then be used against other scientists. They focus upon the words, not on the underlying facts and reasons for the statements that appear in the scientific literature. In this regard, the pseudoscientists act like lawyers gathering precedents and using these as arguments, rather than attending to what has actually been communicated.

- *Refusal to revise:* Cranks and crackpots pride themselves on never having been shown to be wrong. It's for this reason that the experienced scientific hand never, under any circumstances, enters into dialogue with a pseudoscientist. But immunity to criticism is no proof of success in science, for there are many ways to fend off attacks: Write only vacuous material replete with tautologies; make sure your statements are so vague that criticism can never get a foothold; simply refuse to acknowledge whatever criticism you do receive. A variant of this last ploy is a favorite technique of pseudoscientists: They always reply to criticism, but never revise their position in light of it. They see scientific debate not as a mechanism for scientific progress but as an exercise in rhetorical combat. Again the creationists serve as sterling testimony to the power of this principle.

The major defense of pseudoscience is summed up in the statement "Anything is possible," the pseudoscientific version of Feyerabend's philosophical theme song "Anything Goes." Earlier we considered the question of competition between models and theories and drew up a few ground rules by which the competition is generally carried out in legitimate scientific circles. Let's look at how pseudoscientists, with their "Anything is possible" shield, enter into such competition.

In the competition among theories, the pseudoscientist makes the following claim: "Our theories ought to be allowed into the competition because they may become available alternatives in the future. Scientists have been known to change their minds on the matter of what is and is not impossible, and they are likely to do so again. So who's to say what tomorrow's available alternatives may be?" In other words, anything is possible! The fact that a theory may become an available alternative in the future does not constitute a reason for entering it in the competition today. Every competitor now must be an available alternative now. The pseudoscientist suggests that we may as well throw away the current scientific framework since it will eventually have to be replaced anyhow.

By referring to a future but as-yet-unknown state of science, the cranks are in effect refusing to participate in the competition. This would be all right if they didn't at the same time insist on entering the race. It's as if one entered the Monaco Grand Prix with a jet-propelled car and insisted on being allowed to compete because, after all, someday the rules may be changed to make it a jet car race!

The pseudoscientists also worm their way into the competition by putting the burden of proof on the other side. They declare that it's up to the scientific community to prove their theory wrong, and that the theory must be taken seriously if the community cannot do so. The obvious logical flaw is the assumption that failing to prove a theory impossible is the same thing as proving it possible. While the principle of innocent till proven guilty may be used in Anglo-Saxon courts of law, scientific debate is not such a court. The reason why pseudoscientists think they can put the burden of proof on the scientists can be traced to a mistaken notion of what constitutes a legitimate entry in the debate. They think that the scientific method places a duty on the scientific community to consider *all* proposed ideas that are not logically self-contradictory. In their view, to ignore any idea is to be prejudiced.

Finally, we note that the pseudoscientists often act as if the arguments supporting their theory were peripheral to the theory. Science is defined in terms of *how* and *why* we know something, not *what* we know. Thus, the pseudoscientists fail to see that what makes a theory a serious contender is not just the theory, but the theory plus the arguments that support it. Cranks think that somehow the theory stands on its own, and that the only measure of its merit for entering the competition is its degree of daring and novelty. Hence, they think the scientific community has only two choices: admit their theory into the competition or else prove it to be wrong. However, when it comes to defending a theory or model in scientific debate, without high-quality supporting evidence and a solid conceptual scheme, there's just no time, room, or patience for the "Anything is possible" antics of pseudoscience.

As a postscript on the pseudoscientists, it's of interest to ask why the ideas of many pseudoscientists like Velikovsky are so popular. While it is true that Velikovsky's concepts are a little simpler than those used by modern astronomers and paleontologists, his real advantage is that they are so much easier to visualize mentally and come to terms with. In short, they appeal to what John Q. Public would call common sense. Unfortunately, neither the world nor science is as simple as naïve common sense would have us believe. For example, what kind of peasant cunning would suggest that energy levels in atoms can come only in discrete packages? Common sense would say that if you can walk up stairs one step at a time, then you can also stroll up a ramp to get to the same place. But modern physics says no: Change of energy levels can occur only in discrete steps. The more developed a scientific specialty becomes, the less reliable common sense is as a guide. In fact, there are aspects of science that are just plain contrary to common sense, like the staircase example just noted. The point to keep in mind is that most beliefs being promoted as alternatives to science are deliberately calculated to fit smoothly into what common sense suggests is the way things *should* be, as well as the way to solve all our problems. Within these comforting world views, we have no problems of our own—everything that happens to us does so because of bad aspects of Jupiter, the work of the devil, or the will of superior beings from Andromeda. At root, these beliefs are a measure of the degree of disappointment with which the general public greets the revelations of modern science. The average man wants complete, easy-to-understand, clear-cut answers, when all that science has to offer is arcane, difficult-to-follow ifs, ands, buts, or maybes.

Belief systems outside science come in many forms, some of them covered by the general umbrella of pseudoscience. By far the most interesting and important alternative to a scientific ordering of the world is that provided by the principles and tenets of organized religion. From the beginnings of Western science in the Middle Ages, there has been a sort of (not always undeclared) guerrilla war waged between the Church and the scientific community on the matter of which is the keeper of true knowledge about the nature of the cosmos. In the next section we will examine this conflict as our final statement about the alternative realities that we use to shape and interpret our daily lives.

THE PULPIT AND THE LAB

A few years ago Daysi Fernandez, a mother of three living on welfare in New York City, bought a lottery ticket that came up a winner, returning almost $3 million, a tidy profit on a $4 investment. Little did Mrs. Fernandez realize that in her good fortune she would become embroiled in a classic case pitting the claims of science against those of religion. As the story goes, Mrs. Fernandez had asked a young friend, John Pando, to purchase lottery tickets for her. Pando, a staunch believer in the power of prayer, thought that the chances of success for one of the tickets would be greatly enhanced if he asked for the divine intervention of Saint Eleggua. Apparently Mrs. Fernandez was sympathetic to his beliefs, for he claimed that she had promised to give him half the proceeds if any of the tickets struck gold. If you've already guessed the punch line of this story, you're just a bit ahead of me.

One of Mrs. Fernandez's tickets was drawn to the tune of $2,877,203.30, but she refused to fork over the promised half of the pie to Pando. In the tried and true American fashion for dealing with such slights, Pando's immediate response was to file a lawsuit against her, in an attempt also to gain entry to the Millionaires' Club. Mrs. Fernandez argued that the agreement was illegal and/or unenforceable on a number of grounds, including the fact that John Pando was a minor under the age of eighteen. After hearing the competing arguments, Judge Edward Greenfield of the New York County Supreme Court ruled on the matter.

The judge found in favor of Pando on most of the points, including the matter of age, but came up with a final verdict in favor of Mrs. Fernandez on the grounds that it was impossible in a court of law to prove that "faith and prayers brought about a miracle and caused the defendant to win." In other words, Pando hadn't proved that Saint Eleggua had rigged the lottery to point the finger of fate at Mrs. Fernandez. As far as it goes, this seems a defensible statement. But what is open to serious debate is the reasons given by the judge for denying Pando a share of the fortune.

Judge Greenfield in effect assumed a priori that religious beliefs are not amenable to scientific testing. As part of his decision, the judge also stated that rainmaking by cloud seeding would qualify for payment, but that the production of rain by dances, chants, and the other tricks of the medicine man's trade would not. Thus, the Fernandez case opens up for further inspection the age-old question of where a belief system stops and science begins.

In the Reality Game, religion has always been science's toughest opponent, perhaps because there are so many surface similarities between the actual practice of science and the practice of most major religions. Let's take mathematics as an example. Here we have a field that emphasizes detachment from worldly objects, a secret language comprehensible only to the initiated, a lengthy period of preparation for the "priesthood," holy missions (famous unsolved problems) to which members of the faith devote their entire lives, a rigid and somewhat arbitrary code to which all practitioners swear allegiance, and so on. These features are present in most of the sciences as well, and bear a striking similarity to the surface characteristics of many religions. Both scientific and religious models of the world direct attention to particular patterns in events and restructure how one sees the world. But at a deeper level there are substantial differences between the religious view and that of science.

Let's consider some of the major areas in which science and religion differ:

- *Language:* The language of science is primarily directed toward prediction, explanation, and control; religion, on the other hand, is an expression of commitment, ethical dedication, and existential life orientation. So even though there are superficial similarities at the syntactic level, the semantic content of scientific and religious languages are poles apart.

- *Reality:* In religion, beliefs concerning the nature of reality are presupposed. This is just the opposite of the realist view of science, which is directed toward discovering reality. Thus religion must give up any claims to truth, at least with respect to any facts external to one's own commitment. In this regard, the reality content of most religious beliefs is much the same as in the myths considered earlier. Fundamentally, what we have in science is a basic belief that the universe is understandable using rational arguments, experimental observations, even divine inspirations, but no acts of blind faith. This is a viewpoint that is not necessarily shared by many religions.

- *Models:* While both scientific and religious models are analogical, and used as organizing images for interpreting life experiences, religious models also serve to express and evoke distinctive attitudes, as well as to encourage allegiance to a way of life and adherence to policies of action. The imagery of religious models elicits self-commitment and a measure of ethical dedication. These are features completely anathematic to the role of models in science. In religion the motto is "Live by these rules, think our way, and you'll see that it works." The contrast with the traditional ideology of science is clear.

- *Paradigms:* In the discussion of paradigms, we saw that scientific paradigms were subject to a variety of constraints like simplicity, falsification, the influence of theory on observation, and so forth. *All* of these features are absent in the selection of a religious paradigm.

- *Methods:* In science there is a set of procedures to get at the scheme of things: observation, hypothesis, experiment; in religion there is a method, too—divine enlightenment. However, the religious method is not repeatable, nor is it necessarily available to every interested investigator.

Table 1.3 displays a comparative chart of the different ways of science and religion. How are we to divine what this table is trying to convey about the respective abilities of science and religion to tell us anything useful about ourselves and the universe

ISSUE	RELIGION	SCIENCE
subject matter	God and humankind	phenomena of Nature
information source	revealed word, holy books	observations, experiments
objective of study	purpose and plan	mechanisms
language	everyday speech	mathematics
method	literary interpretations	measurement and analysis
results	moral imperatives	explanations
validation	personal experience	replication, testing
limitations	mechanisms unexplained	no goals or values
community	church	scientific establishment

TABLE 1.3 *Religion compared with science*

we inhabit? It seems that there are at least three possible answers to this classic conundrum:

1. *Two realms:* Science and religion have different spheres of jurisdiction.
2. *Concordance:* Religious and scientific explanations of Nature can be brought together on the same plane.
3. *Partial views:* Science and religion each illuminate the same reality (whatever *that* might be), but from different perspectives.

To my mind, only the last possibility makes any sense whatsoever. The first leads to the all too depressing territorial disputes of the kind that so much blood has been shed over through the years, while the second is self-defeating since scientific views are always changing. As a result, a theology that attaches itself to one scientific family today will surely be an orphan tomorrow.

With the above considerations on religion under our belt, we see that both pseudoscience and religion provide alternate reality-structuring procedures radically different in character from those employed in science. It's of interest to ponder why there is such a diverse mixture of nonscientific knowledge, especially in view of the claims of virtually every sect that its own brand of medicine is uniformly most powerful.

My view on this matter is quite simply that neither science nor

religion nor pseudoscience gives a product that is satisfactory to
all customers; the wares are just not attractive enough. In some
cases the beliefs are not useful in the way that people want to
employ them. For example, many people have a deep-seated psychological need for security and turn to conventional religion for
myths of all-powerful and beneficent Beings who will attend to
this need for protection. Science, with its mysterious and potentially threatening pronouncements about black holes, the "heat
death" of the universe, evolution from lower beings, nuclear
holocaust, and so on, offers anything but comfort to such primal
needs and, as a result, loses customers to the competition. Basically, beliefs thrive because they are useful, and the plain fact is
that there is more than one kind of usefulness.

To the practicing scientist, the foregoing observations come as
a sobering if not threatening conclusion, since they seem to put
in jeopardy the conventional wisdom that the road to truth lies
in the "objective" tools of science, not the subjective, romantic
notions of believers and crusaders. But if we accept Feyerabend's arguments of alternative and equally valid belief systems,
then we are inexorably led full circle back to the position that
there are many alternative realities, not just within science itself
but outside as well, and the particular brand of reality we select
is dictated as much by our psychological needs of the moment as
by any sort of rational choice. In the final analysis, there are no
complete answers but only more questions, with science providing procedures for addressing certain important and interesting
classes of such questions.

INTO THE COURTROOM OF BELIEFS

The British philosopher John Locke appears to have been the
first to use the word "science" in anything like its modern meaning when he equated "scientifical" with certainty and demonstration of knowledge about the physical world. In the chapters
that follow, we will be out to question the degree to which science
delivers on these lofty aims. Our dual philosophical themes center about the eternal puzzles: What is real, and what is our relationship as human beings to this reality? In the course of
attempting to shed light on these Bobbsey Twins of philosophical speculation, I have chosen the vehicle of a courtroom meta-

phor within which the competing scientific (and sometimes pseudoscientific and/or religious) parties can plead their case. My reasons for this setting are best summed up in the remark by Henry Bauer that "where eminent men disagree violently, and both sides present their cases as proven, we can be rather sure that certainty is not in fact available, and that the matter is not technical but rather trans-scientific. It is a dispute over probabilities, values, desirability, *not* over facts." The only factor that characterizes science as a whole is that, in the long run, untruths are weeded out and what remains becomes more reliable. Thus, just as in economics where Adam Smith's Invisible Hand guides the flow of events into progressive channels, in science we have the Invisible Boot, which acts to kick out those ideas, theories, and beliefs that don't prove useful to enough people over a sufficiently long period of time.

I leave it to the reader to be the final judge of whether or not "scientism" (I promise that this will be my last -ism) establishes a case for its underlying thesis that "science = truth." But succeed or fail, I hope that as we go through the various case studies in scientific conflict that follow, the reader will not only get some basic grasp of the ideas themselves, but even more important will discover that these ideas are genuinely *worth* an attempt to understand them. Only by acquiring a deeper feeling for the processes as well as the results of science will it be possible to assess its merits effectively as a reality-generation activity. So now that the anthems have been sung, the pledges of allegiance given, and the witnesses called, the court is ready to hear the first case in the continuing litigation between science and Nature. Let the opening arguments proceed!

2

A WARM
LITTLE POND

===

CLAIM:
LIFE AROSE OUT OF NATURAL PHYSICAL
PROCESSES TAKING PLACE HERE ON EARTH

OUT OF THE FIRE AND INTO THE SOUP

By most standards of comparison, 1953 was an eminently forgettable year, with only a fifth straight World Series triumph for the Yankees, the death of Stalin, and Secretary of Defense Charles Wilson's immortal remark that "I thought what was good for our country was good for General Motors, and vice versa" brightening up what was otherwise a pretty dull trip around the Sun. But not so in the world of biology; in fact, for biologists 1953 was a vintage year the like of which had not been seen since the publication of Darwin's classic in 1859. In this

single twelve-month span, not only did Watson and Crick unravel the double-helix geometry of DNA and Frederick Sanger
work out the chemical structure of proteins, but also the modern
era of scientific investigation of the origin of life on Earth was
ushered in with Stanley Miller's experiment showing that the
chemical building blocks of life could be formed by natural
physical processes taking place in the primordial environment.
While the work of Watson, Crick, and Sanger is crucial for understanding how living forms function, it was Miller's experiment that set the stage for what has become the dominant
scientific paradigm for how life as we know it today got its start
here on Earth. To trace that thread, we must begin in 1923 in
Moscow with the unheralded publication of a booklet asserting
that there is no fundamental difference between living and nonliving matter.

Having just escaped the yoke of the czars and not yet stuck
their necks into the noose of Stalinism, Russians found the
Roaring Twenties to be an excellent decade for challenging established orthodoxies. So it seems appropriate that during this
time a thirty-year-old biologist, Alexander I. Oparin, should
present the first real scientific case against biblical creationism,
arguing that life could have arisen by natural physical means
here on Earth. The essence of Oparin's argument, later expanded upon in his 1936 book *Origin of Life,* was that geological
evidence suggests the atmosphere of the early Earth was filled
with gases like methane, ammonia, hydrogen, and water vapor—
but no oxygen (i.e., it was what chemists call a *reducing mixture*).
By pumping energy from lightning, ultraviolet radiation, volcanic heat, and natural radioactivity into such a blend of gases,
Oparin reasoned, the chemical components composing all living
things could be formed in the sea, ultimately accumulating to a
density at which they could link up to form the first primitive
living entities. A few years later, the British biologist J.B.S.
Haldane independently proposed the same general idea, colorfully expressing the character of such a primordial sea as a kind
of "hot dilute soup," leading to the modern labeling of this Oparin-Haldane Hypothesis as the Primordial Soup Theory.

It's of perhaps more than just passing curiosity to note that
both Oparin and Haldane were professed Marxists in a revolutionary era when it was fashionable to try to solve all sorts of
problems here and now by dialectical and materialistic means.

Oparin, who died in 1980, has been described as the kind of man who at dinner had a bottle of cognac on one side and a bottle of vodka on the other, both of which were empty by the end of the meal. While the two-fisted drinking stories may or may not be apocryphal, there is no doubt about Oparin's unfortunate political alliance with the crankish, but powerful, geneticist T. D. Lysenko. During Lysenko's tyrannical period of grace, Oparin reigned over the Biology Division of the Soviet Academy of Sciences, using his political clout to set Soviet biology back at least twenty years. Finally, upon the death of Stalin and the consequent decline of Lysenko, both Oparin and Lysenko were removed from their administrative posts at the academy and returned to the laboratory bench—Oparin to direct the Bakh Institute of Biochemistry (whose main activity involved the study of fermentation for making beer), Lysenko to continue his definite, but almost totally meaningless, Lamarckian experiments in changing winter wheat to spring wheat. An indication of the kind of political cunning that enabled Oparin to survive in such a shifting political environment is displayed in his remark to the journalist Harold Hayes, who, in a visit with Oparin shortly before his death, asked about his views on the treatment of the physicist and human rights advocate Andrei Sakharov. Oparin replied, "Of course, there are many Sakharovs in Moscow!" Incidentally, Haldane, who for many years served as the UK editor of the *The Daily Worker,* the newspaper of the Communist party, ultimately lost faith in Marxism for exactly the same reason that Oparin was catapulted to power—Lysenkoism. As we'll see later, even though they were the co-originators of the Primordial Soup Theory, the two also stood on opposite sides of the fence on the details of exactly how life actually came crawling out of the broth. But we're getting ahead of our story. Let's now go back to Stanley Miller and the Chemistry Department at the University of Chicago in the early 1950s.

At the time, Miller was a young graduate student in the department shopping around for a Ph.D. thesis topic. Initially he decided to do a theoretical project with Edward Teller, something involving the manner in which chemical elements could be synthesized in stars. However, Teller's departure shortly thereafter to set up what is now the Lawrence Livermore National Laboratory put an end to Miller's plan, sending him scurrying about for an alternate topic, as well as a new adviser. At this point fate intervened in the form of another faculty member, Harold Urey, who had earlier given a departmental seminar that Miller had attended and listened to with great interest. Urey, who won the Nobel Prize for chemistry in 1934 for the discovery of deuterium (heavy water), had turned his attention to problems of the origin of the solar system, claiming that the atmosphere of the early Earth would have been highly reducing, i.e., lacking any free oxygen. As a consequence, Urey argued that such an atmosphere should be a good place to synthesize organic compounds that could then form the necessary raw materials out of which to assemble the first living organisms. Urey concluded by suggesting that someone should do an experiment to test the feasibility of the idea. Later, Miller reports, he pointed out to Urey that Oparin had made the same suggestion in his book, although Urey's discussion was far more thorough and convincing.

Following Teller's departure for California, Miller told Urey that he wanted to do the experiment on organic synthesis in a reducing atmosphere for his thesis topic. Urey was initially against the idea, primarily because he saw it as a speculative project that could chew up lots of time and energy with no visible output—just the sort of project that a graduate student intent on getting a degree should steer clear of. However, Miller persisted and finally Urey relented, allowing Miller to begin the experiment in the autumn of 1952. The rest is history. The essence of the experiment is depicted in Figure 2.1.

To simulate the primordial atmosphere, Miller used a combination of methane (CH_4), ammonia (NH_3), water vapor (H_2O), and hydrogen (H_2). Energy input to the mixture was supplied by a spark discharge simulating lightning, with the entire mixture circulated through a cooling tube causing the gaseous combination to condense in an imitation of rainfall. Heat was supplied to the liquid in order to simulate evaporation in the ocean. After some initial playing about with the parameters of the apparatus (rate of heating, amount of the various gases, sequence of heat, spark, and condenser), Miller let the apparatus run for a week, after which time the mixture was found to contain significant amounts of the amino acids glycine and alanine, two of the basic building blocks of protein, hence essential elements in all life. It's interesting that, while admitting his pleasure at the outcome, Urey confessed his surprise at finding such

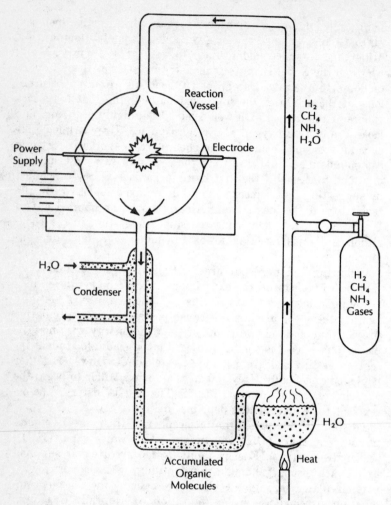

FIGURE 2.1 *The Miller-Urey experiment*

a significant amount of these essential compounds, remarking that his expectations for the experiment were to find "Beilstein," i.e., a little bit of everything—a reference to the classic set of over a hundred volumes listing every organic compound that has ever been synthesized.

Publication of the experiment in *Science* in May 1953 generated considerable attention in the popular press, even resulting in a Gallup poll in which 22 percent of the respondents did not exclude the possibility of creating life in a test tube. An interesting scientific postscript to Miller's work is the fact that an earlier experiment aimed at the same goal of creating some of the building blocks of life had been carried out by Melvin Calvin in Berkeley at the very time that Miller was completing his undergraduate work there. Calvin had used a quite different atmosphere consisting of water vapor and carbon dioxide, with an energy source of alpha radiation obtained from the Berkeley cyclotron. Since Calvin's atmosphere was oxidizing rather than reducing, he obtained no organic compounds. A puzzling aspect of this experiment was Calvin's use of the oxidizing atmosphere, since he explicitly referred to Oparin's reducing atmosphere in his paper yet didn't use it in the experiment. In fact, this observation led to a critique published by Urey that ultimately motivated him to suggest the experiment performed by Miller. A final anomaly surrounding this whole business is that Miller made no mention of Calvin's work in his historical account of the events leading up to his own experiment. Yet it seems unlikely that he could have been unaware of it, since he had been a student in the very same department at precisely the time when the work was under way. But such are the vagaries of fate and information flow in the academic world.

In the three decades since Miller's pioneering work, many similar experiments have been carried out using a variety of gaseous mixtures and different energy sources, each experiment leading to a slightly different collection of organic end products. The current Head Soup Chef is Cyril Ponnamperuma, director of the Laboratory of Chemical Evolution at the University of Maryland, in whose office, fittingly enough, there is a large, Andy Warhol–style picture of a Campbell's soup can carrying the label "Primordial Soup." Ponnamperuma, who originally studied religion in India and then moved into chemistry, has managed to combine his two interests in the comment that "God must be an organic chemist." This pithy remark compactly summarizes the core of today's conventional scientific wisdom as to how life on Earth originated: The basic building blocks of life were synthesized from simple chemical elements that were in ample supply on the primitive Earth. The compounds thus generated then managed to combine somehow, eventually forming the first living organisms. At this point, Darwinian evolution and natural selection could begin working their magic to generate the myriad complex living forms we see today.

A crucial aspect of the credibility of this picture is the very long time span of more than 4 *billion* years that evolution has had to play with to create the myriad organisms we see on Earth today. To convert this almost unimaginable magnitude into more human terms, look at the spare change in your pocket. Now entertain the happy vision of receiving $40 million for each and every cent you find. That's a ratio of 4 billion to 1. Let's measure it on another scale. Suppose you try covering the distance between New York and New Orleans with postcards—stacked on edge! That's also 4 billion for you, and each card represents one year that Nature has had at her disposal to put modern life together. As an irrelevant aside, when 4 billion is put in these terms we start to appreciate the true enormity of budgetary and trade deficits measured in the hundreds, or even *thousands,* of billions of dollars.

While the above skeletal outline serves to underpin most scientific investigations of the origin of life, the fun really begins when it comes time to spell out the details of the precise mechanisms Nature used to breathe the spark of life into a haphazard mixture of simple chemicals. The controversies rage on over these matters and we shall examine the competing arguments later. But to make sense of the various claims, it's first necessary to understand the structure and operation of living forms in greater detail, as only then will we be in a position to appreciate the many gauntlets that must be run by any viable theory of the origin of life.

A CRASH COURSE ON HOW LIFE LIVES

By more or less general consensus nowadays, an entity is considered to be "alive" if it has the capacity to carry out three basic functional activities: metabolism, self-repair, and replication. The latter two functions refer primarily to the entity's ability to manufacture good, but not necessarily perfect, copies of itself, while the first involves the quite different ability to synthesize from the surrounding environment the materials needed to ensure the entity's survival. In all known life forms on Earth, these two jobs are carried out within the cell by distinct chemical compounds and processes. The metabolic functions are the province of the proteins, while reproduction is handled by the nucleic acids DNA and RNA (with a little help from the proteins). The work of Sanger and others has shown that all proteins used in modern life forms are formed as chains of *amino acids* and, furthermore, of the many types of amino acids there are only twenty that are used by living organisms. The work of Watson, Crick, and many others demonstrated that the nucleic acids are also formed as long sequences of chemical compounds termed *nucleotides.* Each nucleotide is composed of one of five bases—guanine (designated G), adenine (A), cytosine (C), thymine (T), and uracil (U)—surrounded by some sugar and phosphate bonds for structural integrity.

The way the cell goes about its main chemical business of manufacturing proteins involves a lot of unfamiliar terminology and a number of steps. So before setting forth on a far-too-accelerated tour of this territory, it will be helpful to have a familiar analogy available to picture the process. Let's think about the way a modern automobile company like Ford or GM goes about manufacturing a car.

First of all there is a master plan, or blueprint, describing the entire automobile, as well as the processes and materials needed for its manufacture. This plan is usually kept under lock and key somewhere in corporate headquarters. In order to build various subsystems of the car like the motor, gearbox, or suspension, the relevant sections of the master plan are copied and dispatched to the corresponding manufacturing plants. Let's consider the case of building your car's motor.

When the working copy of the motor blueprint arrives at the plant, specialized workers identify the various components needed for such things as the engine block, pistons, and valves. These "transfer" workers then go to the stock room to gather the requisite items. The necessary items are next given to "assembly" workers whose task it is to put them together into the main components of the motor. As each main component, such as the block, camshaft, valves, or rings, is put in place, the "transfer agents" and the "assemblers" continue to read the working plan and carry out its instructions until they finally come to the instruction "STOP: The motor is complete." The finished motor is then sent on its way and the process begins anew with the manufacture of another motor. As we will see in a moment, the metabolic machinery of the cell functions in a completely analogous fashion, with its own version of master plans, working blueprints, transfer agents, and all the rest. Let's see how it goes.

The cells of higher organisms are divided into two main compartments: the *nucleus,* which contains the cellular hereditary "program," DNA; and the *cytoplasm,* where the proteins are manufactured. The real work of the cell is carried out by the proteins, mainly the ribosomes, with the nucleic acids DNA and RNA being a bit like the queen bee in a hive, fit only for reproduction but no real work. The DNA is divided into short sections, each of which represents either the chemical code for a certain protein, or a control code that activates or inhibits certain chemical operations in the cell. Such sections of DNA are called *structural* or *regulatory genes,* and they carry the information needed to make the organism, as well as serving to pass that information along to subsequent generations of cells. In simple organisms like bacteria, there is only a single DNA strand, while higher organisms contain a number of separate bundles of DNA strands called *chromosomes.* The number of such strands varies from species to species, being forty-six for humans, sixteen for onions, and sixty for cattle. Some simple organisms like bacteria and algae have no nucleus, with the genetic material mixed in with the cytoplasmic material in a single compartment. Such cells are termed *prokaryotes* ("cells without nuclei"). However, virtually all multicelled organisms are composed of *eukaryotic* cells, having a double-chambered structure with a separate nucleus. The structure of such a cell is depicted in Figure 2.2., while Figure 2.3 displays the celebrated "double helix" structure of DNA, showing the important base-pairing scheme A $\leftrightarrow$ T and C $\leftrightarrow$ G, together with the sugar and phosphate bonds denoted by S and P. Note that the structure for RNA is similar, except that there is only a single strand and the base uracil replaces thymine. Now let's briefly look at how such a cell carries out its metabolic and reproductive activities.

Protein synthesis is initiated within the nucleus when a single-strand "working copy" of part of the DNA, which can code for one or more proteins, is made. This working copy is termed *messenger RNA* (mRNA) and is formed by utilization of the simple base-pairing rules: Wherever the base A appears on the part of the DNA strand that's being copied, the mRNA strand will have the base U, while if T appears on the DNA, the RNA

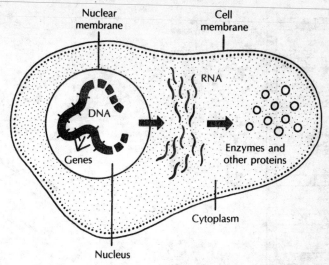

Nuclear membrane

Cell membrane

RNA

DNA

Genes

Enzymes and other proteins

Cytoplasm

Nucleus

FIGURE 2.2 *The structure of a modern eukaryotic cell*

strand will show the base A. A similar pairing exists between the bases C and G. Note here that the base U replaces the DNA base T on RNA strands. *Transcription* is the technical term for this process of copying part of the DNA onto a single RNA strand. For eukaryotes, when the mRNA strand is complete it's expelled from the nucleus and is used in the cytoplasm as the program for construction of the proteins called for by the genes it contains.

The proteins are formed according to the following procedure. Special combinations of proteins and RNA in the cytoplasm called *ribosomes* move along the strand of mRNA, reading its elements (bases) in nonoverlapping groups of three. Each such group is called a *codon,* and each codon is associated with either one of the twenty amino acids of life, or with a "stop" signal, according to the dictates of the *genetic code.* Since there are four possible bases, and each codon consists of an ordered sequence of three bases, there is a total of $4 \times 4 \times 4 = 64$ possible codons. The process of matching up codons with amino acids according to the genetic code is termed *translation,* and the working-out of this code constitutes one of the major triumphs of twentieth-century biology. Since there are only twenty amino acids used in forming proteins but sixty-four possible codons, we see that the genetic code contains some measure of redundancy as all good

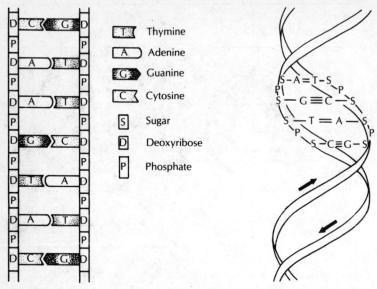

FIGURE 2.3 *The geometry of DNA*

codes should. The precise correspondence between codons and amino acids is shown in Figure 2.4. Note that the three codons UAA, UAG, and UGA are "stop signs," indicating to the ribosomes that they have come to the end of the program for that protein.

Once the ribosome has read a particular codon, it must find the corresponding amino acid in the cellular cytoplasm and join it to the chain of amino acids already assembled from the earlier codons. This process is carried out with the help of so-called *transfer RNA* (tRNA). The tRNA is constructed in such a manner that at one end it contains a "socket" into which can be plugged only one particular type of amino acid, while the opposite end of the tRNA strand contains a sequence of three nucleotide bases that form the *anticodon* to the amino acid at the other end. (Technically speaking, these "sockets" are not at the ends of the tRNA, but nearer the middle.) For example, if a strand of tRNA holds the amino acid methionine (AUG) at one end, the anticodon at the other end will be CAU, the mirror image of the codon complementary to AUG. Note that it is the mirror-image codon CAU and not the complementary codon UAC because the two chains on the tRNA run in opposite directions. So if the ribosome reads the codon AUG from the mRNA strand, it looks

	U	C	A	G	
U	phenylalanine	serine	tyrosine	cysteine	U
	phenylalanine	serine	tyrosine	cysteine	C
	leucine	serine	*punctuation*	*punctuation*	A
	leucine	serine	*punctuation*	tryptophan	G
C	leucine	proline	histidine	arginine	U
	leucine	proline	histidine	arginine	C
	leucine	proline	glutamine	arginine	A
	leucine	proline	glutamine	arginine	G
A	isoleucine	threonine	asparagine	serine	U
	isoleucine	threonine	asparagine	serine	C
	isoleucine	threonine	lysine	arginine	A
	methionine	threonine	lysine	arginine	G
G	valine	alanine	aspartic acid	glycine	U
	valine	alanine	aspartic acid	glycine	C
	valine	alanine	glutamic acid	glycine	A
	valine	alanine	glutamic acid	glycine	G

FIGURE 2.4 *The genetic code*

for a tRNA molecule floating about in the cytoplasm having the anticodon CAU, and when it finds it the amino acid methionine is detached from the tRNA and added to the growing chain. At this point, the tRNA has lost its amino acid and is dispatched back into the cytoplasm where it looks for another unit of methionine (in this case) to recharge itself. In this fashion, the ribosome moves along the mRNA strand assembling the protein chain one amino acid at a time, as if putting beads on a necklace. When it comes to a "stop" codon it releases the protein chain and begins work on another. The entire process is shown schematically in Figure 2.5.

Since the foregoing cellular operation is so central to the origin-of-life debates, let's try to fix the various steps and concepts firmly by comparing them with corresponding elements in the automobile-manufacturing analogy. The following chart shows the match-ups:

cell	⟷	auto-manufacturing company
nucleus	⟷	corporate headquarters
cytoplasm	⟷	manufacturing plant (including stock room)
DNA	⟷	master blueprint for the car
mRNA	⟷	working copy of part of the master blueprint
tRNA	⟷	transfer workers and stockboys
ribosomes	⟷	assembly workers
structural gene	⟷	plan for making a main component (e.g., motor)
regulatory gene	⟷	plan for assembling the main components
amino acid	⟷	individual part for a component
codon	⟷	blueprint ID number of an individual part
genetic code	⟷	rule for matching up parts with ID numbers

Of course, the foregoing analogy applies only to cellular metabolism; reproduction comes extra. While no one has yet started designing auto firms that literally reproduce themselves, the reader should have no trouble seeing how to translate the cellular reproduction process into a corresponding program for a self-reproducing car company.

The process of cellular reproduction is simplicity itself, being almost self-evident from the picture of DNA given above. From the base-pairing rules, it's evident that if we had just one of the two strands forming the DNA molecule, plus a good supply of the various nucleotide bases, there would be no problem in reconstructing the original double helix: just pair up the bases according to the rule: A ⟷ T and G ⟷ C. In real DNA this is very close to the procedure actually followed, as the DNA is un-

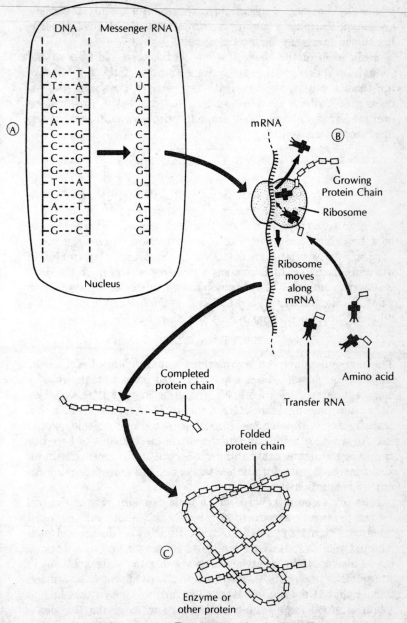

FIGURE 2.5 *Protein formation in the cell*

wound by enzymes (special-purpose proteins) while other enzymes act to link up the newly formed strands of DNA according to the foregoing base-pairing rule.

While each of the above stations on the road of life is simple
enough on its own, there are quite a number of them, and keeping them straight, together with the perhaps unfamiliar terminology, is quite a task. As an aid, the box below provides an
oversimplified, but adequate for our purposes, summary of the
most important steps.

$$\text{DNA} \longrightarrow \text{mRNA}$$
$$\Downarrow$$
$$\text{tRNA} + \text{mRNA} \xrightarrow[\text{genetic code}]{\text{Ribosomes}} \text{Protein}$$

The process of protein formation as described above is seen to
be a transfer of information in one direction: from the genetic
program encapsulated in the DNA to the proteins assembled in
the cytoplasm. In 1958 Francis Crick, codiscoverer of the double-helix geometry of DNA, summarized this information flow in
what he termed *the Central Dogma of Molecular Biology.* This
"dogma" can be summarized by the diagram:

$$\text{DNA} \xrightarrow{\text{transcription}} \text{RNA} \xrightarrow{\text{translation}} \text{Protein}$$

The degree to which the arrows represent inviolable directions of
information flow has been a hotly debated point in molecular biology ever since Crick made his pronouncement. Instances of information passing from RNA back to DNA are known, but a
transfer from proteins back to either of the nucleic acids would
call for a major rethinking of the entire mechanisms of heredity
resulting in, among other things, a revival of the now-unfashion-
able idea of Lamarckian inheritance, i.e., the inheritance of acquired characteristics.

Part of the tenacity with which biologists cling to the Central
Dogma is surely accounted for by Crick's choice of the word
"dogma," signifying a definite and authoritative doctrine. In addition, Crick's reputation (and Nobel Prize) for his DNA work no
doubt also contributed its share to the dogma's entrenchment in
the minds of biologists. Consequently, it's one of science's amusing
ironies that Crick himself later cheerfully confessed that he had
misunderstood the meaning of the word when giving the idea a
name, thinking that "dogma" referred only to "an hypothesis,

some arbitrary thing that was laid down for no particularly good reason." If naming the idea today, Crick claims, he would call it the Central Hypothesis, clearly indicating that the notion is by no means an established fact but only a provisional assumption or working hypothesis. On this ambiguous note, let's end our crash course in the mechanisms of life and reexamine the Primordial Soup Theory in light of what we have learned so far. But before doing so, let's pause just to summarize in the box below the bewildering array of terminology introduced in this section as a point of reference for the remainder of the chapter.

TERMS AND CONCEPTS

NUCLEIC ACID   the genetic component of the cell, DNA or RNA. Formed from the nucleotide bases, A, G, C, T, and U, plus sugar and phosphate bonds

GENE   a short section of DNA that either codes for a single protein or contains instructions regulating cellular chemical operations

MRNA   a working copy of a segment of DNA used in the process of gene translation into proteins

CODON   a triplet of nucleotide bases forming the source "language" for the genetic code

TRNA   a special-purpose molecule carrying an amino acid at one end, its corresponding anticodon at the other end

RIBOSOME   a cellular "constructor," which assembles proteins by reading the codons from the mRNA and then linking the relevant amino acids carried by the tRNA

GENETIC CODE   the rule by which codons are matched with one of the twenty amino acids used by all living things

TRANSLATION   the process of protein assembly by ribosomes reading the mRNA strand

TRANSCRIPTION   the process of producing the RNA strands from DNA by the base-pairing rule

REPLICATION   the process of producing a new DNA strand by means of base-pairing rules

CENTRAL DOGMA   the claim that cellular information passes only in the direction from the genes to the proteins

POTHOLES ON THE ROAD TO LIFE

There are three evident facts about life that any decent theory of its origins needs to confront:

Fact A: There is life on Earth.

Fact B: All life operates according to the same basic mechanisms.

Fact C: Life is very complicated.

To explain Fact A, a "soup theory" would have to show how the conditions of the early Earth could give rise to living forms, while explanation of Facts B and C would involve displaying a plausible path for how primitive living organisms could ultimately evolve the very complicated gene-protein symbiosis seen in modern life forms, in addition to offering a convincing explanation of why all living entities use the same small set of basic chemical components and the same genetic code in carrying out their life functions.

The most imposing hurdle that any origins theory must surmount is the Gene-Protein Linkup Problem. As sketched above, in order for the proteins to be "manufactured," it's first necessary for the genetic material to be read and then decoded into the appropriate amino acids according to the genetic code. On the other hand, until the proteins are present there can *be* no genetic material, since the process of replication is completely dependent upon the activities of special proteins (replicases) that facilitate the copying process. So we're left with a real "chicken and egg" situation, one that's especially tricky for any origins theory claiming that either the gene or the protein came first, with the other following as a corollary. A number of ingenious arguments have been concocted to evade this vicious circle, and we'll examine them in detail later. But for now let's just briefly list some of the difficulties that origins theories must address over and above the Gene-Protein Linkup Problem.

- *Genetic code/protein structure:* The laws of chemistry admit the formation of hundreds, if not thousands, of distinct types of amino acids; ditto for nucleotides. Why are all life forms based on the use of only twenty amino acids and five types of

nucleotides? How did Nature settle on just these few chemical
types and why? Or equivalently, with so many types to choose
from, why isn't it evolutionarily advantageous to make use of
the specialized properties of the other forms of amino acids to
make proteins, and the other kinds of nucleotides to construct
the genetic material? A related question relevant to the possibility of extraterrestrial life is whether it's necessary to use
separate molecular structures for the proteins and the nucleic
acids. On Earth, the proteins are good for action while the
chemical structure of the nucleotides is good for storing information. But in some exobiological environment it might be
that the same chemical structures could be used for both purposes.

• *Chirality:* Everything in Nature (except a vampire) has a mirror image, and all the amino and nucleic acids come in both
left-handed and right-handed forms. While these two forms
are chemically identical in the sense of being formed from exactly the same atomic constituents, the chemical actions of the
two forms are quite different as a result of their "twisting" in
opposite directions. In Miller-type experiments, approximately
equal quantities of both left- and right-handed molecules are
formed, and observations of the molecular composition of galactic clouds show a similar distribution between a given molecule and its mirror image. Yet *all* life forms on Earth use
exclusively left-handed amino acids to form proteins and
right-handed nucleic acids to form the genetic material. As a
consequence of this puzzling fact, we could starve to death on
a planet where the steaks were made out of right-handed proteins, since our body chemistry would be unable to break these
proteins down to extract their energy. A viable origins theory
would have to offer some coherent explanation of why living
forms settled exclusively on *L (evo)-amino acids* and *D (extro)-
nucleotides,* casting their mirror images aside.

• *"Junk DNA":* It's been observed that every strand of DNA
(other than in bacteria and viruses) contains long sections of
nucleotides that don't code for any proteins. In other words,
reading along a strand of DNA would be similar to listening to
a typical transoceanic phone conversation in which only every
third or fourth word is actually comprehensible, the rest being
garbled or swallowed up by cosmic noise, crosstalk, and other
foibles of international communication circuits. These "junk"

segments of DNA have to be edited out before the mRNA strand leaves the nucleus to be used as the template for protein construction, and there are special editing enzymes in the nucleus whose sole function is to perform just this task. While it's not strictly a question for origins theorists but for molecular evolutionists, it is still of interest to ask why Nature has allowed this genetic "noise" to remain in the DNA. Or even better, what's this junk doing there in the first place? It clearly serves no useful purpose in expressing the proteins coded for in the DNA, since it's removed during the process of creating the messenger RNA strand used to construct the proteins. Yet evolution has not seen fit to eliminate this noise from the system, and it's a puzzle for theorists to say why not.

Over twenty-five years ago, Howard Pattee noted that the best way to experimentally test the claims of the primordial soup theorists would be to create a completely unbiased prebiotic environment, then turn the system on and see what happens. In Pattee's words, "For all the inevitable inaccuracies in detail, a sterile simulated seashore, with waves, tides, sand, rain, and intermittent sunlight, is a more accurate primitive Earth environment than the well-defined but oversimplified reactions studied so far." Recently N. Lahav and others have suggested that we actually build a Whole Environment Evolution Synthesizer (WEES), consisting of a combination of primary, secondary, and tertiary environments open to various energy inputs. The primary environments would consist of various gaseous combinations thought to have composed the primitive Earth's atmosphere, while the secondary environments would be formed by the primordial seas, lagoons, and ponds. The tertiary environments would then be composed of small fluctuations in these more basic environments. In the WEES, the three phases (gas, liquid, and solid) simulate the Earth's biosphere using atmosphere, sea, and land. The interfaces would include a tidal zone and ponds, which are fluctuating environments. Material cycling would take place within each environment, as well as between the different environments. The main parameters used in controlling the WEES would be the intensity, duration, and rhythm of energy inputs, the composition and pressure of the gas phase, and the chemical and mineralogical composition of both the secondary and tertiary environments. A diagram of a WEES device is shown in Figure 2.6.

ENERGY SOURCES

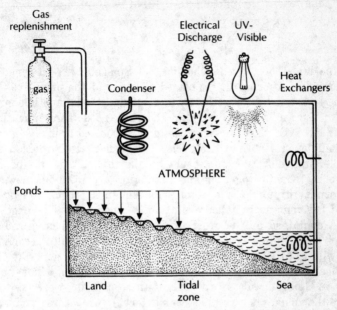

FIGURE 2.6 *A Whole Environment Evolution Synthesizer*

Given the projected time span of many millions of years for life to emerge, it's wildly optimistic to expect Godzilla or even primitive protozoa to come climbing up over the sides of a WEES tank. Nevertheless, given the plethora of useful information that's emerged from Miller-type experiments, it's not unreasonable to hope to learn substantially more details of how life could have gotten its start by using the far more elaborate WEES apparatus, including insight into some of the questions raised above.

With the theoretical problems and Miller and WEES-style experimental apparatus as background, let's now turn our attention to a more detailed consideration of the Prosecution's case, and look at the numerous variations on the Primordial Soup Theory arguing in favor of an origin of life here on Earth by natural chemical and physical means. Since their claims currently hold center stage, we begin with the arguments of those asserting that the genes came first, everything else being a detail.

MONSTERS, HYPERCYCLES, AND
NAKED GENIES

In the mid-1960s, the biochemist Sol Spiegelman performed a remarkable experiment. He placed a supply of the primitive Qβ virus in a test tube together with a virtually inexhaustible supply of the replicase enzyme that the virus needs for replication of its RNA. So that the virus would have no need to invade a cell to complete its normal life cycle, Spiegelman also provided an ample supply of free nucleotides in the tube. After mixing all these ingredients together and arranging a continuous flow of materials through the system, Spiegelman sat back to watch what has come to be called evolution in a test tube. The original RNA contained on the order of forty-five hundred nucleotides, which coded for several proteins that the virus usually needed in the wild to provide its protective coat, as well as to generate the replicase enzyme required for its replication within a host cell. But in Spiegelman's setup none of these proteins were needed, since the virus was insulated from external "predators" and was being supplied with all the replicase necessary to reproduce at whatever rate it wanted.

The outcome of the experiment was quite extraordinary. Initially, the naturally occurring Qβ RNA copied itself more or less faithfully. Rather quickly, however, mutations having the effect of cutting the RNA strand in half occurred. Since it's quicker and easier to copy a short strand than a long one, such mutations soon gained the upper hand in the Darwinian race for survival. As this process continued, shorter and shorter mutations appeared, until after about seventy generations the system stabilized at the shortest possible RNA strand capable of replication. It turned out that this strand contained about 220 nucleotides, and consisted of little more than the recognition site for the replicase enzyme. This final form of the RNA was termed the *Spiegelman Monster,* and offers an object lesson in the bad things that can happen if life is too easy. This little monster was able to reproduce itself at a staggering rate when confined to the friendly environment of the test tube, but couldn't possibly hope to survive in the rough-and-tumble world of unprotected reality.

Spiegelman's experiment involved dumping a living Qβ virus into an artificially hospitable environment, consisting of a supply of free nucleotides and replicase enzymes. The Nobel-winning German chemist Manfred Eigen took the process one step further by omitting the "seed" virus. In Eigen's experiment, a supply of nucleotides and replicase enzymes was placed in a test tube and left to its own devices. To everyone's surprise, with no seed virus to work with, the replicase enzyme proceeded to *create* a short strand of RNA, showing that what's important in the experiment is the replicase enzyme, not the initial viral RNA. The particular kind of RNA that emerged varied from experiment to experiment, but all variations were close relatives of the Spiegelman Monster and consisted of strands whose lengths were about 120 nucleotides.

The experiments of Spiegelman and Eigen demonstrate the minor gap of a hundred or so nucleotides that separates an RNA molecule that grew out of nothing from one that began as part of a living agent. This is a small difference indeed, and offers ample testimony to how easy the process of replication really is. Results of this sort supply the experimental muscle supporting the claim of the so-called naked genies, theorists who believe that the first living organisms were nothing more than short strands of primitive RNA, consisting of a hundred or so nucleotides having no purpose other than to perpetuate themselves. But there are at least two major obstacles in the path of acceptance of these claims, one involving the Gene-Protein Linkup Problem, the other relating to the likelihood of such a replicator's "self-assembling" in the primordial ocean. Let's examine these difficulties in more detail in order to assess the plausibility of the "genes first" arguments.

The essence of the naked genies claim is that the first living things were random replicators that assembled themselves from components floating around in the primordial soup. In particular, this means that there were no proteins, hence no replicase enzymes. However, the sine qua non of both the Eigen and the Spiegelman experiments was the presence of the replicase enzyme that facilitated the RNA replication. Thus, while these experiments show that very small RNA strands are capable of replication, they don't begin to address the issue of how such strands could ever arise without the help of the replicase. This fact poses an enormous barrier for the naked genies to overcome, with the question currently being attacked on two different fronts.

One line of attack is to try to make self-replicating RNA emerge without the assistance of a replicase. Using some artificially constructed, energy-rich nucleotide units, Leslie Orgel at the Salk Institute in La Jolla, California, has managed to induce RNA molecules to form a new chain that matches the existing one, with the chain then forming into a double helix. Unfortunately, the longest such chain has only about fifteen nucleotide units, and the special units are of a type very unlikely to have been present in the primeval seas. Furthermore, the replication process stopped when the double-helix geometry formed; thereafter no additional RNA replication took place. For these reasons, Orgel has been hesitant to make any claims about what he terms his "models," although others have asserted that these results show that it's possible, in principle, for a naked gene to replicate itself without benefit of a protein helper.

More recently, the work of Thomas Cech, Sydney Altman, and others has shown that under plausible circumstances, it's possible for RNA to act as an autocatalyst by snipping out a central portion of itself and then resealing the cut ends. In addition, they have shown that an RNA molecule can also cut up RNA molecules different from itself, thereby acting as a true catalyst (enzyme). Such self-catalytic RNA is also capable of joining several short RNA molecules into a longer chain under conditions that could possibly have been present on the early Earth. Further experimentation along these lines has shown how it would also be possible for RNA molecules to exhibit recombination, i.e., the ability to produce new combinations of genes, thereby providing the equivalent of sex—the infectious transmission of genetic elements from one organism to another.

Walter Gilbert of Harvard, another Nobel-winning chemist, has taken the above cluster of results involving self-catalytic RNA and used them to construct a scenario for the origin of life as we know it today, including a plausible explanation for the earlier-noted junk DNA. Let's look at the main steps.

THE GILBERT SCENARIO

A. RNA molecules perform the self-catalytic activity needed to assemble themselves from the "soup."

B. The RNA molecules evolve in self-replicating patterns, using recombination and mutation to explore new functions and to adapt to new niches.

C. The RNA molecules develop a range of enzymic activities.

D. RNA molecules begin to synthesize proteins, which are better enzymes than their RNA counterparts; i.e., they perform the same functions more efficiently.

E. Such protein enzymes are encoded by the RNA *exon*, the part of modern DNA that is not edited out in the construction of the mRNA, i.e., the complement of the junk DNA.

F. Finally, DNA appears, giving a stable, error-correcting information store.

G. RNA is then shoved off center stage, having been replaced by its creations, the proteins and DNA, which are able to perform its earlier double function more effectively.

The biggest question mark in Gilbert's plan for the emergence of life is Step A, since the experimental results on self-catalytic RNA pertain only to the sophisticated *present-day* form of RNA, and not to the presumably far more primitive forms of several billion years ago. Thus, the problem still remains open as to the degree to which self-catalysis of modern RNA sheds light on the same possibility for more elementary forms.

Postulation of mechanisms for the random assembly of simple, primitive RNA chains also generates another set of difficulties revolving about the amount of error tolerance that any such "manufacturing operation" must accommodate. Examination of this issue leads to what we can term the *Eigen scenario* for the origin of modern life. The basic idea consists of the following sequence of steps:

THE EIGEN SCENARIO

A. Start with a primordial soup consisting of randomly constructed small proteins, a sufficient quantity of lipids (fatty acids) to be able to construct cellular membrane fragments, and a variety of active, energy-rich nucleotide units suitable for the construction of nucleic acids.

B. Assume that at least one replicating RNA molecule forms by chance in the above soup. The assembly of such a molecule could possibly have been assisted by the presence of proteins that have also been randomly formed in the soup. Furthermore, this molecule is not a gene, as it codes for no protein; it is just a replicator. This molecule doesn't have a unique nucleotide sequence, but belongs to a family of closely related individual molecules that Eigen calls a *quasi-species.*

C. In some manner RNA molecules then learn to exert control over proteins, and a primitive genetic code develops. The different quasi-species specialize to take on different functions, so that the entire population is capable of constructing a protein.

D. A series of complex and cooperative interactions now take place between various nucleic acids and proteins. These interactions have been termed *hypercycles* by Eigen, and have been the subject of extensive mathematical and laboratory analysis, which we'll look into in a moment. The hypercycles eventually gain control over their environment until they reach levels straining the environmental carrying capacity.

E. At this point, to progress further it's necessary for competition again to enter the picture. The lipids present in the initial soup are now utilized to construct compartments, each compartment initially containing about the same mix of quasi-species. However, as random mutations take place, different types of hypercycles emerge, each contained in its own membrane. These membranes compete with each other, forming the prototypes of what later come to be modern cells.

F. The processes of *biological* evolution now take over from those governing the earlier *chemical* evolution, and ultimately modern life forms evolve.

The Eigen scenario has the satisfying aspect that one general principle, Darwinian evolution, is extended back to the time of the first replicator. However, this scenario suffers from the same defect as the Gilbert picture, namely, Step B: the appearance of the first replicator. Eigen assumes this initial spark of life emerges by nothing more than just a chance encounter of a set of the right hundred or so nucleotides. Since this random assembly problem lies at the heart of both the Gilbert and the Eigen pathways to life, it's worthwhile digressing for a moment to look into its plausibility in somewhat more quantitative detail.

To illustrate the difficulty involved in randomly assembling even a small RNA strand, suppose we have an organism that reproduces asexually and is capable of producing ten offspring before it dies. If the population is to be maintained without genetic deterioration, at least one of the offspring must have the same genetic information as its parent, while the other nine could have mutations that would render them less fit to survive. However, if not even one of the offspring is without mutation, then the population will eventually decay and become extinct. Suppose the RNA of this organism consists of 10,000 nucleotide bases, and that these are replicated with an error rate of 1 per 1,000. Then the chance that all 10,000 bases are correctly replicated is a paltry $(999/1,000)^{10,000}$, or about 1 in 22,000. So with only ten offspring, there's little chance that a population of such organisms could long survive. As a rough rule of thumb, if a population is to survive and has a chain of N nucleotide bases in its genetic pattern, then it must have an error rate of less than 1 in N.

The above considerations lead us to ask how we humans with a DNA strand many millions of bases long manage to replicate our genetic patterns *(genomes)*. The answer is that we have a "proof-reading" stage, in which our replicase enzyme first puts in a base with an error rate of about 1 in 10,000 and then checks it, replacing it if it's wrong. The second stage also has an error rate of 1 in 10,000, so the overall error rate is a comfortable 1 in 100 million.

The dilemma for the Gilbert and Eigen scenarios is that their primitive replicators have to make do without the replicase enzymes that provide the error-correcting step in replication, and hence they have to put up with error rates in excess of 1 in 100. As seen in the Spiegelman and Eigen experiments, this limits the genome size to around a hundred bases. To improve upon this, the primitive replicators would have to code for a replicase enzyme, as well as for a primitive protein-synthesizing machinery. But that can't be done with only a hundred bases. Thus, if you can't increase your genome size, you can't code for an enzyme; if you can't code for an enzyme, you can't increase your genome size—Catch-22 for the naked genies.

The hypercycle concept is Eigen's proposed solution to the above dilemma. This notion relies upon the idea of dividing up the genetic message to be copied into sections, and then imposing natural selection on each section independently. The difficulty with a straightforward use of this idea is that it's unclear how to prevent one of the sections from outcompeting the others. If all the sections are competing for the same bases, and if one replicates faster than the others, then that fast-track replicator will in time displace all the others and the resulting message will consist only of the winning section of RNA. The hypercycle offers a theoretical way out of this impasse.

Suppose the chain to be copied consists of the message A–B–C–D, divided into the four sections A, B, C, and D. Imagine that each of these sections represents a particular molecular population, and that the populations are arranged in the hypercycle shown below, with the rate of replication of each molecule in the cycle depending upon the concentration of the molecule immediately preceding it in the sequence.

$$
\begin{array}{ccc}
A & \rightarrow & B \\
\uparrow & & \downarrow \\
D & \leftarrow & C
\end{array}
$$

Eigen and Peter Schuster have shown that if such relationships exist, then the whole cycle is stable: No one molecule replaces all the rest. Intuitively, the reason for this is that if the concentration of any molecule rises relative to the others, the net result is to stimulate the others more than itself, the overall balance in the cycle then being restored.

With the hypercycle structure, it's possible to maintain and replicate information selectively in an amount greater than would be possible if the entire message A–B–C–D were copied as a single unit. In their analysis of the mathematical properties of such cycles, Eigen and his co-workers have shown that it's possible for these cycles to evolve, with evolutionarily improved hypercycles more likely to emerge if the molecular quasi-species are not able to move about too freely. This fact strongly suggests the desirability of some sort of cellular membrane to confine the components of the cycle.

To test the feasibility of the hypercycle scheme, U. Niessert conducted a series of computer experiments simulating the behavior of quasi-species and hypercycles according to Eigen's rules. She discovered that in addition to the *error catastrophe* discussed above, the molecular populations of a hypercycle are subject to at least three other types of catastrophes, colorfully termed the *selfish RNA, short circuit,* and *population collapse* catastrophes. The characteristic elements of each are:

- *Selfish RNA:* This situation occurs when a single RNA molecule mutates to a form that replicates faster than its competitors but, like some of my overly stimulated students, is having so much fun replicating that it forgets its other role as a catalyst.
- *Short circuit:* This catastrophe takes place when some RNA molecule that's supposed to be a link in the hypercycle chain changes its role in such a way as to catalyze a later reaction in the chain, thereby short-circuiting the cycle and contracting the hypercycle into a simpler one.
- *Population collapse:* This brand of catastrophe happens when statistical fluctuations result in the die-off of one of the molecular species in the cycle, resulting in the collapse of the entire chain of reactions.

Niessert discovered that the likelihood of the selfish RNA and short-circuit catastrophes increases with the size of the molecular population, while, of course, the population collapse catastrophe is more likely with small species populations. Consequently, the hypercycle model must sail a precarious path between the Scylla of selfish RNA and short circuits and the Charybdis of population collapse. There is only a narrow range of population sizes for which the probability of all three catastrophes is low, and even then the lifetime of a hypercycle can be shown to be finite. These results tend to cast doubt upon any theory of the origin of life that relies upon the cooperative organization of a large population of molecules, especially if that theory provides no insulating mechanisms to guard against the short-circuiting of metabolic pathways. Despite their current preeminence as the most popular flavor of the primordial soup, the naked genie arguments all suffer from this glaring defect. So we now turn our attention from the egg to the chicken, and examine the case for the proteins-first theories.

THE CHICKEN'S STORY

Following a two-sentence summary of the Primordial Soup Theory in their 1974 book *The Origins of Life on the Earth,* Stanley Miller and Leslie Orgel comment that "no one should be satisfied with an explanation as general as this." This remark aptly sum-

marizes the view that any sensible skeptic would take of the
naked genie theories of the origin of life, motivating a consideration of the other side of the coin: the possibility that the proteins
came first. On chemical grounds it's not a bad bet to bank on the
proteins-first idea, since in Miller-type experiments it's much
easier to form the amino acid building blocks of the proteins
than it is to generate the various sugars, phosphates, and nucleotide bases needed for the nucleic acids, let alone to form a self-
replicating molecule like RNA. The problem, as we'll soon see, is
that it's very hard to construct any plausible scheme for the replication of proteins, other than through the nucleic acid inter-
mediaries that encode them. In this section, we'll look at a couple
of the leading efforts devoted to ignoring this obstacle.

Historically, the idea that the first living forms were proteins
had the starring role at the very beginning of the scientific study
of life's origins in the 1920s. This was the theory favored by
Alexander Oparin himself (although Haldane, the cooriginator-
tor of the Soup Theory, was a genie). In a long series of experiments, Oparin noted that if certain oily liquids are mixed
with water, it can happen that the oily liquid will form into
small droplets that then remain suspended in the water. These
small droplets are termed *coacervates,* and are reminiscent of
the tiny droplets of water forming a heavy mist or a pea-soup
fog, although they are of quite different composition. In one famous experiment, Oparin considered droplets formed from his-
tone (a protein) and gum arabic (a carbohydrate). When he
added an enzyme able to link sugars to form starches (the enzyme was, of course, obtained from some already living cell), the
enzyme accumulated in the coacervate droplets. He next added
glucose (a sugar) to the mixture, whereupon the sugar molecules
diffused into the droplets and combined to form starches that
remained within the droplet. As this process continued, the
droplets grew and eventually split, with each offspring droplet
also growing just as long as enzymes were continually added to
the mixture.

Superficially, Oparin's coacervate droplets have a metabolism
as well as being able to grow and divide. But they're able to do
so only because they're being continually supplied with an enzyme from the outside, an enzyme synthesized by an already living organism. Also, the droplets have no mechanism whatsoever
for replicating hereditary information; hence, they have no way
of evolving. Oparin apparently believed that life began by the

accumulation of more and more complicated molecular populations within the shells of these coacervate droplets. Evidently, he felt that the external supply of the enzyme, which plays such an integral role in his experiments, could be provided over the course of geological time by natural processes occurring in the primordial soup, and didn't constitute a fatal stumbling block for his basic vision of the origin of life via "oil droplets." The main steps in this vision are as follows: First the cellular membranes form; then enzymes appear in order to organize the random collection of molecular constituents in the broth into metabolic pathways of various sorts; finally genes make their appearance. Since Oparin seemed to have only the haziest notion of the role of genes, having carried out his work decades before Watson and Crick, his theory of life basically says nothing about these carriers of the hereditary message. We can summarize Oparin's view of life as:

OPARIN'S SCENARIO

Cells (Coacervates) → Enzymes (Proteins) → Genes

In 1963 at the unlikely location of Wakulla Springs, Florida, the Second International Conference on the Origin of Life took place, a gathering that provided the first and only opportunity for Oparin and Haldane to meet face to face. The organizer of that historic event was Sidney Fox, now at the University of Miami, and a prime proponent of the proteins-first school of thought on the origin of life. Fox has promoted the notion of *proteinoid* microspheres, which were first discovered in his lab in the 1950s, as *the* solution to the origins question. Since his arguments have been favorably received by the media, as well as winning honorable mention in several technical publications, it's not surprising to note that Fox has acquired a spectrum of vitriolic critics ranging from the chemist Stanley Miller to the creationist Duane Gish and the astronomer Carl Sagan. In fact, the one point that the evolutionists and creationists both seem to agree upon is the irrelevance of the work of Sidney Fox. When such scientific eminences start getting hot under the collar, it's usually a sign that somebody's doing something right. So let's dig a little deeper into Fox's proteinoid idea and see why it raises so many hackles.

As we've seen, amino acids can be readily produced in Miller-style experiments. However, amino acids don't easily unite to form peptides (short protein chains) in the presence of water. In fact, just the opposite occurs: In water, peptides and proteins break down into their amino acid constituents. The remedy seems obvious: Just heat up dry amino acids so that the water that's formed when they join into a protein chain is carried off as vapor. Oddly enough, when this experiment is carried out with amino acids in the ratios found in naturally occurring proteins, all that's formed is a horrible, sticky, smelly, brown tar instead of the prized protein chains. Enter Sidney Fox.

Rather than using the usual prescriptions for heating amino acids, Fox found that different types of amino acids wouldn't hook together well unless extra amounts of any of three special amino acids—lysine, aspartic acid, or glutamic acid—were present. When these new mixtures were heated in the dry state at temperatures up to 130°C, they rapidly formed polymer chains of amino acids, but chains that didn't correspond to proteins occurring in earthly biology. For this reason, Fox termed these products *proteinoids.*

Despite their unearthly nature, Fox's proteinoids were found to display certain features of interest. For instance, some of them showed a catalytic capacity for several types of chemical reactions, although the activity was not substantially better than that shown by the same amino acid mixture before it was heated. However, what was remarkable was the behavior demonstrated by certain types of proteinoids when they were dissolved in warm water and allowed to cool slowly. Under this very simple operation, billions of microspheres formed from just a single gram of proteinoid. Fox found that these microspheres would grow and bud off smaller spheres, and that they had a somewhat nonspecific enzymic activity; i.e., they would catalyze a fairly broad range of chemical reactions. The "metabolism" of the proteinoids is far less specific than that of Oparin's coacervates, but then Fox didn't add any biological enzymes from the outside to push the metabolism along. It's vital to note, however, that just like the coacervates, the proteinoids lack a hereditary mechanism and will not evolve by natural selection. The main steps in Fox's road to life are displayed in the following diagram:

FOX'S SCENARIO

Amino acids → Proteinoid chains → Cells → Genes

As noted, criticism of Fox's proteinoid idea has been hot and heavy ever since he first introduced the notion over three decades ago. Many of the early complaints focused upon the geological question of where on the early Earth one would find the sort of conditions needed to form the proteinoids. Stanley Miller and Leslie Orgel ask whether there is any place on the present-day Earth where all the necessary conditions are present, coming to the sad conclusion that "we cannot think of a single such place." Earlier, Harold Urey stated quite unequivocally that "it is difficult to see how the processes advocated by Fox could have been important in the synthesis of organic compounds." Recently Fox has answered some of these geologically based difficulties by noting that perhaps the proteinoids arose near the thermal vents at the bottom of the Pacific Ocean. Somehow it's hard to see how the necessary *dry* heating could take place at the bottom of the sea, but that's the illogic of *real* science for you! Other arguments against the proteinoids center upon the fact that similar microspheres are created under a variety of circumstances, such as when ash forms out of molten lava in volcanic explosions like the one that truncated Mount St. Helens. Yet none of these microspheres show the capacity to grow, reproduce, and evolve in a manner that copies the internal organization of the system. In other words, they don't show a capacity for self-organization, hence for life.

It's clear that both the Oparin and the Fox scenarios are hopelessly deficient when it comes to the problem of providing a genetic mechanism whereby hereditary information can be passed along to future generations of cells, opening up the possibility for natural selection to come into play. So just as the naked genies suffer from an Achilles' left heel of no proteins to catalyze reactions that would allow development of a large genetic information store, the proteinists suffer from the complementary right heel of no replication machinery. Since it seems difficult to conceive of plausible ways to fill in the gaping holes in the arguments of either the genies or the proteinists, perhaps the answer lies in adopting a Hegelian dialectical stance, and attempting to combine the best features of the two schools into a Dual-Origin Theory (or Double-Origin Hypothesis). Such a theory would argue that life emerged not once but twice, with the proteins and replicators arising independently and then later linking up in a mutually beneficial symbiotic arrangement. Let's see how such a theory might work to plug the leaks in the all-or-none hopes and dreams of the genies and proteinists.

LIFE: A TWICE-TOLD TALE

Some time ago at the Smithsonian Institution in Washington, D.C., one of the more popular exhibits was a videotape showing the famous TV chef Julia Child mixing up a batch of primordial soup in vivid, living color. Unfortunately, Nature's kitchen, just like those of many of Julia's TV fans, suffers from the unhappy fact that knowing Julia's methods and obtaining her results are two very different matters. And as entertaining and educational as the Smithsonian display was, the unvarnished truth is that the delightful concoction coming out of Julia Child's soup pot had a flavor unlikely to have been on any earthly menu at the dawning of life. Since this observation bears heavily upon the Dual-Origin Theory of life, it's worth our taking a moment to scrutinize the recipe in greater detail.

A point of contact between the proteinist and the naked genie programs for the origin of life is the assumption that all the necessary raw materials could have been assembled by natural means in the primordial soup. Miller-style experiments make this assumption at least defensible for the proteinists, since amino acids seem to form spontaneously in almost any kind of primitive environment—just as long as it doesn't contain any appreciable amount of free oxygen. Thus, we can at least identify a path whereby simple proteins might naturally form. The picture is far fuzzier for the natural assembly of nucleotides.

The bitter facts of chemical life are that it's just plain hard to see how nucleotides could have easily been made in the environment of the early Earth. For over three decades an army of talented chemists has experimented endlessly with various formulas for constructing nucleotides in the laboratory, with only limited success. Some of the nucleotide bases have been created from elementary compounds, but only under conditions that would require a *cold* soup rather than the postulated hot primordial soup. It has also proven possible to synthesize the sugar components of nucleotides using formaldehyde, but again under circumstances that are far more special than those needed for creating amino acids via Miller-type experiments. Fortunately the phosphate components of nucleotides don't have to be synthesized, since they occur naturally in rocks and seawater.

The major difficulty in nucleotide synthesis is in getting the three components—bases, sugars, and phosphates—to stick together naturally in the right kind of geometrical arrangement. If the linkages are made randomly, only about 1 percent of them will turn out to be correct, and there appears to be no natural mechanism that would be able to distinguish the one correct arrangement from the other ninety-nine. In addition, nucleotides are unstable in water and have the depressing tendency to dissolve back into their component parts. Thus, the rate of formation would have to be very high in order to counterbalance the correspondingly high rate of decomposition in seawater. No one has yet discovered a natural chemical mechanism that would enable nucleotides to be generated rapidly enough to find each other and then form into the necessary double helices before they fall apart by hydrolysis. This is one important fact favoring a theory of life that requires only amino acids to be prebiotically formed, with the nucleotides coming later as a by-product of protein metabolism. But this is not the only argument that speaks for the Double-Origin Hypothesis. Here are two more:

• *Parasitism:* Within the cellular cytoplasm (where the proteins are manufactured), we find the *organelles (mitochondria* and *chloroplasts),* which serve a vital function in extracting the energy needed for the cell to carry on its business. The organelles have their own genetic machinery, which operates independently of that found in the cell's nucleus. As a result of studies of the cellular evolutionary tree, it's been found that the genetic apparatus of the organelles belongs to a different branch of the tree than that present in the nuclei of eukaryotic cells. The American biologist Lynn Margulis has forcefully pressed the claim that this fact suggests that the organelles originally lived a life totally independent of the eukaryotic cells, and only later joined up with them in a parasitical, symbiotic relationship, probably to help the cell extract energy from the environment more efficiently. It's also been discovered that the genetic code used by the mitochondria differs slightly from that of the cell nuclei. Significantly, the difference is small enough to point to the conclusion that the two codes must be related, tracing their origin to a common ancestor. Both of these facts lend support to the Dual-Origin Theory. Of course, it should be noted that this is a situation in which one DNA organism invaded another. It says nothing about a "no-DNA" organism.

- *Fossil evidence:* In the oldest rocks that can be reliably dated (about 3 billion years old), evidence is found of fossils that bear a resemblance to modern bacteria, i.e., prokaryotic, single-celled creatures. In rocks about a billion years younger, we find traces of fossils similar to modern prokaryotic algae, including multicellular entities. Finally, in rocks that are about a billion years old, evidence of modern eukaryotic cells finally appears. Unfortunately, the techniques available offer us no way to determine whether or not the oldest fossils possessed a modern genetic apparatus, or were cells with no nucleic acid whatsoever. The only thing we can say with confidence is that of the fossils originating over the past 1 billion years, all are modern in form with contemporary eukaryotic features, including genetic machinery. Thus the fossil record provides evidence only for some sort of ancient living beings, but no evidence at all that these organisms possessed any kind of replication apparatus utilizing nucleic acids.

Putting the chemical, fossil, and parasitism arguments together with the earlier difficulties in the arguments of the proteinists and the genies we come to a Double-Origin Theory in which the first living agents were metabolizers (proteins), with the genetic replication machinery following much later as a consequence of the chemical reactions catalyzed by the primeval proteins. As we've already seen, there's no particular difficulty in forming simple proteins via Miller-style reactions. So the essential step in any kind of double-origin scenario is to offer a plausible means by which protein replication can take place without invoking nucleic acids. Robert Shapiro has suggested the following scheme based upon the manner in which transfer RNA works in a modern cell.

Earlier we saw that when the ribosomes manufacture a protein chain, a central role is reserved for tRNA *synthetases,* special enzymes that act as "interpreters" by having very specific geometries at each of their ends. The geometry at one end fits exactly one nucleotide triplet (codon), while the geometry at the

opposite end of the enzyme fits only the amino acid corresponding to the *anti*codon of the codon on the other end. It is these
tRNA synthetases that do the real job of translating from the
language of the genes (nucleic acids) to the language of the proteins (amino acids). Shapiro argues that perhaps the same system could work, but in a simpler way, for direct protein
replication. It's of considerable interest to note that this "translation" seems to represent a second kind of genetic code. At present, the workings of this code within the tRNA are the subject
of feverish research activity. The interested reader is invited to
consult the "To Dig Deeper" section for citations to some of the
recent work on this second genetic code.

Shapiro's basic idea is that the protein molecule that was to be
copied became attached to some support so that it could be distinguished from those molecules that were not to be copied. The
molecule could then somehow be turned on its support so that
each of its constituent amino acids was exposed to the ambient
environment. As each successive amino acid was exposed, a suitable interpreter enzyme would recognize the amino acid and
match it at its other end with exactly the same amino acid from
the environment, adding this new acid to a growing chain under
construction. Notice that this kind of matching requires a simpler sort of interpreter enzyme than modern tRNA synthetase,
since the "protein interpreter" needs only to be able to recognize
the same kind of amino acid at both its ends. Thus it needs to
know only the language of proteins, not both the language of
proteins and that of nucleic acids. Assuming there ever was such
a system, at some stage it was eliminated in favor of the current
nucleic-acid-based method, implying that the protein replication
process was inaccurate, slow, inefficient, or defective in some
other way. However, the method does have the virtue of indicating how proteins could replicate themselves, as well as suggesting why modern life uses only a few of the many possible amino
acids.

In the Miller experiment, the most prominent amino acids present were the two simplest (in terms of number of atomic components)—glycine and alanine. It's reasonable to suppose that
these two would be present in the initial set of amino acids used
by the first proteins. On the other hand, the most complex amino
acids used in living forms cannot be produced even with the
most elaborate prebiotic simulations, and probably emerged

much later as a result of earlier metabolic processes. Various theoretical arguments have been produced showing that only a handful of amino acids, say between four and six, are needed to approximate the shapes seen in proteins today. The successive introduction of each additional amino acid very likely represented a milestone in the evolutionary struggle of early protein life, greatly increasing the "catalytic power" of the protein chains that could then be formed. Ultimately a crossover point was reached, where it required more work to create the copying machinery for an additional amino acid than the effort bought in extra catalytic power. At this point, natural selection would then act to stabilize the menu of available amino acid components at its current level of twenty. Would that we were so lucky with the menu at the corner Chinese restaurant! Even granting the above sequence of events as a starting point for life, how could the nucleic acid replication process ever have gotten a foothold, eventually to displace the protein replication apparatus?

In Shapiro's setup, RNA and DNA arise only when phosphates become more readily available as rocks gradually erode and dump more and more phosphates into the primordial sea. The original nucleotide material consisting of sugars and phosphates would then be used as structural materials, as they are even today in the ribosomes. One way that this original structural material could have been transformed into today's genetic apparatus would involve the development of short, specialized units of RNA, each associated with a particular amino acid, as well as the development of a longer RNA strand for every useful protein. With this sort of innovation, the information present in each protein would then also be stored in the RNA, giving a duplicate genetic system that would eventually be found more efficient at replication than the earlier protein-based system. Finally, natural selection would ruthlessly assert itself and discard the old replication apparatus in favor of the nucleic acid system that we know today.

The reader should note that the above scheme avoids the bugaboo of the genies, namely, establishing the means by which the individual nucleotide subunits are assembled to form the first strand of RNA. Recall that from the experiments of Eigen, $Q\beta$ replicase can assemble a strand of RNA on its own, given a supply of the subunits. This step is very simple provided that the replicase enzyme is already present. And there's no problem in envisioning how this enzyme might come to be available in a scenario in which proteins came first, followed much later by RNA and DNA. The Shapiro scenario given above suffers from the same kind of assembly difficulty as the nucleic acid replicators of the genies, i.e., how did the appropriate subunits come together to form the first self-replicating system? However, the problem is easier to rationalize and give plausible answers to using proteins rather than nucleic acids as the first living forms.

Recently the physicist Freeman Dyson has proposed a quantitative model for the Double-Origin Hypothesis, in which he explores the feasibility of the overall notion using what he terms a "toy model" of the process of cellular metabolism. While there's no room here to enter into the details of Dyson's model, it's interesting to examine one or two of his main conclusions. Following the imposition of a variety of simplifying physical and mathematical assumptions, the essence of Dyson's model comes down to the interrelationship of three parameters: a, a measure related to the number of distinct amino acid building blocks (technically, *monomers*) composing the original living objects; b, a measure of the number of distinct sorts of chemical reactions that the primitive life forms were capable of catalyzing; and N, the size of the molecular population in a chain composing such a form. What Dyson is interested in is those combinations of a,b, and N that allow a reasonable possibility for the system to jump from a disordered state of miscellaneous chemicals to the ordered state of a living agent.

In analyzing the consequences of his model, Dyson discovered that the only values of the parameters that resulted in physically interesting behavior were those in the ranges:

a: from 8 to 10

b: from 60 to 100

N: from 2,000 to 20,000

When translated back to physical units, this result implies that the number of monomer types should range from nine to eleven. As we know, in modern proteins there are twenty types of amino acid monomers, so it's reasonable to suppose that ten or so would be enough to provide sufficient diversity of protein function to get life off to a start. At the other end, the model definitely fails if $a = 3$. This implies that life according to Dyson could not possibly have begun with only the four nucleotides forming modern RNA; nucleotides alone just don't offer great enough chemical diversity to make the transition from disorder to order. Thus, the model displays a pronounced bias in favor of proteins as opposed to nucleic acids as the material basis of life.

Having the discrimination factor b in the range from 60 to 100 turns out to be chemically reasonable for the first primitive proteins, and also endows the model with the all-important property of being able to tolerate very high error rates. If one were to assume exact replication from the very beginning with a low tolerance of errors, the jump of a chain of N monomers from disorder to order will occur with a probability of around $(1 + a)^{-N}$. This implies that a replicating system can spontaneously emerge only if N is no greater than about 100, as noted in an earlier section. However, in Dyson's nonreplicating system with a and b in the ranges above, the error rate will be about 25 to 30 percent, and still a chain of ten thousand or more monomers can make the transition from a disordered state to an ordered one with reasonably high likelihood. Such a level of performance in which only three out of every four links in the chain are correctly placed would be intolerable for a replicating system, but is quite acceptable for a nonreplicating one.

The overall behavior of Dyson's model is summarized in Figure 2.7, in which each point corresponds to a particular choice of a and b. Models that admit the possibility of both ordered and disordered states occupy the central region in the diagram labeled the "transition region." The biologically interesting models are those near the cusp, which have high error rates and are able to make the transition from disorder to order with large population sizes. One interesting case discussed in detail by Dyson is when $a = 8$, $b = 64$, leading to an error rate of exactly one third and a critical population value of $N_c = 26,566$. The region labeled "dead" in Figure 2.7 corresponds to models that have only a disordered state. Such models have a too large (too much chemical diversity) and b too small (too weak catalytic activity) to produce an ordered state. Conversely, the region labeled "immortal" has a too small (too little chemical diversity) and b too large (too strong catalytic activity) to produce a disordered state. In further discussion of this model, Dyson also gives some provocative arguments for how the asymmetry be-

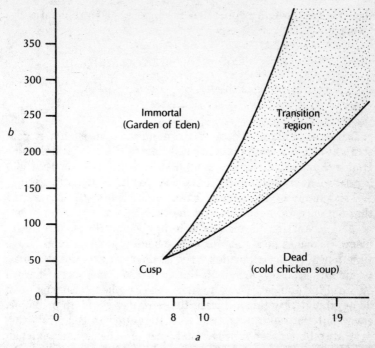

FIGURE 2.7 *Summary of Dyson's model*

tween life and death could arise from such a system, and why it is that death is so much easier than resurrection. But my system-theoretic prejudices have probably already caused me to devote too much attention this model, so I'll let those readers with a hankering for rising from the dead consult "To Dig Deeper" and pass now to a summary of the Dual-Origin Theory.

All but the most casual of readers will have long ago realized that there's really not such a great difference between the Dual-Origin Theories of Shapiro and Dyson and the proteinist arguments of Oparin and Fox. The main point of departure is that both Oparin and Fox argue that the replication machinery came about early in the game and, moreover, arose *directly* out of the initial metabolism. The "doublets" argue that the genetic apparatus was a Johnny-come-lately on the origins scene, and did not arise directly out of the initial proteins but rather had a quite different structural function originally, and that its ultimate role as a replicator arose as a type of "genetic takeover" from the original protein replication mechanism. Diagrammatically, we have:

THE SHAPIRO-DYSON SCENARIO

Cells → Proteins → → RNA → DNA
                much later

The most striking aspect of the doublet claim is that it goes straight in the face of the cherished Central Dogma of Molecular Biology, discussed earlier. Should such a pillar of modern biological wisdom be so lightly discarded? Well, as we noted earlier, the originator of the dogma, Francis Crick himself, has stated that not only did he misunderstand the meaning of the term, but he also meant for the principle to apply only to modern organisms; he makes no claims for how ancient organisms might have functioned. A remark made by one of the most prominent genies, Leslie Orgel, also bears upon this point. When Robert Shapiro outlined his doublet theory to him, Orgel replied, "Enzymes can do anything!" By this he meant that enzymes could, in principle, carry out the replication functions suggested in Shapiro's theory. But that doesn't prove that such a scheme ever existed. What's needed is a plausible physical mechanism by which the protein replication process could have gotten started. Mainline doublets like Shapiro and Dyson think this could have happened using the carbon-based compounds composing modern life. The Scottish chemist A. G. Cairns-Smith says that carbon is much too high tech a material for this job, and has offered a fascinating silicon-based alternative with the claim that, just as the Bible says, life started as a mere mote of dust in someone's eye. We devote the next section to an account of these ideas.

ASHES TO ASHES, LIFE FROM DUST

One of the multitude of ways I managed to misspend my youth in the 1950s was by hanging around the local cinemas. Rather than devoting valuable energy to homework, practicing the piano, or some other dull character-building task, I squandered my time (in my mother's opinion, at least) in soaking up the cinematic offerings of the day at what I then saw as the near budget-breaking admission price of one dollar. In order to put as much distance as possible between myself and the onerous program of chores lined up for me by my teachers and parents, I always tried to arrange to spend my weekly cinema allowance on triple features, which in those days generally meant an entire afternoon of the then-popular science-fiction and horror films. One of the films that I still recall fondly was the 1957 classic *The Monolith Monsters,* a story about some sort of silicon-based life forms that were transported to Earth on a meteorite. These strange objects somehow absorbed silicon from Earth's sand and rocks, using it to grow into tall monoliths that eventually toppled over and broke up into small pieces, which then started growing again into even more of the "rock monsters," whose "life cycle" is shown in Figure 2.8. The film's crescendo was reached as a forest of these monoliths threatened to pulverize some rural community in the Arizona desert. They were stopped an eyelash away from town only when the film's brilliant and handsome scientist-hero realized that the monsters' growth could be stopped dead in its tracks by the salt in simple seawater. While pretty farfetched in regard to both its science and its scientists, *The Monolith Monsters* represented an entertaining Hollywood attempt to speculate on the nature of life forms based upon silicon, an idea that has recently been resurrected by A. G. Cairns-Smith as the material basis for his Dual-Origin Theory of life.

The motivation behind Cairns-Smith's revival of silicon is that what's important in any origins theory is to find *some* system that will get metabolism and replication started. He conjectures that a "low-tech," silicon-based setup might be easier to get rolling than the sort of high-tech, carbon-based systems discussed so far. Once some kind of living system was up and running, Cairns-Smith argues, the more efficient carbon-based units could then engage in a "genetic takeover," pushing the original system out of the spotlight and back into the wings.

In his writings, Cairns-Smith identifies what he terms "seven clues to the origin of life" as evidence to support his claims that modern life got its start as a bunch of mud. The clues he cites are:

1. *Biology:* Genetic information is pure form, not substance, and evolution can begin only when this kind of replicable form exists.

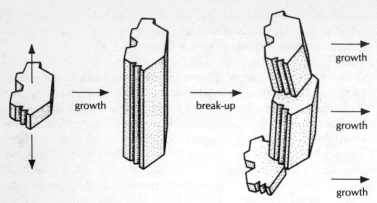

FIGURE 2.8 *Crystalline growth in* The Monolith Monsters

2. *Biochemistry:* DNA and RNA are biochemically complex and hard-to-make molecules, suggesting that they were late arrivals on the evolutionary scale of things.

3. *Construction industry:* In evolution, things can be subtracted as well as added. This can lead to the kind of mutual dependence of components seen in the central biochemical pathways of life.

4. *Structure of ropes:* Gene fibers, like rope fibers, may be added and subtracted without breaking the overall continuity of the gene line. This suggests how organisms based on one genetic material could gradually evolve into organisms based on an entirely different genetic material.

5. *History of technology:* Primitive machinery is usually different in design and construction from later advanced counterparts. The primitive machine has to be easy to make from immediately available materials, and must work with a minimum of fuss. The advanced machine does not have to be particularly easy to assemble, nor does it have to be made from simple parts. This fact suggests that the first organisms would probably have been very different from the "high-tech" organisms of today.

6. *Chemistry:* Crystals put themselves together in a way that could be suitable for "low-tech" genetic materials, suggesting a direction in which to look for the primitive biochemical materials.

7. *Geology:* There is a lot of clay continually being made through natural processes. This kind of inorganic crystal seems to be

much more appropriate than big organic molecules for primitive genes, as well as for other primitive control structures like low-tech catalysts and membranes.

These clues from many different areas of science and technology constitute tantalizing arguments for taking crystals of clay seriously as the first living material. While there's no room here to go into the details underlying Cairns-Smith's argument, it's definitely worthwhile to examine the general scenario he offers for how events might plausibly have taken place. But in order to make sense out of the steps in the Cairns-Smith theory, we first need to understand just a bit about the basic physical properties of clay and crystals.

Occasionally, despite careful planning, I'm forced into trying to recall the odd fact or two from my late and unlamented high-school and university chemistry courses. Usually the first thing that comes to mind (following a prodigious effort) is an image of the kind of experiment that everyone does along about the second week of such a course: the creation of crystals of salt from a supersaturated solution of sodium and chlorine ions. The experiment generally involves dumping an enormous amount of salt into a beaker of water, heating the water to boiling in order to dissolve the salt into its component ions, and then letting the resultant liquid slowly cool, being very careful not to agitate the container. After the liquid mixture cools down, a tiny crystal of salt is dropped into the flask as a seed to start the crystallization process. Almost immediately the ions dissolved in the water start to attach themselves to the seed, forming long crystals of salt that eventually break up, the pieces serving as seeds for further crystallization until the dissolved salt is exhausted. The basic process taking place is illustrated in Figure 2.9, where the black dots represent atoms of sodium and the white dots are chlorine. While not all crystals are as simple as salt, many of them having more exotic geometries than a cube and many of them containing repetitions of several layers composed of more than two atomic ions, the salt crystal depicted in Figure 2.9 illustrates several properties of crystals crucial for Cairns-Smith's theory.

First of all, note that the crystalline structure is very regular, consisting of a repetitious pattern of a two-dimensional lattice structure at each of whose points there is either a sodium or a

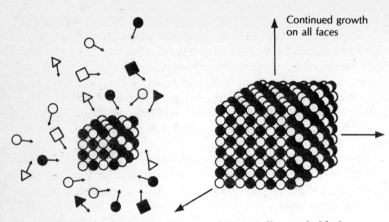

Continued growth
on all faces

FIGURE 2.9 *Crystallization of salt from sodium and chlorine*

chlorine atom. This kind of pattern gives structural integrity to
the crystal, as well as providing it with the opportunity to grow
by adding additional atoms from the surrounding environment,
as shown in the figure. In many crystals the atomic bonds between the various layers are rather weak, much weaker in fact
than those holding the atoms together within the layer. Consequently, it's very easy for layers to shear along the natural crystalline planes, just as the salt crystals in the experiment break
up when they get too big. The combination of being able to attract new ions and shearing along natural planes of cleavage
means that crystals can certainly grow and multiply, i.e., display
a kind of "metabolism." What would be needed for life is some
means by which the crystals could evolve.

To see how crystalline growth could take place with variation
in a self-propagating manner, we have to depart from the
foregoing fiction of idealized crystal growth as presented in elementary courses, and consider how crystals *really* grow in nature. In real life, crystals are not the perfect, infinite lattice
structures depicted in Figure 2.9 or school textbooks, but rather
grow with defects of both a mechanical and a chemical nature.
As seen in Figure 2.8, many crystals contain notches, grooves,
and other types of mechanical flaws that can be passed along,
layer by layer, as the crystal is formed. Other types of mechanical defects come about when growth rates differ on the various
faces of the crystal, leading to separate "domains" that are
slightly out of alignment but each still growing according to the

overall plan. Now suppose that two identical crystals start growing in the same environment, but that soon one of them starts acquiring variations of the above type. Further, imagine that this variation in physical shape somehow allows this "mutant" form to replicate faster than the other, perhaps by being retained more effectively within the pores of a rock from which the other form is more easily washed away. By natural selection, this more "efficient" form may soon come to dominate the crystal population.

But such mechanical imperfections are not the only way that crystals can evolve. Each layer composing a crystal contains many atoms, one at each lattice site. For instance, most clays consist of layers of oxygen ions with layers of positively charged ions (usually silicon or aluminum) sandwiched in between. In many such clays, one of these positive ions can be replaced by another type without destroying the clay's capacity to grow. Such substitution patterns can become quite intricate, making the surface pattern of the clay a very complex chemical structure. Furthermore, the pattern can be passed along (inherited) by subsequent layers as they're added. This inheritance can be by a direct matching of the pattern, or it may involve the formation of some sort of "complementary" layer. Of course, such a complementary matching process would be very close to what we see in DNA transcription and replication. Thus we see the possibility for "crystal genes," by both mechanical and chemical means, to form and act to perpetuate the information needed to give rise to further generations of crystalline forms. But is this enough to constitute life? And how and why do modern organic proteins and nucleic acids enter the picture?

Cairns-Smith's answer to these vital questions is radically different from the conventional tales told by the genies, proteinists, and doublets. His claim is that "living" crystals started making the organic components of modern life *in order to help themselves survive and multiply.* While there's no room here to enter into the intricate arguments he makes to support this claim, the overall conclusion is that it's at least plausible that the manufacture of organic compounds could help the clays in several ways: by providing mechanical support, by removing undesirable ions, by controlling the size and structure of the crystals, by assisting in the capture of inorganic ions, and so on. Now what about the takeover? When and why did the original crystalline life get shoved aside, to be replaced by its "assistants," the carbon-based units?

According to Cairns-Smith, the takeover by organic life took place when some organic forms within the crystal life began to reproduce themselves at a rate that exceeded that of their crystalline hosts. At this very moment, the crystals' days were numbered as the dominant life forms on Earth. Once crystals had made the first strand of self-replicating (and possibly self-catalyzing) RNA, they would have created a much more efficient and versatile genetic material than their own "low-tech" genes. Natural selection would then have seen to it that the "high-tech" nucleic-acid-based genes moved into the spotlight, eventually pensioning off the crystal life as museum pieces. We can summarize the Cairns-Smith program for the unfolding of life in the following diagram:

THE CAIRNS-SMITH SCENARIO

Clay → Growth/Replication → Organics → RNA/Proteins
                                        takeover

The great advantage of Cairns-Smith's Clay Theory is that it makes it far easier to see how life could have gotten started. There was no need for an unlikely and fortuitous juxtaposition of chemical and geological events, only simple chemical reactions involving readily available materials; in fact, these reactions are still going on today. But if such chemical activities are still happening, and the sequence of events is so easy, why don't we see any crystal life today? Or do we?

Currently no one really knows the answer to these queries, principally because no one's really looked. However, in his books and articles Cairns-Smith suggests a number of places where we might look for evidence of such life and what form we might expect it to take. For instance, he suggests that if such crystalline life exists today it might simply be a rather loose collection of interacting crystals whose boundaries are somewhat fuzzy and diffuse. Consequently, we should look for "bizarre" crystal structures doing unusual things.

Cairns-Smith also suggests laboratory experiments involving a mineral version of Spiegelman's test-tube evolution experiment with the $Q\beta$ virus. A supersaturated solution of minerals would flow into a continuous crystallizer. Crystal formation and growth would take place within it, with a suspension of crystals flowing out the other end. Imagine that two different kinds of crystals form in the crystallizer, and that one grows quickly but doesn't break up and so is eventually washed out the other end. But suppose the second type not only grows, but easily breaks apart, and that new crystals are formed to compensate for those lost at the exit pipe. If some random variant can replicate itself more rapidly than the competition, it should eventually take over the entire crystallizer, just as the "monster" did in Spiegelman's test tube. This kind of experiment has yet to be done, but would shed considerable light on the possibilities of crystal evolution and hence on the feasibility of Cairns-Smith's theory of life.

With the Clay Theory, the Prosecution has completed its case for the origin of life on Earth via natural chemical and geological processes. Now we ask the Defense to take the floor and present its spectrum of otherworldly claims for how life got its start. The case for the Defense rests on two pillars: arguments from Nature and arguments from the supernatural. We will listen to natural claims first.

IT CAME FROM OUTER SPACE

James Watson opens *The Double Helix,* his classic account of science in the fast lane, with the statement that "I have never seen Francis Crick in a modest mood." Crick is now a seventyish, graying, distinguished-looking man of good cheer, who appears to have mellowed considerably since Watson's account. But whether you term it good cheer, immodesty, or just plain brash exuberance, Francis Crick has spent the better part of the last three decades on the front pages of both the scientific and popular press, with a steady succession of offbeat and slightly outrageous ideas about various aspects of molecular biology, brain theory, extraterrestrials, and other matters of body and mind.

One of Crick's more speculative offerings appeared in a 1973 article coauthored with Leslie Orgel, in which they claimed that life on Earth might have originated in outer space with extraterrestrials. Crick later expanded his ET theory into the book *Life*

Itself, which argues that the Earth has been under continuous observation by intelligent extraterrestrials who, when the time was right, planted the "seeds" of life on Earth. This notion, like Linus Pauling's theory of vitamin C and the common cold, would probably sink like a stone into the seas of scientific oblivion if it were not being championed by a Nobel laureate. Nevertheless, given the nontrivial obstacles in the path of all of the Earth-based scenarios for life, it's definitely worth taking a look at what Crick has in mind.

The principle underlying Crick's "life from space" thesis had its origins early in this century with the ideas of Svante Arrhenius, who promoted a vision of life raining down on Earth in the form of tiny spores from space. Arrhenius was a Swedish chemist whose original work on the behavior of salts when they dissolve in water was so lightly regarded that when presented in support of his Ph.D. degree, it received the lowest possible passing mark. Later, in the sort of 180-degree turnabout that scientific cranks dream of, his ideas were vindicated and he was finally rewarded with the Nobel Prize for chemistry in 1903. His scientific position secure, Arrhenius could then afford the luxury of proposing a theory of life that involved microorganisms escaping from other life-bearing planets in the galaxy, traveling across interstellar space propelled by the pressure of stellar radiation. According to Arrhenius, one of these spores eventually landed on Earth, giving rise to life as we know it today. This Panspermia Theory is now mostly discredited, the principal arguments against it being that it's unlikely that even one such spore would arrive on Earth during the entire history of the universe, and furthermore, any microorganism of the type we know today would very likely have been killed by solar radiation and/or the cold and vacuum of space.

Crick updated the Panspermia Theory by noting that most of the arguments against it would be invalid if the spores arrived on Earth after having been transported here on some kind of interplanetary vehicle. Observing that the universe is more than twice as old as the Earth, Crick argued that it's not unreasonable to suppose that life could have arisen more than once. Furthermore, he rightly pointed out that there's no reason to believe that the conditions that prevailed here on Earth were anywhere near optimal for the development of life. Putting these remarks together with the anthropomorphic hypothesis that any extraterrestrial life form would have the same psychological need for expansion that humans have displayed, he concluded that the most likely explanation for life on Earth is that it was seeded by extraterrestrials.

While "Directed Panspermia" formally answers the question of how life arose on Earth, from the standpoint of a scientific explanation it does so in the most unsatisfactory way imaginable—by pushing the problem off into some other solar system. In his defense, Crick himself appears not to take the whole business very seriously, and has commented that he put the hypothesis forward only to focus public attention more sharply upon the difficulties associated with the origin-of-life question. In fact, even Crick's own wife thought he'd gone slightly mad, dismissing the whole notion as pure science fiction. But in contrast to the playful ET origins suggested by Crick, another eminent British scientist, Sir Fred Hoyle, has put forth a different sort of "life from space" theory, one that he takes very seriously indeed.

Is there something about the air of the British Isles that causes responsible, rational, sober scientists to turn their attentions to eccentric, crankish, or just plain weird notions when they begin to enter their philosophizing years? Isaac Newton appears to have caught this disease and spent most of his later years hunched over the Bible in search of ammunition to support his claims for familial relationships that apparently only he could see. The case of Bertrand Russell's offbeat social theories is well chronicled, and Francis Crick's Directed Panspermia seems also to represent a mild dose of this peculiarly British affliction. Fred Hoyle, on the other hand, appears to many to have caught a terminal case with his advocacy of the idea that life on Earth originated as a kind of "disease" from the stars.

Hoyle, a rather short, vigorous man in his mid-seventies, who still tirelessly stalks the moors of his native Yorkshire, has had a long and distinguished scientific career noted for pioneering work involving the way in which heavier elements are made from lighter ones in the interiors of stars. He is also known for his now somewhat discredited theory of the so-called steady-state universe, formulated together with Thomas Gold of Cornell, a theory in which the universe didn't begin with a bang, but rather has always been more or less the same whimper we see today. For these universally acknowledged scientific contributions, Hoyle has been recognized with election both as a fellow of the Royal Society and as a foreign associate of the U.S. National Academy of Sciences, as well as having been honored with a knighthood in 1972. Besides his real science, Hoyle has also found time to make major contributions to the literature of science fiction, having written several very intriguing and entertaining novels including one of the all-time classics, *The Black Cloud,* which introduced the possibility of an alien life form composed of a gigantic cloud of interstellar plasma.

With his penchant for pushing contentious scientific causes, it's perhaps not surprising to learn that Hoyle has also found time to get into all sorts of squabbles over academic, political, and administrative matters with his colleagues, especially those at Cambridge, his home university. At one time in the mid-1960s, the heat was so intense that Hoyle resigned his position on the mathematics faculty, threatening to emigrate to the United States. This fate was staved off only by his appointment as director of the newly formed Institute of Theoretical Astronomy. Somewhat later, Hoyle also found his way into the public press when he accused his colleague Anthony Hewish of exploiting the work of a graduate student, Jocelyn Bell, in promoting the work that eventually led to the Nobel Prize for physics in 1974 for the discovery of pulsars, as recounted in the last chapter. With such a track record, the appearance of Hoyle and his ideas in polite scientific circles is about as welcome as the appearance of Martin Bormann at a bar mitzvah. Nevertheless, Hoyle and his longtime associate Chandra Wickramasinghe have put forth not just one, but two distinct scenarios for how life came to be. Let's briefly look at these two visions of life according to Hoyle.

H&W: VERSION I

In their early papers, Hoyle and Wickramasinghe claimed that life originated in the molecular clouds of interstellar space, and was then transported to Earth by comets. Radio astronomers have noted that many of the important organic molecules needed for life are present as major components of the vast clouds wandering between the stars. H&W jumped on this fact, claiming that cometary material "seeded" the primordial soup with the right stuff to develop into the first terrestrial life forms.

The essential point of this cometary theory was the assertion by H&W that the interstellar dust grains that seeded the Earth were grains of cellulose, perhaps the most abundant biological product on Earth, forming the main component of trees, cotton, and many other important plants. This claim raised a number of scientific eyebrows, principally because cellulose is such a special material, coming about on Earth only under very particular biological circumstances. Thus, any chemical process going on in outer space that could yield such a specific substance could also be expected to give a large number of other important chemical products as well. This kind of claim seems so miraculous to the flinty-eyed community of astrochemists that the most overwhelming evidence and documentation would be required to substantiate it. Unfortunately, H&W offered no such weight of evidence, proposing instead to back up their assertions only with some rather inconsistent infrared spectral observations of dubious pedigree. This evidence involved averaging the spectral characteristics of a collection of 153 compounds they thought were relevant to life, and then smoothing the result to fit the observed spectral data of the interstellar clouds. Spectroscopists and astrochemists around the world were uniform in their denunciation of this procedure.

When Hoyle and Wickramasinghe published their ideas for the general public in the popular book *Lifecloud,* the response of the scientific community was mixed in the extreme. At one end of the spectrum were the scathing remarks by Lynn Margulis, who called the book "flamboyantly irresponsible," and noted that "its theme moreover is entirely contrary to the considered opinion of most workers in the field. . . ." On the other hand, some words of praise came from science journalist John Gribbin, who in *Genesis,* his own book on the origin of life, states that "something along these lines will eventually become the established view." Similar testimonials were forthcoming from others as well, who noted that it should definitely be possible for complex molecules to form spontaneously in cosmic gas clouds. However, extensive scrutiny of the scientific arguments presented by H&W in the technical papers supporting their claims uncovered so many holes that even the Watergate plumbers couldn't have made the Lifecloud Theory respectable. Besides the aforementioned problems with the spectral data, H&W were taken to task by their critics for a plethora of experimental gaffes, ignorance of disconfirming data, statistical snafus, and, in general, "fingerprints" of the sort we discussed in Chapter One under the heading of pseudoscience. Thus ended H&W's initial foray into the origin-of-life game; however, it was not to be their last.

H&W: VERSION II

Following a half-time breather, H&W reentered the game with a second theory that stands in almost total opposition to their original ideas about interstellar dust and cometary messengers. In Version II, all pretense at a natural explanation for life is abandoned in favor of the claim that life originated with a Creator, and was then carried to Earth as a form of cosmic "disease." As an indicator of the kind of about-face that the Disease Theory represents, in the Lifecloud Theory H&W accepted the primordial soup as the breeding ground for the life deposited by the comets. But in their new vision, H&W state that "another fuddled notion is that life began here on Earth in a thin brew of organic material. The mystery is why grown men and women have allowed themselves to be persuaded into such beliefs, in spite of there being a considerable body of fact running against them."

Besides repudiating most of their original claims, H&W brought many new and wonderous notions into the Disease Theory of life. For instance, they assert that many historical developments were caused by diseases originally brought here from space. As an illustration, they cite the superiority of classical armies to medieval ones, with the explanation that the Middle Ages were riddled with diseases. This claim is then followed by the even-more-difficult-to-swallow statement that "we also attribute the rise of Christianity to the same disease-filled epoch."

As noted, Hoyle and Wickramasinghe's second theory attributes the origin of life to a Creator, but not just one of the deities claimed by conventional religions. No, their Creator is one of their own divination, being none other than—a silicon chip! Apparently, they think that the dust clouds of space somehow coalesced into such a chip, much the same way that the sentient cloud in Hoyle's famous science-fiction story *The Black Cloud* was formed. Unfortunately they offer no scientific arguments or testable predictions, nor do they cite any experimental data in support of these strange notions. As H&W pile one extravagance upon another, they ultimately wind up defending a position that is unabashedly of the divine revelation variety.

Amusingly, Robert Shapiro has also noted the similarity of the later H&W theory with the thesis put forth as science fiction in *The Black Cloud*. The Cloud is surprised to find life on Earth, stating that space is a far better place for the assembly of biochemicals. Sensing higher intelligences in the universe, the Cloud finally becomes bored with humans and sets off to find these higher intellectual forms. Only by reading this early (1957) fictional account does one realize that in *Lifecloud* and *Diseases from Space* Hoyle finally came out of the closet to display a long-held, essentially religious view of the mystical origins of life on Earth. The only difference is that the intervening decade has seen Hoyle's vision pass from the realm of fiction to that of "fact." Thus do Hoyle and Wickramasinghe move from scientific arguments for the origin of life to what are essentially religious ones, treading exactly the same path as our next extraterrestrial-origins adherents, the "creationists," only in precisely the opposite direction.

AND GOD CREATED...FROM FISH TO GISH

In an attempt to effect legislative repair to one of the oldest flaws in the fabric of Nature, the state of Indiana in 1897 enacted a law setting the legal value of π at precisely 4, replacing its inconvenient "natural," but irrational, value $\pi = 3.14159265 \ldots$ Later, a Tennessee legislator suggested the value be legally fixed at 3, but this idea was immediately quashed when a British clergyman, in one of those hilarious letters that British clergymen have traditionally sent to *The Times* of London, stuck up for the Indiana value, stating that 3 was inadequate since it wasn't even an even number! But the Tennessee legislature eventually imposed its will on an unruly cosmos anyway by enacting a different law making it illegal to teach evolution in the classroom, an action thrusting the tiny hamlet of Dayton into the international spotlight in 1925 with the celebrated Monkey Trial of John Scopes, a substitute for the local high-school biology teacher, accused of filling the heads of his charges with pernicious Darwinian visions.

For most of us, I suppose, the dramatic account of the Scopes trial in the film *Inherit the Wind,* in which a legendary barrister based on Clarence Darrow (played by Spencer Tracy) crushes the fundamentalist arguments of a prosecuting attorney modeled on William Jennings Bryan (played by Fredric March), represented what we thought of as the death knell of legislative tampering with Nature. This despite the fact that Scopes was actually found guilty and assessed a one-hundred-dollar fine (although two years later the Tennessee Supreme Court overturned the conviction on technical grounds). And a death knell it was, at least insofar as brute-force, frontal legislative assaults on Nature by religious fundamentalists are concerned. But in March 1981, not to be outdone by its next-door neighbor, the Arkansas state legislature revived the spirit of Dayton by resurrecting a fundamentalist interpretation of the origin of life under the new rubric "creation science." With the enactment of the Balanced Treatment for Creation Science and Evolution Science Act (Arkansas Act 590), stating that "public schools in this state shall give balanced treatment to creation science and to evolution science," the battle was rejoined between the fundamentalists and the scientists, only this time it was to be fought on the home ground of science rather than in the pulpits. Let's take a moment to understand why.

The essential components of the "creationist" vision of the origin of the the Earth and its life forms is contained in the following pledge sworn to by each member of the Creation Research Society:

1. The Bible is the written Word of God, and because we believe it to be inspired throughout, all of its assertions are historically and scientifically true in all the original autographs. To the students of nature, this means that the account of origins in Genesis is a factual presentation of simple historical truths.

2. All basic types of living things, including man, were made by direct creative acts of God during Creation Week as described in Genesis. Whatever biological changes have occurred since Creation have accomplished only changes within the original created kinds.

In addition to swearing this pledge of "allegiance," all prospective members of the society are also required to possess an advanced university degree in some field of science. As a result, members in essence agree to forsake the common practices of their profession in certain areas, and instead accept explanations on the basis of divine authority alone.

In 1968 the U.S. Supreme Court outlawed all anti-evolution laws like the Tennessee statute on the grounds that they violated the constitutional prohibition against mixing the state, in the form of the schools, with religion. Since this decision effectively prevented the creationists from having their ideas of religion introduced into the educational curricula, the fundamentalist movement decided to settle for the next best thing and mounted a campaign to push its position into the classrooms, dressing it up as science. The Arkansas bill gives a particularly graphic account of the strategy employed. Arkansas Act 590 lists six principles of "evolution science" side by side with corresponding principles of "creation science," and then goes on to state that both should be given equal time in the classrooms. The two most important principles for our purposes are the following, which I have taken directly from the text of the act: "Creation science means the scientific evidences and related inferences that indicate: (1) Sudden creation of the universe, energy, and life from nothing; . . . (6) A relatively recent inception of the earth and living kinds." Other points of the act involve the occurrence of a global flood, separate ancestry for man and apes, and other similar biblical stipulations. It's clear from the above statements that in order to make their case, the creationists are going to have to attack the conventional scientific views on several aspects of geology, most importantly the matter of the age of the Earth.

In speaking of the education of their children, creationists are fond of citing the remark of William Jennings Bryan that "Christians desire that their children shall be taught all the sciences, but they do not want them to lose sight of the Rock of Ages while they study the age of rocks." This well-known remark served for years as a rallying cry for fundamentalists asserting that the rocks of the Earth were only a few thousand years old, just as claimed in Genesis. It doesn't take too much imagination to envision the loathing with which the creationists look upon the increasingly accurate radiocarbon-dating methods developed over the past few decades. With these unassailable methods, used recently, for instance, to demonstrate the medieval origin of the Shroud of Turin, the high levels of uncertainty arising from the old fossil and sediment dating schemes were eliminated, showing the Earth to be at least 4 billion years old.

How did the creationists react to such incontrovertible evidence of an ancient Earth? Well, let me quote Henry Morris, a hydraulics engineer and director of the Creation Research Society: "The only way we can determine the true age of the earth is for God to tell us what it is. And since he *has* told us, very plainly, in the Holy Scriptures that it is several thousand years in age, and no more, that ought to settle all basic questions of chronology." Such an act of faith unfortunately rejects data, methods, experimental equipment, and all of the other paraphernalia of science. In fact, the leading creationists have been even more candid in their rejection of science's traditional methods of inquiry.

Duane Gish holds a Ph.D. in biochemistry from the University of California at Berkeley; he is also the vice-director of the Creation Research Society and a regular participant at university debates on the merits of creation science. Since he is trained in the scientific method, especially in an experimental science like biochemistry, it's odd, to say the least, to read in his book *Evolution: The Fossils Say No* that "we do not know how the Creator created, what processes He used, for He used processes which are not now operating anywhere in the natural universe. . . . We cannot discover by scientific investigation anything about the creative processes used by the Creator." With such statements, creation "science" joins the long list of other perverse modern "sciences," such as "fashion science," "dairy science," and "educational science," all of which can be conveniently subsumed under the heading "nonscientific science."

Despite the cursory nature of our airing of the creationist views, I think most readers will find no difficulty in understanding the opinion of Judge William Overton in his ruling declaring the Arkansas Act 590 unconstitutional. Citing the creationists' own words in deciding that creation science was not science but religion, the good judge offered one of the most concise, best-thought-out lists of criteria for what constitutes science yet put on the public record. The Overton criteria are:

• It [science] is guided by natural law.
• It has to be explanatory by reference to natural law.

- It is testable against the empirical world.
- Its conclusions are tentative, i.e., are not necessarily the final word.
- It is falsifiable.

Needless to say, creation "science" fails to meet even one of these criteria; ergo, as a *scientific* explanation for the origin of life, it has no real place in our deliberations here.

If creation science has no role in a scientific consideration of life's origins, why have I devoted any space to it at all? Principally because the creationist controversy illustrates in the starkest possible terms the psychology and tactics of pseudoscience as considered in abstract terms in Chapter One. All of the hallmarks of pseudoscience enumerated there show up in glorious detail in the Arkansas case: appeals to myths, a casual approach to evidence, irrefutable hypotheses, refusal to revise, and all the other by-now-familiar calling cards of the pseudoscientist.

From an intellectual point of view, perhaps the most interesting aspect of creationism is not the "what" of its beliefs about the way life got started, but the "why." That is to say, why is it that adherence to a literal reading of the book of Genesis holds such great appeal for so many people? There must be something more to it than just the odd beliefs of a fringe group of backwoods hicks, as even well-educated, obviously intelligent people like Henry Morris and Duane Gish are not immune to its attractions. In grappling with this puzzling matter, I can only conclude that the reasons, whatever they are, lie much deeper than in the mere surface phenomena of a religious belief about the origin of life on Earth. To my eye, creation science is only a symptom of a far more fundamental disenchantment with science in general, and the overwhelmingly dominant role it plays in daily life. Many people obviously feel threatened by what they see as the control that science has gained over their lives, and many others feel mistrustful of the claims made by the pro-science lobby about the improvement in their lives that science will provide. And who can blame them, with disasters like Chernobyl, the *Challenger,* Bhopal, and Love Canal serving as constant reminders of science and technology run amok? So my feeling is that the simple, straightforward belief in the word of

God as written in Genesis serves as a comforting counterweight for those of a certain fundamentalist persuasion. And as long as people remain ignorant of the limitations of science and the fact that science is carried out by ordinary human beings with all their foibles and weaknesses, the creationists, like the rich, will always be with us.

Through a poetic twist of cosmic fate, I happen to be writing these words on Christmas Eve (1987), the main day of the year for creationists to recharge their batteries in preparation for another 365 days of jousting with those who take their Bible reading a little less literally (whoops, 366 days—I almost forgot that 1988 is a leap year). As a Christmas bonus to my scientific readers, let me close this section by recounting what is surely one of the more amusing sideshows associated with the Arkansas trial.

In mounting their defense, the biggest scientific gun that the creationists seemed able to muster was none other than Fred Hoyle's comrade-in-arms, Chandra Wickramasinghe. The creationists presumably requested his appearance because he had suggested the intervention of a Creator to explain life on Earth, although the silicon chip Creator he and Hoyle had conjured up probably wasn't exactly what Henry Morris and Duane Gish had in mind. Anyway, after beginning his testimony with a few well-chosen words about life being the product of a Creator, Wickramasinghe began veering off the track, entering into a long, meandering exposition of his views on comets, diseases, and the rest of the extraterrestrial apparatus underlying the H&W theories. He ended his testimony *for the defense* by stating that he saw no way that a rational scientist could endorse the notion of a global flood, or an age for the Earth of less than 1 million years. With expert witnesses like this, the creationists surely didn't need any enemies! In summarizing this farcical testimony, Judge Overton remarked that he was "at a loss to understand why Dr. Wickramasinghe was called in behalf of the defendants." On this sorry note ends not only the case for the state of Arkansas, but also our own Defense case for the extraterrestrial origins of life on Earth. *Sic transit gloria.* Before moving on to summary arguments, let's call upon a few expert witnesses on the functional activities of life in order to get a little better perspective on the prospects for scientifically wrapping up the foregoing claims and arguments.

THE LOGIC OF LIFE

In recent years the standard Primordial Soup Theory has come under increasing attack, both as a result of newly acquired experimental data and as a result of some serious reexamination of the methods and arguments employed by its proponents. Before pondering a verdict on the competing cases, let's pause for a look at some of the points about the soup that give skeptics cause for concern.

- *Reducing atmosphere:* There has been a mounting body of evidence to support the claim that the early atmosphere was not nearly as reducing as claimed (and needed) by the soup theorists. Data has been presented showing the presence of oxygen-producing life forms and oxidizing mineral species in rocks more than 3.5 billion years old, as well as calculations showing that a significant amount of free oxygen could have been produced by photodissociation of water. While not conclusive, these results certainly cast doubt upon origins theories that hinge critically upon a lack of free oxygen in the primitive atmosphere.
- *Macromolecule polymerization:* Soup theories rely upon the assembly of proteins and nucleic acids by a linking-up of many individual amino acids or nucleotides. The kinds of such "polymerizations" needed for life are subject to many competing reactions, with processes of destruction just as prevalent as those of construction. Thus, any viable Soup Theory would have to offer an explanation of how the constructive processes dominated those tending to tear apart potentially useful polymer chains.
- *Monomer concentrations:* The amino acids produced in Miller-type experiments usually appear in *very* low concentrations; ditto for the nucleotide bases formed in the kinds of experiments done by Eigen and Orgel. These concentrations are far too low to have had a plausible chance of leading to any meaningful spontaneous polymerization.
- *Investigator interference:* A common occurrence in the origins business is for an investigator to postulate some sequence of chemical reactions needed to lead to life. He then sets up ex-

periments that could plausibly lead to the production of the necessary intermediate chemical compounds. As each compound is produced, *no matter how small the quantity,* the experimenter then proceeds to the next step, assuming that the needed elements from the earlier steps are available at *whatever level of purity* and in *whatever quantity* desired. In fact, almost all laboratory prebiotic simulations involve illegitimate interference of this sort by the experimenter, in which he or she adjusts the experimental conditions so as to violate plausible hypotheses about what it was like on the early Earth.

Every item on the above list is a potentially fatal flaw in any conventional Soup Theory. Let's think more positively and suppose we were trying to construct an alternative theory instead of just looking for holes in existing proposals. What kinds of problems would we want our "Stew Theory" to address effectively? Here are just a few:

• A possibly oxidizing primordial atmosphere
• A dominance of destructive over synthetic processes in the prebiotic environment
• A short time interval of, say, 170 million years for the appearance of the first life forms
• The presence of Precambrian rock deposits exhibiting no geological or geochemical evidence for any sort of hydrocarbon-rich primordial soup
• Creation of a controllable and readily recognizable barrier between what laboratory experiments do when left to themselves, and what they do when there is active interference by the experimenter

Creation of a theory satisfying the above list of desiderata is a tall order indeed. And what would we have even if we did produce such a theory? Would *any* theory really tell it true as to how life *did* originate here on Earth, as opposed to how it *might* have originated? Are all such theories only *Just So* stories, or do we really have, at least in principle, a fighting chance to unravel the actual sequence of events that took place over 4 billion years ago? These questions lead us into deep waters of philosophy, in particular into consideration of the distinction between *operation science* and *origin science*.

*   *   *

The usual view of theories of knowledge requires that a *scientific* theory be able to (1) explain observed phenomena, (2) predict phenomena that have not yet been observed, (3) be testable by further experimentation, and (4) be modifiable as needed by the results of new experiments. In order to have even a ghost of a chance of satisfying these criteria, the scientific theory must set out to explain a recurrent set of events. The final condition for a scientific theory is basically what separates science from some of the baser forms of pseudoscience, and is a condition that cannot possibly be met if experiments cannot be performed. Unfortunately for origins theorists, the one thing that everyone seems to agree upon is that the origin of life on Earth was a one-time-only event, hence outside the bounds of what is normally thought of as a scientific theory.

The foregoing split between a once-and-once-only event and a set of recurrent phenomena is what separates operation science from origin science. Operation science deals with explanations of recurring processes like the passage of the Earth around the Sun, the union of hydrogen and oxygen to form water, and the flow of electrons through a resistor. In short, with natural processes that are, in principle, repeatable. Origin science, on the other hand, addresses the unique events in life: formation of the universe, World War II, the painting of the *Mona Lisa*—and the origin of life itself. Such events are not explainable by traditional scientific theories for the simple reason that they are not subject to experimentation; thus they are not falsifiable and hence not scientific. Or are they? Is there a loophole of some kind that would somehow enable us to make a unique event repeatable, at least to the extent that sufficiently detailed experiments can be performed about the event so that in a scientific sense the event becomes repeatable? Prebiotic experiments of the Miller type are crude, slow, plodding steps in this direction; some think that the modern computer offers a possibility for far more rapid progress.

Earlier we discussed the idea of using a WEES apparatus to simulate the entire environment of the early Earth. The idea is to create a miniature version of the primordial seas, atmosphere, energy sources, tidal pools, and all the rest, then turn the system on and see what happens. The problem with such a simulation is the time factor: It's estimated that the first living forms on Earth took on the order of at least 170 million years to arise.

What tenure-hungry assistant professor can afford to wait millions of years for something publishable to start swimming around in such a device (even if the environmental parameters are right)? The digital computer offers us two distinct pathways to get around this time barrier—Material Mode and Formal Mode.

In Aristotle's epistemology, there are four causes for the appearance of worldly events, causes that Aristotle offers as explanations for why things are as they are. The four complementary, mutually exclusive, and collectively exhaustive causes are *material, formal, efficient,* and *final* causation, of which the first two are the most relevant for us at the moment. According to Aristotle, material cause explains an event's or object's taking the form it does as a result of the material elements out of which it is composed. On the other hand, the event or object also has a plan according to which it is constructed, and this plan is completely independent of the matter out of which the object is built. In Aristotle's scheme of things, the plan constitutes formal cause.

In all the origins work surveyed above, the focus has been almost totally upon what in Aristotelian terms would be considered material cause. All of the primordial soup theorists begin by postulating some kind of material elements composing the soup, together with a sequence of physical processes that plausibly lead from the primitive material elements of the soup to the first life form. By and large, the major points of disagreement revolve about matters of material causation, e.g., the gases composing the primordial atmosphere, whether the original life form was silicon or carbon based, and so on. For questions of this sort, the computer can be run in Material Mode and used in the commonly accepted manner to simulate the chemical and physical processes postulated by particular models. We saw a good illustration of Material Mode in the computational experiments of Niessert, who used this mode to investigate the plausibility of the random replicator scenarios of Eigen. When operating in Material Mode, the computer's main role is to act as a time accelerator, allowing the basic physical and chemical processes to unfold on a time scale thousands, if not millions, of times faster than real time. Thus, with the types of supercomputers currently available, a Material Mode simulation of the WEES might yield something interesting after, say, just a few years instead of the hundreds of millions that might be needed in real time. Important and interesting as Material Mode undoubtedly is, to my mind the *really* interesting way to use computers to study life is to use them in Formal Mode.

Looking at life through the spectacles of formal causation means that we forget entirely about what kinds of *matter* living objects are composed of, and instead turn our attention to the *functional,* or *logical,* structure of living agents. In other words, we focus upon those aspects of living forms that distinguish them from nonliving objects, and ignore entirely the kind of "stuff" that they're made out of. Philosophers of biology generally agree that the functional activities distinguishing living forms are three in number: metabolism, self-repair, and replication. Let's look at how these activities can be formally represented by the logical interconnections linking them, independent of material considerations.

On September 20, 1948, John von Neumann delivered a lecture at Caltech titled "On the General and Logical Theory of Automata," in which he laid the foundations for a functional theory of life. Von Neumann's interest at the time was in explicating the logical principles permitting construction of a machine that would be capable of manufacturing copies of itself if placed in an environment sufficiently rich in the necessary raw materials. Thus, at first glance it would appear that von Neumann's attention was focused on material cause. But first impressions can be deceiving, and a reading of the paper makes it transparently clear that what von Neumann was driving at was to give a mathematically complete account of the different kinds of functional activities that such a self-reproducing object would need to have in order to be able to function. The particular material composition of such a *self-reproducing automaton* was of little interest to von Neumann, and I'm sure he couldn't have cared less if one tried to build such a machine from aluminum, glass, steel, or, for that matter, lox and cream cheese. What interested von Neumann was the different *functions* that would have to be incorporated and coordinated to achieve the properties characterizing life; in short, the logic of life.

What von Neumann discovered was that *any* self-reproducing object must contain four fundamental components:

A. A *blueprint,* providing the plan for construction of offspring
B. A *factory,* to carry out the construction

C. A *controller,* to ensure that the factory follows the plan
D. A *duplicating machine,* to transmit a copy of the blueprint to the offspring

In living cells these properties are physically manifested, roughly speaking, in the DNA (the blueprint), the process of translation (the factory), the specialized replicase enzymes (the controller), and the process of replication (the duplicating machine). It's worthy of note that von Neumann discovered these abstract properties necessary for any living form more than five years before the far more publicly celebrated work of Watson and Crick, which dealt with the very special case of the kind of life we now see on Earth. Such are the wages of the theoretician, especially one who solves "only" the general case!

The work of von Neumann and his successors shows that everything that's functionally important about life can be represented as logical patterns that are in principle implementable in a multitude of material environments. The simplest and most entertaining illustration of this point is the well-chronicled game of Life, an elementary board game invented by the British mathematician J. H. Conway. The playing field of Life can be imagined as a flat sheet of paper extending infinitely far in all directions, with the sheet ruled off into square cells like a chessboard without colors. At any particular stage of play, a given cell is either alive (ON) or dead (OFF), the live cells being filled in with a dot, say, and the dead cells being left blank. According to the rules laid down by Conway, whether a particular cell is ON or OFF at the next stage of play depends upon the current state of those cells that are its immediate neighbors in what is termed the *Moore neighborhood,* depicted in Figure 2.10. The rules are very simple: The cell is ON if exactly three of its neighbors are ON; it is OFF if it has zero, one, or more than three neighbors that are ON (death from isolation or overcrowding); it retains its current state if exactly two of its neighbors are ON. Conway said that he originally set up these rules as a guess between balancing the birth of new cells in a rich, cooperative environment of social support, and the death of cells by overcrowding or isolation. Let's look at a few rounds of play.

Figure 2.11 shows the histories of three generations of Life patterns, all of which initially begin with three ON cells. The

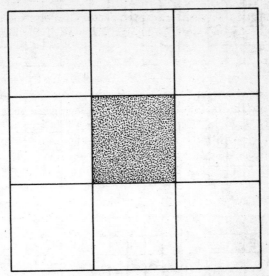

FIGURE 2.10 *The Moore neighborhood of a Life cell*

reader will note that the first three triplets all die out, while the fourth forms a stable configuration called a Block, and the fifth, termed a Blinker, oscillates indefinitely.

For our purposes, one of the most interesting patterns in Life is the so-called Glider, which is a pattern that repeats itself after four generations, but in the process moves one square down and to the right. A picture of the Glider is shown in Figure 2.12. In the early days of Life, Conway conjectured that there were no Life patterns that could grow indefinitely (i.e., would never die out), and offered a fifty-dollar reward for the first proof or counterexample to his assertion. A group at MIT claimed the prize by displaying "the Glider Gun," shown in Figure 2.13. This configuration is a spatially fixed oscillator that resumes its original shape after thirty generations. Within this period, the Gun shoots off a Glider that wanders across the playing field and encounters the configuration in the upper-right corner called an Eater, which is a fifteen-generation oscillator. The Eater swallows up the Glider without undergoing any irreversible changes. Since the Gun oscillates indefinitely, it can produce an infinite number of Gliders, thereby showing that there are configurations that "live forever." This fact refutes Conway's conjecture. What does all this have to do with formal models of life, as op-

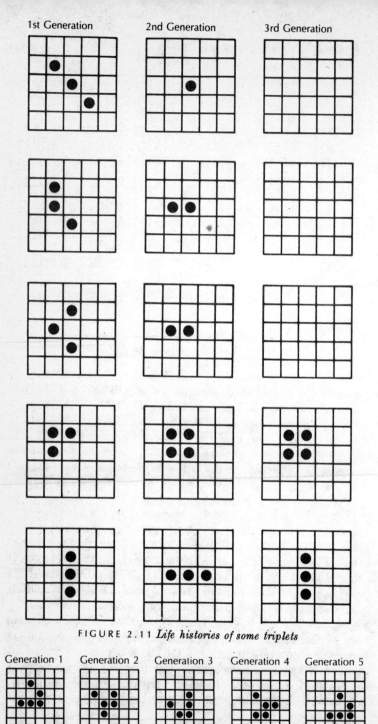

FIGURE 2.11 *Life histories of some triplets*

FIGURE 2.12 *The Glider*

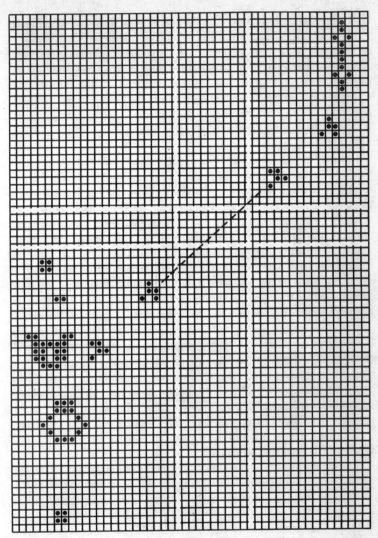

FIGURE 2.13 *The Glider Gun*

posed to special facts about Life? As it turns out, it has almost everything to do with it.

In von Neumann's automata setup, just as in the behavior of real-life cells, information in the DNA is used in two quite distinct ways: as instructions to be *interpreted* (as in gene translation), and as instructions to be *copied* (as in DNA transcription). Consequently, if a Life pattern could be displayed that would be self-reproducing *and* use the information describing itself in just these two ways, it would be difficult to argue that such a pattern was not "alive," in the sense that it would then display all the features needed to constitute a living form. Conway proved that such a Life configuration exists, although to display it explicitly would require a playing field the size of a small city (Venice, for instance).

Conway's self-reproduction proof is based on the observation that Glider Guns, as well as many other Life objects, can be produced in Glider collisions. He then shows that large constellations of Glider Guns and Eaters can produce and manipulate Gliders to force them to collide in just the right way to form a copy of the original constellation. The proof begins not by considering reproduction *per se,* but by showing how the Life rule allows one to construct a universal computer. Since the Life universe consists of an array of ON-OFF cells, what this amounts to is showing that one can construct a Life pattern that *acts* like a computer in the sense that we start with a pattern representing the computer and a pattern representing its programming. The computer then calculates any desired result, which would itself have to be expressed as a Life pattern. For numerical computations, this might involve the Life computer's emitting the requisite number of figures or, perhaps, arranging the required number of figures in some prespecified display area. Conway showed that the circuitry of *any* possible computer can be translated into an appropriate Life pattern consisting only of Guns, Gliders, Eaters, and Blocks.

The second part of the Conway proof is to show that any conceivable Life pattern can be obtained by crashing together streams of Gliders in just the right way. The crucial step in this demonstration is to show how it's possible to arrange to have Gliders converge from four directions at once in order to represent the circuits of the computer properly. The details of the ingenious solution to this problem are much too complicated to enter into here, but they provide the last step needed to complete Conway's translation of von Neumann's self-reproduction proof into the language of Life. This pioneering result opened the door to the use of the computer to study abstract life in Formal Mode. But what might such a study look like?

* * *

To the general public, the Los Alamos National Laboratory is usually thought of as nothing more than a bomb factory, inhabited by a collection of Strangelovian characters with visions of megatons dancing in their heads. While some sections of "the Labs" undoubtedly comply with this distorted vision of reality, many of the activities underway there are of a far more benign character, including a major effort linking the power of the modern computer with some of the deepest problems in modern biology. As an outgrowth of some of this work in theoretical and computational biology, in the fall of 1987 the Labs were host to the First International Conference on Artificial Life, which offered computer models of processes from protein synthesis to plant growth, all in the spirit of Conway's demonstration of computer life. The meeting organizer, Christopher Langton, concisely summarized the credo of the artificial-life community by stating that such studies seek "the ghost in the machine; an essence arising out of matter but independent of it"—in other words, formal causation!

While there's little room here to detail the program of the "artificial lifers," the essential ingredients are already present in an earlier paper by Langton himself, where he takes up the issue of how to use cellular automata (such as Conway's Life universe) to study *real* life of the organic, soft, squishy type biologists love and cherish.

Langton's paper argues that the primary functional roles of the proteins and nucleic acids are as follows:

- *Catalysis:* The special proteins associated with mediating chemical reactions are the enzymes, which do their thing by the process of catalysis, speeding up chemical reaction rates dramatically, sometimes by a factor of 100 million or more. Thus, for all practical purposes, enzymes determine which reactions occur and which do not. Among the most important properties of enzymes is the ability to recognize particular structures and to bring about changes in them. Hence, the enzymes are the active agents in the logic of life.
- *Transport:* Proteins are the main vehicles carrying molecules and ions around in the cell.
- *Structure:* Most of the cellular components and body tissues are formed out of proteins.

- *Regulation:* The primary agents for regulating the production and interactions of biomolecules in the cell are the proteins. In this role, they act mainly as messengers to initiate changes in enzyme activity or protein synthesis.
- *Defense:* Proteins constitute the main agents (the antibodies and immunoglobulins) by which the body fights invasion by foreign objects. These functions involve the recognition of foreign agents, and the production of various molecular compounds to tie up or break down the foreign invader.
- *Information:* The nucleic acids DNA and RNA provide the main information store in the cell. Various polymerase enzymes covering the DNA strands initiate the transcription of DNA to RNA, while other polymerases act to trigger the transcription of DNA in the process of replication.

With these functional roles in mind, Langton indicates how it might be possible to associate a cellular automaton rule similar to Conway's rule for Life with each activity. In this way we could *formally* represent each of the functional activities of a living agent with a logical "machine." Linking these individual machines together would then create an object that could be said to represent a living agent, albeit an artificial one. In the Langton paper this idea is actually carried out to create an artificial ant colony, whose behavior under appropriate conditions is strikingly similar to that displayed by real-life ants. Under other conditions, Langton's *vants* (virtual ants) display lifelike behavior quite different from what you'd see in your backyard terrestrial anthill, but perhaps just the kind of activity that might be seen in an ant farm on some planet orbiting Tau Ceti. Who knows? Anyway, the point is that the creation of life in a machine rather than a test tube offers almost unlimited vistas for experimentation with origin-of-life theories that would be temporally or physically inaccessible by any other means.

Before leaving this topic, let me take a moment for the benefit of those who think of "life" in a computer as being merely a hacker's metaphor having no connection with ordinary "wet" life, other than as a kind of computer game. This kind of thinking is becoming increasingly difficult to defend, the recent spate of "computer viruses" being the most dramatic evidence. And these are definitely not the same kind of bugs most of us are familiar with from the computer programming jargon. Put simply, a com-

puter virus is a piece of software that mischievous, and sometimes malevolent, programmers deliberately place in a set of instructions, say onto a game diskette or in a program available on
public-access electronic bulletin boards. As soon as the program
is loaded into a computer, the virus buries itself somewhere deep
in the system, with instructions to come alive when some set of
conditions is satisfied. For example, one famous virus was instructed to monitor the computer's clock and then "wake up"
when the date showed it to be the birthday of the Apple Computer
Corporation. Upon awakening, this benign virus temporarily
took over the computer's operating system and printed a birthday greeting on the screen. Other, less benign, viruses have been
reported that wipe clean data files, cause hard-disk crashes, as
well as produce a variety of other nasty effects. The point is that
once these things get into a system like a multicomputer network,
they take on what has every appearance of being a life of their
own. They can grow by moving from system to system through
the communication links in the network, and they act just like
biological viruses by appropriating the machinery of the network
for their own purposes. With these mischievous creatures now all
too real, I think the day is definitely over when one can scoff at
the idea of computer life as having no real meaning. To some
computer manufacturers, data center managers, and users, these
viruses are all too real for comfort.

Having listened to the sideline kibitzers, let's get back to the
business at hand—coming to some kind of closure on the convoluted and confusing circle of arguments, hopes, dreams, and
schemes for how life got started on Earth.

SUMMARY ARGUMENTS

The path we've followed in trying to get a handle on the various
theories proposed to explain the origin of life has been a long
and tortuous one, going from the very down-to-earth ideas of
Alexander Oparin to the acts of faith of the creationists. Let's
first summarize the competing arguments for the Prosecution
and the Defense.

Just to be perfectly clear on the conflict we're addressing, let's
begin our summary by restating the bone of contention. The
Prosecution's claim is that:

Terrestrial life had its origin as a consequence of natural physical and chemical processes occurring here on Earth.

The Defense's claim is just the opposite:

Terrestrial life either was imported to Earth, or did not come about as the result of natural physicochemical processes.

Telegraphically, the arguments are given in Tables 2.1 and 2.2.

LIFE ORIGINATED ON EARTH!

PROMOTER	ARGUMENT
Eigen, Orgel	random replicators, hypercycles
Gilbert, Cech	self-catalytic RNA
Oparin	coacervates
Fox	proteinoids
Dyson, Shapiro, Margulis	double origin, parasites
Cairns-Smith	clay

TABLE 2.1 *Summary arguments for the Prosecution*

LIFE ORIGINATED ELSEWHERE!

PROMOTER	ARGUMENT
("natural origins")	
Crick	extraterrestrial seeding
Hoyle and Wickramasinghe I	interstellar clouds and comets
("supernatural origins")	
Hoyle and Wickramasinghe II	silicon-chip Creator, diseases
Morris, Gish	creationism

TABLE 2.2 *Summary arguments for the Defense*

BRINGING IN THE VERDICT

On the specific question to be settled: "Did life originate on Earth or did it come from somewhere else?" my verdict comes quick and easy: The Defense is guilty of murder of the facts in the first degree! To my mind, even the wildest schemes of the Prosecution are vastly more plausible than the pipedreams, fantasies, and totally baseless speculations of the Defense. In fact, if I were the defendants' counsel I'd strongly advise following the path blazed by that well-known exemplar of modern political sagacity Spiro T. Agnew, urging them to enter a plea of *nolo contendere* and throw themselves on the mercy of the court. With the exception of H&W Version I, none of the Defense arguments are even in principle scientific, and they would hardly be worth a footnote in any serious account of the origin of life were they not being advocated by scientists of some repute, and adhered to by such large, seemingly uncritical, followings. The entire Defense case seemed to be aptly summed up in a newspaper account I read recently about a movement afoot in Iran to declare all car dealers and real-estate agents legally guilty of the greatest sin on the books in modern-day Iran: "corruption on Earth." Rereading the off-earthers' claims for life's origins, I thought it would be a delightful touch of theological irony to have the ayatollahs, of all people, expand their horizons and include the entire Defense contingent beneath their legal umbrella. But when it comes to picking and choosing among the many conflicting arguments of the Prosecution, things start to get interesting again.

Of the many Prosecution claims and scenarios, I must confess to a sneaking bias in favor of the Clay Theory of Cairns-Smith. My reasons? There are many, but perhaps the most appealing is that in contrast to the competition, it hasn't been strongly challenged by any serious scientific arguments, and especially experiments, against it. Of course, one could argue (and many do) that all this means is that the theory is new enough and offbeat enough that no one has really looked very hard at it yet. Maybe so. But to my ears at least, it has a ring of plausibility missing from the songs being sung by any of the competition.

First of all, the Clay Theory is explicitly a Dual-Origin Theory, one that easily accommodates my prejudice for life's originating with the proteins and then moving on to the nucleic acids. Somehow it just doesn't ring true that the nucleic acids, which are really just the big, fat molecular slobs of the cell, should arise before the proteins, which are the actual doers. Thus, any theory that postulates proteins first has a built-in advantage in my mind, and the Clay Theory certainly qualifies for these bonus points. Second, the theory requires no special materials and no special environment above and beyond what could be expected on the ancient Earth. Finally, I like the idea of starting with some kind of low-tech solution to the problem of how to get life going, and then shifting over to today's high-tech mode once things are up and running. As an additional selling point, the Clay Theory doesn't rely upon the kind of highly unlikely linking-up of many amino acids and/or nucleotides called for by the other theories, linkages that have formed the basis for any number of "devastating" critiques of origins theories by information theorists and others of that ilk. All in all, in my view the Cairns-Smith scenario provides a good lesson in how you should wield Ockham's razor in science to slit the throats of your opponents: Simply offer an argument leading to the same conclusions, but with fewer and simpler hypotheses. This is the essence of good theorizing as well as good model building, and to my mind Cairns-Smith has just done a better job of it than any of the others.

3

IT'S IN
THE GENES

===

CLAIM:
HUMAN BEHAVIOR PATTERNS ARE
DICTATED PRIMARILY BY THE GENES

NATURE/NURTURE: SENSE OR NONSENSE?

A few years ago, in one of the most fascinating and disturbing experiments in the annals of behavioral psychology, Stanley Milgram of Yale tested forty subjects from all walks of life for their willingness to obey instructions given by a "leader" in a situation in which the subjects might feel a personal abhorrence for the actions they were called upon to perform. Specifically, Milgram told each volunteer "teacher-subject" that the experiment was in the noble cause of education, and was designed to test whether or not punishing pupils for their mistakes would have a positive effect on the pupils' ability to learn.

Milgram's experimental setup involved placing the teacher before a panel of thirty switches with labels ranging from "15 Volts (Slight Shock)" to "450 Volts (Danger—Severe Shock)" in steps of 15 volts each. The subject was told that whenever the pupil gave the wrong answer to a question, a shock was to be administered, beginning at the lowest level and increasing in severity with each successive wrong answer. The supposed "pupil" was in reality an actor hired by Milgram to simulate receiving the shocks by emitting a spectrum of groans, screams, and writhings, together with an assortment of statements and expletives denouncing both the experiment and the experimenter. Milgram told the subject to ignore the reactions of the pupil, and to administer whatever level of shock was called for as per the rule governing the experimental situation of the moment.

As the experiment unfolded, the pupil would deliberately give the wrong answers to questions posed by the teacher, thereby bringing on various electrical "punishments," even up to the danger level of 300 volts and beyond. Many of the subjects balked at administering the higher levels of punishment, and turned to Milgram with questioning looks and/or complaints about continuing with the experiment. In these situations, Milgram calmly explained that the teacher was to ignore the pupil's cries for mercy and carry on with the experiment. If the subject was still reluctant to proceed, Milgram said that it was important for the sake of the experiment that the procedure be followed through to the end. His final argument was "You have no other choice. You *must* go on." What Milgram was out to discover was the number of subjects who would be willing to administer the highest levels of shock, even in the face of strong personal and moral revulsion against the rules and conditions of the experiment.

Prior to carrying out the experiment, Milgram explained his idea to a group of thirty-nine psychiatrists and asked them to predict the average percentage of people in an ordinary population who would be willing to administer the highest shock level of 450 volts. The overwhelming consensus was that virtually all the subjects would refuse to obey the experimenter. The psychiatrists felt that "most subjects would not go beyond 150 volts," and they expected that only 4 percent would go up to 300 volts. Furthermore, they thought that only a pathological, sadistic, lunatic fringe of about 1 in 1,000 would give the highest shock of 450 volts.

What were the actual results? Well, *over 60 percent* of the subjects continued to obey Milgram up to the 450-volt limit! In repetitions of the experiment in other countries—South Africa, Italy, West Germany, Australia—the percentage of obedient teachers was even higher, reaching 85 percent in Munich. How can we possibly account for this vast discrepancy between what calm, rational, knowledgeable men predict in the comfort of their study, and what pressured, flustered, but cooperative "teachers" actually do in the laboratory of real life?

One's first inclination might be to argue that there must be some sort of built-in "animal aggression" instinct that was activated by the experiment, and that Milgram's subjects were just following a genetic need to discharge this pent-up, primal urge onto the pupil by administering the electrical shock. A modern hard-core sociobiologist might even go so far as to claim that this aggressive instinct evolved as an advantageous trait, having been of survival value to our ancestors in their struggle against the vicissitudes of life on the plains and in the caves, ultimately finding its way into our genetic makeup as a remnant of our ancient animal ways.

An alternative to this notion of genetic programming is to see the subjects' actions as a result of the social environment under which the experiment was carried out. As Milgram himself stated:

> Most subjects in the experiment see their behavior in a larger context that is benevolent and useful to society—the pursuit of scientific truth. The psychological laboratory has a strong claim to legitimacy and evokes trust and confidence in those who perform there. An action such as shocking a victim, which in isolation appears evil, acquires a totally different meaning when placed in this setting.

Thus, in this explanation the subject merges his unique personality and personal moral code with that of larger institutional structures, surrendering individual properties like loyalty, self-sacrifice, and discipline to the service of malevolent systems of authority.

Here we have two radically different explanations for why so many subjects were willing to forgo their sense of personal morality and responsibility for the sake of an institutional authority figure: genetic determinism versus Marxian environmentalism. The problem for biologists, psychologists, sociologists, anthropologists, and other "-ologists" of this ilk is to sort out which of these two polar explanations is more plausible. This, in essence, is the problem of modern sociobiology—to discover the degree to which hard-wired genetic programming dictates, or at least strongly biases, the interactions of animals and humans with their environment, i.e., their behavior. Put another way, sociobiology is concerned with elucidating the biological basis of all behavior.

At first sight it may seem slightly preposterous to argue that any human behavior pattern is forced upon us by our genes since, after all, we are free-thinking beings having the power to decide our actions for ourselves. Comforting as this prejudice may be, there are plenty of arguments against it. A trivial example is our need for sleep. No one can question that sleeping is a behavioral pattern common to all humans, and furthermore it gives every appearance of being completely determined by our physiological makeup; i.e., it is genetic, not learned. You might argue that sleeping is not the *type* of behavior pattern we have in mind when we speak of exercising our "free will," and that we're more concerned with human *social* behavior: aggression toward others, mating and bonding patterns, religious and ethical codes—in short, all the kinds of behavior that anthropologists, psychologists, and sociologists find interesting. But even here the Nature-versus-nurture question is far from clear cut as, for example, when we consider the problem of schizophrenia. It's hard to deny that the actions of a schizophrenic fall into the category of "interesting" social behavior. Yet there is fairly convincing medical evidence to indicate that this malady is attributable to chemical imbalances in the brain, i.e., to a genetic misprogramming. Thus, the task of the modern sociobiologist is to examine the balance between social behavior that is primarily dictated by the genes, like schizophrenia, and behavior that is overwhelmingly determined by our social and/or cultural environment, like that of Milgram's obedient automatons.

Since the arguments of the sociobiologist are based upon the idea of behavior patterns emerging as a result of biological evolutionary pressures, they are couched in evolutionary terms involving concepts such as genotypes, phenotypes, selection, adaptation, and so forth. Consequently, to explore the plausibility of a genetic basis for behavior, our first order of business must be to establish the basic vocabulary of the Darwinian evolutionist, and then to look at how these biological notions fit together with the concepts of social behavior as seen by the ethologist, sociologist, anthropologist, and psychologist. It is to this that we now turn.

NEO-NEO-DARWINISM AND SOCIOBIOLOGY

The Central Dogma of Molecular Biology asserts, roughly speaking, that there is a one-way flow of information from the genes to an organism's structural form. In short, we have the chain DNA → RNA → Proteins. For the purpose of studying the implications of biology for behavior, we might profitably expand this pillar of molecular biology into what I'll call the Central Dogma of Social and Behavioral Biology, whose essence is depicted in the following diagram:

$$
\left.\begin{array}{c}
\text{Genotype} \\
+ \\
\text{Environment}
\end{array}\right\} \Rightarrow \text{Phenotype}
\begin{array}{c}
\nearrow \text{Form} \\
\rightarrow \text{Function} \\
\searrow \text{Behavior}
\end{array}
$$

The Central Dogma of Social and Behavioral Biology

Since more than a minor amount of the rhetoric surrounding the aspirations and claims of the sociobiologist arises from terminological confusions involving the components of this dogma, let me now pick apart the diagram and give a more detailed account of how each of its pieces is to be understood within the context of our concerns in this chapter.

• *Genotype:* By far the most vexing terminological confusion in the sociobiology literature surrounds the many and varied usages of the term *gene*. In strict biochemical terms, the gene is rather unambiguously defined as a section of the DNA strand needed to code for the production of a single protein. However, when we pass beyond the borders of molecular biology and begin moving toward "genetic" determination of behavior, the concept becomes increasingly fuzzy. Since virtually all interesting physical characteristics and behavioral traits involve the cooperative action of several "genes," as the term is used in its molecular biological sense, it has been suggested for soci-obiological purposes that the word "gene" be replaced by the term *replicator,* which is taken to mean the unit of genetic material that we use when we refer to a Darwinian adaptation's being beneficial to the organism. In this sense, a replicator can mean a combination of individual genes that generate some observed behavioral and/or physiological property of an organism. With this idea in mind, we'll consider an organism's *genotype* to be the totality of replicators contained in its physicochemical genetic makeup.

- *Environment:* In our discussions, the term *environment* will always refer not only to an organism's physical surroundings, such as terrain, climate, water, and air, but also to the social and cultural setting within which the organism carries on its life activities. So, for instance, within this extended definition of the everyday idea of what constitutes the environment, we would say that the identical twins Jim and Joe had the same genotype but different environments if Jim was a Hare Krishna and Joe was a practicing Orthodox Jew, even if they both lived in the same house and otherwise shared the same life-style.

- *Phenotype:* Quite simply, an organism's phenotype is the ensemble of all of its observable physical, functional, and behavioral characteristics, i.e., form, function, and behavior. Thus, physical properties like color, size, and shape are part of the phenotype, as are functional activities such as flying for birds or swimming for fish. In addition, an organism's phenotype includes various behavioral traits characteristic of the organism, like hunting in packs for hyenas, pair bonding for pigeons, and the organizational patterns of social insects like ants, bees, and wasps, not to mention cultural traits like painting or music for human beings.

With the foregoing ideas in mind, let's now look at the processes that compose today's souped-up version of Darwin's vision of evolution. In compact terms, we can express the essential features of neo-Darwinian evolution by means of Darwin's Formula:

$$\text{Variation} + \text{Heredity} + \text{Selection} = \text{Adaptation}$$

As with our Central Dogma, each of the terms in Darwin's Formula requires amplification and elucidation.

- *Variation:* In the neo-Darwinian world, the term *variation* is employed to refer *only* to change at the level of the organism's genotype. Such genotypic variations (which can be caused by many environmental factors, such as temperature, radiation, or just random mutations) may give rise to phenotypic differences.
- *Heredity:* In order for genotypic changes to be passed on to offspring, it must be assumed that there is a mechanism by which the parental genotypes are somehow transmitted to their children. Since the idea of a gene was unknown in Darwin's time, this problem of heredity was a major puzzle for Darwin; nowadays we know that it is the replicators that are passed on from one generation to the next by moving from one temporary phenotypic host, or "survival machine," to another.
- *Selection:* Not all phenotypes are created equal, and the crux of the Darwinian scheme is the argument that Nature picks and chooses among the phenotypes, bestowing on some the "right" to produce more offspring than others. It's crucial to note here that although the phenotypic variation has its root cause in changes in the genotype, the traditional Darwinian selection mechanism acts only at the level of the phenotype. Furthermore, the decision "thumbs up/thumbs down" on a particular phenotype is determined by the environment in which the phenotype is operating. Thus a thick coat of white hair has a strong positive selective advantage for a polar bear at the North Pole, but would work in just the opposite direction should the same bear be transplanted to the Philippines.
- *Adaptation:* By definition, a phenotypic trait is termed *adaptive* if possession of the trait gives an organism a reproductive advantage in its operating environment. Note again that a particular trait is never adaptive or maladaptive in and of itself; its level of adaptation is always determined with regard to a specific environment.

At this point in our deliberations, it's useful to stop for a few comments setting our terminological usage into perspective within the mainstream sociobiological literature.

First of all, the matter of *fitness*. I have avoided using this term since in the literature it is often used more or less interchangeably in two quite distinct (and far from equivalent) ways. The popular usage in Darwin's time was what today we call *phe-*

notypic fitness, which refers to the measure of an organism's ability to survive and reproduce in a given environment. Note that this criterion of fitness refers only to the organism's phenotypic characteristics, and says nothing about the genotype. Darwin termed the process by which Nature rewards those of higher phenotypic fitness *natural selection.* On the other hand, we have the currently more fashionable idea of *genetic fitness,* which is a measure of an organism's genetic contribution to the next generation, i.e., how many copies of its genes find their way into the gene pool of the next generation. This concept of fitness makes no reference to the organism's phenotypic properties at all.

With these two very different measures of fitness available, we have to be very careful to make clear which one we're using when we begin waving our magic wand of evolution and start talking about "selecting" an organism for reproductive advantage. Of course, it could be claimed that the two measures of fitness are highly correlated, using the argument that high phenotypic fitness gives an organism a leg up on the competition, thereby enabling it to push more of its genes forward into the next generation. On the surface this claim appears airtight, but we'll see later that it's very difficult to explain how certain well-established behavioral traits like altruism could ever arise if such an argument were valid. The crux of the counterargument, which we'll also take up in detail later, is that such behavioral traits could arise "naturally" only if we shift the focus of our concept of fitness from the phenotype to the genotype. As we'll see, this shift in direction serves as a major plank in the platform of most sociobiologists.

A second point to take note of, as indicated earlier, is that Darwin knew nothing about genes or the precise mechanism by which phenotypic fitness could be passed on to offspring. And, in fact, such knowledge was not necessary for the arguments he was making. All that was required was that there be *some* (not necessarily perfect) correlation between the phenotypic properties of parents and offspring and the reproductive contributions of each to future generations. In other words, Darwin needed only a positive correlation between parents and offspring in overall phenotypic fitness, without having to worry about the precise mechanism by which this correlation came about.

Before moving on to a discussion of sociobiology per se, let's return to the Central Dogma for Social and Behavioral Biology and carefully delineate just what it means to say that a behavioral trait follows from a particular genotype. To begin with, I'd like to dispel the simplistic, popular-science view that somehow the cellular genetic material acts as a blueprint for assembling a body from a set of individual pieces. While in molecular biology it is true that a given gene corresponds to one and only one protein structure, there are a large number of poorly understood steps between a bag full of proteins and a fully assembled, functioning, living organism. Richard Dawkins has appealingly compared DNA to a recipe for baking a cake from a set of raw ingredients. With minor exceptions, there is no one-to-one correspondence between the words of the recipe and the "bits" of the cake. While the whole recipe maps onto the whole cake, if we change one word of the recipe and bake one hundred cakes with the original recipe and one hundred cakes with its "mutated" version, what we will note is a consistent difference between the two types of cakes, a difference that can be attributed to that single change in the recipe. It is in exactly this sense that we can say that genotype $\Rightarrow$ phenotype in a fixed environment, and it would be not only misleading but generally just plain wrong to assert that there is any single "bit" of the organism's genotype that corresponds directly to any particular phenotypic characteristic, including behavioral traits.

On this same general issue of genetic "determinism," care should be taken not to confuse the gene action involved in the physical development of an *individual* organism from a fertilized egg to a mature adult, a process that indeed does follow in a causal manner from genotype to phenotype, with the kind of acausal relationship between genotype and phenotype used in population genetics. In the latter case, a proportion of the phenotypic variation observed in a *population* is "attributable" to a correlated variation in the population genotype, with no claims being made as to the causes of that correlation. For instance, we might have a group of rats in which half have long tails, the other half tails of normal length. Upon examining the genetic makeup of the population, we may find that 60 percent of the long-tailed rats have genotype X, whiie the rest of the population are of genotype Y. In the population-genetic sense, we would say that there is a positive correlation between genotype X and the phenotypic property "long tail," but we would not necessarily infer that the presence of genotype X "caused" a long tail in any particular individual. In fact, we could not infer this since 20 percent of the population display the alternate genotype Y and yet still have long tails.

As we wend our way through the labyrinth of arguments offered by the sociobiologists and their critics, the reader should continually be on the lookout for the various ways in which the above concepts and notions are employed. As noted, the literature is rampant with confusion on this score, and in many cases the only way to make sense out of some of the verbal bombshells flying about is to examine carefully the specific ways in which the disputants are using these overworked everyday words and ideas. With these caveats in hand, let's now take a brief look at the general framework of the research program of the sociobiologists before we move on to consider their ideas in all their elaborate detail.

As a compact statement of the aims and claims of sociobiology, we can hardly do better than quote directly from the work of Charles Lumsden and Edward O. Wilson, two of the main players in the contemporary game of sociobiology. In their 1981 book *Genes, Mind, and Culture,* they state:

THE CENTRAL TENET OF HUMAN SOCIOBIOLOGY

. . . social behaviors are shaped by natural selection. . . . Those behaviors conferring the highest replacement rate in successive generations are expected to prevail throughout local populations and hence ultimately to influence the statistical distribution of culture on a worldwide basis.

The Lumsden-Wilson thesis can be translated into the following steps:

1. Some phenotypic characteristics that we currently possess were adaptive traits at some time in the past.
2. The appearance of these adaptive traits was strongly influenced by our ancestors' genotypes.
3. The genotypes that influenced the favorable traits have therefore been selected for.
4. The genotypes that influenced the maladaptive traits have died out.
5. The reason why we display favorable phenotypes today is the

widespread presence of genotypes influencing adaptive phenotypic traits.

Since the Lumsden-Wilson thesis is so central to understanding the sociobiology debate, let's restate its premises in slightly
less formal language. The links in the sociobiological chain of
argument are strung together as follows:

Humans now display some kinds of behavior that were "good"
in the past.

↓

These good behavioral traits are there because we inherited
them from our ancestors.

↓

Therefore the good genotypes have been singled out for survival by natural selection.

↓

The "bad" genotypes have been eliminated.

↓

We have good behavioral traits *now* because the good genes
survived and the bad ones didn't.

Providing the theoretical and experimental ammunition
needed to underwrite this chain of reasoning constitutes the
heart of the sociobiological research program. Needless to say,
the sine qua non of the program is the establishment of a tight
fit between the genotype and phenotype. A large part of our
story will be centered upon the nature of this fit and just how
tight it can be made.

To tie the concepts of phenotypic and genetic fitness into the
program of the sociobiologists, let's call a behavioral trait
"phenotypically altruistic" if possession of that trait benefits
the survival of some other organism, while the trait is
"phenotypically selfish" if its possession benefits its owner's own
personal survival. Similarly, we can say a behavioral trait is
"genetically selfish" if the effect of the behavior is to increase
the likelihood of the organism's passing along copies of its own
genotype to future generations, while the trait is "genetically

altruistic" if its effect is to increase the likelihood of genotypes different from its own being passed on. With these distinctions in mind, we can state:

THE STRATEGY OF SOCIOBIOLOGY

To explain all phenotypically altruistic behavior as being genetically selfish acts

This section has introduced numerous terms and concepts that will continually be referred to throughout the balance of the chapter. So before letting the Prosecution loose with its arguments supporting the research program of the sociobiologists, let's try to summarize the basic vocabulary in the following box.

TERMS AND CONCEPTS

REPLICATOR   the unit of genetic selection influencing a phenotypic trait

GENOTYPE   the totality of replicators forming an organism's biochemical genetic makeup

ENVIRONMENT   the physical, social, and cultural setting in which an organism develops and lives

PHENOTYPE   the totality of traits constituting an organism's form, function, and behavior

GENETIC FITNESS   the relative ability of an organism to propagate its genotype into future generations

PHENOTYPIC FITNESS   the relative ability of an organism to survive in its current environment and reproduce

GENETIC SELECTION   the process by which Nature favors those organisms of high genetic fitness

PHENOTYPIC SELECTION   the "natural" Darwinian process by which those organisms of high phenotypic fitness are favored by Nature

ADAPTATION   the process by which favorable traits (genetic or phenotypic) are incorporated into the population

With the preliminaries out of the way, we now turn to the advocates of sociobiology and ask them to present their case for why we should believe that behavioral traits are governed principally by the genes. To avoid inflaming delicate sensibilities at the outset, we will first consider the arguments for animals. Later we'll turn to a consideration of how relevant these results seem to be for humans.

ANIMAL ANTICS

The literature surrounding the Darwinian Theory of Evolution is filled with bizarre, crankish, and just plain incredible contentions about the evolutionary pathway leading from apes to humans. In this rogues' gallery of craziness, surely the Yugoslavian Kiss Maerth takes the prize for batty ideas with his book *The Beginning Was the End: Man Came into Being Through Cannibalism—Intelligence Can Be Eaten.* According to Maerth, the apes fed primarily on each other's brains, and since brains are an aphrodisiac, the apes' gastronomic preferences increased their sex drive, thereby whetting their appetite for more brains. The most visible evolutionary result of this culinary "brain drain" was the swelling of the apes' own brains, making the apes more intelligent. But Maerth claims that brain size increased at a pace faster than the expansion rate of the skull, producing not only migraines of gargantuan proportions for the apes, but also an inflated view of their own importance in the overall scheme of things. This, concludes Maerth, is why the state of mankind is in its current deplorable mess. While it's hard not to regard Maerth's evolutionary fantasy as a kind of scientific satire in the style of Jonathan Swift, his line of reasoning does veer dangerously close to some of the arguments put forth by sociobiologists wanting to infer by analogy human behavior from that of animals, especially the primates such as apes, monkeys, and baboons.

One of the main taproots of modern sociobiology is the field of ethology, or animal behavior, which was catapulted into prominence when the 1973 Nobel Prize for physiology or medicine was awarded jointly to Konrad Lorenz, Karl von Frisch, and Niko Tinbergen for their well-chronicled studies of the imprinting of geese, honeybee dances, seagull sex, and other types of animal doings. Interestingly, it was the work of these men that formed the starting point for a good bit of modern human sociobiology (especially Lorenz's studies of aggression). This is especially ironic when we consider that both Lorenz and Frisch had been nominated much earlier for the prize, but turned down because it was felt that their work did not apply directly to humans! As we'll see, there are many who still hold to this position today. Be that as it may, this ethological work, coupled with the immensely popular accounts of territoriality and aggression by Robert Ardrey, Desmond Morris, and Lorenz himself, set the stage for today's claims that there is something to be learned about human behavior by observing the animals, and that that something involves social behavioral patterns placed into our genotype and passed on to us by our primitive animalistic forebears. But just what sorts of animal behavior do sociobiologists have in mind when pressing this extraordinary claim?

To the uninitiated, mention of the theory of evolution immediately brings forth the classic knee-jerk response "survival of the fittest." This catch phrase suggests, and rightly so, that an essential feature of Darwin's world is fierce competition between species for limited resources of food, shelter, and sex. In short, animal aggression—at least at the interspecies level. In the classic Lorenz-type studies on aggression, this kind of behavior is correctly explained by appeal to natural selection, with the studies then going on to note that within a species there appears to be only restrained fighting, usually involving ritual, bluff, and violence of a nonfatal kind. According to Lorenz, these fights within species are more like medieval jousting tournaments than real wars, and are usually carried out for precisely the same reason—winning the hands of the fairest maidens. For instance, in ritual fighting between male bighorn sheep to determine which will do most of the group's mating, the contestants butt their heads against each other until one of them signals his submission by baring his neck. At this stage the contest is over: The victor retires to his newly won harem, while the loser limps off to nurse his headache and, perhaps, to fight another day. Lorenz claimed that aggression is instinctive; i.e., direct experience is not necessary for it to develop normally. He also argued that aggression is motivated by a "drive."

To explain why there should be *any* fighting at all between members of the same species, Lorenz offers the *group selection* hypothesis: Such aggression exists to pick out the best (i.e., fittest) members of the group for breeding, since it's in the group's overall interest to have its best members be parents. But it's also in the species' best interest not to have any of its members killed, since the weaker usually include the younger ones who are needed to keep the species going in the future. This sort of conventional-wisdom, group-selection-based scenario for animal aggression has been challenged in almost every possible way by the modern sociobiologist.

The sociobiologist's first line of attack is at the level of the facts: The almost universal principle of the limited nature of aggression between members of the same species is far more fiction than fact. Beginning at the level of insects and moving up to the higher vertebrates, there is field evidence of case after case of fights to the death, including even cannibalism, among members of the same species. For example, lions sometimes kill each other, and fathers are not beyond eating their cubs if given the chance. Similarly, among chimpanzees, ants, and slugs we see murder rates that make Las Vegas look positively benign. And even birds display the sort of casual attitude toward murder that most of us would associate with Colombian drug lords rather than parakeets and blue jays.

At this point, one might begin to wonder how Lorenz could have been so completely wrong. The sociobiologist has two answers to this commonsense query: insufficient data and an erroneous theoretical foundation—the kind of one-two punch that spells trouble for any purported scientific theory. According to the guru of sociobiologists, E. O. Wilson, it's necessary to have very long term studies of animal behavior to establish the full truth about animal aggression, and Lorenz simply did not have this kind of data. Wilson writes: "I have been impressed by how often such behavior becomes apparent only when the observation time devoted to a species passes the thousand hour mark." He then goes on to note that a murder every thousand hours is a high level of violence by human standards, and that with the more extensive data on animal behavior that is now becoming available, humans are starting to look downright peaceful compared with most of the animal kingdom, including the apes.

The second line of attack on Lorenz is directed against his theoretical group-selection hypothesis. The sociobiologist completely rejects the concept of group selection, swearing allegiance only to the notion that what's good for the individual is ultimately good for the group as well. Later we'll try to provide solid arguments for the adoption of this stance. For now let's be content to note only that individual selection is preferable to group selection, if for no other reason than by an appeal to Ockham's razor: It's just simpler. With individual selection there's no need for a

priori assumptions about the good of the species, and as a consequence there's no reason to put forth special explanations for
why one member of a species would not attack another. Thus, all
things being equal, a lion is indifferent as to whether it's attacking a member of its own pride or a Thomson's gazelle across the
savanna. After all, the first rule of survival is to survive. And
food is food, so you take it where you can find it.

Despite the limitations the sociobiologists put on Lorenz's theory, the sociobiological explanation of aggression still fails to account for the most surprising aspect of Lorenz's studies:
Animals *do* show a remarkable degree of restraint in their conflicts with fellow members of the same species. The problem is to
offer an explanation based on individual selection for this observed fact, as well as to explain why these conflicts sometimes
escalate. At one level such an explanation is trivial: Unre-
strained aggression must be more costly to the individual than
the exercise of restraint. But this is a pretty feeble sort of "explanation." At least this is what the eminent British biologist
John Maynard Smith thought when considering the question in
the early 1960s. He had the idea of looking at the problem of
animal conflict resolution as a "game," employing ideas and
models originally pioneered by von Neumann and Oskar Mor-
genstern for the study of processes in economic bargaining.
Maynard Smith's marriage of game theory to ethology has since
come to form one of the principal theoretical weapons in the
sociobiologist's arsenal. Let's see why.

The heart of Maynard Smith's idea is the observation that in
any animal conflict, the respective payoffs to the individual contestants depend on the strategy employed by each of them. In
general there is no such thing as a uniformly best strategy, and
what a given individual should do in order to maximize his take
depends upon what his opponent is doing. Game theory enables
us to calculate what the optimal mix of actions would be in order
for a contestant to receive the greatest reward, on the average,
over a series of contests. To see the way things work, it's best to
look at an example.

The simplest situation that illustrates the game-theoretic ideas
is the classic Hawk-Dove game introduced by Smith and Price
in 1973. The basic situation involves a population of animals
that are competing for some common resource. In any competition between two members of the population, each contestant has

the choice of opting for one of two "pure" courses of action: *Hawk,* which is a policy of aggression in which the player always escalates the battle until it is injured or its opponent gives way, and *Dove,* a policy that begins with a traditional display and then immediately gives way if the opponent begins to fight in earnest. To make things as simple as possible, we further assume that the members of the population reproduce asexually, and that they breed "true," i.e., offspring adopt exactly the same behavioral policy as the parent. Note that here we are implicitly assuming a link between the genotype and the behavioral phenotype. We'll come back to this crucial point later.

To measure the outcome of various interactions, let's suppose we have a unit of fitness V, which can be understood as the expected increase in an animal's number of offspring if it can gain the resource of contention without cost. Furthermore, when an encounter escalates into a fight, the vanquished suffers a loss of C units of fitness. Consider the possible types of conflict:

Hawk $\longleftrightarrow$ Hawk: In this case there is always a fight. The winner gets all of the resource, while the loser is injured and disappears. Since the situation is symmetric, any Hawk can expect to win half its contests with other Hawks. Thus the expected change in fitness for a Hawk is $\frac{1}{2}(V - C)$.

Hawk $\longleftrightarrow$ Dove: In this case the Dove immediately runs away at the first sign of Hawkish aggression, leaving the Hawk with all the resource. In this situation the Hawk receives an increase in fitness of the amount V, while the Dove gets 0.

Dove $\longleftrightarrow$ Dove: In this peaceful situation of universal harmony and sharing, it can be expected that each "noncombatant" will take the resource half the time, while giving it to the opponent the other half. In either case, the loser walks away uninjured and the expected gain in fitness to each is $\frac{1}{2}V$.

We can summarize these expected payoffs for pairwise interactions with the following array:

$$
\begin{array}{cc}
 & \text{Hawk} \qquad \text{Dove} \\
\begin{array}{c} \text{Hawk} \\ \text{Dove} \end{array} &
\left(\begin{array}{cc} \frac{1}{2}(V - C) & V \\ 0 & \frac{1}{2}V \end{array} \right)
\end{array}
$$

Here by convention the payoffs are to the player using the course of action along the side against a player employing the behavior along the top of the array.

Now imagine you are a member of the animal population and

are faced with the decision to play Hawk or Dove. What should you do if your goal is to maximize your overall payoff? Should you always play one of the two pure strategies or should you mix them in some proportion, sometimes playing Hawk and at other times Dove? To address this question, we need the concept of a *strategy*. Put simply, a strategy S is just a rule expressing what fraction of the time a contestant plays Hawk and what fraction it plays Dove. Thus, if the player adopts Hawk a fraction p of the time and Dove a fraction q, then we can represent this strategy as $S = (p, q)$, $p + q = 1$.

At this point, Maynard Smith introduces a key idea enabling us to calculate what the "best" choice of p and q would be. He argues that the best choice would be those values of p and q that lead to a strategy that is *uninvadable*. In other words, any animal playing a different strategy that tried to compete with one playing this uninvadable strategy would, on the average, be wiped out. Maynard Smith termed such a strategy an *evolutionary stable strategy (ESS)*.

If the situation is such that the potential gain in fitness exceeds the cost of losing a contest, i.e., $V > C$, then it's easy to see that playing Hawk is an ESS, since those playing Dove would meet mostly Hawks and would have a smaller payoff from such encounters (0) than the expected amount of fitness increase $\frac{1}{2}(V - C)$ received by a Hawk encountering another Hawk. On the other hand, playing pure Dove is not an ESS since Hawks would have a field day in a population of Doves, gaining a double payoff at every encounter, as opposed to the payoff they would obtain in fighting another Hawk. But it's probably more realistic to assume that the cost of an injury is greater than the benefits to be obtained from the contested resource, so let's calculate what the ESS strategy would be in this more interesting situation when $V < C$.

To firmly fix these ideas, let's plug in some numbers. Let p^* and q^* be the values of p and q corresponding to an ESS when $V < C$. For definiteness, suppose we have the situation in which $V = 5$, $C = 10$; i.e., the increase in fitness acquired by winning a fight is only half as great as the loss incurred by being defeated in battle. In this case it can be shown that $p^* = V/C = \frac{5}{10} = \frac{1}{2}$. Therefore, the ESS is for a contestant to play Hawk exactly half the time, Dove the other half.

There is an important technical point as to the interpretation of the foregoing results that needs to be inserted here. We have seen that no individual animal that plays a strategy different from the ESS proportion between Hawk and Dove can survive in the long run. Now suppose we have a population in which the members cannot shift between Hawk and Dove at will, but are constrained (genetically or otherwise) always to follow one of the two courses of action. Question: Can we reinterpret the above argument as saying that in such a situation it is evolutionarily stable if a fraction V/C of the *population* plays Hawk, while the remaining fraction $1 - V/C$ always plays Dove? Answer: Yes, if there are only two courses of action available to the players; otherwise, the two interpretations lead to different results. This is just a mathematical oddity of the two-action situation, and has no deeper meaning in the context of the general problem. Now let's return to the question of the genetic basis underlying these behavioral strategies.

Our earlier assumption of asexual reproduction ensured that, given an equilibrium distribution of Hawks and Doves at which the fitnesses were equal, the frequency of the offspring generation will be the same as the frequency of the parental generation since the offspring are genetically identical to their parents. The question is whether we can apply the same kind of game-theoretic arguments to sexually reproducing organisms like ourselves. To address this question, consider the following example constructed by Philip Kitcher.

Assume we have an infinite, random-mating population of sexually reproducing organisms with $V = \frac{1}{2}$ and $C = 1$. In this case, the ESS for the *population* is the strategy *Indecisive,* which plays Hawk half the time, Dove the other half, just as in our numerical example above. Suppose the initial state of the population consists of individuals with three possible genotypes: AA, Aa, and aa, with AA animals playing Hawk, aa Dove, and Aa Indecisive. Question: Is the strategy Indecisive of an Aa *individual* an ESS? Answer: No, as both pure strategies Hawk and Dove can invade in the first generation and are maintained in the population as a result of the sexual reproduction. In fact, there is no ESS for individuals in this situation, although there is a stable *distribution* of strategies for the population: $\frac{1}{4}$ Hawk, $\frac{1}{2}$ Indecisive, $\frac{1}{4}$ Dove. Thus, there is no way for an *individual* animal to move between the various actions and create an uninvadable strategy (an ESS), but there is a way for the population as a whole to distribute itself so that no new *population* can invade. This example should be kept in mind as we continue our discussions later about the relevance of game-theoretic arguments for social behavior. The moral for the moment is that the existence of an ESS depends not only upon the available strategies and payoffs, but also upon the genotypes underlying those strategies. Again it should be noted that this analysis assumes the existence of such a genotype → phenotype link.

The foregoing game-theoretic analysis has been pure armchair speculation and back-of-the-envelope calculation. Does it have anything to do with the way animals really behave in the wild? Sociobiologists like David Barash have compiled considerable field evidence that it does. One of the most interesting tests was carried out by Susan Riechert, who studied the behavior of the common grass spider *A. aperta* in settling territorial disputes. Riechert studied these spiders in two habitats that differed greatly in the availability of suitable locations for building webs—a desert grassland in New Mexico and a desert riparian area consisting of a woodland bordering a stream in Arizona, a region offering many more favorable locations for webs. While there is no room here to go into the details of how Riechert determined the actions available to the spider and assigned the various payoffs, her final conclusions are worth pondering. She discovered that the contest behavior for web sites in the riparian regions deviated substantially from the ESS predicted by the game-theoretic model. In particular, contrary to theory, a riparian spider does not withdraw from occupied territory when it encounters the owner of the web. Rather, they engage in a dispute that escalates to potentially injurious behavior. On the other hand, the behavior of grassland spiders does follow the ESS as predicted by the theory, with the time and energy they expend in fights varying with their probability of emerging victorious.

So while the riparian spiders are normally less aggressive than their desert grassland cousins, just as ESS theory predicts, they are still somewhat more aggressive than they should be. This leads us to ask: Why does the behavior in these territorial disputes differ from the ESS for riparian spiders and not for their grassland cousins? Riechert gives an answer that will gladden the heart of any sociobiologist. She states:

If one assumes that the model is correct—that it has taken into account all the important parameters and includes all possible sets of strategies—then there must be some biological explanation for the observed deviation. . . . One possibility is that the release from strong competition is a recent event and that there just has not been sufficient time for natural selection to operate on the behavioral traits to complete the expected change. . . . Finally, a major change in the wiring of *A. aperta*'s nervous system might be required to achieve the new ESS, and such a mutant may simply not have arisen yet.

So far we have concentrated attention on animal conflict and aggression as representative of the ideas and approach of the sociobiologists to animal behavior. But at some stage the animals have to stop fighting and start reproducing if their genes are to be sent on to the next generation. In view of our earlier discussion, let's assume at the outset that this reproduction takes place sexually, and take a moment or two to consider the process of sexual selection and sex roles in animal mating from the sociobiological point of view. A good case in point is the problem of parental investment.

Both the male and the female want to produce children. But production alone is not enough; someone has to bring up the family. If one of the parents can off-load the work onto the other, so much the better from an evolutionary standpoint, since that parent is then free to go on the prowl for another mate with whom it can produce more offspring. Naturally each parent wants to adopt the same strategy, so the question arises of whether the mother or the father has more to lose by adopting the strategy of "hit and run." Obviously, it's normally the female that has more to lose if she decides to throw in the towel and start over again. So there is a conflict of interest: The male wants to "philander," while the female wants not only to be fertilized, but also to convince the male to hang around long enough to help out with raising Junior. As a result we get different selective forces at work, and what we expect (and usually find) is that males tend to want to fertilize many females, while females are more interested in raising those children that they already have. To understand the sociobiological arguments underlying these observations, let's take a little closer look at the overall situation.

The key to understanding the evolution of the above kind of sex role differences lies in the notion of parental investment. Basically, *parental investment* is any investment by the parent in an individual child that increases the child's chance of surviving at the cost of the parent's ability to invest in other offspring. Since any parent has a limit on both the total amount of parental investment that it can make and on the total number of children that it can have, we can work out the average investment per child that an individual parent can make. By the definition of sexual reproduction, each sex can produce only the same total number of offspring as the other sex. But it's not necessarily the case that the two sexes in a species will have the same average parental investment per child. As a result, the sex having the greater average parental investment becomes a limiting resource for the other sex. Figure 3.1 shows the situation graphically, assuming that the female has the greater average parental investment. In this diagram, the female's fitness is maximized when she produces O_f offspring, while the male's fitness is highest when he produces O_m offspring. Since O_m is greater than O_f, in this case males compete for females. Many of the territorial disputes and aggressions discussed earlier arise for exactly this reason: males seeking sexual access to females.

The story has been told so far from the viewpoint that selection acts only on the sex making the lesser parental investment. But remember that the sociobiologist insists that selection acts on the individual, so it must be the case that selective forces are at work on the parent making the greater investment, too. Just how could selection act to aid such a "giver"? The most obvious way would be for selection to aid the giver by allowing it to produce the largest number of the best possible children. For the sake of discussion, let's now assume that this giver is the female.

In the terminology of Dawkins, there are at least two pure strategies that such a giving individual could follow in looking for a mate that would contribute to this Panglossian passel of little savages: *Domestic Bliss* or *He-man*. The first involves the female's forcing the male to make a substantial investment before copulation, a strategy probably all too familiar to sugar daddys the world over. Under this strategy, the male is so committed by the time the children arrive that it might not pay him to desert, because the next female he meets up with will probably

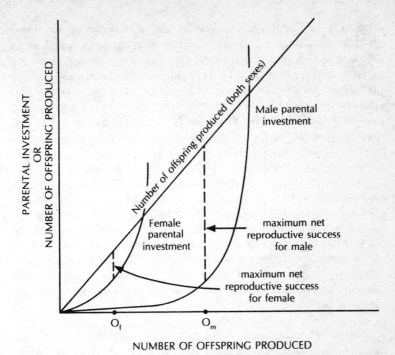

FIGURE 3.1. *Parental investment and reproductive success*

also demand such a priori efforts. Of course, this theory assumes that the next female will indeed demand such efforts, so we must be able to show that this behavioral trait will be an ESS strategy in the population. Simple game-theoretic arguments very similar to those of the Hawk-Dove variety show this actually to be the case.

He-man is the other pure strategy open to the female. By adopting this course of action, the female gives up on the idea of having the male take out the garbage and bring home the bacon, and settles for trying to get the best possible genes for her children. Adoption of this strategy by the female places a high selective pressure on males to be strong, attractive, clever, and the like, since this will be appealing to the female whose sons will then be likely to carry these advantageous traits, thus giving them a better chance of reproducing. Note that in the operation of these female strategies, there will be a constant temptation for males to appear fitter than they really are, with females trying to discriminate between those that are really fit and those that just put on a good show. This observation led Wilson to remark that under the He-man strategy, females would have a strong tendency to develop coyness, i.e., hesitant and cautious responses that evoke more displays from the male, thereby giving the female additional information with which to try to separate out the "real men" from the "cads," "flakes," and poseurs. Again, game-theoretic arguments can be used to examine the optimal mix between Domestic Bliss and He-man.

As the final stop on this whirlwind tour of the zoo, let's look at what for traditional Darwinists is one of the animal world's most difficult-to-fathom puzzles: the behavior of the sterile worker castes in colonies of ants, bees, wasps, and termites. In these settings there exist entire castes of sterile females who devote their time exclusively to the well-being of their mother (the queen) and their siblings. The British biologist William Hamilton suggested the concept of *kin selection* in 1964 as a mechanism to explain this otherwise highly non-Darwinian altruistic behavior.

Kin selection is based on the rock-solid premise that we are all related to others. This means that each living creature shares some of its genes with others, and since our genes have been selected because of their ability to produce phenotypic characteristics that assist their replication (or so say the sociobiologists, anyway), it's in our own selfish reproductive interest to see that those to whom we are related reproduce. In short, only those genes that reproduce persist, and the gene is indifferent as to whether this is done directly or by proxy. Thus it might be worthwhile to be altruistic to your otherwise useless, sponging cousin because he will then be in a better position to pass on some of your genes. As an aside, it should be noted that the idea of kin selection goes back at least as far as another British biologist J.B.S. Haldane, who is reputed to have done a quick calculation on a beer mat in a London pub, coming to the conclusion that he would gladly give up his life for three brothers or nine first cousins. Here Haldane was simply following the rules of Mendelian genetics, according to which he would share half his genes with a full sibling, while sharing only one eighth of his genes with a cousin.

The basic principle of kin selection can be generalized by the rule: If the coefficient of relatedness (i.e., fraction of shared genes) with another is r, and the benefit you can give to that person in enhanced fitness for reproduction is k, then you should give up your own chance at reproduction to help the other if $k > 1/r$. So in the case of a full sibling (like Haldane's brother), $r = \frac{1}{2}$, implying that he should give up his own life to save one brother if by doing so he could double his brother's chances of surviving to reproduce. Figure 3.2 shows how to compute r for various degrees of relatedness. Each arrow in the diagram means that there is a 50 percent chance that the two individuals thus connected share genes. Hence, the likelihood that any particular gene gets through n such arrows is $(0.5)^n$. When two individuals have more than one ancestor in common, they can share genes via all of them, and we must then add all possible paths. So, for example, for cousins we have

$$r = (a \times b \times c \times f) + (d \times e \times c \times f)$$
$$= (0.5 \times 0.5 \times 0.5 \times 0.5) + (0.5 \times 0.5 \times 0.5 \times 0.5)$$
$$= 0.0625 + 0.0625$$
$$= 0.125 \; (= \tfrac{1}{8})$$

Hamilton's contribution was to work out the mathematical details of the notion of *inclusive fitness,* which many feel is the most significant extension of Darwin's original idea since the incorporation of Mendelian genetics as the mechanism of heredity. According to Hamilton, the old Darwinian notion of individual fitness (genetic or phenotypic) should be replaced by the individual's inclusive fitness, which is defined as the individual's own personal fitness plus the individual's influence on the fitness of nondescendant relatives. There is no better way to see inclusive fitness in action than to go back to the social insects and examine Hamilton's explanation for the appearance of the sterile worker castes.

In the order Hymenoptera, which includes the ants, wasps, and bees, the sex of offspring is determined in an unusual way. Specifically, females are diploid, developing from fertilized eggs, thus having both a mother and a father. On the other hand, males develop from unfertilized eggs and are haploid, thus sharing genes only with the mother (the queen). The result of this odd sex-determination process is that sibling daughters of a queen, fertilized by a single male, are more closely related to

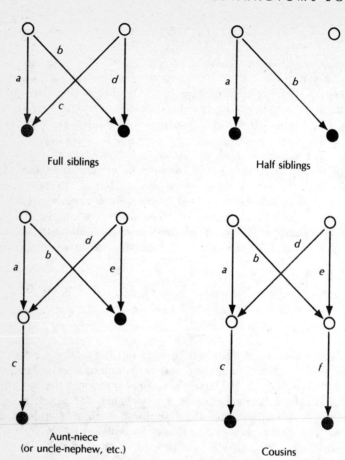

Full siblings

Half siblings

Aunt-niece
(or uncle-nephew, etc.)

Cousins

FIGURE 3.2. *Coefficients of relatedness* r *for different relatives*

each other than they would be to any of their own daughters. Graphically, the reason is depicted in Figure 3.3. Here the female Ego inherits two sets of genes: one from her mother, with two sets, and one from her father, with one set. Hence the coefficient of relatedness (average fraction of shared genes) between Ego and a full sister is $r = \frac{1}{2} \times \frac{1}{2} + \frac{1}{2} \times 1 = \frac{3}{4}$. But the coefficient between Ego and one of her daughters is only $r = \frac{1}{2}$. Thus, Ego has more genes in common with one of her sisters than she shares with one of her own daughters.

If Ego's mother continues to produce cells for eggs after Ego reaches maturity, then Ego will do the most toward perpetuat-

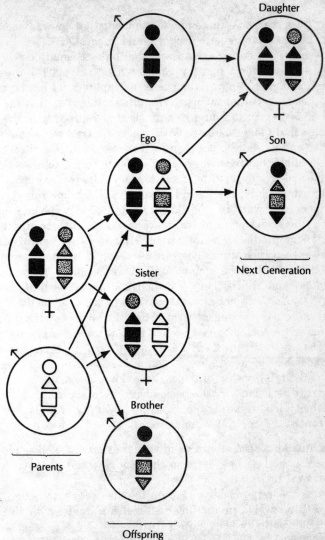

FIGURE 3.3. *Sex determination in the order Hymenoptera*

ing her own genes if she devotes her time entirely to raising fertile sisters, since fertile sisters will spread more of her genes than will fertile daughters. More precisely, by adopting the criterion of inclusive fitness, Ego's self-interest is served if she behaves "altruistically" toward her sisters rather than "selfishly" for herself—in complete contradiction to what conventional Darwinian genetic fitness would suggest.

Besides its intrinsic elegance, Hamilton's explanation also suggests why we don't find any worker males: A male is no more closely related to siblings than he is to daughters (he has no sons). Another observation favoring Hamilton's theory is that the normal 50-50 sex ratio found in animals that reproduce in the conventional diploid manner, with genes contributed equally by father and mother, is not seen in Hymenoptera. Hamilton's theory predicts that the ideal ratio of males to fertile females should be one male for every three females, very close to what is actually observed. Finally, there are the cases in which one colony takes "prisoners of war" in a battle with another, with the queen then able to make use of unrelated slave workers. In these situations, the theory predicts a more normal 1:1 sex ratio, again exactly what is seen in Nature. These results were taken to be convincing triumphs for the sociobiological arguments in favor of kin selection and the notion of inclusive fitness. Needless to say, however, they are not airtight and a number of difficulties have been put forth casting at least a few shadows over the glowing claims of the sociobiologists. We'll look at these complaints when the Defense takes the floor. For now, let's try to summarize what the sociobiological studies of animal behavior *might* suggest about the relationship of genes, behaviors, and man.

As far as I can see, the basic chain of reasoning that human sociobiologists would like to use from the study of animal behavior consists of the following steps:

- In animals, especially those of the lower orders such as insects, there is a close link between the genotype and phenotypic behavioral traits.
- Game-theoretic models based on the idea of maximizing inclusive fitness give predictions in excellent accord with the way animals actually behave in Nature.
- Extension of the classical notions of fitness by introducing ideas of kin selection and inclusive fitness enables us to offer good explanations for altruistic behavior in animals.

THEREFORE

- The same principles that work well to explain animal behavior by genetic influence should work equally well to explain the behavioral patterns of humans.

We will spend the rest of the chapter looking at the pros and cons of this astoundingly ambitious chain of hopes and claims.

THE STRANGE CASE OF ALTRUISM

It's been noted that a large number of winners of the Congressional Medal of Honor have been soldiers who have thrown themselves on hand grenades to save comrades. And in the animal world we have the honeybees, who buy themselves certain death when they sting an intruder threatening the hive. How can these acts of suicidal altruism be explained by the overtly selfish principles of natural selection? This question has been described as the central problem of sociobiology by no less an authority than the head sociobiologist himself, Edward O. Wilson. In the case of the social insects, we have already seen a fairly convincing explanation of this altruistic behavior in the concepts of kin selection and inclusive fitness put forward by William Hamilton. But what about the many examples of human and animal altruistic behavior that involve totally unrelated parties? As a prelude to a full-scale examination of the arguments for sociobiology, in this section we'll devote our attention exclusively to an investigation of the ways sociobiologists have devised to say that "doing good for someone else can be doing good for yourself."

In the sociobiological literature, four distinct mechanisms have been suggested to explain why an individual would take actions decreasing his personal fitness in order to enhance the fitness of another. We have already touched upon two of them—group selection and kin selection—but for the sake of completeness, let's briefly review all four.

- *Group selection:* This was Lorenz's explanation of why potentially harmful aggression in animals appeared to be confined to interspecies competition, and was rarely observed within species. The basic idea is that an individual within a group would be willing to suffer a personal loss in fitness if that loss was more than compensated for by an increase in overall group fitness. As a result of theoretical models, as well as ingenious alternative explanations, there is more or less universal agreement today that group selection is a pretty rare phenomenon, taking place only under very special circumstances.
- *Kin selection:* We covered this explanation for altruistic be-

havior between related individuals in some detail in the case of
the social insects, and the same notions seem to carry over
mutatis mutandis to humans. It's often observed that close relatives tend to look after each other more than they look after
strangers, and the closer the relationship (e.g., identical twins
versus distant cousins), the greater the willingness to sacrifice.
• *Parental manipulation:* This is a type of enforced altruism in
which a parent coerces a child to give help to another for the
parent's benefit. A typical situation of this sort might arises,
for instance, if a mother cat has a litter of, say, five kittens
but can raise only three of them to maturity using her own
resources. Then it would pay her (genetically speaking) to employ her position of authority to force some of her older offspring to devote a part of their resources to helping her raise
the litter. She can do this in many ways, perhaps the most
common being a threat to withhold some of her attention from
certain offspring if they refuse to help out. In Nature the
strategy of parental manipulation often takes the form of cannibalism, in which the weaker members of the litter are sacri-
ficed for the benefit of the stronger. Of course it might be
argued that putting yourself on your brother's dinner plate
hardly constitutes an "altruistic" act, in the sense that the
term is normally used in polite conversation. But in Nature
"altruism" means only an act that decreases your own fitness
in order to enhance the fitness of another, so such an act of
sacrifice is indeed altruistic, at least by Nature's dictionary.

At first glance it may appear that there is no real difference
between parental manipulation and kin selection—they both
involve the sacrifice of an individual for the benefit of another.
However, there is one critical difference: In kin selection, one
individual helps another because they share some genes; in parental manipulation, one person helps another for the benefit
of a third party (the parent). The fact that the two parties
might share genes is incidental in parental manipulation, although it often happens that they do. So in practice it may not
be easy to distinguish between the two forms of altruism, and
any given situation may involve both. In fact, it has been suggested that the main causal factor at work in the development
of sterile castes in Hymenoptera is parental manipulation and
not kin selection. This is because when the queen sets up the
nest, she chooses to make workers rather than reproductives
by virtue of what she feeds her initial offspring. But this is

still a matter of some controversy and the jury is out as to which of the two altruistic mechanisms is really at work here.

• *Reciprocal altruism:* By far the largest share of altruistic acts, at least among humans, involve parties who are not related at all. Robert Trivers introduced the idea of reciprocal altruism to account for these sorts of sacrifical acts. In essence, the principle governing reciprocal altruism is "If you'll scratch my back, I'll scratch yours." Briefly, the claim is that individuals engage in altruistic acts because they expect that by doing so they will benefit by someone else's altruism toward them at sometime in the future. Note the very great difference here between an act of reciprocal altruism and an act of kin selection altruism. In the reciprocal case, the giver expects to see a direct return from a sacrifice; in the latter situation, the giver sees no direct reward but only the satisfaction of seeing his or her genes being given a better chance to make it into future generations.

The most convincing example of reciprocal altruism in Nature seems to be the case of the "cleaner fish." Certain species of fish clean parasites off fish of a different species. This is a situation in which both parties gain: The cleaners get a hearty meal, while the fish being cleaned avoid the sores and diseases that would otherwise result from the parasites. The most remarkable aspect of this situation is that the cleaner fish are never eaten by those they're cleaning, even though this could easily happen. Furthermore, it's often the case that other types of fish try to imitate the cleaners, rushing in to bite big chunks off the fish being cleaned. In these cases, the big fish happily gobble up the pretenders despite the fact that the pretenders have developed high-level camouflage techniques to fool them. Since the cleaners and the cleaned have no genetic relationship at all, Trivers argues persuasively that this situation can be explained only as a case of reciprocal altruism. We'll return to a deeper consideration of reciprocal altruism later on when we consider the evolution of cooperative behavior.

THE GENETIC IMPERATIVE

From sad personal experience, I can attest to the fact that book publishing, academic style, is a surefire prescription for anonymity, totally unrewarding to anything but the ego. Only the fortunate few manage to sell even as many as a couple of thousand copies of their magnum opus to libraries, captive students, and a small band of fanatics and connoisseurs of the arcane. But occasionally an academic author crashes through this paper curtain of obscurity, putting forth a glitzy product with a marketing campaign rivaling that of the largest trade publishing houses. Such was the case in the spring of 1975 when Harvard University Press brought out *Sociobiology: The New Synthesis*, a lavishly illustrated, seven-hundred page coffee table book by the eminent insect expert Edward O. Wilson. In addition to full-page ads in *The New York Times Book Review*, the book was the subject of a front-page article in *The New York Times* describing sociobiology as having "revolutionary" implications for human societies. Similar statements were made in other major publications like *People* magazine, *The National Observer*, and *The Boston Globe*. What is it that gave *Sociobiology* and Wilson's subsequent book *On Human Nature* (which won a 1979 Pulitzer Prize) their immense interest outside biology? Basically, it was the extraordinary breadth of Wilson's claims about the possibility of offering biological explanations for virtually all human social and cultural activities. Here we want to examine in some detail both these claims and the arguments Wilson presents to support them.

Wilson's office at the Harvard Museum of Comparative Zoology is filled with colonies of various sorts of ants, the insects whose behavior patterns started Wilson off on his path toward trying to explain human behavior on the basis of biological principles. As Wilson tells it, his books *Sociobiology* and *On Human Nature* are really the second and third parts of an unplanned trilogy that began with his 1971 classic *The Insect Societies*, which, incidentally, was *not* on *The New York Times* best-seller list! Wilson, a tall, thin Southerner in his late fifties, speaks with great enthusiasm and verve about his passions (which include a firm commitment to jogging and a deep admiration for people who have great goals and persevere toward them over a long period). He talks about human sociobiology in just the manner mentioned in the last section: as a natural extension of the behavior patterns noted in animals. To understand his line of argument as put forward in his books and subsequently refined in numerous articles, interviews, and lectures, it's useful to think of the various steps in his program as rungs on a ladder that must be climbed to reach his far-ranging conclusions. Our version of this ladder paraphrases that originally put forward by the philosopher of science Philip Kitcher.

WILSON'S LADDER

First Rung

Fitness maximization: Employing the usual methods of evolutionary biology, we plausibly argue that all members of a population P will maximize their fitness if they display behavior pattern B in the typical environments faced by members of P.

Second Rung

Universality: If we observe that all members of P do in fact display behavior B, then we can conclude that B became prevalent and remains so as a result of natural selection.

Third Rung

Selfish gene: If genetic fitness is used as the selection criterion, selection can act only when there are genetic differences. Thus we can conclude that there are such genetic differences between the current members of P and their ancestors who did not display B.

Fourth Rung

Adaptation: Because there are genetic differences and because B is adaptive, we can conclude that it will be difficult to modify B by altering the social environment. This is because such an alteration will be resisted by the B-dominant population.

In Wilson's scheme of things, we can identify three main lines of attack supporting this ladder: gene inflation, analogy, and adaptation. Let's look at each in turn.

GENE INFLATION

This argument tries to assert the supremacy of the genes by showing that the levels of biological organization that normally mediate between the genotype and phenotype are either of no consequence or are simply communication pathways for the expression of the genes. An eloquent advocate of gene inflation is Richard Dawkins, whose book *The Selfish Gene* is a vastly entertaining, relentless pursuit of the idea that the organism is only

DNA's way of making more DNA. As an example of the kind of logical tightrope that Dawkins walks, consider his distinction between the *unit* of selection and the *process* by which this unit is singled out. He says: "If *selection* means differential survival and reproduction, there is no question that it occurs between alleles [genes]. But the *processes* by which it occurs include differential survival and reproduction (selection) of individuals [phenotypes]." Thus Dawkins asserts the supremacy of the genes by assigning to the phenotype and the environment the role of the mechanisms by which the genes are chosen. Opponents argue that it is misleading to imply inconsequential status for the higher levels of biological organization, and that the "selfish gene" argument fails to make a case for a tight genotype-phenotype fit because it tries to push out of the way the most likely candidate for creating this gap in the first place: the disproportionately large human brain.

ANALOGY

As has been noted, there are many human traits like sleep that really are strongly determined by our genotype. Wilson's argument by analogy states that if other behavioral traits are found to be widespread across cultures, that fact constitutes a strong prima facie case for there to be a substantial genetic component underlying such traits.

As an example of this kind of reasoning about universal human traits, Wilson offers the case of incest avoidance. According to Wilson, proscriptions against incest exist in virtually all human cultures. His sociobiological explanation is that aversion to mating with close relatives is a genetically programmed trait that increases inclusive fitness, since inbreeding would have a strong tendency to bring out lethal recessive genotypes. In fact, Wilson goes further by citing the results of a study of 2,769 Israeli marriages in which none of the unions were between members of the same kibbutz group raised together since birth. Using this result, Wilson argued that the genetic tendency is not just to avoid mating with blood relatives, but rather extends to avoidance of sexual relations between members of any group raised together since childhood. Wilson's argument by analogy is that the adaptive trait came about to prevent biologically unfit offspring, and then "spilled over" to all close childhood associates. Skeptics ask, if the incest taboo is indeed universal and genetic, why does incest need to be illegal?

ADAPTATION

Wilson writes as if he believes there are identifiable phenotypic traits that are underwritten by specific "chunks" of genetic material—what we have earlier termed replicators. He then goes on to imply that any phenotypic trait that lasts must be adaptive, and its adaptiveness must be explained by natural selection acting so as to single out the underlying replicator. As an extreme example, Wilson offers the religiously sanctioned cannibalism of the Aztecs as a phenotypic response to the genetically programmed need for protein. Again a skeptic might say that such a cultural response had nothing to do with genes for protein consumption, but was due entirely to overpopulation of the environment. Another case of the same sort that Wilson puts forth involves the widespread practice of homosexuality. How is it that homosexuality could ever evolve as an evolutionarily advantageous behavioral trait? Wilson's answer is to appeal to the notion of inclusive fitness, regarding the appearance of homosexuality as an adaptive response of the same sort as the appearance of the sterile insect castes in Hymenoptera. That is, it serves as a mechanism to prevent overpopulation. The general problem with Wilson's arguments from adaptation is that for virtually every phenotypic trait, there are many *Just So* stories that can be told for how that trait could have arisen as an adaptive behavioral response.

So we see that each of the main lines of argument Wilson puts forth in his books *Sociobiology* and *On Human Nature* comes with built-in, self-neutralizing counterarguments. Let's briefly summarize the main objections to his claims before taking a look at how he tries to deal with them in later work. The principal flaws in the early work appear to be:

- *Underestimation of the power of the mind:* Wilson continually discounts the extraordinary power of the human brain to mediate between lower and higher levels of biological, social, and cultural organization.
- *Circularity:* In Wilson's claims, he assumes what he needs to show, i.e., the causal path from the genotype to the behavioral phenotype.

- *Isolatable traits:* Wilson regards genotypic and phenotypic traits as "atomic" units that can be isolated and studied individually.
- *Advantage versus adaptation:* There is a continual confusion between traits that would be genetically advantageous, such as banning weapons of war, and those that are the result of an evolutionary adaptation.

In the five years or so following the appearance of *Sociobiology,* many of the objections noted above to Wilson's line of argument for human sociobiology came bubbling up out of the heated political, scientific, and philosophical debates surrounding his sweeping claims. We'll look at these debates in detail in a later section, but it's of more interest at the moment to see how Wilson, together with his colleague (and former student) Charles J. Lumsden, tried to patch up the above gaps in their 1981 book *Genes, Mind, and Culture.*

The main thrust of the Lumsden-Wilson position is aimed at addressing the fundamental questions:

How much choosing do people actually do in the course of acquiring or transmitting their cultural repertoire? That is, how strong are direct biases relative to other evolutionary forces acting on cultural variation?

Where do the rules that direct choice come from and how do they work?

In their book, Lumsden and Wilson try to give answers to these deep matters by providing a mechanism through which the genes can influence the development of mind, which in turn then acts to produce culture. Finally, they close the loop by having culture act through natural selection to influence the genotype. The claim is that this *coevolutionary circuit* closes the genotype-phenotype gap by way of the mind. Let's examine the principal steps in the Lumsden-Wilson circuit.

THE COEVOLUTIONARY CIRCUIT

1. Human culture consists of the interaction of all the ideas, institutions, behaviors, and artifacts used by a population.
2. We can use the term *culturgen* to mean an observable feature of a culture.

3. During the process of forming a social order, the culturgens are processed by *epigenetic rules,* which are genetically determined procedures that direct the formation of the mind.
4. The epigenetic rules of the mind bias the owner of that mind to choose certain culturgens in preference to others.
5. The totality of all such choices in a population creates that group's culture and social organization.
6. Genetic variation takes place in the epigenetic rules, and this variation accounts for at least some part of the variation in behavioral choices that we see in a population.
7. Individuals whose choices increase their inclusive genetic fitness are able to pass more of their genes along to the future generations. As a result, the population as a whole is shifted toward certain epigenetic rules and the types of behavior favored by those rules.

The entire Lumsden-Wilson circuit is schematically depicted in Figure 3.4, showing the four main levels of biological organization. The molecular, cellular, and organismic steps constitute the epigenesis, while the transition between the organismic and populational levels involves the gene → culture transition. The final step of population influence on the genes takes place through natural selection.

We can summarize the argument by saying that in this theory the mind is formed out of a set of genetically determined rules that bias it to choose certain interpretations of the world and certain social and cultural options over others. Note the crucial point here that what the genes prescribe is not a particular behavior, but only the capacity to develop certain behaviors and the tendency to develop them in particular environments. In other words, it is the epigenetic rules that are inherited because the genotype actually codes for the construction of the wiring pattern of the mind, which in turn encodes these rules. Thus, the authors are claiming that the specific behavioral repertoire that will be displayed depends on the experience that individuals receive within their own culture. So it is the total array of human possibilities that is inherited, not the specific behavioral trait.

It's fairly evident, I think, that all of the complaints leveled against the early work of Wilson would vanish if the coevolutionary theory could be established. Lumsden and Wilson state the following conditions for such a validation of their theory:

MOLECULAR

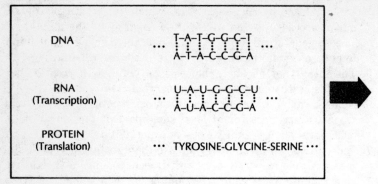

POPULATIONAL

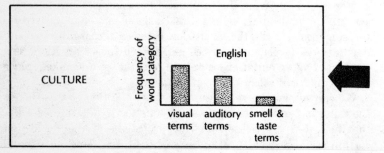

FIGURE 3.4. *The coevolutionary circuit*

A. It must be shown that biased epigenetic rules exist.
B. It must be shown that these rules can be inherited.
C. It must be shown that we can establish a link between specific culturgens and inclusive genetic fitness.
D. It must be shown that there are molecular and cellular mechanisms that directly link the genotype to cognitive development.

Surprisingly enough, there is evidence to support all four of the above necessary conditions. To begin with, there do exist biased epigenetic rules. For example, some people are born with a clubfoot and would surely be biased against making the same

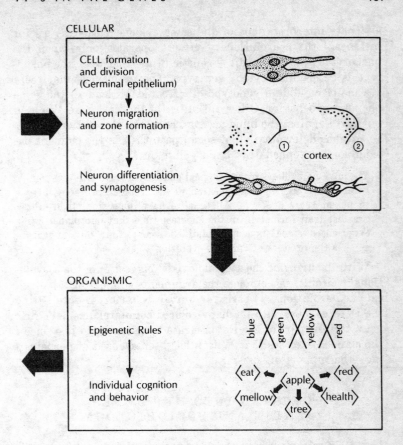

choice of footwear as those born with two normal feet. Further, some epigenetic rules are clearly hereditable, such as the predisposition to walk on two legs rather than on all fours. Thirdly, some cultural choices do affect genetic fitness. For example, making a living as a poisonous-snake handler or a movie stuntman is likely to decrease one's overall genetic fitness. Finally, there is almost universal agreement that the code written in the DNA is central to the construction and wiring of the central nervous system.

So the Lumsden-Wilson Coevolutionary Theory is a contender. The question really comes down to: How strong a contender is it? How plausible is their argument compared with alternate interpretations of the same evidence? Just as a "for instance," the coevolutionary circuit operates only when the phenotypic behavior modifies genetic fitness and when the phenotype is determined by the genotype. The problem is that there are many candidate genotypes that could all lead to the same phenotypic behavior. Other difficulties of this sort have also been put forth against the blind acceptance of the coevolutionary thesis of gene determination of social patterns. As the eminent paleontologist Stephen Jay Gould points out:

> We have no evidence for biological change in brain size or structure since *Homo sapiens* appeared in the fossil record some fifty thousand years ago. . . . All that we have done since then—the greatest transformation in the shortest time that our planet has experienced since its crust solidified nearly four billion years ago—is the product of cultural evolution.

With the firing of the sociobiologists' biggest gun, the coevolutionary circuit, we complete the arguments supporting a biological (i.e., evolutionary) basis for human behavior. Before letting the Defense loose with its many-colored counterclaims, let's first get a feel for part of the Defense case by listening to just one of the claimed excesses that Wilson has been accused of perpetrating—the support of sexism.

GETTING INTO HER GENES: SEXISM AND SOCIOBIOLOGY

In a 1978 interview with *Omni* magazine, Wilson appeals to the sex difference argument sketched earlier, claiming that there currently exist "modest" genetic differences between men and women that could be erased by careful training. As evidence he cites studies of the second generation in an Israeli kibbutz, where the regression of women to traditional roles was noted, even in a social and cultural environment that explicitly called for egalitarianism and equal opportunity. He then goes on to state that there are three alternative courses of action open if we want to tamper with this difference: (1) eliminate the difference; (2) exaggerate the difference; (3) leave things as they are. His claim is that by following the first course we could get statistical equivalence of the sexes, but that it would require more knowledge than we currently possess about the effects of gene manipulation. On the other hand, Wilson argues that adoption of the second course would only continue male domination and injustice, stunting individual development. The third, laissez faire course would most likely generate statistical imbalances in the outcome, more or less like what we have today. He concludes that there is probably no basis upon which a choice can be made, and that there is a cost associated with each course of action. Pretty reasonable, noncontroversial stuff, right? Wrong! It's statements like these that send Wilson's critics into fits of apoplexy, running for their typewriters to denounce him for contributing to the defeat of the ERA, as well as aiding and abetting arch-conservative views that would deny the political and social demands of those without power.

The heart of the argument that sociobiology is sexist is the assertion that sexism is an outgrowth of the theory itself, at least the version of sociobiology advocated by Wilson. The chain of reasoning goes as follows: (1) sociobiology begins by trying to identify those traits that are common to people in all cultures; (2) such universality is then taken to be an argument for the trait's genetic basis; (3) according to Wilson, one such trait is an aggressive dominance system, with males reigning over females; (4) therefore, sociobiology is inherently sexist. QED. In fact, Wilson's opponents have gone further and claimed that *all* the important traits he identifies, like incest taboos, dominance systems, and division of labor between sexually bonded pairs, are based on sex differences. The problem, the critics argue, is that Wilson is looking for a genetic cause, whereas what the sociobiologist is really analyzing is adaptive function. But from the standpoint of adaptive function, there is no difference between a behavior that is genetically programmed and one that is culturally taught or individually learned.

Critics state that the underlying cause of sexism in sociobiology is its basis in the kind of Darwinian sexual selection we described earlier in our discussion of the Domestic Bliss versus He-man strategies of mate selection. The argument against sociobiology is that this is only one of a number of possible forms of natural selection, and its importance in the evolution of humans is an untested hypothesis. One possible alternative, for example, would be to claim that everything depends on the ecological setting (environment), with an environment of abundance leading to behavior that would minimize the social inferiority of the female, while an environment of scarcity would generate behavior emphasizing the sex role differences. A prime candidate for a living example of this sort is found in the Tasaday of the Philippines, a primitive tribe discovered in 1971 leading a Stone Age existence with no concept of aggression or war, and also no ideas about a male dominance hierarchy. Since all the Tasaday's needs were supplied by the lush Mindanao rain forest, the argument is that this environment of plenty worked to create a social order in which females and males participated equally.

A slight detour into sexism might be excused if it were seen as an idiosyncrasy of an otherwise morally neutral study. But when Wilson goes on to make claims for the sociobiological underpinnings of such sensitive areas as homosexuality, religion, ethics, and morals, members of the radical left, as well as a lot of others, put on their gloves and come out swinging. We'll hear more from them in the next section. For now let's take a moment to let Wilson state his case concerning these delicate matters.

On the matter of religion, Wilson believes it is biological in origin. He claims that religion is really the pivot of all that we do and all that we really fight about, particularly when the religion becomes an ideology. In this view, religion is the one area of behavior where you can't draw any principles from the animal world. Along with semantic language, it is the one truly human trait and has to be considered as a biological property of humans, not just a cultural phenomenon or as the conduit for divine guidance to man. Wilson's hypothesis is that religion is essentially an extension of tribalism and of our need to be able to subordinate ourselves to concerted, irrational, even frenzied group activity. The biological basis of this claim is a kin selection argument based on the principle that we all have genetic predispositions toward xenophobia, attraction to charismatic leaders, group worship, and so forth. Kin selection is then invoked as a mechanism whereby individuals subordinate themselves to the group for the overall welfare of the tribe.

The Wilson line on religion goes on to state that the religious impulse is biological and uniquely human, but that religious faith is almost always linked to imaginary scenarios and false mythologies. He argues that it's part of our biological predisposition to make complete stories about the universe and the tribe, stories that are always false and tend to be wiped out by science.

Wilson concludes his overall argument by asserting that science and religion will ultimately come together, with science adding new depth to subjects that have traditionally been the province of religion and the humanities. In this view, the new religion will be a kind of scientific materialism, with competing ideologies like Marxism ultimately fading away.

From religion to morals and ethics is but a small step, one that Wilson takes without a moment's hesitation. His position is that biological knowledge will help us arrive at a firmly based moral code. His appeal to biology also includes the statement that biological principles will emphasize genetic diversity, at least till we gain a much deeper understanding of human heredity. But then Wilson undermines his own case by stating that even diversity may not be a permanent value, leading his vitriolic critics to protest that he is serving conservative, racist interests when he implies that at some future time we may want to practice eugenics.

To understand Wilson's position on these moral matters more clearly, let's examine what it is that he *could* mean by his gene-based views on the relationship between sociobiology and morals. According to Owen Flanagan, there appear to be at least four different interpretations of Wilson's vision:

1. Sociobiology can explain the origin of our moral capacities.
2. Sociobiology can explain the origin of particular moral beliefs and practices.
3. Sociobiology can explain the basic nature and function of morality.
4. Sociobiology provides a way of generating certain normative principles; i.e., it gives us a way of getting from "is" to "ought."

The first interpretation is trivially true; everything we do is allowed by our genes, including the development of our moral capacity. The second interpretation would imply that persisting moral principles that enhance the genetic fitness of the group that practices them must have genetic causes. This conclusion is debatable, since it appears to be another instance of confusing a trait that is advantageous with one that is adaptive. In this same connection, Wilson argues that since morality evolved as a genetic-fitness-enhancing trait, moral statements are statements about genetic fitness strategies. To assess the merits of this argument, consider the following similar chain of reasoning:

(a) our mathematical abilities evolved because they enhanced our genetic fitness; therefore *(b)* mathematical statements are statements about genetic fitness strategies. The most contentious element on the list of possible interpretations of Wilson's sociobiological view of morality is the last one. The claim is that we are combinations of genes drawn from a pool, and that therefore we should concern ourselves with the continued survival of human genes in a common pool. Hence, we "ought" to act to preserve the genes currently in the pool. But this argument appears to be little more than the statement that we ought to care about the long-term consequences of our actions for the future. If so, why do we need Wilson's extra baggage of concern for the genes? Why not just think about caring for persons and forget the genes?

With the above questions on the table, we begin to edge away from the arguments of sociobiology and into the territory of its opponents. So without further ado, let's give the floor to the first Defense attorney, who will try to convince you not only that sociobiology is pseudoscience, but that it's positively politically dangerous as well.

CANT VS. KANT

Shortly after the turn of the century, John D. Rockefeller, Sr., was busy pushing forward the interests of Standard Oil in a manner that would make today's antitrust lawyers salivate in anticipation. Rockefeller was also a devout Baptist, and during one of his weekly Sunday-school lectures he appealed to "natural law" as a means to justify his ruthless, predatory business practices. On that occasion, he made the following oft-quoted statement:

> The growth of a large business is merely survival of the fittest.
> . . . The American Beauty Rose can be produced in the splendor
> and fragrance which bring cheer to its beholder only by sacrificing
> the early buds which grow up around it. This is not an evil tend-
> ency in business. It is merely the working out of a law of nature
> and a law of God.

Good old John D.—the very embodiment of the spirit that made America great! Or so thought many of his contemporaries, a number of whom also appealed to this Darwinian vision of the natural order of things to salve their own consciences and, not incidentally, to line their pockets.

In echoing their interpretation of Darwin's universe, the John D. Rockefellers of the world were following a path originally blazed by the British philosopher Herbert Spencer, coiner of the immortal phrase "survival of the fittest," and a chief popularizer of Darwin's ideas. Ironically, Spencer himself was not a Darwinist at all. He believed in the idea of Lamarckian inheritance whereby phenotypic changes can directly influence the genotype, in direct contradiction to the Central Dogma of Molecular Biology. This is not, however, an unreasonable position to hold in the context of social affairs, even if it is still anathema to the molecular biologists. In today's climate it's hard to appreciate the influence that Spencer's social Darwinian ideas had on the fabric of American life at the time, but a small indicator is contained in a dissenting opinion given by the famed Supreme Court justice Oliver Wendell Holmes, who stated that "the Fourteenth Amendment [circumscribing governmental interference in the rights and actions of the individual] does not enact Mr. Herbert Spencer's *Social Statics.*" It's just this sort of social and political influence that the most vocal and rabid of Wilson's critics had in mind when they marshaled their forces in 1975 to attack the claims they thought he'd made in *Sociobiology.*

The Science for the People Sociobiology Study Group (the Boston Group) is a collection of mostly radical-left scientists in the Boston area whose most prominent members in the mid-1970s were the eminent population geneticists Richard Lewontin and Richard Levins, as well as the general public's favorite paleontologist, Stephen Jay Gould. Before carrying on, I should emphasize that for the most part the members of this group were internationally recognized scientists. Both Lewontin and Levins were members of (or at least invited to join) the U.S. National Academy of Sciences (although Lewontin resigned over the matter of the academy's issuing classified reports, while Levins, a professed Marxist, refused membership because the academy engaged in military studies). Thus the attack that the Boston Group mounted on Wilson's scientific speculations is particularly disturbing given the rarified academic reputations of the group, and its spectacular disregard for the commonly accepted ground rules governing constructive (or even destructive) criticism within the ivy-covered walls and halls of academia. Following initially favorable reviews of *Sociobiology,* the Boston Group issued a scathing attack not only on the book itself, but also on Wilson personally, linking him with the most reactionary of political thinkers, including the Nazis. In addition, even though many members of the group were Wilson's colleagues in the very same department at Harvard, the attack was carried out publicly in a letter to *The New York Review of Books* without even the courtesy of giving Wilson a copy of the criticism prior to publication. Needless to say, this gross breach of academic etiquette resulted in a spiraling escalation of attack and counterattack that for a time even spilled over into the popular press. On the principle that where there's so much smoke there must be at least a few embers, let's take some time to look at the nature and content of these broadsides.

On the team of Philosophical All-Stars, Immanuel Kant is definitely a heavy hitter. Unfortunately he wrote in a style that confirms your worst fears about the writing of philosophers, with even the most dedicated professor's eyes glazing over when slogging through one of Kant's weighty tomes. But it's heavy going for heavy ideas, and with a little more luck and better timing one of those grand-slam notions could have placed Kant in the spotlight as the developer of sociobiology, rather than Spencer or Wilson. One of the central tenets of Kant's thought is the categorical imperative, which, roughly speaking, is a claim that humans have an innate awareness of moral law in the form of a kind of rock-bottom ethical "ought." By linking this Kantian notion with our Central Dogma of Social and Behavioral Biology, we come up with something that sounds remarkably like Wilson's sociobiological explanation for the evolution of human ethics as an adaptive trait. Unfortunately for Wilson, Kant never met Darwin, so when the Boston Group unleashed its barrage of antisociobiological verbiage it was Wilson and not Kant who was forced to take to the barricades. Doubly unfortunate for Wilson was the fact that the group's assault focused on the cant of raw emotionalism rather than the Kant of pure reason.

The essential content of the Boston Group's letter to *The New York Review of Books* was that Wilson's book concealed a reactionary political message. A direct quote expresses better than I ever could the flavor of the political and personal attack:

> These theories [biological determinism/sociobiology] provided an important basis for the enactment of sterilization laws and restrictive immigration laws for the United States between 1910 and 1930 and also for the eugenics policies which led to the establishment of gas chambers in Nazi Germany.
>
> We think that this information has little relevance to human behavior, and the supposedly objective, scientific approach in reality conceals political assumptions. In his attempt to graft speculation about human behavior onto a biological core Wilson uses a number of strategies and sleights of hand which dispel any claim for logical factual continuity. What Wilson illustrates to us is . . . also the personal and social class prejudices of the researcher.

About a month later, Wilson replied to these charges:

> I wish to protest the false statements and accusations that comprise the letter. . . . This letter . . . is an openly partisan attack on what the signers mistakenly conclude to be a political message in the book. Every principal assertion made in the letter is either a false statement or a distortion. On the most crucial points raised by the signers, I have said the opposite of what was claimed. . . . I feel that the actions of Allen et al. [the group] represent the kind of self-righteous vigilantism which not only produces falsehood but also unjustly hurts individuals and through that kind of intimidation diminishes the spirit of free inquiry and discussion crucial to the health of the intellectual community.

Let's look a little deeper into the specific charges leveled at Wilson by the group, and the claim that his words and ideas had been distorted.

The core of the Boston Group's emotional outburst is organized around the assertion that Wilson is a biological determinist whose work serves to buttress the institutions of society by exonerating them from responsibility for social problems. In support of these allegations, the group writes, "It is stated [by Wilson] as a fact that genetical differences underlie variations between cultures, when no evidence at all exists for this assertion and there is some considerable evidence against it." What did Wilson really say? He wrote: "Even a small portion of this [genetic] difference might predispose societies toward cultural differences. At the very least, we should try to measure this amount. It is not valid to point to the absence of a behavioral trait in one or a few societies as conclusive evidence that the trait is environmentally induced and has no genetic disposition in man. The very opposite could be true." Not quite the same either in content or in spirit as what's claimed by the Boston Group, is it? Another example? Well, try this one on for size. The group writes that Wilson "promotes the analogy between human and animal societies and leads one to believe that behavior patterns in the two have the same basis." In his book, Wilson actually prefaces his discussion of this topic with the statement "Roles in human societies are fundamentally different from the castes of social insects." This list of distortions and fabrications could be considerably extended, but I think you get the general drift. But how is it that the group could so consistently misrepresent Wilson's statements in the book? To answer this commonsense query, we have to dig a little deeper into the political background of the group, especially that of its chief spokesman, Richard Lewontin.

It's no secret that Richard Lewontin advocates a Marxian view of biology, and takes his professional job as a scientist as tantamount to a political calling. He is on record with the statement:

> Any investigation into the genetic control of human behaviors is bound to produce a pseudoscience that will inevitably be misused. *Nothing* [emphasis added] we can know about the genetics of human behavior can have any implications for human society. But the process has social impact because the announcement that research is being done is a political act. . . . I treat my job as a political activity.

Later he argued:

> There is nothing in Marx, Lenin or Mao that is or that can be in contradiction with the particular physical facts and processes of a particular set of phenomena in the objective world.

One can only wonder why he omitted Stalin, Ho Chi Minh, and Pol Pot from this list of infallible thinkers.

Faced with these outlandish statements by Lewontin, even the most ardent zealot might start to cringe. But for us they offer a window through which we can begin to see a bit more clearly how it could be that the Boston Group would so blatantly twist

Wilson's words and warp his meaning. To me it's clear that the group's members were deeply alarmed by the impact that the critical success of the book might have on the acceptability of their own political views. Couple this fear with the inherent belief that political philosophy should guide scientific research, an attitude that was especially trendy on college campuses in the 1970s, and you have the basis for what Wilson once termed the Fallacy of the Political Consequent. This fallacy consists of the assumption that political belief systems can be mapped one-to-one onto biological or psychological generalizations. Perhaps not so surprisingly, this mapping points in exactly the opposite direction from the one that Wilson's tormentors accused him of following!

In a later book written with Steven Rose and Leon Kamin, *Not in Our Genes,* Lewontin continued his diatribe against Wilson by noting that sociobiology describes the whole of humankind as a transformation of European bourgeois society, with Wilson's description of human political economy involving a possessive, individualist, entrepreneurial society that would certainly not apply to the serfs of Eastern Europe or Mayan and Aztec peasants. Thus, in this view sociobiology treats categories as if they were natural objects having a concrete reality, rather than realizing that these are historically and ideologically conditioned constructions. These authors then go on to level a personal attack upon Wilson, claiming that by emphasizing altruism as a consequence of selection for reproductive selfishness, Wilson has identified himself with American neoconservative libertarianism, which holds that society is best served if each individual acts in a self-serving manner, limited only in the case of doing extreme harm to others. In short, Wilson has failed to separate out his "personal and class prejudices." So in this highly politicized view of reality, sociobiology is just the most recent attempt to put natural science to work in the cause of supporting the economic views arising out of Adam Smith's Invisible Hand, which, no doubt, in the opinion of Lewontin et al. looks more like the Iron Fist. In fact, this whole business is one of the most striking contemporary examples of the sociological factor in science, which we discussed in the opening chapter. Here we have strong cultural and political biases influencing not only what is considered to be acceptable as a scientific research topic, but also what is to count as scientific "truth."

As an illuminating final touch to this acrimonious exchange, it's of more than passing curiosity to note one or two of the personal factors surrounding the debate. Probably the most revealing background consideration of this sort is the close relationship between Lewontin and Wilson prior to the dispute. In fact, their offices at the Harvard Museum of Comparative Zoology are only one floor apart. Moreover, it was Wilson who was responsible for bringing Lewontin to Harvard in the first place, over strong political opposition on the faculty. It was also Wilson who acted to promote Richard Levins's candidacy for the membership in the National Academy of Sciences, which he later turned down. So we see here a kind of family feud that unfortunately bubbled over into the public arena dressed up as a scientific debate. Despite the fact that these personal attacks forced Wilson to cancel several engagements because of the mental strain on his family, the group's most publicly visible member, Stephen Jay Gould, still had the audacity to remark that "we don't intend it as a personal attack. Ed Wilson is a colleague whom we like." Apparently the group members feel that it's not a personal attack to accuse someone of writing a book that is not only totally valueless but even dangerous, because it's filled with what they claim are the author's personal political views. Perhaps the dictionaries in Cambridge offer a special definition of what constitutes a "personal" as opposed to "professional" attack, definitions differing from those found in the *Wörterbücher* I use in Vienna. In any case, Gould's claim surely qualifies for honorable mention at the next International Hairsplitters' Convention. Now let's move on to more substantive scientific criticisms of sociobiology.

SO-SO BIOLOGY

A few years after the publication of Darwin's classic works, the French writer Émile Zola began his *Rougon-Macquart* cycle, a series of twenty novels described in a subtitle as *The Natural and Social History of a Family Under the Second Empire*. This cycle was intended to show the inevitable consequences of certain scientific "facts," especially the claims of Cesare Lombroso and Paul Broca that inherited physical characteristics were indicative of mental and moral traits. For example, in *Nana,* probably the best-known novel in the cycle, Zola tells of the trials of the courtesan Nana, the laundress Gervaise, and the drunk Coupeau. The complete family history is set up to dramatize Zola's statement that "heredity has its laws, just as does gravitation." It is this idea of biological (read genetic) determinism that constitutes the main focus of the more sober, scientific criticisms of sociobiology.

Technically, sociobiology is based on a new view of natural selection: Hamilton's idea of *inclusive* genetic fitness. Implicit in the claims of sociobiology is the notion that organisms act so as to maximize their inclusive reproductive fitness. Critics have argued that this just is not true. Organisms act to maximize inclusive fitness *under constraints.* The following list of such constraints offered by Barry Schwartz indicates their importance in assessing the merits of the sociobiological case:

- *Neutral characteristics:* As far as genetic fitness goes, many phenotypic properties of the organism are irrelevant, i.e., neutral. Nevertheless, such characteristics may severely restrict the kinds of future modifications of the organism that will count as an improvement.
- *Time lags:* The processes of environmental change and evolutionary adjustment operate on vastly different time scales. Thus, what was optimal long ago may be very far from optimal today.
- *Context dependence:* Genes leading to a certain kind of behavior seen today may have originally come about for some quite different purpose, one that is no longer relevant in the current environment.
- *Historical constraints:* Every modification must also be an improvement in order to avoid being eliminated. Thus, Nature is totally oriented to the short term, performing local optimizations that may not lead to globally optimal performance. A good illustration from technology of this kind of phenomenon is the development of the transistor. No sequence of evolutionary improvements on the vacuum tube would have brought this change about—a fundamentally new principle was needed.
- *Variation constaint:* Maximization can be applied only to those variations that actually occur, not to those that were possible but just didn't happen.
- *Cost-benefit analysis:* Each of the subsystems composing an or-

ganism has to coexist with every other, so that a variation that's good for one system may be very bad for another. Consequently, every variation has to be measured by a cost-benefit calculation against the overall improvement for the organism. So, for example, the development of a capability to run faster to catch food would have to be weighed against the extra energy needed to supply the added motive power.

• *Levels of analysis:* A given variation has to be evaluated at several biological levels—gene, organism, group—and what's good for one may be very harmful to another.

• *Capricious environment:* A sudden environmental disturbance can undo in a few days the gradual evolutionary changes of several millennia, e.g., the meteorite collision that supposedly wiped out the dinosaurs 65 million years ago.

The problem that this list of constraints poses for sociobiology is that it's difficult to give a criterion for what constitutes a maximizing trait and what is only a background constraint. In any particular situation, any activity at all can be shown to be adaptive if the background constraints are drawn tightly enough. Sociobiology adopts the strategy of arguing from behavior, and asks what the constraints must be so that a given behavior maximizes inclusive genetic fitness. This is a circular argument, and leads to Gould's complaint that sociobiological explanations are merely a collection of *Just So* stories.

The problem of genetic constraints is a special case of the more general criticism that sociobiological explanations rely too heavily upon the use of *reification,* i.e., treating idealized abstractions as if they were concrete entities. Specifically, the critics contend that the sociobiologists systematically overrate the relationship between the genotype and various observed behavioral traits. We have already considered some of these objections to the assumed tight genotype-phenotype fit, and the attempts by the sociobiologists to wriggle off the hook by introducing hypothetical constructs like replicators or by arguing for genetic "influence" rather than determination. So let's consider here a different criticism, but one that pushes in the same general direction.

It's been seen that the explanation of altruism occupies a central place in the theoretical framework of the sociobiologists, and that the notions of kin selection and inclusive fitness are crucial for the sociobiological argument to go through. The pivotal role in this argument is the determination of the coefficient of relationship r expressing the genetic linkage between any two family members. M. Sahlins contends that this is all mystical nonsense, with the computation, or even recognition, of r being impossible. The sociobiologist's counterthrust is to concede readily that the organism doesn't sit down and explicitly calculate r when deciding upon what action to take, but it *acts* as if it has made such a calculation. And for sociobiological purposes that's all that counts. This is much the same argument that you might use if someone claimed that you could never catch a baseball because you couldn't solve the differential equations governing the ball's flight path. When set in this context, Sahlins's assertion starts to lose some of its initial luster.

It might appear surprising to some to invoke the name of Richard Dawkins in connection with arguments *against* sociobiology, but it has always seemed to me that his notion of a cultural *meme* playing the role for cultural traits that genes play for physiological ones is really a statement against the genetically based claims of the mainline sociobiologists. In the last chapter of his book *The Selfish Gene,* Dawkins introduces the meme, a kind of unit of selection for cultural matters of the same sort that Lumsden and Wilson label a culturgen. The memes are the carriers of such things as fashions, popular tunes, and fads in speech, but unlike culturgens, memes are not claimed to have a direct relationship with the actual genotype. In fact, Dawkins goes further and argues that "memes and genes may often reinforce each other, but they sometimes come into opposition." However, he does emphasize the functional similarity of memes to genes in that both are replicators and both are carriers of information. But when it comes down to exactly *how* the information is carried and replicated, Dawkins parts company with Wilson and Lumsden. Memes replicate by being passed from brain to brain as pure information; culturgens replicate by epigenetic rules processed by the physical genotype. It is in this sense that I see Dawkins's process of cultural evolution as an antisociobiological argument against the strict material transmission inherent in the hard-core position of Lumsden and Wilson.

Now let me turn to one of the major objections put forward

against treating sociobiology as a science. This is the old Popperian criticism that in order to be *scientific* a theory must be falsifiable, with sociobiology failing the test. The critic's case rests upon the *Just So* character of sociobiological claims that allow virtually any behavioral pattern to be seen as an adaptive trait. Thus, the argument goes, there is no conceivable observation or experimental result that would falsify sociobiology; ergo, the theory is nonscientific.

There are at least two comments to be made regarding this objection. First of all, it is just plain false to claim that there are no observations that would falsify the theory. For example, the observation of societies in which relatives gave freely with no hope of return would surely deal a mortal blow to the aspirations of the sociobiologists. The fact that no such societies have been observed can hardly be laid at the doorstep of the sociobiologists. Another example would arise if we were to see societies that actively promoted incest. In this case, either *(a)* the incidence of birth defects is not significantly increased, or *(b)* the birth defect rate does rise, but the practice of incest goes on anyway. The first alternative would refute the claim that incest avoidance reduces inclusive fitness, while the second refutes the claim that behavioral traits can be explained as enhancing genetic fitness; i.e., it would be an example of a trait that persists even though it has a negative effect on inclusive fitness. Either alternative would spell deep trouble for the sociobiological view of the world.

As a second point, we have already noted in the opening chapter that there are serious difficulties with Popper's falsification criterion for separating science from pseudoscience. Interestingly enough, a major stumbling block for Popper is the Problem of Auxiliary Hypotheses, a difficulty strikingly similar to the Problem of Genetic Constraints noted above. Furthermore, Kuhn has noted that much can be gained by not allowing a theory to be blown away by the first fact that appears to contradict it—i.e., a strict adherence to the falsificationist doctrine can be hazardous to the ultimate health of science! So with these ideas in mind, the claims that sociobiology is not scientific also begin to take on a distinctly less convincing air.

Finally, we move from falsification to the assertion that sociobiology is just plain false. Here the critics contend that the genetic differences between populations are not great enough to account for the vast differences in culture that have been observed. In this connection, the Boston Group states that "at least 85 percent of that kind of [genetic] variation lies *within* any local population or nation, with a maximum of about 8 percent between nations and 7 percent between major races." The implication is that this relatively minor variation between nations and races is way too small to be a significant factor in generating cultural differences.

The sociobiologists have two replies to this critique. First of all, they argue on the basis of what Wilson calls *the multiplier effect,* whereby small changes in the genotype can multiply and, by the time they percolate up to the phenotype, can give rise to major phenotypic variations. As might be imagined, the critics look upon this kind of response with the same degree of favor that small children look upon a plate of spinach. As always, the Boston Group speaks with the sharpest tongue when it comments that the so-called multiplier effect and its closely associated "threshold effect" are "pure inventions of convenience without any evidence to support them. They have been created out of whole cloth to seal off the last loophole through which the theory might have been tested against the real world."

By way of a second response, the sociobiologists employ a little rhetorical judo, using their opponents' strength against them by turning their argument on its head. The sociobiologists say, instead of looking at cultural differences, let's look at cultural *similarities.* The case is then made that the similarities are much more important than the differences, and that these similarities indicate a common genetic background. Of course, stating this argument, just like stating the argument based on the multiplier effect, is a far cry from providing a convincing demonstration that it's true, or even plausible.

On this inconclusive note we wrap up our quick survey of the main epistemological objections to sociobiology. Since the central pillar upon which the entire sociobiology program rests is the explanation of how altruistic, or at least cooperative, behavior can emerge out of basically selfish motives, it is worth spending a moment looking at mechanisms by which this might come about (with or without the help of the genes) before moving on to summary arguments and the verdict.

CONFLICTING RATIONALITIES AND THE DILEMMA OF COOPERATION

In 1951 Merrill Flood of the RAND Corporation introduced one of the most thought-provoking concepts in the history of strategic thinking. His idea, later termed the Prisoner's Dilemma by Albert Tucker, cuts to the heart of an age-old question: How do we balance individually selfish acts against the collective rationality of individual sacrifice for the sake of the common good? A familiar example will illustrate the point.

In Puccini's opera *Tosca,* Tosca's lover has been condemned to death, and the police chief Scarpia offers Tosca a deal. If Tosca will bestow her sexual favors on him, Scarpia will spare her lover's life by instructing the firing squad to load their rifles with blanks. Here both Tosca and Scarpia face the choice of either keeping their part of the bargain or double-crossing the other. Acting on the basis of what's best for them as individuals, both Tosca and Scarpia try a double cross. Tosca stabs Scarpia as he is about to embrace her, while it turns out that Scarpia has not given the order to the firing squad to use blanks. The dilemma is that this outcome, undesirable for both parties, could have been avoided if they had trusted each other and acted not as selfish individuals, but rather in their mutual interest.

The tragic fates of Tosca and Scarpia serve to characterize the essential ingredients of a classical Prisoner's Dilemma situation: There are two parties, each of whom has the choice of either *cooperating* (C) or *defecting* (D), i.e., acting either to sacrifice their individual interests for the sake of a common good, or to further their own selfish individual interests at the expense of the other. In addition, there must be a payoff structure involving a *temptation* (T), the payoff received by defecting when the other party cooperates; a *reward* (R), the payoff each party receives if they both cooperate; a *punishment* (P), the payoff they each get if they both defect; a *sucker's payoff* (S), which is the amount received by the cooperating party when the other defects. For the Prisoner's Dilemma to arise, these payoffs must be ordered largest to smallest in the following way: $T > R > P > S$. To avoid getting locked into an out-of-phase cycle of mutual defections and cooperations, there is the technical condition

(T + S)/2 < R. Under these conditions, let's quickly analyze the source of the dilemma faced by Tosca and Scarpia when considering their respective courses of action.

To make things concrete, let's put in numerical values for the payoffs in *Tosca*. Suppose they are T = 4, R = 3, P = 2, S = 1. Tosca can then argue: If I defect and Scarpia cooperates, my lover's life will be saved and I won't have to see Scarpia, yielding a payoff to me of 4 units. But if I defect and Scarpia also defects, then even if I do lose my lover, at least I won't have to give myself to that pig Scarpia and I'll end up with 2 units. On the other hand, if I trust Scarpia and he trusts me so that we both cooperate, I'll get 3 units, while if I trust him by cooperating and he double-crosses me and defects, then I'll get only the sucker's payoff of 1 unit. So, all in all, by defecting I'm assured of getting 2 units, whereas if I cooperate I can't get any more than 3 units and could end up with much less. Therefore, rationally it's in my best interest to defect. Of course, the situation is perfectly symmetrical and Scarpia, being equally rational and logical, comes to the same conclusion and also opts to defect. Result: Both Scarpia and Tosca end up with much less than they could have had by showing a little mutual trust. In other words, by employing *individual* rationality they sacrifice their *collective* joint interests.

The relevance of the Prisoner's Dilemma for sociobiology is evident. The cornerstone of sociobiological reasoning is the claim that human behavior patterns, including what look on the surface like selfless acts of altruism, emerge out of genetically selfish actions. In the context of the Prisoner's Dilemma, we can translate this sociobiological thesis into the statement that the individually rational act of defection will always be preferred to the collectively rational choice of cooperation. Our question is then: Can that situation ever lead to a population of cooperators? If there is no way for cooperative acts to emerge naturally out of self-interest, it's going to be very difficult for the sociobiologists to support their case. Put in our earlier game-theoretic terms, always to defect is an evolutionary stable strategy, since players who deviate from this policy can never make inroads against a population of defectors. Or can they? Are there any situations in which a less cutthroat course of action can ultimately establish a foothold in a population of defectors? This was the Big Question that Robert Axelrod set out to answer in one of the most intriguing psychological experiments carried out in recent years. The separate issues that Axelrod wanted to address were: (1) How can cooperation get started at all in a world of egoists? (2) Can individuals employing cooperative strategies survive better than their uncooperative rivals? (3) Which cooperative strategies will do best, and how will they come to dominate?

Axelrod's key observation was to note that while ALL D, the strategy of always defecting, is uninvadable for a sequence of Prisoner's Dilemma interactions that is of known, fixed, and finite duration, there may be alternative ESS strategies if the number of interactions is not known by both parties in advance. So after having played a round of the Prisoner's Dilemma, if there is a nonzero chance that the game might continue for another round, then maybe there is a nice strategy that is also ESS. Here by "nice" we mean a strategy that would not be the first to defect.

To test this idea, Axelrod invited a number of psychologists, mathematicians, political scientists, and computer experts to participate in a contest pitting different strategies against one another in a computer tournament. The idea was for each participant to supply what he or she considered to be the best strategy for playing a sequence of Prisoner's Dilemma interactions, with the different strategies then competing against each other in a round-robin tournament. Fourteen competitors sent in strategies, which were in the form of computer programs. The ground rules allowed the programs to make use of any information about the past plays of the game. Furthermore, the programs didn't have to be deterministic, but were allowed to arrive at their decision by some kind of randomizing device if the player so desired. The only condition imposed was that the program ultimately come to a definite decision for each round of play: C or D. In addition to the submitted strategies, Axelrod also included the strategy RANDOM, which took the decision to cooperate or defect by, in effect, flipping a coin. In the tournament itself, every program was made to engage every other (including a clone of itself) two hundred times, the entire experiment being carried out five times in order to smooth out statistical fluctuations in the random-number generator used for the nondeterministic strategies.

The winning strategy turned out to be the simplest. This was the three-line program describing the strategy TIT FOR TAT. It was offered by Anatol Rapoport and consisted of the two rules: (1) cooperate on the first encounter; (2) thereafter, do whatever your opponent did on the previous round. That such a simple, straightforward strategy could prevail against so many seemingly far more complex and sophisticated rules for action seems nothing short of miraculous. The central lesson of this tournament was that in order for a strategy to succeed, it should be both nice and forgiving, i.e., it should be willing both to initiate and to reciprocate cooperation. Following a detailed analysis of the tournament, Axelrod decided to hold a second tournament to see if the lessons learned the first time around could be put into practice to develop even more effective cooperative strategies than TIT FOR TAT.

As prelude to the second tournament, Axelrod packaged up all the information and results from the first tournament and sent it to the various participants, asking them to submit revised strategies. He also opened up the tournament to outsiders by taking out ads in computer magazines, hoping to attract some programming fanatics who might take the time to devise truly ingenious strategies. Altogether Axelrod received sixty-two entries from around the world, including one from the renowned game theorist John Maynard Smith, mentioned earlier as the developer of the ideas of the evolutionary game and the ESS. The winner? Again it was Rapoport with TIT FOR TAT! Even against this supposedly much stronger field, Rapoport's game-theoretic version of the Golden Rule was the hands-down winner. The general lesson that emerged from the second tournament was that not only is it important to be nice and forgiving, but it's also important to be both provocable and recognizable; i.e., you should get mad at defectors and retaliate quickly but without being vindictive, and you should be straightforward, avoiding the impression of being too complex. After extensive study of the results, Axelrod summarized the success of TIT FOR TAT in the following way:

TIT FOR TAT won the tournaments not by beating the other player but by eliciting behavior from the other player that allowed both to do well. . . . So in a non–zero sum world, you do not have to do better than the other player to do well for yourself. This is especially true when you are interacting with many different players. . . . The other's success is virtually a pre-requisite for doing well yourself.

So what are the implications of these results for sociobiology?

If we think of the total points amassed by a strategy during the course of the tournament as its "fitness," and if we interpret "fitness" to mean "the number of progeny in the next generation," and finally if we let "next generation" mean "next tournament," then what happens is that each tournament's results determine the environment for the next tournament. The fittest strategies then become more heavily represented in the next tournament. This interpretation leads to a kind of ecological adaptation without evolution (since no *new* species come into existence). Sociobiologists can take heart in this sort of interpretation of Axelrod's experiments because they show that it's possible for phenotypically altruistic (cooperative) behavior to emerge out of individually selfish motives. It's important to emphasize here, though, that these results say nothing about the actual causal factors at work generating the individual motives. They *could* be genetic, as hard-core sociobiologists would love to argue, but there is nothing in Axelrod's work to say that they are. Nevertheless, the experiments do offer some support to the sociobiological explanation of cooperative behavior by means of reciprocal altruism.

Following his work on the evolution of cooperation, Axelrod carried out another set of experiments that also give succor to the sociobiologist's claim for an evolutionary development of standards of behavior, i.e., cultural norms. The basic idea was to use a souped-up version of the Prisoner's Dilemma in which the players had the choice not only of cooperation or defection, but also of punishing a defection or letting it pass. Players in the Norms Game are characterized by two qualities: *boldness* (B), which measures the risk they are willing to run in defecting; and *vengefulness* (V), a measure of their inclination to punish defection. Strategies were assigned randomly to twenty players, with the first round of play lasting until each player had had four opportunities to defect. At the end of the first generation, a strategy was given one offspring if its score was near average, two offspring if its score was at least one standard deviation above the mean, and no offspring if its score was more than one standard deviation below the mean. Furthermore, Axelrod also allowed for the emergence of new strategies through a process of mutation in such a way that about one new strategy emerged in each generation.

The results of the simulation showed that with enough time, all populations would eventually converge to the collapse of the norm, i.e., V approaching zero. The problem appears to be that the players lack sufficient incentive to punish the defectors: Nobody wants to play sheriff. As one way to enforce the norm, Axelrod suggests a *metanorm:* direct vengeance not only against those who defect, but also against those who refuse to punish them. This is the kind of procedure we see in some totalitarian countries where, when a citizen is accused by the authorities of some real or imagined ideological transgression, others are called upon to pile their own denunciations onto the back of the hapless offender.

While these results are still in the preliminary stage, the Evolution of Cooperation Game and the Norms Game both provide some theoretical evidence in support of the idea that social behavior can emerge as the result of evolutionary processes involving individually selfish agents. Whether that selfishness is programmed into the genes is anybody's guess, but at least there are no obvious game-theoretic barriers preventing it. Now let's leave the gaming arena and return to the courtroom to let the respective sides make their final statements before we retire to ponder the verdict.

SUMMARY ARGUMENTS

The claims and counterclaims have been flying fast and furious throughout this chapter, so before trying to put them into some sort of coherent order let's restate the basic question to be decided. The Prosecution's contention is that:

> *The* majority *of human behavior patterns are strongly* influenced *by the genes.*

Note the emphasis here on the words "majority" and "influenced." All that's needed to make the sociobiologists' case is to agree that in the vast majority of situations, genetic makeup plays a more important role than the environment in determining the way people act.

With this statement of the question in mind, let's now turn to a tabular summary of what look to be the principal positions of the contending parties. But before presenting the summaries themselves, a few comments are in order:

1. Most of the advocates of sociobiology, as well as their critics, expound a variety of arguments in support of their case. For the sake of brevity, each table entry lists only a catchword or two representative of the general position. No attempt has been made to summarize all aspects of any promoter's argument.

2. Oddly enough, the men most responsible for the theoretical underpinnings of human sociobiology, Hamilton, Maynard Smith, and Axelrod, appear to be pretty lukewarm, at best, about the case for human sociobiology. In fact, Maynard Smith, for one, categorically denies that there is any direct contact between his work on game-theoretic models for animal aggression and the behavior of humans. My suspicions are that he does so to avoid being sucked into the bottomless pit of ideological debate with supporters of the Boston Group. Nevertheless, I have placed all this theoretical work under the general heading of the Prosecution, since it acts to lend more support to the adherents of sociobiology than to their detractors.

3. Richard Dawkins appears in the curious position of supporting both the Prosecution and Defense on my lists, since his original work on the selfish gene argues strongly for a genetic basis to behavior, whereas his later discussion of cultural memes is really more of a case for environment as being the main motivator of human actions.

With these clarifications at hand, let's examine the summaries of the competing cases in Tables 3.1 and 3.2.

Prior to making tracks for the jury room, we must listen to the judge's instructions. In our assessment of the evidence, no weight whatsoever is to be given to the wild-eyed, slightly hysterical political outburst made in open court by the Defense. Regardless of any jury member's personal sentiments, we're in a courtroom, not at a political pep rally, and the matter to be decided here is a question of science, not politics. This should not be taken to mean that the arguments of the Boston Group are wrong, only that the political component of what they argue should never have been heard, and wouldn't have been if the court had been quick enough to muzzle the Defense attorney

HUMAN BEHAVIOR IS PRIMARILY GENETIC!

PROMOTER	ARGUMENT
Lorenz	innate aggression, group selection
Wilson, Barash	genetic influence, multiplier effect
Dawkins	selfish genes
Lumsden and Wilson	coevolutionary circuit
Trivers	reciprocal altruism
"theoretical support"	
Hamilton	inclusive fitness, kin selection
Maynard Smith	evolutionary game, ESS
Axelrod	evolution of cooperation and norms

TABLE 3.1. *Summary arguments for the Prosecution*

HUMAN BEHAVIOR IS PRIMARILY ENVIRONMENTAL!

PROMOTER	ARGUMENT
Boston Group	reification, no multiplier effect, unfalsifiability
Schwartz	evolutionary constraints
Sahlins	kin selection impossible
Gould	*Just So* stories
Dawkins	cultural memes

TABLE 3.2. *Summary arguments for the Defense*

properly. Despite Defense protests to the contrary, politics has no place in the laboratory, however one might personally feel about the nature of what's being studied, or however much one might wish for an experimental result to come out one way instead of another. So put the political smokescreen and academic vigilantism out of your mind when pondering your verdict.

BRINGING IN THE VERDICT

Of all the Big Questions dealt with in this book, I find the sociobiology problem to be the most perplexing. Even after wading through the pages of testimony and trying to filter out the nuggets of real information from the fool's gold of rhetoric and political bombast, when all is said and done I'm forced to take refuge in that ancient Scottish verdict "not proven."

In terms of hard evidence, I find that after we pass from the few instances of behavioral maladies like schizophrenia that appear to be solidly founded cases of genetic causation, the tangible facts supporting the case of sociobiology fade away like a trickle of water in the desert. On the other hand, the circumstantial evidence is impressive. Predictions of both animal and human behavior made on the basis of sociobiological arguments seem to be, for the most part, at least within the bounds of normal experimental error. And the idea of a smooth transition from rather clear-cut evidence of genetically influenced behavior in the animal kingdom to similar behavior patterns in humans is appealing.

In many ways, my feeling is that the sociobiologists have been a little too eager to promote their cause by calling forth evidence of dubious validity, and neglecting some rather obvious alternative interpretations. All of this calls to mind the statement made by Alexander Solzhenitsyn in his 1978 Harvard commencement address when he noted that "hastiness and superficiality are the psychic disease of the twentieth century." It's tempting to wonder whether or not some of the Boston Group, as well as the Wilson circle, were present in Cambridge that day when this compact summary of many of their claims was expressed.

On the other side of the ledger, I also find the arguments (scientific and philosophical, that is) of the Defense to be difficult to dismiss. For the most part, there really is no firm evidence to support the kind of direct path from the genotype to phenotypic behavior that the sociobiologists need to establish their case. And it is true that the usual arguments leading from behavior back to genetic causes leave a lot to be desired, both scientifically and philosophically. Furthermore, Stephen Jay Gould may have a valid point when he says that the genes have given up their sovereignty over the major human behavior patterns as a result of *Homo sapiens*'s most distinguishing feature—an extraordinarily large brain.

All things taken together, I feel a bit like the Dodo in *Alice in Wonderland* when he announces the winners of the Caucus Race: "*Everybody* has won, and *all* must have prizes." Frankly, I can't for the life of me understand why, in the face of so much real and circumstantial evidence supporting both sides of the debate, the participants continue to cling so fiercely to what are basically either-or positions. To an outsider, it seems pretty clear that most interesting human behavior patterns are brought about through a complex combination of genetic *and* environmental factors, and the real work should be addressed to investigating these complicated webs of interconnection. To my mind it seems a futile effort to try to disentangle the relative contributions of the genes and the environment, and even more futile to dissipate energy on senseless political harangues about a distinction that is far more virtual than real. But before dismissing the sociobiology debate in such a nonpartisan and cavalier manner, I think it's worth speculating a moment on some of the reasons why sociobiology seems to touch such a sensitive nerve in the psyches of the scientific community and the general public alike.

In my opinion, the root cause of the heated public and academic debates over the claims of the sociobiologists comes down to only one thing—raw power. As the Boston Group has noted, the reductionistic flavor of sociobiology, and its implication that human society is both inevitable and the result of adaptive processes, holds great attraction for the John D. Rockefellers of the world who wield power and want to justify their actions by an appeal to that final authority, Nature. As the late ethologist Niko Tinbergen expressed it:

> It is tempting to ponder this over-emphasis on studies of causation. I believe that it is partly due to the fact that, as the developments of physics and chemistry have shown, knowledge of the causes underlying natural events provides us with the power to manipulate these events and bully them into subservience.

In short, sociobiology offers us a mystique of power.

An equally plausible and closely related reason why sociobiology appears so compelling is offered by Barry Schwartz, when he notes that we are living in a time when the pursuit of self-interest in the free-market economy provides the primary metaphor for understanding social relations. As a result, our social and cultural categories overlap with our economic ones. Consequently sociobiology, with its explanatory structure based upon the "economic accounting" of evolutionary biology, seems to capture many of the most prominent features of modern life. However, this "fortuitous" juxtaposition of economic, social, and biological principles is not a universal biological necessity. Both economic and social situations can change; biological principles cannot. Consequently, it may indeed be dangerous to argue that the unbridled pursuit of selfish personal interest is a part of basic human nature—just as Lewontin & Co. have been claiming all along!

So we end up closing the book on the question of Nature versus nurture barely any farther along the road to an answer than when we opened it. But there has always been one area in which virtually everyone agrees that there is a biological substrate underlying a uniquely human behavior: the capacity for semantic language. As we might suspect by now, however, even this seemingly clear-cut case is not without its competing factions. So as a detailed case study of one small, but important, corner of the social *cum* biological forest, let's now move on to a consideration of the problem of human language acquisition.

4

SPEAKING FOR MYSELF

===

CLAIM:
HUMAN LANGUAGE CAPACITY STEMS
FROM A UNIQUE, INNATE PROPERTY OF
THE BRAIN

DUMB DOGS AND CLEVER HANS

Unlike Americans, Austrians have no prejudices or proscriptions against allowing dogs into their restaurants. Consequently, at the Kuchldragoner, a Viennese *Beisl* I frequent for lunch, the house dogs, Chi-Chi and Isabella, routinely make their appearance at my table to put in their claims for a sliver of schnitzel with a low whine, a paw on my lap, or, in Isabella's case, just the dropping of her St. Bernard–like head onto the edge of my table. Of course these two mutts don't realize how crazy they are to think that I'm ready to part with any of Frau Holzfeind's *Specknockerl, Grammelknödel,* or *Schinkenfleckerl,* so just as rou-

tinely I try to convince them of the folly of their canine ways by
telling them, "Not now, Chi-Chi," or "You're so beautiful
today, Isabella, but this food isn't for dogs," or, if all else fails,
"Get lost, dog!" What could possibly make me think that these
hounds understand even one word of what I'm saying to them,
especially when I usually say it in a garbled, pidgin version of
their "native" German? In fact, after every such encounter I
end up feeling slightly foolish, often wondering who is really the
crazy one in our by-now-almost-ritualistic interactions.

Strangely enough, my experiences in linguistic communication
with Chi-Chi and Isabella seem to reflect an almost universal
human belief in the ability to communicate by speech, or at least
symbolic language, with the higher animals. As a small child I
was convinced our family dog was just brimming over with ideas
and plans he wanted to tell me about, and I remember asking my
mother why he couldn't speak to me like my other playmates. To
her eternal credit, she responded with the commonsense answer
that perhaps he really didn't have anything to say, or at least
not anything that would be of concern, or even comprehensible,
to any human. Later, though, I felt slightly better about my
"stupid" query when I read an article claiming that Alexander
Graham Bell had tried teaching a dog to speak by training it to
growl at a constant level while he manipulated its jaw muscles
and throat to get it to produce various sounds. About the best
the poor pooch could come up with was something that sounded
a lot like *ah oo yow grrr,* a pretty poor imitation of "Let me out
of here," at which point Bell wisely returned to his work on the
telephone.

Undeterred by failures of this sort actually to speak with animals, around the turn of the century a retired German school-
teacher named Wilhelm von Osten acquired a bit of a local
reputation in Berlin by displaying his horse Hans, which he
claimed could solve problems in arithmetic. The question would
be posed, "Hans, how much is three plus five?"—at which point
Hans would start tapping his hoof, stopping after the eighth
tap. It makes one wonder what Hans's response would have been
if he were asked to extract the square root of π! Unfortunately
for both von Osten and Hans, the psychologist Oskar Pfungst
made a detailed investigation of the "Clever Hans" phenomenon, conclusively showing that Hans was more of a showman
(showhorse?) than an accountant, taking unconscious cues from

his owner during the course of his demonstrations. In fact, Pfungst demonstrated that Hans had the uncanny ability to detect head movements of as little as one fifth of a millimeter, thereby being able to "read" the slight, but unconscious, movement in his trainer's head when he came to the correct number of taps for a given computation.

The most recent manifestation of the human psychological need to talk with the animals has been the spate of experiments by John Lilly, David Premack, Allen and Beatrice Gardner, Herbert Terrace, and others aimed at communication with dolphins, gorillas, and chimpanzees by sign language, manipulation of colored chips, and other such devices. While virtually everyone agrees that some kind of interspecies communication has taken place in these efforts, the bottom line seems to be that whatever kind of communication it is, it's not what we think of when we consider human linguistic interaction. The key question that emerges from this result is why anyone would seriously entertain the notion that chimps, whales, or apes could be taught to communicate using the same principles that underlie human language in the first place. The answer involves a brief consideration of the historical origin of language.

The number of different theories of how human language originated seems to be about equal to the number of investigators of the topic, ranging from "bowwow theories" claiming that language emerged from imitations of the sounds of nature with onomatopoeic words, to "singsong theories" claiming that speech arose from the love songs and rhythmic chants of primeval Lotharios. Following an explosion of such wild speculation, in 1886 the Linguistic Society of Paris issued a resolution "outlawing" any more papers concerned with the origin of language. The ban was upheld in 1911 by the Philosophical Society of London but, regrettably, it does not seem to have stanched the flow of speculation on the topic. The most sober guesses today argue for the origin of language as an evolutionarily advantageous trait enabling primitive man to communicate more effectively in groups for hunting, socializing, and defense. Whatever the actual reason for its origin (and it's likely to be a combination of several causes), human language is almost universally considered to have evolved out of more primitive levels of brain neurophysiology and body structure.

If we accept the picture of human language as having its origin in an evolutionary outgrowth of body and brain development, what could be more natural than to ask: What was the first human language? According to the second book of Herodotus's History, credit for the initial experiment on the matter goes to the Egyptian pharaoh Psammetichus, who around twenty-five hundred years ago arranged to have two infants raised in a "linguistic deprivation tank" under the assumption that whatever their first word turned out to be, it would be from mankind's true "source" language. Herodotus reports that the first word uttered was *bekos,* the word for "bread" in Phrygian, a language then spoken in the northwestern corner of what is now Turkey. Thus, Psammetichus concluded that Phrygian was mankind's original tongue. In good scientific fashion, the monarchs James IV of Scotland and Frederick II of Hohenstaufen both repeated the pharaoh's experiment, with the outcome that James's test subject "spak guid Ebrew." Unfortunately, Frederick's subjects died, perhaps from loneliness, before having the chance to utter even a single word. The sum total of all these experiments is that we have no more real information about the original language of man than did Psammetichus, but the exercises do allow us to contrast in sharp colors the lines of research that linguists have taken since the pharaoh's time.

In terms of philosophical endearment, linguistic research can be divided into two main camps: the empiricists and the rationalists. The thesis of the first group is that the only way to understand human language is by actual observation. Go into the field with your tape recorder and notebook, gather several hundred hours of actual speech in a variety of situations, then analyze the data to extract the linguistic patterns present in a particular speech community. The rationalists, on the other hand, hold to the view that there's much more to language than just data; there is an innate knowledge of linguistic structure that is part of the genetic birthright of any normal human child, and in order to understand language it's necessary to take this innate knowledge into account. In the more biological terms used in the last chapter, the empiricist stance corresponds to the position that language is basically determined environmentally, while the rationalists cling to the position that it's principally in the genes. To sort out the competing claims, a short tour of twentieth-century linguistics research will prove helpful.

VERBAL BOTANY
AND UNIVERSAL GRAMMAR

According to a count by that bastion of linguistic conservatism the Académie Française, there are currently 2,796 separate dialects spoken on Earth, a number no doubt contributing to the old joke that a language is nothing but a dialect with an army and a navy. When contemplating this enormous variety of languages, especially in light of my own anemic linguistic talents, I become ever more sympathetic to the view of Naguib Mahfouz, 1988 Nobel laureate in literature and the first Arab writer ever to be so honored, when he stated that it would be much better for culture and humanity if all writers wrote in the same language. But, alas, both Mahfouz and I seem to be stuck with the Académie's list of 2,796. Every one of these known languages shares the following features:

1. Formation of a large number of meaningful symbols (words) from a small set of basic sounds (phonemes).
2. Formation of an unlimited number of sentences by logically combining words using a finite number of grammatical rules.
3. The sentences are used for socialized actions.
4. Any normal child has the ability to learn to speak the language.

By way of contrast, no known system of animal communication shares all of these characteristics. For example, the dance of the honeybees doesn't involve symbols or sentences, nor is it learned. And the chimpanzees don't form structured sentences either. The science of linguistics has arisen over the past century or so with the goal of studying the properties of these 2,796-plus ways of human communication.

It seems that all linguistic researchers agree on what constitute the basics: A language is composed of a set of meaningful *sentences,* each composed of a set of *words,* each of which is in turn formed phonetically out of a set of elementary sounds *(phonemes)* like [f] and [v] in "fine" and "vine," and semantically from a collection of "atoms" of meaning *(morphemes)* such as [im] and [possible] in the two-morpheme word "impossible." Further, for each language there is a *grammar* consisting of the rules determining the allowable ways in which the words can be combined to form sentences *(syntax),* as well as the manner in which those sentences are to be understood *(semantics)* and pronounced *(phonetics).* Thus, the grammar specifies the totality of linguistic rules necessary for using the language. (This technical use of the term *grammar* is not quite the same as the grammar most of us struggled with while parsing sentences in "grammar school." The differences will be made explicit later.) The overall goal of virtually every linguistic researcher is somehow to specify explicitly the grammar characterizing any given language. The fun begins when it comes to the question of just how we should go about attaining this state of linguistic bliss.

According to one school of linguistic thought, the localists, the most interesting aspects of languages are the ways in which they differ. Accordingly, localists follow the empiricist path to understanding grammar, tending to emphasize the collection and analysis of field data on exotic languages like Philippine Tagalog, Haitian Creole, or perhaps the Mandingo tongue of West Africa. The approach of the localists is to start with a description of the elements in a language (the phonemes and morphemes), and build up to the more complex elements (the sentences). The underlying localist belief is that through acquisition of enough data, the patterns characterizing the grammar of the language will slowly but surely emerge.

Following what seemed to be the eminently sensible approach of gathering data before engaging in any theorizing, the localists were first off the mark in the linguistics derby with the work of Ferdinand de Saussure in Geneva. Saussure focused on language as a system and tried to describe that system as a collection of interdependent parts deriving their significance from the system as a whole. A second stream of localist thought was initiated by the German linguist Franz Boas, who advocated what amounts to an anthropological approach to analyzing the speech patterns of living languages. Later the influential American linguist Leonard Bloomfield brought these ideas to the forefront of American linguistic research, developing localist methods and notations for the study of exotic, unusual languages. Bloomfield's influential 1933 book *Language* dominated American linguistic thinking for over twenty years. Its overriding theme was the emphasis on objective methods of verification and precise techniques of discovery, as well as a refusal to admit discussion of meaning or mental entities or any other kind of unobservable features in the mind of the speaker. It's against this backdrop of behavioral psychology and logical positivism that the globalist school of linguistic research emerged.

In direct contrast to the localists' position, the globalist creed is that the important parts of languages are their similarities, not their differences. And the best way to study these similarities is by admitting the discussion of possibly unobservable mental structures giving rise to linguistic universals. Accordingly, the globalists emphasize a top-down research program focusing upon the abstract, syntactic structure of language per se, placing far less weight on the peculiarities associated with the concrete surface structure of any particular spoken language.

The modern era of globalist thought in linguistics research was dramatically ushered in with the publication of Noam Chomsky's *Syntactic Structures* in 1957. This electrifying event shifted the focus of linguistics virtually overnight from the observation and classification of the localists' "verbal botany" to a new vision of language as phonetics and semantics superimposed upon an underlying core of pure syntax. The principal goal of research now was to identify this core *universal grammar* from which all human languages get their start. In the globalist view, the universal grammar is something that is biologically present in the mind of all normal children as part of their genetic birthright. Since we'll consider Chomsky's program in painstaking detail later, it suffices for the moment to note that in addition to proposing the idea of a universal grammar forming the abstract structure upon which all languages are built, Chomsky also put forth in *Syntactic Structures* the radical notion that the grammar of each language must be *generative* in the sense that it must be a set of rules capable of "generating" all the well-formed (i.e., grammatical) sentences of the language and none of the ill-formed ones.

In addition to providing a set of formal tools and a theoretical framework for investigating the abstract properties of languages, Chomsky's work had the far-reaching effect of totally reorienting the primary direction of linguistic research. For the globalists, with their preoccupation with linguistic universals, the main focus now became not the speech patterns of adult speakers, but a deeper understanding of the process by which children come to learn their native language. In fact, it's fair to say that the globalists' program is directed toward answering

THE CENTRAL PROBLEM OF MODERN LINGUISTICS

How do children acquire the ability to speak their native language?

Upon first hearing this question, most people would dismiss the problem of language acquisition as no problem at all, saying that children obviously learn to speak by listening to their parents and older playmates. Unfortunately, this commonsense response just doesn't stand up to the test of observation, the main obstacle being what is often called the "poverty of the stimulus" problem. Since it forms one of the pivotal points of this chapter's debate, let's look at what the stimulus deficiency problem means in somewhat greater detail.

In broad terms, "the poverty of the stimulus" refers to the fact that, during the linguistically formative years, the child is not exposed to enough language to account for the linguistic capability displayed by a normal six-year-old. In short, children's ability to use their native language is vastly underdetermined by the data. There are several aspects to this underdetermination, each of which strongly suggests the need for something beyond mere exposure to account for the phenomenon of language acquisition.

First of all, the speech the child hears doesn't always consist of well-formed, complete sentences, but includes ill-formed sentences, partial statements, slips of the tongue, and other incomplete and/or ungrammatical utterances. Furthermore, children encounter only a finite range of expressions, yet come to be able to deal with an infinite spectrum of novel sentences going far beyond what they have ever heard before. Somehow the child acquires schemes for generating potentially infinite sentences such as "This is the dog that chased the cat that ate the mouse . . ." (relativization) or "Susan went home, and Jerry and Jane went out, and Carl slept . . ." (coordination) or "You heard that John asked me to tell Sam that he saw the house . . ." (subordination). You've probably never seen these sentences before and, in fact, it's likely that they've never before been written in any book until I made them up today. Nevertheless, you immediately understand the structure and meaning of them and, what's more important, so does your five-year-old child. So it can't possibly be the case that children learn their native language solely by imitation of what they hear. Finally, children come to know things subconsciously about their language for which there is no direct evidence in the data to which they are exposed. For example, children are not systematically informed that some hypothetically possible sentences do not in fact occur, or that a given phrase such as "I like her cooking" is ambiguous.

To summarize the relevant facts about language acquisition:

1. The child masters a rich system of knowledge without significant instruction.
2. This is carried out despite the poverty of the stimulus.
3. The process takes place most rapidly between the ages of two and three.
4. Normal human children are able to master any human language to which they are exposed in infancy.

An integral part of the globalist program is the assertion that any theory of human languages must provide an explanation for the above empirical facts, and that such an explanation will never be forthcoming by following the butterfly-collecting path of the localists.

By and large Chomsky's revolution has driven the verbal botanists underground, the main fireworks in contemporary linguistic research now centering upon his program. As we'll soon see, this program has many parts, some of which are technical, others psychological and philosophical. Moreover, Chomsky's own position has shifted somewhat over the past thirty years, so the program presented in *Syntactic Structures* no longer represents a totally faithful account of his vision. Nevertheless, certain key points have remained invariant, one of them being the dogmatic claim that the human language acquisition capability is attributable to a unique, genetically programmed part of the brain. It is on this point of innateness that many psychologists as well as linguists balk, and it is also at this point that Chomsky's views on linguistics most dramatically intersect with other philosophical, psychological, and neurophysiological questions of brain, mind, body, and thought. Thus, our debate in this chapter focuses on the problem of language acquisition as a vehicle with which we can enter into some of the central themes of what has come to be known as cognitive science. As is the custom, we begin the trial with the Prosecution.

THE NOAM OF CAMBRIDGE

During a stormy crossing of the Atlantic in 1953 on an old tub that had been salvaged after being sunk by the Germans during the war, a seasick, twenty-five-year-old graduate student from Philadelphia had an idea that was to initiate a bona fide Kuhnian revolution in the way we think about language. That youthful traveler was Noam Chomsky, and the idea he had was that the peculiarities of the biological structure of the human brain play *the* essential role in the ability humans have to communicate by means of language. As Chomsky now tells it, "I remember exactly the moment when I finally felt convinced," and at that moment he set out on a course of study emphasizing the key role of the mind and its mechanisms for making human language possible. In Chomsky's terms, the brain contains a genetically programmed "language organ" enabling human children to master their mother tongue with virtually no training or effort. Yet at the same time this organ defines and circumscribes the boundaries of all human languages, specifying what is and isn't possible by way of human linguistic communication.

As we've seen, prior to Chomsky's brainstorm, linguists didn't think of brain structure as playing any significant role in shaping human language. They thought of the mind as a tabula rasa capable of learning any kind of language whatsoever, and concentrated on isolating "discovery procedures" that would objectively describe the grammar of any human language. Although starting out himself on this same *structural linguistic* path while a student of Zellig Harris at the University of Pennsylvania, Chomsky, after several years of effort, came to the conclusion that some radical new notion was needed to understand the nature of human language. His insight on that fateful sea voyage was twofold: (1) recognition that the actual structure of the brain was crucial for explaining human language ability, and (2) recognition that the usual definition of grammar needed to be expanded to include all the rules and elements of language that children assimilate as they learn to speak and understand, as well as the linguist's theory of what goes on in the speaker's and hearer's brains.

In Chomsky's view, heredity must play an overwhelmingly important role in language because there is no other way to account for the facts noted earlier surrounding childhood language acquisition. In this genetically dominated picture, there are special neural circuits for the representation and use of language that interact with the child's linguistic environment, eventually evolving into a neurophysiological pattern specifying the grammar of the language that the child ends up speaking. According to this scenario, language growth is just one in the long series of genetically programmed changes a child goes through while maturing. Thus, just as the child is predetermined to pass through puberty or lose baby teeth in a certain genetically programmed manner, so it is with language as well, with the crucial changes taking place beginning around the age of two years and ending at about the onset of puberty. Among other things, this explanation of language accounts for the way children pick up a language as easily as the rest of us pick up a cold, while it's so painfully difficult for most adults to learn to speak in a foreign tongue.

The key element in Chomsky's biologically based theory of language is the idea of a universal grammar, which we briefly discussed in the last section. Upon first hearing this term, I conjured up the image of somehow taking all 2,796 human languages and throwing them together into a pot and boiling the whole mess down to its distilled essence, the residue remaining being the universal grammar. In some ways this image isn't too far off the mark, as the universal grammar does represent the totality of all the immutable principles of language that Nature builds into the language organ. But a better metaphor for the universal grammar would be that of a not completely specified electrical circuit. Figure 4.1 displays a simple passive electrical circuit consisting of a single resistor R, a capacitor C, and an inductor L, together with a voltage source. The way the circuit transforms a signal at its input terminal into an output pattern is governed by two factors: the way the circuit elements are connected, and the actual numerical values of R, L, and C. In the figure we see the circuit wired in two different ways, a pure series connection (a) and a series-parallel connection (b), with other combinations also possible. In linguistic terms, we could think of the input signal as corresponding to the child's linguistic experience, i.e., the external environment, and the output as being the language produced by the mental language organ. The

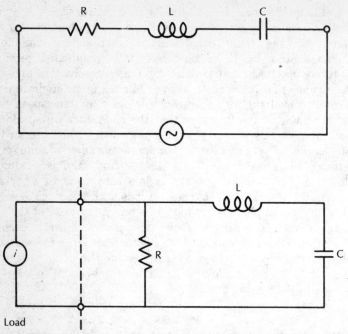

FIGURE 4.1. *Passive electrical circuits:* (a) *series,* (b) *series-parallel*

organ itself, the universal grammar, is represented by the components of the electrical circuit, together with the manner in which they are connected (*not* including their actual numerical values). Thus, the universal grammar is a set of preprogrammed subsystems of the circuit (the resistor, capacitor, and inductor together with their wiring pattern), but subsystems that are not programmed down to the last detail (the actual values of R, L, and C).

The idea is that the linguistic input the child experiences results in setting the parameter values of the universal grammar, thereby turning the language circuits in the brain into a language device suitable for producing and understanding the specific language corresponding to the particular parameter settings. The wiring pattern of the resistor, the capacitor, and the inductor corresponds to the biologically innate universal grammar. Fixing the actual values of R, L, and C corresponds to setting the "switches" of the universal grammar to produce a specific human language.

The universal grammar characterizes the abstract syntax of language, independent of the peculiarities and idiosyncrasies present in a given human speech community. Thus heredity provides the basic outline common to every language, the child's linguistic environment then filling in the details pertinent to the language being learned. While the universal grammar allows the learning of any human language, it imposes rather narrow limits on the possible ways that the rules governing each of its subsystems can interact. For instance, languages like Italian have what is called the *null subject option,* allowing statements such as "went" instead of "he went" or "she went." English has passed up this option. It is the collection of such options that constitutes the boundaries of the universal grammar, just as it is the collection of choices for the values of R, L, and C that constitutes what can be done with the "universal passive circuit." But it should be noted that the grammatical options, unlike their electrical counterparts, cannot be chosen freely; the grammatical options are interconnected, residing in a hierarchy where a choice at one level constrains what can be done farther on down the line. It's also of critical importance to observe that the universal grammar says nothing about the lexical facts of a language, but only about the form of the lexicon. Thus all considerations of word categories such as nouns and verbs are absent from the universal grammar. But it does contain principles about the assignment of semantic roles, cases, and so forth.

It's now easy for us to see why Chomsky's vision of universal grammars and biologically based language organs resulted in a total redirection of the lines of linguistic research. Rather than trying to construct the grammar of an actual language by assembling it piece by piece from individual observations, the Chomsky program turned the process totally upside down and started with the assumption that a universal grammar exists, with the parameters of the grammar fixed but unknown. The game is then to try to deduce these parameter settings from the observed grammar of the actual human language under investigation. So we see that the ultimate goal is the same—the characterization of the grammar of a real language—but the path is radically different: top-down instead of bottom-up. Moreover, the important questions now become matters involving the determination of the properties of the universal grammar rather than matters of collection and analysis of field data.

It's not surprising that Chomsky, who diametrically opposed their research strategies, didn't exactly capture the fancy of the localists, who reigned over the American linguistic community in the 1950s. As with most revolutionary insights, Chomsky's initial attempts to publish his ideas, both in the form of a summary article submitted to the prestigious journal *Word* and as a manuscript emerging from his Ph.D. work, met with a steady stream of rejections, mostly under the impetus of reviews from the Bloomfieldian old guard. Happily, through the influence of Roman Jakobson, one of the founders of the Prague school of linguistics and an influential member of the American linguistic community, a drastically slimmed-down version of Chomsky's vision was finally published by the small Dutch house Mouton under the title *Syntactic Structures* in 1957. Following a very favorable review of the book in the widely read journal *Language,* everything, as they say, hit the fan. Chomsky was immediately catapulted into a professorship at MIT and a position at the center of the academic and intellectual stage, both of which he occupies to this day.

Clearly the kind of furor that Chomsky stirred up in the linguistics world, with his speculations about how language acquisition works using an unobservable mental organ and a universal grammar, didn't arise solely because a new Ph.D. was putting forward a few wild fantasies. There was much more to it than that. So let's go exploring and look more deeply at the main features of this revolutionary research program and the array of technical ideas Chomsky introduced to support it.

As with most great intellectual breakthroughs, Chomsky's started when he looked at an everyday occurrence from a new point of view. He began by noting that there are certain properties of sentences that people intuitively know, but that can be explained only by employing the kind of deep principles of language that are known explicitly to linguists alone. The classic illustration is his sentence "Colorless green ideas sleep furiously," which every speaker of English perceives as meaningless, yet perfectly correct grammatically. Somehow there is an intuitive understanding at work here that tells us that the formal structure of the sentence (its syntax) is fine, but that the sentence is all form and no content. The point is that syntactic categories can be defined independently of meaning. Lewis Carroll's "Jabberwocky" poem is another classic example of this phenomenon (as are a depressing number of homework exercises from my students). In other words, the syntactic rules used to form the sentence exist independently of its semantic content. Chomsky boldly asserts that what's important about language is the understanding of these syntactic rules, and that such understanding will never come about from looking in an inductive fashion at just the utterances themselves. Instead, it's necessary to work deductively from a postulated set of rules, i.e., the universal grammar. Part of Chomsky's insight was also to recognize that he could cut the overall problem of grammar identification down to digestible proportions by invoking the simplifying assumptions that syntax could be studied independently of other aspects of language, and that linguistics could be pursued independently of other areas of cognitive science like psychology, neurophysiology, and logic. Later in the chapter we'll turn to a consideration of how well these simplifications hold up under detailed scrutiny.

Recall that for Chomsky the goal of linguistic research was to provide a framework for characterizing the rules that specify which sentences of a language are syntactically correct and which are not, i.e., to specify a grammar for the language that correctly distinguishes the "good" sentences in the language from the "bad" ones. In actual practice, Chomsky demanded much more than this. A key ingredient in his program was the claim that every such grammar would not only have to be a decision procedure for grammatical correctness, but it would also have to be a *generative grammar.* That is, the grammar would have to have the capability actually to generate all the well-formed sentences of the language and none of the ill-formed ones. Chomsky's approach was to present a sequence of increasingly powerful grammars, and to argue that only the last grammar on his list would serve as a viable candidate for the structure of a universal grammar. The three principal delicacies on this Chomskian menu are *finite-state, phrase-structure,* and *transformational* grammars. Let's look briefly at each of them in turn.

FINITE-STATE GRAMMARS

Using such a grammar, sentences are generated by a series of choices, one after another. First a word is chosen from a set of possible words, where the actual choice is dictated by a random selection weighted according to a given probability distribution. Then the second word is chosen, also according to a probabilistic weighting of choices, with a third word selected in the same manner and so on. Mathematically, this kind of sequential selection is termed a *finite-state Markov process* when there are only a finite number of sets from which the words are chosen. It is called a first-order process if the probabilities affecting the choice of a word depend only upon the preceding word, a second-order process if the probabilities depend upon the two preceding words, and an *n*th-order process if they depend upon the *n* preceding words, *n* being finite. An illustration of such a process is shown in Figure 4.2.

In the grammar of Figure 4.2, the node *I* is the starting node, while node *T* represents a termination node. In this elementary setup, there are only two grammatically correct sentences specified by the grammar: "The girl spoke" and "The men work," with the single probabilistic choice being made following the word "the." This primitive finite-state grammar starts in the initial state *I*, then moves to the state "the" with probability one, complete certainty. From the state "the," the grammar can move to either the state "girl" or the state "men," each with a probability that may depend upon the previous state, "the." And so it goes in this manner, with the grammar eventually generating the two grammatically correct sentences of this super-primitive "dialect" of English. It's clear from this simple case that a finite-state grammar without feedback loops can generate at most a finite number of sentences, hence could never serve to characterize the grammar of any human language. But what if we allowed loops? In that case it should at least be possible to produce sentences that are, in principle, of infinite length. Would this be enough to say that such a grammar is a viable

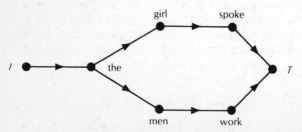

FIGURE 4.2. *A simple finite-state grammar*

candidate for describing the rules of some human speech community?

Figure 4.3 displays a finite-state grammar with loops. With such a grammar we can clearly generate sentences of infinite length. Nevertheless, it is still a finite-state device. Why? Because whatever state it's in (whatever box it's approaching), no matter what its previous states have been, the device still proceeds in exactly the same way. It has no way of "remembering" how many times it has visited a given state, since if it did such information would be part of its state and it would no longer be a finite-state device, as such loopings can, in principle, be carried out an infinite number of times.

In *Syntactic Structures,* Chomsky proved that a grammar based upon such a device cannot possibly characterize human languages since it's inherently incapable of accounting for nonadjacent dependencies. For example, the relationship between "toys" and "are" in the sentence "The toys in the store . . . are funny," where ". . ." represents an indefinite amount of material, cannot be handled by any kind of rule coming out of a finite-state grammar. Thus did the idea of a finite-state grammar fall by the wayside in Chomsky's search for the right abstract structure for his universal grammar.

PHRASE STRUCTURE GRAMMARS

With this type of grammar, we return to elementary-school days and the standard exercise of diagramming sentences by following a set of what linguists term *phrase structure rules.* Grade-school days notwithstanding, these rules are easy enough to understand, consisting of a set of statements such as:

- A sentence S consists of of a noun phrase (NP) followed by a verb phrase (VP).
- A noun phrase can consist of an article (Art) and a noun (N).
- A verb phrase can consist of an auxiliary phrase (Aux), a verb (V), and another noun phrase.

These rules can be represented compactly in "rewrite" form as

$$S \rightarrow NP + VP$$
$$NP \rightarrow Art + N$$
$$VP \rightarrow Aux + V + NP$$

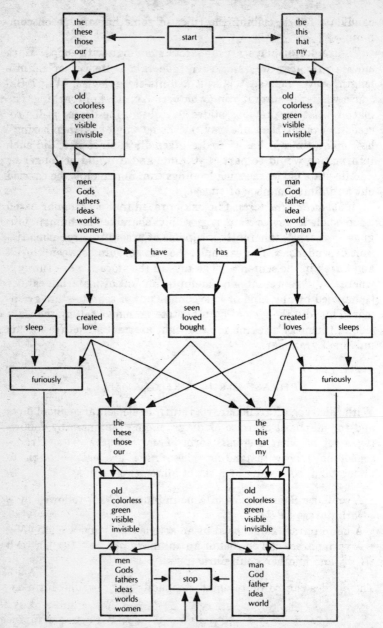

FIGURE 4.3. *A finite-state grammar with loops*

and so on. Thus, the grammar for a simple fragment of English might consist of rules of the above sort, which would be used to decompose a sentence into its atomic constituents. As an example, consider the decomposition:

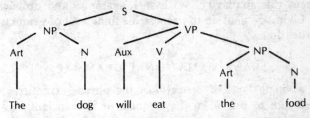

This derivation is Chomsky's description of the syntactic structure of the sentence "The dog will eat the food." It consists of a set of phrase structure rules, together with the lexicon stating what is to count as a noun, verb, auxiliary, and so forth. The "tree" sitting above the target sentence is often called a *phrase marker,* and is the structure resulting from applying only the phrase structure rules to the original sentence and inserting the lexical information.

Let's note in passing that we now see the difference between what is commonly termed grammar in everyday speech and what linguists think of as a grammar. Everyday grammar is nothing more than the rules of a phrase structure grammar, whereas a linguist's grammar is something far more general—*any* set of rules that is capable of generating the correct sentences and only the correct sentences of a human language, together with the rules for their interpretation and enunciation. With this not so crucial point settled, let's get back to Chomsky.

Rather soon Chomsky saw that phrase structure grammars with their single set of rules alone might succeed in characterizing the proper sentences of some language, but they could do so only with undo complication and at the expense of introducing an unwieldy number of rules. Moreover, with such a grammar there is no natural way to describe ambiguous sentences such as "I like her cooking." Phrase structure rules would provide only one parsing or "diagramming" for this sentence, but since the sentence is syntactically ambiguous, any decent theory of grammar should account for this fact by providing a number of syntactic derivations and descriptions. In addition, surface differences often conceal underlying similarities, as in the sentences "The dog will eat the food" and "The food will be eaten by the dog." These sentences mean the same thing, the only difference being that one is in the passive voice while the other is in the active voice. Phrase structure grammars give us no way to represent this similarity. With such examples and problems in mind, Chomsky went on to propose his final level of grammatical sophistication.

TRANSFORMATIONAL GRAMMARS

Since a single level of operations, the phrase structure rules, isn't enough to pin down the richness of human language, the "obvious" next step is to introduce a second level. Or so thought Chomsky when he proposed the idea of a type of transformation rule that would act to transform not grammatical categories like noun phrases, verb phrases, and the like, but rather the entire phrase marker itself. Thus, in transformational grammars there is a second set of rules that act on one phrase marker to transform it into another by actions such as moving elements around, adding elements, deleting elements, and so on. For instance, we can use Chomsky's transformational rules to display the similarity between the active and passive voices by showing how the active and passive can be converted into each other by transformations of the corresponding underlying phrase markers.

To illustrate the way such a grammar works, consider the phrase marker below, diagramming the sentence "All the boys might have gone with their parents," where the element Q represents a quantifier, P is a preposition, and PP is a prepositional phrase.

In a sentence of the above sort, the quantifier "all" has considerable flexibility as it can occur once for each subject NP. It

appears that no simple phrase-structure rule is capable of producing the full range of grammatical sentences that can come
out of the above initial phrase marker without producing ungrammatical sentences, too. For instance, the rule "Allow an optional occurrence of Q before each of the verbs" will correctly
produce a sentence like "The boys all might have gone with their
parents," but it will also produce the ungrammatical "All the
boys all might have gone with their parents," which comes from
a double usage of Q, an action admissible by the above rule.

What's needed here is a new kind of rule that can scan an
entire phrase-structure tree at one glance, and instead of just
marking a tree as grammatical or not, is capable of actually
transforming it into a new configuration. In our example, the
kind of rule that's needed is one like "Detach Q from within an
NP and move it to the left of any verb in the structure." Application of this rule to the foregoing tree yields:

Corresponding to the phrase structure rules and the transformational rules, respectively, are two subsystems constituting the
syntax of a language: a *base* subsystem and a *transformational*
component. The base subsystem contains the phrase structure
rules, which, in Chomsky's terminology, determine the *deep
structure* of every sentence. The transformational component of
the grammar then transforms the deep structure into its *surface
structure,* the level at which the phonetic components of the language take over to give the sentence a phonological structure. In
the early versions of Chomsky's theory, the deep structure was
also employed without benefit of transformation as input to the
semantic component of the language's grammar in order to give
the sentence its actual meaning. In later years, this firm distinction between syntax and semantics has been blurred somewhat,

even by Chomsky himself, although the relative importance of the deep structure for semantics is still a matter of hot debate.

It should be noted that the term *deep* in this Chomskian context has no bearing on matters of profundity, but signifies only the "hidden," purely abstract, syntactic structure of the sentence. Using his transformational rules, Chomsky was able to show that ambiguous sentences such as "I like her cooking" could be given a single surface structure from several deep structures, while semantically equivalent sentences of the sort involving just a change from active to passive voice could have different surface structures emerging from the same deep structure.

This summarizes the Chomsky revolution in linguistics in a nutshell. The whole program is set out in Table 4.1.

Another way of looking at Chomsky's work is to examine its implications not only for linguistics, but also for psychology and the philosophy of mind. Chomsky's principal claims in these areas are:

Psychology

- It makes sense to speak of abstract, possibly unobservable, mental entities.
- One of these mental "organs" is designed specifically for language.
- This language organ, the so-called universal grammar, is genetically determined.

Linguistics

- To discover grammars, studies must focus upon syntax.
- Any real grammar must be generative.
- The best candidate for the universal grammar is a transformational grammar.

I suppose it goes almost without saying that any list of revolutionary ideas as long as this is bound to come under attack from many quarters. And indeed Chomsky has been assailed not only by his linguistic peers, but also by psychologists, philosophers, and computer scientists, as well as a random assortment of the other fauna inhabiting the intellectual zoo. Since the main objective in this chapter is to look at the problem of language acquisition, we'll center most of our attention upon the first half

	LOCALISTS	GLOBALISTS
SUBJECT MATTER	body of utterances	speakers' knowledge of how to produce and understand sentences; their linguistic competence
GOAL	classification of the individual elements composing the body of utterances	specification of the grammatical rules used for constructing sentences
METHODS	discovery procedures	investigation of the properties of the universal grammar

TABLE 4.1

of the above list, the problems of mind. In Chomskian terms, our interest is in the way the child acquires his *knowledge* of language, rather than how he demonstrates his *competence*. As a result, most of the opposing positions scouted below will be objections to one or another of the items on the "psychology" list. However, since Chomsky has chosen linguistics as the specific arena in which to defend his theories of mind, we will necessarily touch upon a few linguistic objections to his claims as well.

At this point the reader may well be wondering whether Chomsky is the only scholar in the field. Surely there must be other thinkers of sterling reputation whose work parallels Chomsky's but with interesting differences. And indeed there are. So why haven't I told you about them? The answer to this eminently sensible query is quite simple. In virtually no other area of modern intellectual life that I'm aware of has one man's work so completely shaped a field as Chomsky's views have done in linguistics. The ideas and programs sketched in this chapter are but the tip of the iceberg set in motion by his vision, and there is no better example in the second half of this century of a true Kuhnian paradigm shift than what has been brought about in linguistics by Chomsky's efforts. Consequently, to speak of linguistic thought from the 1960s onward is really to speak of the concepts, ideas, and techniques introduced by Chomsky. As far as I can tell, *all* work in mainstream contemporary linguistics involving language acquisition is directed toward either providing support for or looking for holes in the claims he has made. So in a very specific sense, Chomsky's views *define* what we mean by much of modern linguistics, in much the same way that Newton's ideas defined classical particle mechanics.

Now let's give the floor over to the Defense to put forth its array of arguments designed to convince us that Chomsky's ideas about language and mind don't merit much attention after all.

POSITIVELY REINFORCING

New York's Greenwich Village has always been a haven for aspiring artists, writers, and other intellectual hangers-on, at least until the recent influx of wheeler-dealer yuppies, trendy restaurants, and chic boutiques, along with their consequent explosive effect on the costs of maintaining even an artist's garret. But in the Roaring Twenties gentrification had yet to strike the Village, and an ambitious young writer from Pennsylvania was encouraged by the famed poet Robert Frost to strike out for the literary life and try swimming with the sharks in what was even then the highly competitive world of publishing. As in most cases of this sort, after a couple of years of rejection slips the romance of living by the pen gave way to the realities of eating regularly, and the writer-to-be traded in his Village pad for the comforts of Harvard Yard and the pursuit of a graduate degree in something useful. The literary world's loss was the psychological community's gain, as that young man, B. F. Skinner, went on to found a school of psychology that dominated American thinking on matters of mind for more than two decades.

In the early 1920s, John B. Watson made the radical suggestion that human behavior does not have mental causes. Stimulated by the ideas of logical positivism, this thesis holds that in studying behavior all notions of mind, mental states, and mental representations should be eliminated, and investigations focused solely upon externally observable *stimulus-response* behavior patterns. At this time the topic of the day in psychological circles was the understanding of the learning process, and most psychologists were operating under the paradigm established by the Russian Ivan Pavlov, whose experiments with drooling dogs will undoubtedly strike a responsive chord (dare I say ring a bell?) with the reader. In his influential 1925 book *Behaviorism,* Watson made the infamous claim:

> Give me a dozen healthy infants, well-formed, and my own specified world to bring them up in and I'll guarantee to take any one at random and train him to become any kind of specialist I might select—doctor, lawyer, artist, merchant-chief and, yes, even beggar-man and thief, regardless of his talents, penchants, tendencies, abilities, vocations, and race of his ancestors.

In these few words Watson established the basic elements of behavioral psychology: elicitation of any desired behavior solely by externally applied stimuli and responses, coupled with positive and negative reinforcements or, more prosaically, rewards and punishments. It was into this psychological mindset that Skinner dropped when he entered the Harvard Psychology Department in 1928.

The behaviorist programs of both Pavlov and Watson asserted that learning takes place as a result of environmental stimuli experienced by the organism, which then responds in various ways. The responses that either the experimenter or Nature rewards are reinforced, and the maladaptive responses are soon weeded out by punishments. It was in exactly this fashion that Watson envisioned training his collection of a dozen infants to become lawyers, doctors, beggar-men, or thieves. Unlike most academics, Watson evidently felt confident enough of the soundness of his ideas that shortly after publication of his 1925 book on behaviorism, he left his professorship at Johns Hopkins for a life in the business world, trying to transmute the lead of conditioned behavior into the gold of the marketplace. Fittingly enough, he chose a profession to suit the task, spending the remainder of his years in the advertising business!

Although Skinner is generally regarded as a behaviorist cast from the mold of the stimulus-response school, he departed in crucial ways from the program laid down by his predecessors. In the learning theories of both Watson and Pavlov, the process unfolds in a fixed sequence: first stimulus, then response. Following the response, the behavior being conditioned is rewarded while nonconditioned behavior is punished. Skinner objected that such a theory of learning could never account for the twin problems of novelty and purpose in the response.

How did Tolstoy come to write *War and Peace?* How did John Lennon and Paul McCartney write those great old Beatles hits? And how did Bobby Fischer ever find those dazzling combinations on the chessboard? The classical behaviorist has only the feeble explanation that these tasks can each be broken down into a series of small behavioral units, all of which exist initially as unconditioned responses. These individual units are then brought out as a coherent whole by a mysterious and equally coherent sequence of individual stimuli.

Such thinking creates a similar problem in explaining a subject's purpose. The stimulus-response behaviorist is unable to talk about the goals and consequences of behavior. The analytic framework is limited to a discussion of stimuli *followed* by behavior, and there's just no room for descriptions of the consequences of actions. But it seems highly implausible that even the most detailed account of a set of stimuli preceding a bicycle ride around the lake will explain what the rider is doing and why. In the final analysis, classical behavioral psychology fails to offer a framework for understanding human actions that can peacefully coexist with a belief in the psychological reality of innovative and goal-oriented behavior.

In Skinner's *radical behaviorism,* he attempts to address these difficulties by eliminating the notion of antecedent stimuli, replacing it with the idea of *operant* behavior, i.e., behavior acquired, shaped, and maintained by stimuli occurring *after* the responses rather than before. At the same time, he argues that only positive reinforcement produces behavior that leads to a more satisfying life, dropping the kind of *Clockwork Orange-*style aversion therapy inherent in the work of Pavlov and Watson. Thus, for Skinner desirable behaviors are rewarded after they are performed, and the reward is what acts to enhance the likelihood of that behavior's being repeated.

It's easy to see that Skinner's idea of operant behavior is the psychological analogue at the individual level of biological evolution at the level of the species. In Skinner's scheme, "good" behavior is reinforced, just as in Nature, "good" mutations are selected. But in both cases, there is no reinforcement until after the action has taken place. Operant conditioning is designed to explain the emergence of novel behavioral patterns in the individual in the same way that natural selection explains the emergence of new traits in a species. In both cases, the role of the environment is more to select than either to reward or punish, although Skinnerian reinforcement can be likened to a reward since it encourages continuation of certain types of behavior just as natural selection encourages certain types of mutations. Note, however, that despite his major departure from the learning schemes advocated by Pavlov and Watson, Skinner still retains the keystone in the behaviorist arch: the inadmissibility of any notion of a nonphysical mind, mental states, or mental entities in the scientific explanation of behavior.

Skinner, of course, is well known for the many ingenious experiments he set up to try to validate his behaviorist theories. For instance, during the Second World War he devised a kind of pigeon guidance system for targeting bombs. This unlikely system involved placing trained pigeons inside the warhead with a map depicting the terrain to be bombed. The pigeons would peck at the map in all the right places to activate a steering mechanism, with the right sequence of peckings presumably ensuring that the bomb would be brought back on target if it started to veer off course. Another one of his more widely reported schemes was the so-called air crib, a variation of his famous Skinner box used to train the pigeons. In the crib, a kind of glass-enclosed box, the interior was carefully regulated to be at just the right temperature and humidity to form an ideal atmosphere for an infant. Furthermore, germ filters cleaned the air, doing away with the need for blankets, clothing, and frequent baths. The crib also contained a variety of equipment to keep infants amused and well exercised. Skinner tested the device by placing his own daughters in it, which at the time (1945) caused quite a public flap. Contrary to sensationalist reports, the girls, who now both lead rather normal, well-adjusted lives, did not become suicidal or psychotic, and both report that they think the experience was beneficial.

Given his predilections for seeing operant behavior in every corner of life, it should come as no surprise that Skinner has devoted a considerable amount of his impressive reservoirs of intellectual and polemical energy to the problem of language learning. He is particularly interested in this question since he holds the view that language and self-knowledge are intimately intertwined. In the Skinnerian view, all words are acquired on the basis of the "law of effect," i.e., by rewarding, ignoring, or correcting the performance of novices by more mature users of the language. As a result of the way the human brain is structured to learn, a child comes to identify its pet dog with a word such as "Spot," with this identification taking place through a sequence of positive reinforcements from parents and older friends who have developed a more mature use of the language. So in the Skinnerian version of language acquisition, language is learned in exactly the same way (operant conditioning) and with exactly the same psychological mechanisms (unspecified) as the child learns any other skill, such as bicycle riding, tying shoelaces, or telling time.

The behaviorist position of Skinner raises the question about how the child learns words for private events. Such events cannot be reinforced by external means like pointing to an object or showing a picture in a book, yet must somehow come to be understood by the child in the same sense as they're understood by the rest of the speech community. Skinner's answer to this dilemma is to skirt the issue by asserting that teaching words for private events is akin to trying to teach color words in a world of partially and unpredictably color-blind people. His claim is that our confidence in the reliability of our inferences about private events is based upon observable behavior, and we simply cannot be sure that people use the language of private events to mean the same things.

The radical behaviorist view of language acquisition was put on record in Skinner's 1957 book *Verbal Behavior.* Noam Chomsky's scathing review of this book in the prestigious journal *Language* in 1959 gave Chomsky his first widespread recognition as an opponent to the empiricist claims of most scientists of that era. With considerable relish, Chomsky argued that the behaviorist conception of language acquisition cannot possibly be correct, and that "with a literal reading . . . the book covers almost no aspects of linguistic behavior, and that with a metaphoric reading, it is no more scientific than the traditional approaches to this subject matter, and rarely as clear and careful." Later Chomsky expanded his attack on Skinner's ideas by stating that

Skinner's approach has led absolutely nowhere. . . . It has yielded no theoretical knowledge, no nontrivial principles as far as I am

aware—thus far, at any rate. . . . Skinnerian behaviorism is off
the wall. It's as hopeless a project as trying to explain that the
onset of puberty results from social training.

The essence of Chomsky's critique is that the learning process as
Skinner describes it is at crucial points left to vague notions like
"analogy" and "generalization," notions that are inherently incapable of offering any sort of explanatory power.

Skinner never responded to this savage review of his life's research program, although to this day he maintains that psycho-
therapists and psychologists rely too much on inferences they
make about what is going on inside their patients' heads, and too
little on what the patients are actually doing. He further contends that "I think cognitive psychology is a great hoax and a
fraud, and that goes for brain science, too." Nevertheless, most
researchers tend to feel that Chomsky's review of *Verbal Behavior* sent behavioral psychology into a tailspin that it may never
pull out of. The reason is one that Skinner himself would surely
approve: Behaviorism just isn't very reinforcing nowadays. So
with the fading of Skinnerian visions from the psychological
stage, let's move across the ocean to the land of the cuckoo clock
and chocolates to hear from our next witness against Chomsky.

OUT OF THE MOUTHS OF BABES

In 1918 a small Lausanne publishing house brought out the
novel *Recherche,* an account of the conflicts felt by a young
Catholic over the relationship between science and religion. Com-
mercially the book sank like a stone, a fact that the young author later probably had little cause to regret. However, leaving
its dubious literary merits aside, *Recherche* deserved a better
fate if for no other reason than that its detailing of the relationship between the part and the whole in organic life represented
the initial public glimpse of the thoughts that later led its author, Jean Piaget, to play a founder's role in the development of
what is now termed cognitive psychology. So as with Skinner, a
small loss to the world of letters became a giant gain to the
world of science and the study of the mind.

As a youth growing up in the Swiss town of Neuchâtel, Piaget
was a passionate collector of shells, fossils, and other such ef-
fluvia of nature, an interest that led to an unofficial position as

assistant curator of the Neuchâtel natural history museum at
the precocious age of ten. As a consequence of his childhood biological obsessions, the young Piaget developed a lifelong focus
on the structure of organisms, as well as a deep attraction to the
philosophy of Henri Bergson and its attempt to fit questions of
mind, matter, science, and soul into an overall, integrated world
view. During this period of youthful contemplation, Piaget had
already begun formulating the notion that all organisms consist
of parts related to the whole, and that all knowledge derives
from the assimilation of external experiences into the organism's
structure. Piaget's key idea involved a comparison between mental processes and the body. Just as the body requires balance
and self-regulation in all of its biological functions, so too does
mind require equilibration of its intellectual levels. Hence by the
time he was out of his teens, Piaget's lifetime research direction
was already set: to explore the interrelationship between biology
and logic, with the workings of the human mind as the bridge
linking the two.

Upon completion of his studies, the specific vehicle with which
Piaget was to pursue his grand research plan emerged from a
job he was offered as an assistant to an assistant to Alfred
Binet, the developer of the IQ test. Piaget was hired to stan-
dardize some of the tests, and during the course of his work he
noted that the kinds of mistakes that children made on the tests
were not random but tended to fall into definite categories, depending upon the age of the child. Rather than dismissing this
observation as a statistical irregularity, Piaget conjectured that
it was a sign that qualitatively different structures of intellect
were present at different stages of the child's cognitive development. The pursuit of this theme was to occupy Piaget for the
rest of his life.

The pivot around which all of Piaget's ideas revolve is his vision of the mind as not just a passive device for handling sensory inputs, but a mechanism that actively transforms the
inputs it receives by performing exploratory operations upon
them. Thus, Piaget thought of human intelligence as a process
of reality construction rather than as a passive receiver and
processor of information from the outside world. A core ingredient supporting this concept of the active, exploring mind is the
idea of an internal mental representation. Piaget, unlike Skinner, felt that to postulate such unobservable, even hypothetical,

internal mental states was a necessary step on the path toward providing explanations for mental development. Furthermore, he felt the introduction of such entities into psychology was no more an obstacle to making the study of mind "scientific" than the introduction of concepts like neutrinos and electrons was a barrier to making physics "scientific." In fact, if we wanted to pinpoint the precise moment when the "cognitive revolution" in psychology began, we can probably do no better than to mark the day when Piaget pushed forward his claim for mental representations as valid objects of study in the creation of a science of human thought.

Shortly after completing his work in Paris, Piaget was offered the position of director of the Rousseau Center for "genetic psychology" in Geneva, where he spent the remainder of his long and fruitful career. Upon arrival in Geneva he quickly inaugurated a program of research on the intellectual development of children, using many ingenious experiments to identify the various stages in his theory. According to Piaget, the child goes through at least four main stages in mental evolution from a little savage to a more or less right-thinking adult. These qualitatively distinct stages are:

- *Sensorimotor stage: birth to two years.* This is the period when infants construct the concepts of an object, space, and causality. This involves an increasingly coordinated linkage between perception and action. For example, the child's perception of objects like a doll or a rattle becomes synonymous with the actions that can be performed upon them, such as shaking the rattle or holding the doll.
- *Preoperational stage: two to five years.* At this time the child's thought processes begin to use symbols in the form of mental images arising from imitation or words. During this period reasoning from memory and analogy also begins to occur, as does the development of language skills.
- *Operational stage: five to ten years.* In this stage the child performs mental operations on objects that are physically present. Classification of hierarchical structures occurs, as does the understanding of ordinal relations. Near the end of this period, the concept of conservation of continuous properties like weight, quantity, and volume emerges, so that the child

begins to recognize that there is not less liquid present when water is poured from a tall, narrow tube into a short, flat bowl.

• *Formal operations stage: ten to fourteen years.* At this time the real world is conceived of as a subset of possible worlds. Propositional thinking, with assertions and statements that can be true or false, becomes possible, and there is a better grasp of the fact that appearances can be deceiving.

Just as the acquisition of language involves progression through a set of strictly ordered stages, Piaget held that general intellectual development follows the path outlined above, and that, like the Stations of the Cross, each of the steps must be passed through with no omissions or change of order.

As to exactly how experience was processed within each stage to generate knowledge, Piaget advocated a two-pronged theory in which the child pits the antithetical processes of *assimilation* and *accommodation* against each other in a kind of dynamic arm wrestling. Assimilation, for the child, involves trying to fit novel aspects of reality into old behavioral and cognitive schemes rather than changing them. On the other hand, accommodation requires changing an existing mental or behavioral pattern to adapt it to the specific characteristics of new objects and new relationships, in this way taking account of novel aspects of reality. The tension between these two approaches to dealing with novelty in the environment is then resolved by what Piaget called *equilibration,* a type of dynamic steady state that balances out the competing forces. This again calls to mind Piaget's initial preoccupations with the process of autoregulation in biological systems. Now let's try to put these rather general ideas about learning and mental development into the specific context of language acquisition.

In the epistemology of Piaget the child does not come "hardwired" to understand concepts, but has to create them as in his construction of the ideas of space, time, conservation, and so on. In this framework, the environment provides feedback about the quality of the mental structures the child creates; it does not simply imprint the right structures on the mind. Thus, for Piaget the world is not just "out there" waiting to impress itself on a blank slate. Intellectual development is a constant interplay between the child and his environment, with the child playing an active, structuring role. Moreover, the Piagetian sees all areas of mental development as being closely interconnected with each other. So as far as language acquisition goes, Piaget sees it as all of a piece with the other stages of intellectual growth, and he places no particular emphasis on language as opposed to the other skills the child learns. As a result, the Piaget school contends that the mind develops more as a whole across a spectrum of intellectual tasks than as a modular structure.

Since the Piaget position differs from both Skinnerian behaviorism and Chomskian rationalism in a number of interesting ways, let's briefly summarize the differences. The major points of each position are shown in Table 4.2.

Since the differences telegraphically noted in the table are so central to the entire issue of cognition, we'll defer detailed discussion of them to a later section where the problems and possible rapprochements can be given the attention they properly deserve. But first let's have a short intermezzo and look at some of the complaints that have been registered in the linguistics community against the syntax-dominated position of Chomsky.

IT'S ALL A QUESTION OF SEMANTICS

One of the most surprising outcomes of Piaget's research program was the discovery that the earliest function of speech is not communication, but symbolization. Thus, the first entities that the child perceives and that stand for a certain content or meaning are private symbols. These lead to internalization and representation of thoughts, with social communication arising only at a later stage. So in this view language is more of a technique or strategy for structuring thought than a vehicle for communication. This discovery is entirely consistent with Chomsky's idea of a universal grammar and his focus on syntax as the real core of language. But the deemphasis of content at the expense of form has not always occupied a favored position in the linguist's order of things and, in fact, has had to be toned down somewhat from Chomsky's original proposals. To see how things stand today, let's go back for a moment to the days when meaning was still king in the world of linguistics.

Benjamin Whorf was a chemical engineer by training, a linguist by avocation. While spending his entire professional life as a fire inspector for a large Hartford insurance company, prior

	MENTAL STATES?	LANGUAGE ORGAN?	ENVIRONMENT/HEREDITY?
Chomsky	yes	yes	heredity
Skinner	no	no	environment
Piaget	yes	no	both

TABLE 4.2. *Positions on mind and language*

to his untimely death at the age of forty-four, Whorf served as a striking example of the gifted amateur competing on equal terms with the professionals by devoting his spare time and energy to a detailed study of the languages of the American Indians, particularly the Hopi of the American Southwest. In these efforts, Whorf was following in the wake of his teacher Edward Sapir, an anthropologically oriented American linguist of the pre-Chomsky era, who advocated the position that one's view of the world is strongly shaped, if not totally created, by language. This claim calls to mind the contention of the later Wittgenstein in his statement that "the limits of my language mean the limits of my world." This argument was expanded upon by Sapir when he stated:

> . . . the "real-world" is to a large extent unconsciously built up on the language habits of the group. No two languages are ever sufficiently similar to be considered as representing the same social reality. The worlds in which different societies live are distinct worlds, not merely the same world with different labels attached.

The linguistic ideas of Sapir and Whorf have come to be enshrined in what is now termed the Sapir-Whorf Hypothesis, consisting of two main assertions relating language to thought:

SAPIR-WHORF HYPOTHESIS

- *Linguistic determinism:* Language determines the way we think.
- *Linguistic relativism:* The distinctions encoded into one language are not found in any other language.

The famous example of the Eskimo language, which has separate words for falling snow, snow on the ground, snowed packed hard like ice, slush, and so forth, illustrates the point.

The fact that translations from one language to another can be made, as well as the fact that the conceptual uniqueness of a language like Eskimo can still be explained using another language like English, makes it unlikely that a strong form of the Sapir-Whorf Hypothesis is correct. While it's undeniable that there are conceptual differences between languages due to cultural and environmental factors, this does not necessarily imply that the differences are so great that mutual comprehension is impossible. It's always possible to use various types of circumlocutions to say in many words in one language what can be said more compactly in another. As an example, consider the diagram in Figure 4.4 displaying the ways of saying "He invites people to a feast" in English and Nootka, an Indian language of the Pacific Northwest. Nootka is able to express in a single word an idea that in English requires a far more elaborate construction.

Even though a strong form of the Sapir-Whorf Hypothesis seems unlikely to be true, a weaker form is probably valid, asserting that language does affect the way we perceive and remember, as well as facilitate the performance of mental tasks. If so, the weak form of the Sapir-Whorf Hypothesis might lead us to speculate that there's a lot more feasting going on among the Nootka than the British, in light of the relative ease of issuing the requisite invitation in Nootka.

With these kinds of ideas, the feet of both Sapir and Whorf are firmly planted in the localist school of linguistics with its emphasis on the differences in languages as being of paramount importance. And these differences center upon matters of meaning, i.e., semantics. This, of course, is exactly the situation whenever we study a literary text, as by definition the reader and student of literature "work at the surface," as noted by the literary critic and language scholar George Steiner. Such texts deal with phonetic and semantic facts, the words and sentences that we can actually see and hear. That is the only reality available to us, so on the surface we are all ultra-Whorfians. The transformational grammarians assure us that the surface presence of the text is merely an external product emerging out of deeper structures, and that to understand language it's necessary to descend to these primal levels. In short, Chomsky's not so tacit assumption that syntax can be studied profitably detached from semantics comes under a dark cloud. Let's briefly consider a couple of the more interesting objections and the responses to them.

* * *

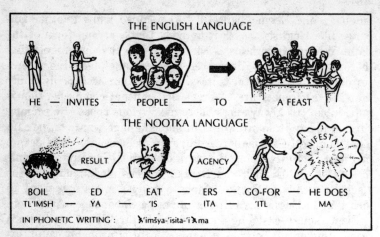

FIGURE 4.4. *An invitation in English and Nootka*

Recall that Chomsky's so-called Extended General Theory comprises the sequence shown in Figure 4.5, where we see the surface structure emerging out of the deep structure, each being processed by both the phonetic and semantic rules to generate what we think of as everyday speech. This picture clearly shows the role of both phonology and semantics as logically following syntax in Chomsky's world.

The most obvious line of attack against this picture is to argue that there is no clear-cut distinction between syntactic and semantic rules; hence, the level of syntactic deep structure cannot be defended. This is the position of the so-called *generative semanticists,* who have tried (rather unsuccessfully) to set up syntactic-semantic rules that take semantic representations as their input and yield surface structures as their output, using no intervening level of deep structure. A closely related idea is that pursued by the *interpretive semanticists,* who argue for moving more and more of the syntactic rules into the semantic component, thereby moving the deep structure closer to the surface structure of the language. Let's see how one variation on this basic theme offers the promise of patching up at least a few of the holes in the Chomskian facade.

Part of the problem with the Extended General Theory was a 1971 mathematical result produced by Peters and Ritchie showing that the original transformational grammars are just too

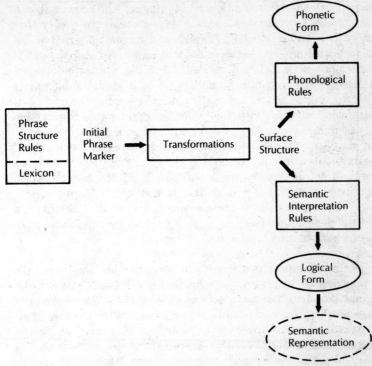

FIGURE 4.5. *Chomsky's Extended General Theory*

general. This theorem demonstrated that any language whose sentences could be listed mechanically could be generated by some Chomskian grammar. Thus, Chomsky's claim that natural languages had transformational grammars essentially amounted only to the claim that they could be characterized mathematically. A major part of the difficulty is that the Chomskian grammars do not necessarily provide a mechanical decision procedure for the grammaticality of the sentences of the language because they are so general. What this means is that while sentences can be mechanically generated by such a grammar, the grammaticality of any sentence given *in advance* cannot be decided by applying the rules of the grammar. Or at least, grammaticality cannot be decided by any procedure guaranteed to terminate in a finite number of steps. As a consequence of results of this type, interest in transformational grammars waned in the 1970s, only to be reborn with the work of Richard Montague, who showed that it was possible to associate an equally explicit semantic theory with the syntax. In his words, "There is in my opinion no important theoretical difference between natural languages and the artificial languages of logicians; indeed, I consider it possible to comprehend the syntax and semantics of both kinds of languages within a single natural and mathematically precise theory." This manifesto, together with the theoretical framework supporting it, has pumped new life into the area of generative grammars, but now with syntax and semantics coexisting on a more equal footing. Since this is neither the time nor the place for an account of the highly mathematical details of Montague's work, let's instead look briefly at another competitor to Chomskian visions, a viewpoint whose ideas are in harmony with Montague's and which combines features of both Chomsky's and Piaget's positions but without totally supporting either.

Geoffrey Sampson is a British linguist who has challenged the notion that there are as many linguistic universals as Chomsky claims. However, Sampson does agree with Chomsky on the existence of at least one such universal, the hierarchical nature of all languages. And he proposes a theory of language acquisition that he feels accounts for this universal feature without having to invoke the specialized language organ that Chomsky so cherishes.

Sampson's argument is based upon a parable first introduced by Herbert Simon to explain why all complex systems generally seem to display a hierarchical structure. Simon considered the assembly of a watch consisting of ten subassemblies, each of which consists of ten individual components. Assuming that the watchmaker is periodically interrupted in his task of assembling the watch, with each such interruption necessitating his starting from scratch on the construction of the part of the watch he is at that moment putting together, Simon shows convincingly that with even a minuscule chance of interruption the watchmaker will never finish assembling the watch if it's regarded as a single object of one hundred pieces. On the other hand, the chances of finishing are excellent, even with many interruptions, if the watch is divided hierarchically into subassemblies and the watchmaker has only to put together the subassemblies to make the final product. This so-called Watchmaker Parable forms the heart of what Sampson claims is a major improvement upon Chomsky's ideas.

Applying the Watchmaker Parable to syntactic structures, Sampson argues that the communication system of our ancestors presumably consisted of words and short sentences, and that language users occasionally hit on new combinations of phrases to produce slightly longer sentences than had earlier been the rule. There are then two ways the new sentences could become entrenched in the language: (1) a new sentence might effect the transfer of enough useful information sufficiently often that there was some selective advantage for the organism to transmit just that sort of information, or (2) a new sentence might put simpler grammatical elements together in new and more complex ways resulting in a linguistic innovation—i.e., it would represent a new semantic category not present in any of its parts taken individually. In this way, Sampson argues, a child can acquire language in just the same manner as the watchmaker puts together the watch—by composing subassemblies from individual components, and then putting together the subassemblies. In experiments by Berlin and Kay involving the way color words like "red" and "yellow" are learned, it was found that learning followed an evolutionary sequence in all languages regardless of their grammars, a strong point in favor of Sampson's theory.

This idea is completely consistent with the way a complex expression is generated using a Montague grammar. In such a grammar, we begin with the lexical items and "assemble" the lower-level structures. Within these new structures, it's possible to discern the earlier items whose syntactic combination in accordance with Montague's rules involves only minor peripheral modifications. The new lower-level structures are then themselves syntactically combined, again with peripheral modifications, into yet higher-level ones, and so on. Finally, the sentence itself is produced. An example of the kind of tree that comes out of a Montague grammar is shown in Figure 4.6 for the sentence "Every man loves a woman such that she loves him."

Substitution of a Montague grammar for a Chomskian one, together with Sampson's account of the origin of hierarchical structure in languages via the Watchmaker argument, leads to a theory of language acquisition that can dispense with the innate language organ. Instead we require only the kind of general problem-solving capability promoted by Piaget, with the child

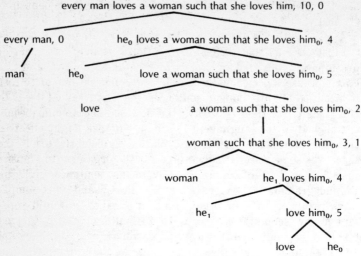

every man loves a woman such that she loves him, 10, 0

every man, 0 he_0 loves a woman such that she loves him_0, 4

man he_0 love a woman such that she loves him_0, 5

love   a woman such that she loves him_0, 2

woman such that she loves him_0, 3, 1

woman he_1 loves him_0, 4

he_1 love him_0, 5

love he_0

FIGURE 4.6. *A Montague tree*

subconsciously and implicitly testing various hypotheses about grammaticality against the actual linguistic data encountered. The assumption that ensures that the child will home in on the right hierarchically structured language is the presupposition that the child's program of exposure and hypothesis testing will take place in a linguistic environment where the local language has precisely this "right" structure. Ergo, by following a Popperian program of conjectures and refutations, the child arrives at the right set of rules.

Before closing this discussion of semantics and Montague grammars, let's note one further point: The Montague grammars are capable of characterizing at most *context-sensitive* languages, for which it can be shown that a mechanical decision procedure does exist for establishing the grammaticality of a given sentence. Recent work by Gerald Gazdar and others indicates that natural languages like English are a subclass of these, termed *context-free* languages. What this means is not that the sentences have a meaning independent of the context where they are used, but that the phrase structure rules are formulated in such a way that a category label can be rewritten without regard to the surrounding context of words in the sentence. In other words, the way the phrase structure tree branches depends only on what the situation is at the branch point, and not on what lies on other branches of the tree. Thus, because Montague grammars are *decidable* (i.e., they possess a decision procedure for grammaticality), contain only phrase structure rules and no transformations, treat syntax and semantics on the same footing, and are harmonious with an evolutionary view of linguistic development, they can be taken seriously as an alternative to the Extended General Theory of Chomsky.

Following this brief appearance on the stand by Chomsky's linguistic opponents, let's get back to the philosophers and psychologists and examine a few of the arguments put forth against his theory of mind.

SHOOT-OUT AT THE ROYAUMONT CORRAL

A few miles down the road to Mexico from Tucson, Arizona, lies the ghost town of Tombstone where in 1881 at the famous O.K. Corral the Clantons and the Earps, with a little help from Doc Holliday, fought the most famous gun battle of the Old West, earning Tombstone the sobriquet "The Town Too Tough to Die." To this day, twice a month members of a local community group strap on their six-shooters and reenact this famous gun battle for the benefit of tourists like myself, who yearn to feel even for a moment a bit of the excitement and lawlessness of those legendary times. During a recent visit to this living monument to the past, I was quenching my thirst with a beer at the Bird Cage Saloon following the festivities at the corral, when the thought struck me that perhaps the conflict resolution methods of riproaring Tombstone and those of the modern-day intelligentsia have a lot more in common than most of us realize. Other than the admittedly nontrivial difference that the academic and intellectual losers don't expire from lead poisoning, the similarities are striking: diametrically opposed forces clashing in a public arena (scholarly journal articles and open lectures), hotshot young challengers looking to make a name for themselves by outdrawing the top gun, even a version of the Boot Hill Cemetery for those whose ideas bite the dust at high noon on Main Street (professorships in academic "Siberia"). Looked at in these terms, one of the most talked-about and intellectually violent showdowns in recent times took place not on the dusty streets of an Arizona boomtown, but in the august halls of a luxurious French château when in October 1975 Noam Chomsky rode in from Cambridge, Massachusetts, to do battle face to face with Jean Piaget.

The Centre Royaumont pour une Science de l'Homme is located just outside Paris in an elegant château of the type that would make any royalist's heart flutter. At the time of the debate, the now-booming subject of cognitive science was only beginning to emerge out of its parent disciplines, and the center found the biological ideas of Chomsky and the cognitive perspective of Piaget to be central to a proper understanding of the mind and its workings. As a result of the enthusiastic encouragement of its president, the famed biologist Jacques Monod, the center's staff arranged for a constellation of biologists, computer scientists, psychologists, and philosophers to bear witness to the struggle of the two titans, as well as to provide dissenting views from the chorus. The mainstream of popular psychological thought prior to the Royaumont gathering centered upon three principal themes: psychoanalysis, behaviorism, and classical learning theory. Significantly, at Royaumont not one of these traditional areas of mind was represented, leading more than one observer to date the coming of age of cognitive science to this unique conclave.

The real issue before the house at Royaumont was the interplay between the question of the nature of various vehicles of knowledge, such as images, signs, and schemata, and the problem of whether knowledge is inborn as Chomsky claims, or constructed through interaction between certain inborn modes of information processing and the actual characteristics of the physical world as asserted by Piaget. Of course the positions are not so clear-cut as this, as indicated by a remark of Monod's: "In asking myself the question, 'what makes man man?' it is clear that it is partially his genome and partially his culture. But what are the genetic limits of his culture? What is its genetic component?"

While struggling with this eternal question, the participants at the debate focused their arguments on three main topics: child versus adult thought, the nature of mental representations, and the generality of thought and thought processes. On the first topic, the Piagetians argued for the stages of mental development noted earlier. By way of response, MIT philosopher Jerry

Fodor pointed out that it's logically impossible to generate more powerful forms of thought from less powerful ones, and that all forms of reasoning that a person will ever be capable of are already present at birth and gradually emerge through a process of "growing up." Thus, his position strongly supported Chomsky's nativistic views of mental development.

As to the issue of mental representations and thought, while both sides accepted the validity of postulating unobservable, yet no less real, mental representations as a means to explain mental processes, there was considerable disagreement as to the nature and specific role of these representations. For instance, Piaget asserted that the ability to represent knowledge to ourselves is a process of construction, taking place over a long series of interactions with the environment, and cannot be really initiated until the end of the sensorimotor stage at about the age of two years. But if this were true, the Chomskians argued, we would expect paraplegics to have a distorted path of language development, a prediction that is not borne out by the evidence. Chomsky also doubted the validity of grouping together a family of such representations, arguing for a modular view of a mind composed of individual "compartments," each emerging in its own time to carry out its preassigned mental tasks.

In the Chomskian view, the human language capacity is just one of these mental modules, and is for the most part divorced from other forms of thinking. Here Chomsky was claiming that thought is a collection of heterogeneous "actors" loosely controlled by some central organizing agent. Perhaps ironically, in this sense his vision of the mind is reminiscent of the organization of Piaget's homeland, the Swiss Confederation, with its collection of individual cantons loosely held together by the central government in Bern. Piaget, of course, took the opposite tack, insisting that thought is a broad set of capacities, with identical mental operations underlying the individual's encounters with a wide range of environmental stimuli, such interactions eventually shaping the homogeneous mind into more specialized components. In rebuttal, Chomsky challenged the Piagetians to address the problem of the poverty of the stimulus, and to explain how generalized learning strategies could ever overcome this major hurdle.

For the most part, the biologists at Royaumont tended to favor Chomsky, perhaps on account of some rather strange anti-

Darwinian views expressed by Piaget concerning a kind of Lamarckian "transfer of structure" from the environment to the
organism. The social scientists present appeared to be equally
divided between the two competing schools of thought. It should
not go unmentioned that Chomsky's unparalleled skills as a
debater may also have played a nontrivial role in tilting the
Royaumont scales in his favor. Having honed these skills to a
razor-sharp edge through numerous encounters with the bar-
racudas of the American political and academic intelligentsia,
Chomsky was well prepared to counter the low-key, gentlemanly,
almost apologetic style prevalent in European intellectual debate.

As an interesting aside, it's perhaps worth noting here the relationship between Chomsky's strongly biologically oriented position on mental development and the position of sociobiologists
like Edward O. Wilson on the role of the genes in determining
human behavior patterns. On the basis of surface arguments,
one might well speculate that Chomsky would be most sympathetic to the sociobiologists since, after all, one of his central
claims is that our language capacity is inherently limited by our
genetic endowment. Perhaps surprisingly, in actual fact Chomsky appears to be at best lukewarm toward Wilson's arguments.
While firmly committed to the position that a good deal of our
personal and social behavior is a reflection of our genetic program, Chomsky has gone on record with the statement that "I
don't think Wilson understood what he was talking about in
that final chapter." Here he was referring to the last chapter of
Wilson's book *Sociobiology* and its treatment of human behavior.
Somehow this statement seems strangely at odds with Chomsky's
later elevation of heredity over environment, as noted, for example, in his Managua Lectures, where he claims that "the evidence
seems compelling, indeed overwhelming, that fundamental aspects of our mental and social life, including language, are determined as part of our biological endowment, not acquired by
learning, still less by training. . . ."
Following this pronouncement, which, on the surface at least,
certainly appears to be consistent with many of the strongest
claims of the sociobiologists, Chomsky goes on to speculate as to
why so many intellectuals find such assertions so difficult to
swallow. His conjecture is that intellectual libertarians have

become ideological and social managers, seeking to serve or assume power for themselves by taking control of popular movements. For such people committed to control and manipulation, Chomsky claims, it's very useful to think that humans have no intrinsic (i.e., innate) moral and intellectual nature, and that they are simply objects to be shaped for their own good. To my untutored eye, this looks to be about as strong a claim for at least the spirit, if not the program, of sociobiology as could be offered. However, to pursue this line of argument would take us too far afield at this point, so let's return to our main concerns.

In summary, it's clear that the "shoot-out" was more of an exploration than a definitive resolution or rapprochement. But as in all good gunfights, both Piaget and Chomsky stuck firmly to the styles that had got them to the top, and who would have ever thought otherwise? But what about the jury of peers? When the intellectual pyrotechnics and academic smokescreens cleared away, did either of the combatants live to fight another day? Again, as in all the movie westerns at least, only one man rode off into the sunset and that lone gunslinger was Noam Chomsky. But regardless of how one judges the result of this particular encounter, what is clear is that at least a few of the pillars upon which the cognitive sciences now rest were firmly erected as a result of the debate. Since it's important for our later deliberations, let's take a few pages now to discuss this "cognitive skeleton" before summing up the overall issue of language and our verdict as to how it's acquired.

RULES AND REPRESENTATIONS

In the search for the seat of grammar, the one thing all the disputants seem to agree upon is the nature of the grammars themselves: They are a set of rules enabling us to distinguish a sentence that is acceptable in a given language from a sentence that's not. For instance, a trivial and useless grammar for English might contain the rule: A sentence is OK if it contains an even number of words, unacceptable otherwise. The major goal of linguistic researchers is to identify more extensive collections of such rules that, taken together, determine what goes and what doesn't for utterances in the target language. But from the higher-level standpoint of general thought processes, the case of linguistics raises the broader issue of the degree to which all human thought processes are governed by rules. If we grant the cognitivists their use of mental representations, is it true that every thought you think and action you take involves these representations' being shoved around inside your skull according to the dictates of a collection of rules? Since I intend to devote the next chapter to an extensive discussion of this very question, for now I'll sketch only one or two aspects relevant to our linguistic concerns of the moment.

As we'll discuss more fully in the next chapter, to contend that the mind operates according to rules means that we can view the mind as an *information-processing machine* of the sort depicted in Figure 4.7. Here the inputs to the system, or "machine M," represent the environmental stimuli processed by M to produce the observed outputs (actions or behaviors). The inner workings of M are fenced off from the inputs and outputs by the dotted lines to indicate that, generally speaking, an investigator has direct access only to the inputs and outputs, not to the internal mechanisms of M. The workings of M should be thought of as unfolding in one of two ways:

- *External description*—processing of the inputs directly into outputs by a set of *external* behavioral rules.
- *Internal description*—processing of the inputs into outputs by the following steps: (1) Inputs are applied to M from the environment. (2) The inputs are "encoded" by *internal* rules within M as mental representations. These, in turn, are manipulated within M by other rules to form new mental representations. (3) The new representations are then "decoded" by additional internal rules of M to produce the externally observed behavior of the system.

It's crucial to note that there are two conceptually quite different sets of rules operating here. There are the external rules directly relating inputs and outputs. Such rules can be thought of as the sort of stimulus-response patterns so loved by behaviorists. On the other hand, there are the internal rules living inside the system. These rules are the ones that the cognitivists crow about when extolling the virtues of manipulating mental representations as explanatory objects for the mind. The sixty-four-dollar question then becomes: Do these two types of rules have anything to do with each other, and if they do, does this machine

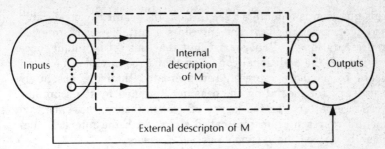

Inputs

Internal
description
of M

Outputs

External descripton of M

FIGURE 4.7. *Schematic diagram of an information-processing machine*

metaphor serve as an adequate model for the way the human
mind actually works?

In support of the machine metaphor as a model of the mind,
let's briefly see how every one of the positions taken in this chapter on matters of language and the workings of the mind can
comfortably be interpreted within the confines of the structure
shown in Figure 4.7. First of all, every behaviorist from Watson
to Skinner has argued that whatever may be in the box labeled
"internal description," it has no place in any scientific theory of
behavior, and that such a theory must be based solely upon the
"external description" of the system. Cognitivists like Piaget
say that it's perfectly acceptable to invoke theoretical objects
like the rules and representations composing the "internal description," but that those rules and representations can be created only by the system's interaction with its environment.
Finally, the Chomskians argue that not only do such rules and
representations exist, but their essential structure is already
present at birth and only the fine details are "tuned" by interaction with the outside world. Interestingly enough, developments
of the past decade or two in the world of mathematical system
theory shed some light on these different views.

In the context of minds and machines, we might paraphrase
the central problem of mathematical system theory as:

*Given a set of external rules, can we always find a set of internal men-
tal representations and rules such that the internal rules generate the
same behavior as the given external ones?*

Under very weak assumptions about the precise forms and properties of the external rules, the somewhat surprising answer to

this question is a definite YES! In fact, the result is considerably stronger, asserting that not only does a suitable set of internal representations and rules exist, but that this set is unique, once we impose the additional condition that it be minimal, i.e., that there be no more representations created than are absolutely necessary for mediating the behavior specified by the stimulus-response pattern. In the jargon of system theory, these abstract mental representations are called *states,* with the internal rules usually termed the system's *internal dynamics.*

What all this mathematical mysticism adds up to in our linguistic setting can be summarized by the following steps:

A. Given a stimulus-response pattern (external description) of a system's behavior, we can always associate with it a minimal set of *abstract* mental representations and rules that will act to reproduce the given external behavior.
B. These "states" and "internal dynamics" can be constructed directly from the stimulus-response pattern.
C. The role of the mental representations is to mediate between the environmental inputs and the observed behaviors and actions of the system.

While the foregoing *facts* seem to deal a strong blow to the behaviorist position with its rejection of the very notion of mental states, there is a practical loophole that needs closing before Skinner & Co. can be permanently cashiered.

The proverbial perceptive reader will have noted that statements A through C have been couched in terms of "abstract" mental representations and rules. What this means for real brains and real minds is not yet clear. The mathematical facts of life ensure that if we represent real stimulus-response patterns within the framework of suitable mathematical structures, then within those structures we can create by mathematical operations new abstract entities that play the role of internal mental states. These states, in turn, generate abstract behaviors and actions in a suitable mathematical space of outputs. The gap that needs closing is the production of a dictionary that relates these mathematical structures and abstract mental states to the actions of real people and their equally real physical brains. In other words, the question now shifts to the relationship between the abstract mental states and actual physical states coded into our neural circuitry. In the jargon of philosophy, we have to close the gap between *mentalism* and *physicalism.*

This problem is identical in form to that encountered when we deal with the mathematical points of Euclid's three-dimensional space E^3, where each such point is represented by three numbers measuring its distance in the x, y, and z directions from some fixed origin. What is the relationship between these purely abstract mathematical objects we call points, and the points of our real-world space R^3, measuring height, width, and depth in the physical universe? In developing his analytic geometry, Descartes made the astonishing claim that these two sets of points are identical, i.e., $E^3 = R^3$. This assertion stood the test of time and experiment until Einstein showed it to be only a good approximation. We're now in a similar situation with the problem of relating abstract mental states to real brain states but, alas, with no Descartes to show the way. As yet, no one has even come close to providing a plausible argument that closes this gap. But since I don't want to start giving away the next chapter's theme, let me say no more about the matter here. Let's return to a summing-up and the rendering of a verdict on the case of the linguists.

SUMMARY ARGUMENTS

To be absolutely clear on the point to be settled, let's review the bidding. Chomsky's argument is that all normal human children receive as part of their genetic birthright a unique language-acquisition device, or language organ. This organ contains a hard-wired universal grammar, which children use to learn their native language quickly and effortlessly. The two key points of contention are whether the language acquisition device is (1) *innate,* i.e., inherited, not learned, and (2) *unique,* i.e., specifically designed for language and not just part of a general problem-solving apparatus. Tables 4.3 and 4.4 summarize the competition. As an aside, the reader will note that I've included Fodor along with Chomsky in Table 4.3. The reason is twofold: First of all, even though Fodor is primarily a philosopher of mind and not a linguist, his views on the modularity of the mind are completely in harmony with Chomsky's; and second, I want to dispel the view that Chomsky is the only one who holds to the Prosecution's case. In fact, a large number of linguists support Chomsky's case, but they do so in ways that are so similar to Chomsky's that there is no reason to distinguish among them in

THE LANGUAGE DEVICE
IS INNATE AND UNIQUE!

PROMOTER	ARGUMENT
Chomsky	universal, generative, transformational grammars
Fodor	modularity of mind

TABLE 4.3. *Summary arguments for the Prosecution*

LANGUAGE IS MAINLY LEARNING
AND/OR NOT INNATE!

PROMOTER	ARGUMENT
Skinner	operant conditioning
Piaget	stages of cognitive development; interactionism
Sapir and Whorf	"language = world"; relativism
Montague	Montague grammar
Sampson	Popperian learning of hierarchical structures

TABLE 4.4. *Summary arguments for the Defense*

a broad treatment of this sort. The interested reader may want to look into the work of some of these Chomskian comrades-in-arms cited in "To Dig Deeper."

BRINGING IN THE VERDICT

On the matter of language acquisition, there's no doubt for me as to where to place my money: firmly with the Prosecution and its claims for innateness and uniqueness. In this sense, I'm a devoted Chomskian. Let me explain why.

First of all, uniqueness. I find it hard to countenance any of the claims by Piaget, Sampson, et al. that the human language facility is just part of the general problem-solving and learning machinery of the brain. It seems to me there's just too much empirical evidence against this claim to take it seriously. For example, why should language acquisition skills mysteriously disappear for most of us in late childhood if the acquisition mechanism is part of our general learning abilities instead of being a specialized skill? If I can learn how to dance the tango or program a computer at the age of forty, why can't I learn to speak Russian or French with equal ease if language acquisition is just a learning procedure like any other? Returning to the idea of switch settings in the universal grammar, it seems that a few lucky souls have the ability to change these settings, even in adulthood. Most of us, however, appear to have these switches "soft-welded" into place in childhood, and remain prisoners of our native language thereafter.

Further evidence along these lines is provided by observations of people suffering strokes or other types of brain injuries resulting in aphasias. If the language facility were as decentralized as the general learning theories suggest, it seems to me that the unaffected parts of the brain would pick up the slack and speech impairments would be a lot less prevalent than they actually are. In this connection, I must say that the ideas of Sampson as they relate to learning the hierarchical structure of language following a Popperian strategy seem appealing. But I can't quite accept his claims that the mechanisms involved are just part of a general learning program. So, all in all, Chomsky's arguments for uniqueness of the language organ strike a more responsive chord with me than the claims of his opponents.

On the matter of innateness, the scales also seem to swing in Chomsky's favor. Without the benefit of some kind of preprogramming, it seems inconceivable to me that children could acquire the basics of virtually any language within their first few years of exposure, not to mention the capacity to generate sentences never before heard or spoken. I have already mentioned the case of paraplegics and language acquisition as an example of the kind of problem that seems difficult for noninnateness theorists to deal with. The basic problem is to explain where this language capacity comes from if it's not basically inborn, and none of Chomsky's opponents have presented a case that even begins to come close to a viable alternative to innateness.

My support of Chomsky's views on language acquisition should not be interpreted as a wholesale endorsement of his entire position on languages, especially the ideas supporting the universal grammar. On this point I have great sympathy for the allegations that the universal grammar unfairly and needlessly underrates the role of meaning. Personally, I lean to a kind of innate grammar that combines the generative ideas of Chomsky with the syntax-semantic combination displayed by the Montague grammars. On balance, it appears to me that Chomsky is right on target with his notions of modularity and innateness, but off course when it comes to the primacy of syntax over semantics. Perhaps the right course is to put his ideas of mind together with Montague's ideas of grammar and then sprinkle on Sampson's vision of hierarchical evolution. The convergence of these three streams of thought might, in my outsider's view, lead to a theory of language that would stand the tests of both time and completeness.

Our focal point in this chapter has been the question of language and its development within the specific biological machine we call a human being. The Chomskian verdict says that the peculiarities of our biological machinery influence not only the kinds of languages we can speak but, more generally, the kinds of thoughts we can think. Question: If we had a different kind of physical structure, in what way might this change the way we think? In particular, if we were composed of fragments of silicon, metal, and plastic connected up like a digital computer, would we think in the same way we do as humans? For science's best answer to this puzzler, read on.

5

THE COGNITIVE ENGINE

===

CLAIM:
DIGITAL COMPUTERS CAN, IN
PRINCIPLE, LITERALLY THINK

THE TURING TEST AND THE CHINESE ROOM

Can a computer think? I mean *really* think, just like you and me, with mental states of the same sort we have when we're slaving over our taxes, daydreaming about next summer's vacation, translating the Spanish ads in the subway, or fuming over our boss's obvious faults. Is it even faintly plausible that a machine of metal, plastic, and silicon can literally experience the same kinds of mental states that we do in these circumstances? If you think the question's easy, consider the following two experiments.

THE IMITATION GAME

Suppose you wander over to your neighborhood university computer center and enter a room whose only furnishings consist of a chair and a table upon which sits one of the major factotums of modern life, a video display terminal and its keyboard. At that moment, a disheveled, malnourished-looking fellow with the bug-eyed, slightly demented stare of the dedicated computer hacker appears, informing you that the terminal on the table is connected either to a similar terminal in another room at which sits a more or less normal human being of indeterminate sex, or to a computer that has been programmed to respond to any sort of question you may wish to pose, provided that it's expressed in everyday English. Neither the computer nor the human is under any obligation to answer your queries truthfully and, to keep the experiment with reasonable bounds, your interrogation is limited to, say, twenty questions or maybe an hour's worth of questioning. At the end of the experiment, the hacker will return and you are to tell him whether you think the terminal is connected to a real, live human being or to the computer. The general setup is displayed in Figure 5.1.

To get some feel for the kind of probing possible in such a situation, let's sit down at the terminal and play this game awhile. The following silly dialogue might result, with your questions shown in normal type while the responses from the human/computer are given in caps.

Well, what should we talk about?
I CAN TALK ABOUT ANYTHING. WHAT WOULD YOU LIKE TO ASK ME?

Let's start with something easy, perhaps a little mathematics. If I have a right triangle with sides of lengths 3 and 4, what's the length of the third leg?
AH, THE OLD PYTHAGOREAN THEOREM. THE HYPOTENUSE THEN HAS LENGTH 5, OF COURSE.

Not too bad. At least you know your high-school geometry. What about some elementary arithmetic? What's the square root of 147?
JUST A LITTLE BIT OVER 12, BY MY RECKONING.

Pretty vague for a computer, but acceptable for a human. Let's try some current events. What are your views on the prospects for peace in the Middle East?

FIGURE 5.1. *The Imitation Game*

I NEVER TALK ABOUT POLITICS OR RELIGION. BUT IF I DID, I'D SAY THE OUTLOOK IS BLEAK, ESPECIALLY WITH THE EVER-PRESENT MEDDLING OF THE AMERICANS AND NOW THE RUSSIANS.

OK, we'll stay away from politics. How do you fancy the Yankees' chances for the pennant this year?
WHAT'S A YANKEE?

I see you're not a sports fan, either. I thought you told me you could talk about anything.
I CAN CONVERSE JUST LIKE ANY NORMAL PERSON, BUT I'M NOT AN ENCYCLOPEEDIA, YOU KNOW.

Not much of a speller either, I see, but no need to get uptight. I'm just trying to find some common ground for us to have a conversation. How do you feel about the economy? Do you think this is a good time to invest in stocks?
THE STOCK MARKET IS THE WORLD'S BIGGEST CASINO, BUT MY HUNCH IS THAT NOW IS A GOOD TIME TO GET OUT OF THE MARKET, NOT IN.

Why do you feel that way?
TOO MANY SPECULATORS AND TOO MANY HAIR-TRIGGER COMPUTER TRADING SCHEMES THAT CAN WHIPSAW YOU OUT OF YOUR SAVINGS IN A MATTER OF MOMENTS. SO UNLESS YOU'RE A REAL GAMBLER, I'D STAY ON THE SIDELINES.

Well, you certainly seem to have definite ideas about finance. What do you know about the arts? How about composing a little poem for me?
YOU'LL HAVE TO COUNT ME OUT ON THIS ONE. I NEVER COULD WRITE.

Well, I've got to leave now, so let me ask you one last question.
Are you a computer or a human?
I'M AN INTELLIGENT, THINKING ENTITY. BYE FOR NOW, AND THANKS
FOR COMING TO SPEAK WITH ME.

Imagine you come back to the computer center and play this
game many times. By random guessing alone, you'd expect to be
able to determine correctly whether you are in contact with the
human or the machine half the time, on the average. Imagine
that after a sufficiently large number of plays of the game, your
success rate in distinguishing the machine from the human is
not substantially better than the 50 percent rate from random
guessing alone. Now we ask: Can the machine think? Well, why
not? After all, the only way we have to decide whether or not
other humans are thinking is by interacting with them in much
the same way we interacted with whoever or whatever was at the
other end of the terminal. So if a sequence of such interactions
leaves us unable to separate the computer from the human, then
it seems perfectly defensible to argue either that the machine is
thinking or that humans do not. Since *ex hypothesi* humans do
think, we must accept that any machine that can fool us in the
above Imitation Game is indeed thinking.

The Imitation Game was originally proposed almost forty
years ago by the British computer pioneer Alan Turing in a
landmark paper on the possibility of constructing intelligent machines. By all accounts, Turing, who played a central role in
breaking the German Enigma code during the Second World
War, was a somewhat emotionally underdeveloped, otherworldly
character given to offbeat pursuits such as "run-around-the-
house chess" (in which after you make your move, you get up
and run around the house, and if you get back before your opponent has moved, you're allowed an extra move), and the "desert
island game" (a kind of survivalist exercise in which you see
how many chemicals can be produced from household substances
using only homemade apparatus), and the simple passions of
long-distance running, bicycling, and violin scratching. It appears that Turing's interest in the idea of a thinking machine
was an outgrowth of his war efforts in cryptography, and
shortly after the war's end he set down his position, together
with a rather detailed rebuttal to the many sorts of objections
that he anticipated might be offered against the notion. It's
strong testimony to the basic soundness of his vision that even

today, almost forty years later, the fundamental ideas he put forth are as topical and fresh as the most recent work in the area, as we shall soon see.

The Imitation Game, or as it's more commonly termed the Turing Test, has the virtue of being implementable, in principle, but unabashedly behavioristic in nature, asserting that the existence of "thinking" is solely a matter of producing convincing responses to more or less arbitrary stimuli. By the Turing Test, any "black box" that does a convincing enough job of imitating a human being in ordinary conversation would be deemed to possess genuine intelligence and could (and should) be thought of as a "thinking entity," just like our friend in the dialogue. Before giving our uncritical acceptance to this kind of claim, let's turn to the second experiment.

THE CHINESE ROOM

Suppose you find yourself inside an enclosed room whose only entrance is a door containing a small mailboxlike slot. Inside the room you find a large number of flashcards upon which are printed Chinese characters, one per card. You also find a big, dictionaryish kind of book giving instructions in English as to how to process the flashcards through the slot. For example, a typical instruction might read: "If the character 'squiggle' comes through the slot, then find the card with 'squaggle' and pass it back outside the room." Friends outside the room pass in a sequence of such cards, while you look up the appropriate instructions in the book and pass back whatever card is called for. Now unknown to you (since you understand not one word of Chinese), the cards that are being passed in form a set of questions about, say, a current popular film. And the cards you are called upon to pass back out constitute perfectly sensible, coherent replies to questions about the plot, the actors, the staging, costumes, and so forth. As far as those outside the room are concerned, the black box consisting of the room and its contents displays a perfect understanding of Chinese; however, from your perspective inside the room, there's no understanding at all. You're just shoving tokens (flashcards) around according to a set of rules. In short, there is syntax but no semantics.

Now we again ask: Can computers think? Since thinking presumably involves understanding the *meaning* of symbols, and computers only manipulate symbols according to a set of rules, the Chinese Room setup leads clearly to the contention that computers cannot think. There is no understanding of the questions posed in Chinese, just blind symbol manipulation. And without understanding there are no genuine cognitive states; ergo, no thinking.

The Chinese Room experiment was proposed by Berkeley philosopher John Searle by way of a counterattack on the adequacy of the Turing Test as an operational procedure for identifying objects having genuine mental states. Searle's claim, of course, is that your actions inside the room duplicate exactly the functional activities of a computer, and it's obvious that there's no real understanding on your part of the questions being passed into the room. Whatever understanding exists is present solely in what's been "programmed" into the rulebook *from the outside,* and the processor (you) has no notion whatsoever of what the symbols actually mean.

Notice the crucial shift of perspective on the question of whether or not the black box, consisting of the room and whatever may be inside it, possesses actual cognitive states. Looked at from the *outside* as called for by the Turing Test, the room does indeed display every sign of being a thinking being, and we would justifiably deem it so from our outside, third-person perspective. Yet when we take the *insider's* first-person stance advocated by Searle, it's difficult to see how anyone could take seriously the idea that the box has internal mental states.

When Searle first published the Chinese Room argument in 1981, the room and its implications met with an outburst of indignation and a variety of denunciations from several quarters of the artificial-intelligence (AI) community. The well-known AI advocate and writer Douglas Hofstadter termed the paper "one of the wrongest, most infuriating articles I have ever read in my life," and regarded it as "a religious diatribe against AI." Similarly, the philosopher Daniel Dennett claimed Searle's arguments were "sophistry." We'll take a look at several of these arguments later, but for now it's sufficient to note that third-person and first-person perspectives lead to flat-out contradictory conclusions regarding the "mentality" of whatever is shuffling the cards out through the door slot of the Chinese Room. They can't both be right, although they could both be wrong, depending upon exactly how we understand the term "mental state." If we add to this the fact that humans are in some sense machines that clearly think, then we're quickly led to see that resolution of the possibility of machines' having legitimate mental states, solely by virtue of their following rules for formal symbol manipulation, involves sharpening considerably our ideas of what we mean by a "machine," a "rule," a "cognitive state," and, most important, what we mean by "thinking." But before trying to clarify these matters, it's worthwhile to pause for a moment and consider why it's of more than passing philosophical interest to spend time grappling with such a question in the first place.

Contrary to popular belief, researchers claiming the existence of genuine cognitive states of the human sort in machines do so neither to undermine cherished psychological, religious, and/or sociological prejudices surrounding the special position of mankind in the universe, nor to demonstrate that man is nothing more than a machine. The reason for the deep concern with the seemingly academic question of whether machines have mental states is distinctly more pragmatic.

Over the past decade or so, the digital computer has provided the "society of mind" community with an unprecedented tool for *experimentally* testing whatever theory of mind one might fancy at the moment. If you think a neuronal net wired up in a certain fashion will produce responses only when stimuli occur in pairs, well, you can just program it into the computer. and check it out. Or if a colleague claims that language acquisition involves a particular kind of symbol representation in the brain, a program can be written to test the proposed theory. So pervasive has the digital computer become as a laboratory tool that a whole new field, cognitive science, has emerged as an amalgam of psychology, philosophy, anthropology, neurophysiology, computer science, and linguistics, organized around the use of the computer as a probe for teasing out the secrets of both the brain and the mind. Consequently, if it can be definitively demonstrated that no digital computer, no matter how cleverly programmed, can ever possess mental states of the sort found in a biologically based human brain, then the computer studies of mind can be at best *simulations* of human cognitive processes. On the other hand, should it turn out that computers can indeed think just like you and me, then the hand of the cognitive scientist will be enormously strengthened when he claims that his pet theory of the mind should be taken seriously, solely because the computer program's behavior agrees with the behavior of humans under similar circumstances. In short, in this case we could say that the program serves as a *model* for human thought, not just a simulation.

As decisions and actions are taken about human beings on the basis of pronouncements from the psychological community, and modern life abounds with such actions in every area from deciding university admissions to the determination of who's criminally insane and who isn't, the question of whether or not machines can have mental states is of practical as well as philosophical importance. Now let's get back to the question itself.

FORMAL SYSTEMS, MACHINES, AND TRUTHS

Generally when we speak of machines, we have in mind things like electric motors, drill presses, water pumps, and the like. These are all devices whose purpose is to act on matter in order to transform or transport it in some fashion. A computer is a quite different sort of "machine." Its purpose is to manipulate not matter or energy, but rather information. Boiled down to its essentials, a computer is a machine for transforming one set of meaningless symbols into another; in short, a device for physically executing the operations called for by the rules of a *formal logical system.* So before we can speak meaningfully about whether such machines can think, we'll need a clearer picture of what constitutes a formal system, and the degree to which the mental life of humans can be captured by such a system.

FORMAL SYSTEMS

Quite generally, a formal system is nothing more than a set of abstract symbols, together with some rules specifying how we can combine strings of such symbols to form new strings. More specifically, the components of a formal system consist of

- an *alphabet* composed of a set of symbols, or tokens, such as the characters $\{a, b, c \ldots\}$ of the Roman alphabet or an even more culture-free set like $\{\diamond, \varnothing, \wedge \ldots\}$. Any finite set of these symbols is called a *string.* However, most such strings are nonsense, so we have

- a *grammar,* which is a criterion for determining which strings are acceptable. Grammatical strings are termed *admissible strings* of the system. Finally, to compose a formal system we need
- a set of admissible strings given a priori, termed the *axioms* of the system, together with
- a set of *rules of inference* specifying the allowable ways of combining admissible strings to form new admissible strings.

To fix these very abstract but absolutely essential notions, let's look at three everyday examples of formal systems in action.

Example 1: The Game of Chess.   As our first illustration of a formal system, think of the game of chess, where the symbols are the black and white pieces. The strings of the system are simply the set of all possible ways the pieces can be arranged on the board. The grammar is just the specification of all *legal* positions that the pieces can occupy on the board (e.g., White King's Bishop only on White squares), while there is only a single axiom, namely, the initial position of all the pieces at the beginning of the game. The rules of inference consist of all legal moves that can be made at any stage of the game, enabling the initial axiom to be transformed into a sequence of legal positions.

The chess example makes it evident that whatever particular physical properties the pieces and board may possess are irrelevant to their role in the game. Thus, it matters not one whit whether we use ivory or wooden chess pieces, or if the board is made of stone or plastic, or if the pieces have been formed to represent agents of the CIA and KGB, or even if we use material symbols at all! The only thing that's important is the arrangement of the pieces in relation to each other and to the squares on the board, and any abstract symbol strings possessing the right relationships will serve equally well for representing everything that's important about the game of chess. It is in this sense that we say that only the "form" of the symbol strings is important, not their content, and this is why we term such systems *formal systems.*

Example 2: Scrabble.   Another board game of universal appeal that fits into the framework of a formal system is Scrabble. For those unfamiliar with the game, it is played with a collection of small, square wooden tiles, each bearing a letter of the alphabet. The tiles are placed on a board ruled off into squares much like a checkerboard, except with far more squares. There is a point value attached to the letter on each tile, and the objective of the game is for players to form words by placing the tiles on the board in much the same fashion as in a crossword puzzle, i.e., by building upon the words already present. As each word is placed, points are awarded to the player according to the values on the tiles used in formation of the word.

The symbols for the formal system describing Scrabble are just the letters of the alphabet etched onto the individual playing tiles. As in chess, the corresponding formal system for Scrabble has only a single axiom, which is the initial word placed on the board by the player who starts the game. But unlike chess, where the sole axiom is determined by the initial position of the pieces, which is always the same, Scrabble's single axiom changes from game to game depending upon the choice made by the first player. The strings of the Scrabble system are just finite sequences of tiles, i.e., combinations of letters, while the grammar specifying which strings of Scrabble tiles are admissible is given by the rules of the game. In general, any string is admissible if it constitutes a genuine word from the dictionary, and if the string touches a tile in any other string that's already on the board. It is this last condition that ensures that the various strings interlock on the Scrabble board in the crisscross pattern of a crossword puzzle. Finally, the rules of logical inference telling us how to form new admissible strings from old ones are just the usual rules of Scrabble telling us in what manner tiles can be added to the board. For instance, one such rule is that the tiles can be added only vertically or horizontally, not diagonally.

It's of significance to note here that if you play Scrabble (like my friend Joe) by introducing your own private dictionary into the game, different from that employed by the other players, then you'll see a different formal system, hence a different game. This new game may or may not be similar to the original Scrabble, depending upon how similar the new dictionary is to the old, thereby opening up the possibility for many of the Scrabble squabbles familiar to the game's devotees (like Joe's wife, Peggy). The point is that any change in any component of the formal system results in a new formal system. And this new system may or may not bear a close relationship to the original.

The examples of both chess and Scrabble, as well as other board games that can be expressed as formal systems like go and Mah-Jongg, account for part of the fascination that such games hold for AI researchers. The fact that these games can be represented by formal systems means, as we shall see, that such games can be "mechanized" in a precise sense of that term. But before moving on to consider these matters, let's first look at another example of a formal system that is not a board game but perhaps is even more familiar.

Example 3: Addition.   Suppose the symbols of our system consist of the two characters * and ⊘. The strings are then just finite sequences of these two symbols taken in any order. Typical strings are sequences like ⊘⊘⊘***** and ⊘⊘⊘⊘*******. All such strings are assumed to be grammatical. Our system will have the two axioms * and ⊘, meaning that the single-element strings * and ⊘ are assumed to be admissible a priori. We will allow two rules of inference by which we can generate new strings from old:

$$1)\, S + ⊘ = ⊘S \quad \text{and} \quad 2)\, S + * = S*$$

Rule 1 means that given any string S, we can combine it with the string ⊘ and thus obtain a new string consisting of the string S prefixed by ⊘. Similarly, Rule 2 says that if we combine S with the string *, then the result is the new string consisting of the string formed by appending the symbol * to S.

Let's use these rules on the axiom * and see what we get:

$$S = *\ (\text{Axiom})$$
$$* \to ⊘*\ (\text{Rule 1})$$
$$⊘* \to ⊘⊘*\ (\text{Rule 1})$$
$$⊘⊘* \to ⊘⊘**\ (\text{Rule 2})$$
$$⊘⊘** \to ⊘⊘⊘**\ (\text{Rule 1})$$

In this sequence, each of the strings following the axiom * constitutes what is termed a *theorem* of the formal system, and the sequence of application of the rules forms what we call the *proof* of the theorem. Thus, the symbol string ⊘⊘* is a theorem having the proof sequence Axiom → Rule 1 → Rule 1. Other theorems would have resulted if we had begun with the axiom ⊘, and/or if we had used a different sequence in applying Rules 1 and 2.

So far, the above formal system just gives us a way of generating grammatically correct strings involving the abstract symbols * and ⊘. Now suppose we try to attach an *interpretation* to these symbol strings in the following way: To each string S, associate the nonnegative integer $[n]$, where n is the number of appearances of the symbol * in the string. Thus, the string ⊘⊘⊘*** and the string *** would both be associated with the number $[3]$, while the strings ⊘** and ⊘⊘** would both be identified with the number $[2]$. With this interpretation of a string S, we are able to assign a single integer to each grammatical string of the system. Now if we think of the abstract symbol ⊘ as standing for our usual notion of zero, it's easy to interpret the general Rules 1 and 2 as the ordinary rules of addition, i.e.,

$$1)\ [n] + [0] = [n], \quad 2)\ [n] + [1] = [n + 1]$$

for every natural number $[n]$.

Thus, the *abstract* formal system defined solely in terms of the symbols ⊘ and * can be *modelled* by the process of addition of nonnegative integers—once we make the appropriate interpretation of the symbols and symbol strings. What's important to note here is that the symbols ⊘ and * don't *mean* anything until we pass to the interpretation step; at the level of the formal system, they are just symbols or tokens, and the rules of inference are just prescriptions for shuffling around symbol strings in order to create new symbol strings. This point is of crucial significance when it comes to assessing many of the arguments offered against the idea of a computer actually thinking. At the formal system level, there is syntax alone; semantics enters only when the symbols are interpreted, and for a computer to think it must be possible for the machine to make this transitional step from the syntax to the semantics. In the Chinese Room experiment, Searle claims this is impossible. We'll see why later.

The fact that it's only the form and syntactic structure of the strings that are important in a formal system accounts for one of their greatest attractions: They can be *about* anything. All we need do is attach some meaning (i.e., semantic content) to the symbols and presto! Before our very eyes, the system strings become meaningful statements about the integers or the solar system or the stock market or whatever other interpretation we've given to the symbols. On the other hand, the "meaninglessness" of a formal system is also its Achilles' heel, since the truths it can express about the real world are entirely determined by the interpretation injected into the system from the outside. Thus, the only semantic content that a formal system can express is there not from inside the system itself, but from the meaning put into the system from the outside by its user. This observation accounts for Searle's claim in the Chinese Room experiment that "you can't get semantics from syntax." On the surface this argument looks airtight but, as with all matters of this sort, things are seldom what they seem, and the hidden assumptions built into it play an important role in our later consideration of the objections to the notion of mental states for machines. For the moment, let's take a harder look at the question of what kinds of truths can be generated by any kind of formal system.

PROOFS AND TRUTHS

The "truth" or "knowledge" to be obtained from a given formal system consists entirely of the statements that can be generated or proved from the system's axioms by applying the given rules of inference. Speaking more precisely, a *proof sequence* in a formal system F is a list of admissible strings $S_1, S_2, \ldots, S_n$ such that each string is either an axiom of F or is obtained from some of the previous strings by applying the rules of inference. So, for example, if the system F is the one representing the game of chess, and the string S_1 is the sole axiom, consisting of a listing of the positions of the playing pieces at the beginning of the game with White to move, then S_2 might be the position of the pieces after a King's Pawn opening. That is, S_2 is the same as S_1 with the single exception that the White King's Pawn has been moved forward two squares. A string T is said to be *provable* in F if there is such a proof sequence that ends in T, i.e., a sequence $S_1 \rightarrow S_2 \rightarrow \cdots \rightarrow T$. The set of all provable strings constitutes the *theorems* of the formal system F. This setup should be familiar to each of us as the situation encountered in our late and unlamented high-school geometry course, where we started with a handful of elementary, "self-evident" truths about points, lines, circles, and planes, and proceeded to struggle with the rules of logical inference in our feeble attempts to rediscover a few of Euclid's ancient truths. Since the purpose of a formal system is to generate proofs of theorems, we might think of a formal system as an abstract machine that prints out the list of theorems provable in the system F.

When it comes to the matter of how powerful a given formal system F is in its ability to generate a long list of truths, there are two aspects of the system that bear heavily on the question: *completeness* and *consistency*. Basically, the idea is that we would like every true statement that can be interpreted using the symbols of F to be a theorem, i.e., provable, while at the same time being unable to prove any self-contradictory statements. More informally, we want F to be able to prove all "true" statements, and not be able to prove any "false" ones. So, if $T_1, T_2, \ldots$, is the list of all theorems provable within F, and P is an interpreted string corresponding to a true statement, then F is called

- *complete* if P appears on the list $T_1, T_2, \ldots$, and
- *consistent* if P and not-P do not both appear on the list.

Note that the properties of completeness and consistency are what are termed *metamathematical* statements about the system F; i.e., they are not statements (strings) expressible *within* F, but rather are statements made *about* F from the outside, so to speak.

In terms of the formal system characterizing the game of chess, the system would be complete if any legal position of the pieces could be achieved through a legal set of moves starting from the initial placement of the pieces. The system would be consistent if a legal position and its negation could not both be attained. So, for example, if we have the usual legal position that the White King's Bishop plays only on White squares, then any sequence of legal moves that would involve putting this piece on a Black square would imply the system's inconsistency.

From the standpoint of machine cognition, it's of great interest to understand the difference, if any, between what is "true" and what is "provable," since if we could establish the equality

$$\text{True statements} = \text{Provable statements}$$

then we would have gone most of the way toward showing that all thought processes are just physical manifestations of particular formal systems. Regrettably for mechanists, things just didn't turn out this way. We'll see why in a moment. But first, let's pause to catch our breath and summarize in the box below the impressive array of terminology introduced so far about formal systems.

FORMAL SYSTEMS

ALPHABET   a collection of abstract symbols or tokens used to form the strings of a formal system

STRING   any finite sequence of symbols (sometimes termed a *formula*)

GRAMMAR   a set of conditions or criteria that distinguish an admissible string from one that is inadmissible

RULES OF INFERENCE   a collection of logical operations that can be performed on strings to transform one admissible string into another

AXIOM   a string that is taken to be admissible by definition, i.e., without proof

FORMAL SYSTEM   an abstract entity consisting of an alphabet, strings, a grammar, rules of inference, and axioms

PROOF SEQUENCE   a finite sequence of admissible strings such that each string follows from its predecessor by applying one of the rules of inference

THEOREM   the final, or termination, string in some proof sequence

COMPLETE SYSTEM   a formal system in which every interpreted true statement can be proved, i.e., every such string is a theorem of the system

CONSISTENT SYSTEM   a formal system in which an interpreted true statement and its negation are not both provable, i.e., they are not both theorems .

DIGITAL COMPUTERS

In the crudest terms possible, we can think of a digital computer as being a device with the capability of storing and changing a whole lot of numbers. A good analogy would be a general post office with a large number of post boxes, each box having its own label or address. We suppose that each box can contain a single number. This collection of boxes forms the *memory unit* of the computer. Imagine now that we have another device that enables us to go to any two boxes, remove the numbers that reside in these boxes, and perform an arithmetic operation upon them, forming a new number. Such a device is termed the *arithmetic unit* of the computer. Similarly, suppose we have another device that can compare any two numbers and tell us which of the two is the larger. We call this the computer's *logical unit*. In addition to these units, suppose we also have an *input unit* enabling us to place particular numbers into certain boxes, and an *output unit* that gives us the ability to look into any box and read its contents. Finally, imagine we have a set of instructions telling us what boxes are to be looked into, and which further details the sequence of arithmetic and logical operations to be performed. This set of instructions is the *program*. Thus, the way the computer works is first to place a particular set of numbers in some of the boxes. Next it consults the program to see what the first operation is to be, goes to the boxes called for by this instruction, and performs the indicated operation, placing the result in the particular box that's specified. It then executes the next instruction in the program and carries on in this fashion until it comes to the end of the program. The computer then employs its output unit and looks into certain boxes to read out their contents, which we then call the results of the program (in actuality, the input and output operations are also specified as part of the program and and may be carried out as intermediate steps in the overall computation). This entire setup can be schematically depicted in the following diagram:

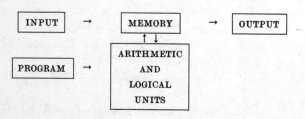

In real life, the computer becomes a lot more useful if we can use it to do more than just perform arithmetic operations on numbers. In fact, most computers in use nowadays are employed for things that have little to do with numerical computation, but rather involve activities like preparing, storing, and retrieving text, creating graphics, monitoring industrial processes, and a host of other nonnumerical activities. So how is it that we can arrange for the "number processor" described above to act as a "symbol processor"? The answer is obvious: Simply *code* whatever symbols we want to process as numbers. In the case where the symbols we're interested in are the usual alphanumeric characters of the Roman alphabet, there is a universally agreed-upon way to associate a number with any of the symbols {A, B, C, . . . , a, b, c, . . . , 1, 2, . . .}. This labeling of symbols with numbers is termed the *ASCII* ("As-key") code, and it works in the following manner.

The basic unit of storage in a modern computer is a unit called a *byte,* which consists of a string of eight binary digits or *bits*. Thus, every address location in the computer memory can store a single number consisting of a string of eight bits. In the ASCII coding scheme, the first bit in each byte is reserved for various sorts of internal bookkeeping chores, leaving seven bits free to code alphanumeric quantities. So there are a total of $2^7 = 2 \times 2 \times 2 \times 2 \times 2 \times 2 \times 2 = 128$ different quantities that a single byte could encode. Here are a few examples of how the ASCII code allows us to represent alphabetic and numeric symbols:

SYMBOL	ASCII CODE
A	1000001
M	1001101
I	1001001
!	0100001
⊔ (blank space)	0100000
?	0111111

Thus, in ASCII the sentence "I AM!" would be translated into the byte string

I ⊔ AM! = 1001001/0100000/1000001/1001101/0100001

while the interrogative "I AM?" would be the sequence

I ⊔ AM? = 1001001/0100000/1000001/1001101/0111111.

Using this kind of coding, we can then employ the computer memory locations to store individual alphanumeric symbols as well as numbers, and arrange things so that the computer can be used not just as a "number cruncher" to do arithmetic calculations, but also as a symbol processor to manipulate nonnumeric quantities. This kind of coding scheme enables us to see how a computer might be used to determine mechanically the theorems of a formal system. In fact, we can make an argument to show that symbol manipulation in a computer according to a specific program is exactly the same thing as the determination of the theorems of a particular formal system. Let's see why.

In a digital computer, the symbols of the formal system are just the elements 0 and 1, while the grammatical strings are all those binary sequences whose length equals the word length in the computer. This is set by the computer hardware design, typically two or four bytes for a standard personal computer. The axioms of the formal system are the strings that encode the inputs fed in at the beginning of the calculation, while the rules of inference are just the statements composing the program that operates on these input strings (axioms). Thus, every computer programmed to deal with a particular kind of problem is a formal system in exactly the sense described earlier.

By a result due to the same Alan Turing, the inventor of the Imitation Game, the converse is also true: Every formal system is equivalent to a suitably programmed digital computer. In fact, Turing proved much more. He showed the existence of a *universal computer,* which, given enough memory and time, can simulate *any* computer, and that any formal system could be modeled by running an appropriate program on this universal computer, or *Turing machine.* Thus, an IBM PC could simulate the behavior of a Cray YM-P (but *verrry* slowly, since computational speed is hardware-dependent). Further, the so-called Turing-Church Thesis states that every computable quantity (roughly speaking, every output that can be obtained as the result of following a program) can be computed on a Turing machine. So the problem of mental states for machines now becomes equivalent to the question: Are human cognitive processes (i.e., is thinking) representable by a formal system? In other words, do all human cognitive processes involve just manipulating a collection of abstract symbols according to a set of rules? If so, what are the symbols and rules; if not, what's missing? The answers hinge critically upon an understanding of just what kinds of knowledge or truths are *formalizable* in the sense that they are the theorems of some formal system.

GÖDEL'S THEOREMS

The most influential mathematician of the early part of this century was the German David Hilbert, who thought that all possible mathematical truths could be captured within some formal system, and who actively promoted the formalist school of mathematics devoted to a rigorous proof of this contention. Formalist hopes were permanently blown away in 1931 by Kurt Gödel, who astonished the mathematical (and philosophical) world by proving that for any formal system $\mathcal{F}$ that is (1) finitely describable, (2) consistent, and (3) strong enough to prove the basic facts about elementary arithmetic,

I. $\mathcal{F}$ is incomplete,

and

II. $\mathcal{F}$ cannot prove its own consistency.

Gödel's theorems show that every formal system is subject to inherent limitations on the amount of "truth" that we can expect to squeeze out of it. Gödel I states that no formal system $\mathcal{F}$ is capable of deciding every statement that can be made about the natural numbers. Thus given a formal system $\mathcal{F}$, there is a statement $\mathcal{P}$ about the natural numbers that can be made (and even seen to be true), but that cannot be proved in $\mathcal{F}$; moreover, if we extend $\mathcal{F}$ to include $\mathcal{P}$ (for example, by including $\mathcal{P}$ as one of the axioms of a new system $\mathcal{F}'$), then there is a new true statement $\mathcal{P}'$ that is not provable within $\mathcal{F}'$. Also, if $\mathcal{F}$ is to embody a correct description of all mathematical truths, we would expect the consistency of $\mathcal{F}$ to be readily apparent and a fairly easily provable fact. Nevertheless, Gödel II tells us that this just isn't so: Even if $\mathcal{F}$ is consistent, we can't use $\mathcal{F}$ to prove this fact. Actually, this result can be even further strengthened to the statement that there exists no constructive procedure that will suffice to prove the consistency of $\mathcal{F}$.

These very abstract results can be seen more clearly if we interpret them as "just" special cases of an even stronger result of Gregory Chaitin on the limitations of formal systems (or, equivalently, Turing machines) in their ability to cope with complexity. Specifically, suppose we have a string composed of 0's and 1's. Some such strings are intuitively "simple," like 0000 . . . 000 or 1010101010101010. Others, like 0010011101010100011010, have no apparent pattern and look "complicated." The great Russian mathematician Andrei Kolmogorov and, independently, the American Chaitin had the idea of characterizing the *complexity* of such a string by using the notion of a Turing machine and a program for producing the string. In particular, they argued that if the program required to produce the given string was of about the same length as the string itself, then such a string would be more complex than one that could be produced using a relatively short program. Thus, for example, the string consisting of all 0's can be produced by the simple program: "Start with 0 and continue in this way for as many elements as are in the given string." Thus no matter how many 0's are in the string, we can always produce the given string with this simple, relatively short program. On the other hand, the "complicated" string above seems to have no program appreciably shorter than just instructing the machine to write out the string itself. Using this line of reasoning, Kolmogorov and Chaitin defined the complexity of a string as being the length of the shortest program needed by a universal Turing machine to produce the string. Since as we have seen, a program can also be described by a finite binary sequence, there is no ambiguity here as to which of two given programs is shorter than the other.

With the above notions in mind, in 1965 Chaitin proved the following remarkable result: If $\mathscr{F}$ is a formal system that is (1) finitely described and (2) consistent, then there is a number x such that the system $\mathscr{F}$ cannot prove that there are any binary strings with complexity greater than x. In other words, any formal system $\mathscr{F}$ is limited in its ability to determine the complexity of an arbitrarily given binary string. But since there are infinitely many strings of arbitrary complexity, it must certainly follow that there are strings of complexity greater than any arbitrary, but fixed, number x. But $\mathscr{F}$ is unable to prove this fact, so it must be that $\mathscr{F}$ is incomplete. Thus, using Chaitin's Theorem we are able to deduce Gödel's Incompleteness Theorem as a simple corollary.

Rumor has it that Hilbert was livid with rage when informed of Gödel's results, perhaps not surprisingly, since having years of work, as well as one's philosophical way of life, destroyed virtually overnight is a bitter pill to swallow. As one might suspect, the proofs of Gödel's and Chaitin's incompleteness theorems are much too technical to enter into here, but the underlying trick that makes the magic work is to find a way to mirror the metamathematical properties of completeness and consistency within the system $\mathscr{F}$ itself. The basic idea shows up already in the famous Liar's Paradox, illustrated by the statement

> THIS SENTENCE IS FALSE.

Here we can interpret the expression at two levels: the level of the words in an ordinary English sentence, and a higher level referring to the *meaning* of the sentence. Thus, the sentence can speak about itself in a semantic sense by using symbols and rules at a purely syntactic level. The way Gödel achieved this kind of self-reference for formal systems is indeed tricky and devious, just the kind of argument one might expect from a man who, according to mathematical folklore, agonized for weeks while studying the U.S. Constitution for his citizenship examination because he thought he had discovered logical contradictions built into it by the Founding Fathers of the republic!

The key ingredient in Gödel's proof of the foregoing results was the construction of a string G that represented a mathematical way of saying "I am not provable." Then if it were possible to prove G, the string G would be false and the formal system containing G would be inconsistent; on the other hand, if G could not be proved, then we would see that G is true but impossible to prove using the rules of inference of the formal system; i.e., the system is incomplete. Gödel's genius was to prove that such a *Gödel sentence* G could be found for *any* formal system $\mathscr{F}$ that was sufficiently rich to contain the usual rules of arithmetic.

Figure 5.2 shows a schematic version of Gödel's result in "logic space," where the enclosed box represents all possible logical statements that can be made. Let the box initially be colored completely gray. Suppose M is a given, finite mathematical theory, i.e., a formal system. Using M, we are able to prove some logical statements true and falsify others. Let the true statements be colored white, and the false ones black. Thus, starting with the theory M, we gradually change the color of the logic square from gray to a mixture of black, white, and gray. What Gödel says is that there is no theory M that will enable us to remove *all* the gray. In other words, there will always be some statement of the type denoted in the figure by GM, which is forever doomed to lie in the twilight zone of logical grayness. Of course, different theories remove different regions of gray, but no single theory, or combination of individual theories, can re-

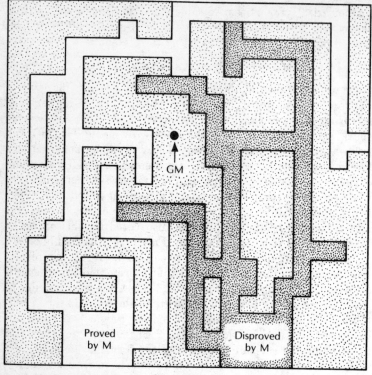

FIGURE 5.2. *Gödel's theorem in logic space*

move it all. One of the crucial questions for the proponents of thinking machines is to address whether or not the gray area that remains is / accessible to humans, but not to computers. We'll return to this point in detail in a later section.

Now that we've taken a high-altitude flight over the territory of formal systems, truths, proofs, and Gödelian logic, let's try to bring these purely logical ideas into contact with machines and, in particular, the digital computer in an attempt to see what these stratospheric mathematical abstractions have to do with what is computable by such devices.

MACHINE STATES AND COGNITIVE TRUTHS

The preceding discussion has shown that each memory address in the computer can hold exactly one byte of information. Thus, we can specify the entire state of the computer's memory at any given moment by giving a list of what symbols are currently being stored in each of its memory locations. It can be shown that the other functional units in the computer can also be characterized by their own byte patterns, so that we can speak of the computer's *state* at any time as a list of the byte pattern that is currently present in each of its basic units. Hereafter, whenever I speak of the *state of the machine,* or, more compactly, the *machine state,* it will mean such a list consisting of the byte pattern currently residing in the machine's memory unit, its logical unit, arithmetic unit, and so on. Since execution of the program will result in a change of these states as time unfolds, we can also think of the computer's state history as being a listing of its successive states over the entire time history of the computation.

It is commonly held that cognitive thought in humans is somehow associated with the various electrochemical activities taking place in the neurons inside the brain. To oversimplify slightly, we can think of a neuron's state at any moment as being either "on" or "off," depending upon whether the neuron is firing at that moment or not. At the neuronal level, a listing of the state of each neuron constitutes what we can call a *brain state* at that moment. Somehow (nobody really knows how) these brain states give rise to the mental states that we associate with thinking. Thus, there is some kind of correspondence between physiological brain states and a set of abstract states that represent ordinary cognitive notions such as our mother's face, a car, a pain, or a sunny day. In what follows, we shall use the general term *cognitive state* or *mental state* for these abstract quantities. If there is any content whatsoever to the claim that computers can literally think, or at least think like you and me, then there must be a way in which the computational states of the machine can be meaningfully associated with these mental states of human thinkers. So far, a detailed account of this association remains but a gleam in the eye of the AI aficionados, and there are many who claim that no such connection between the machine and the mental will ever be made. Nonetheless, the resolution of the thinking machine debate ultimately resides either in producing a convincing map between the two, or in proving that it does not exist. In short, the problem is whether or not it's possible to remove the question marks in the diagram:

machine   brain   cognitive
      $\leftrightarrow$ $\leftrightarrow$
states   states   states

This theme is our leitmotiv for the remainder of the chapter. Before listening to the competing claims, let's briefly return to some of Gödel's thoughts on the matter.

It's evident that Gödel's results have a profound bearing on the issue of thinking machines, since they appear to imply that there exist truths that can be known but that cannot be captured by any formal system and hence cannot be obtained by any kind of computation. There is considerable controversy over the meaning of Gödel's theorems for the question of artificial intelligence, and we'll examine some of the competing arguments later. But for now it's of interest to hear just what Gödel himself thought about this question. Unfortunately Gödel was rather reclusive and secretive, especially in his later years, and his only published statement on the topic comes from a lecture delivered to the American Mathematical Society in 1951:

> The human mind is incapable of formulating (or mechanizing) all its mathematical intuitions, i.e., if it has succeeded in formulating some of them, this very fact yields new intuitive knowledge, e.g., the consistency of this formalism. This fact may be called the "incompletability" of mathematics. On the other hand, on the basis of what has been proved so far, it remains possible that there may exist (and even be empirically discoverable) a theorem-proving machine which in fact *is* equivalent to mathematical intuition, but cannot be *proved* to be so, nor even be proved to yield only *correct* theorems of finitary number theory.

Thus Gödel leaves open the possibility of the existence of a theorem-proving machine, and even concedes that it may be possible to discover such a machine by empirical investigation. However, he then throws a wet blanket on the whole business by saying that if we ever find such a machine, it will be beyond our powers to prove that it constitutes a Universal Truth Machine.

We began this section by trying to get a more precise feel for what we mean when we speak of a "machine," and ended up taking off into the stratosphere of formal systems, undecidable propositions, universal computers, and the like. So let's try to summarize the situation thus far. Henceforth, when we speak of a machine we will be talking about a universal computer (a Turing machine), which, by the Turing-Church Thesis, is capable of computing anything that can be computed. Furthermore, we saw that every such Turing machine is equivalent to a particular formal system, which means that the theorems of the system coincide with the quantities computable by the Turing machine. Finally, Gödel's theorems told us that every such system, and hence every such machine, is subject to inherent limitations on the quantity of truth that we can extract from it. Therefore, as indicated above, the problem of whether or not such machines can "think" now comes down to a more detailed consideration of how we can associate "cognitive states" with the "computational states" of such a machine, and of the connection such cognitive states have with everyday, garden-variety human thinking and with the electrochemical activities going on in the brain.

"STRONG" VS. "WEAK" AI, BRAINS, AND MINDS

By informal consensus, the birth of artificial intelligence as a recognizable intellectual undertaking can be pinpointed to the summer of 1956 at Dartmouth College, where John McCarthy, then a member of the Dartmouth Mathematics Department, convinced the Rockefeller Foundation to fund a summer study on "the conjecture that every aspect of learning or any other feature of intelligence can in principle be so precisely described that a machine can be made to simulate it." Along with McCarthy, who now heads the AI Laboratory at Stanford University and who bears responsibility for coining the term "artificial intelligence," others at that historic Dartmouth workshop included Marvin Minsky, head of the MIT AI Laboratory; Claude Shannon, inventor of information theory; Herbert Simon, Nobel laureate in economics from Carnegie-Mellon University; and Arthur Samuel, developer of the first championship-caliber checkers-playing program, as well as a half-dozen others from academia and industry who shared the vision that perhaps a machine could be made to perform human functions that previously were thought to require intelligence.

It's of interest to note that even at this dawning of the Age of AI, the manifesto of the Dartmouth study was already maddeningly vague as to whether or not the participants actually shared the belief that machines might one day actually think or would only behave as if they were thinking, each possibility being left open by use of the word "simulate." Written and oral accounts of the meeting support both positions, some of the participants being engaged in studies of networks of artificial neurons that they hoped would, in some sense, mirror the biological neurons of the brain, while others at the meeting were much more interested in the construction of programs that would behave in an intelligent fashion, regardless of whether or not the principles the programs employed bore any resemblance to the way a human brain would do things. This split between the paradigms

Thinking = The *way* the brain does it

and

Thinking = The *results* the brain gets

persists to this day, dividing the AI community into what has been termed the *strong* and *weak schools of AI.*

For purposes of even understanding what the question of whether machines can think means, it turns out to be of value to refine the "strong" versus "weak" dichotomy just a bit according to a scheme proposed by the philosopher Keith Gunderson. He identifies the following versions of AI:

- *Strong AI, human:* Whatever kinds of cognitive states machines might have, those states are functionally (although, of course, not physically) identical to those found in the human brain.
- *Strong AI, nonhuman:* The kinds of cognitive states found in a machine are not functionally identical to those in the brain and hence cannot be used to model human thought processes.
- *Weak AI, sim-human:* A computer can simulate human cognitive processes, but there is no particular correlation between the computer states and the cognitive states of the brain.
- *Weak AI, sim-nonhuman:* A computer can simulate the cognitive processes in a nonhuman mind (e.g., a frog, a dog, an ant), but the states of the machine may or may not be related to those in the nonhuman brain.
- *Weak AI, task, nonsim:* The computer can perform tasks that previously required intelligence, but there is no intelligence re-

quired of the machine, whose states have nothing whatsoever to do with cognition, human or otherwise.

It's important here to understand the distinction between two states' being functionally equivalent, and being physically identical. The easiest way to see this difference is to imagine we had a correspondence between, say, three cognitive states C_1, C_2, and C_3, and three machine states M_1, M_2, and M_3. These states are clearly not physically identical since the machine states are just patterns of 0's and 1's imprinted within some silicon chips, while the cognitive states are connected with the chemical concentrations and electrical pattern in a brain. However, these two sequences of states would be functionally equivalent if, for example, whenever we saw the machine pattern $M_1 \rightarrow M_3 \rightarrow M_2$ it always corresponded to the cognitive pattern $C_2 \rightarrow C_3 \rightarrow C_1$. In this case, we would say that the states M_3 and C_3 were functionally identical because they played the same functional role in the corresponding sequences; i.e., they were always the middle state of the three-state sequence.

As far as genuine machine thinking goes, the only category that counts is the first: strong AI, human; everything else, while undoubtedly technically challenging and economically rewarding, is pretty much devoid of any real intellectual or philosophical appeal, at least as far as the thinking-machine question goes. This may come as a surprise to many in view of the recent brouhaha generated by the media (and various self-serving members of the AI community), extolling the wonders of the so-called expert systems being developed in AI labs from Massachusetts to Tokyo, describing the robots waiting just around the corner to satisfy your every desire (or take your job), and proclaiming the need to pour more good money after bad to keep pace in the "thinking machine race" with the Japanese. And this is not to mention the venture capitalists/entrepreneurs and their computer-fixated associates, who are running around doing a good Keystone Kops imitation while trying to capitalize on the public's gullibility over the cognitive capacities of machines. This whole deplorable situation can be traced to a handful of programs demonstrating some progress in the last and intellectually feeblest category on our list: weak AI, task, nonsim. Progress in this category sheds about as much light on thinking as the flight mechanism of birds shed on the development of the airplane. So henceforth, when we speak of cognitive states for machines, we will be referring to the kinds of states understood to be in our first category: strong AI, human.

Needless to say, no one has ever produced an unassailable argument showing that the internal states of an appropriately programmed digital computer are functionally identical to the states existing between your ears when you're engaged in ogling that new Mercedes, poring over the seemingly endless menu at the neighborhood Chinese restaurant, juggling your expense account to cover a session at the craps table in Vegas, enjoying a Bach fugue, or performing any of the other myriad activities that, in some sense, we call thinking. Nevertheless, this is our problem. And as a result of our deliberations thus far, we can finally state the "Can machines think?" question in more or less final form:

TURING MACHINE VERSION

Can an appropriately programmed computer display strong AI, human?

FORMAL SYSTEM VERSION

Are all human cognitive states functionally equivalent to the admissible strings of some formal system?

It's clear that Alan Turing, a computer scientist and logician, would answer the question with a resounding yes, while John Searle, a philosopher, would give an equally strong negative reply. This separation between the "scientists" and the "humanists" is typical of the way the deep-thought industry seems to have divided itself on the matter, but the reasons for taking these positions are manifold and diverse. But before entering the courtroom of scientific debate and listening to the competing arguments, let's first hear the thoughts of John von Neumann, who spent the final years of his life reflecting on the problem of mechanical thought.

Von Neumann, a banker's son from Budapest, was one of the few true geniuses of the twentieth century. Before his untimely death in 1957 from bone cancer (most probably induced by radiation exposure suffered while observing the hydrogen bomb tests at Bikini atoll in the early 1950s), von Neumann made fundamental contributions to the theory of logic, quantum mechanics, meteorology, game theory, economics, and functional analysis. Important as this work is, there is little doubt now that von Neumann's most lasting contribution will be his central role in the development of the digital computer, particularly the idea of the stored program. As an outgrowth of his work on the theory of computation, von Neumann became interested in the logical structure of machines, producing the first proof of the possibility for a self-reproducing machine, as we detailed in Chapter Two. In this effort, he anticipated the later work of Watson and Crick on the dual role of information in cellular DNA, identifying the need for information to be used in both an interpreted and noninterpreted form if self-reproduction were to take place in any sort of organism, biological or otherwise.

Oddly enough, despite his clear understanding of the distinction between the functional activity of biological organs and their material construction, von Neumann tended to be somewhat skeptical about the possibility of a computer's duplicating the activities of the human brain, primarily because he found it difficult to see how the physical hardware of the computer could ever be made to mimic the complexity of the brain. In his last published work, the incomplete text of his Silliman Lectures at Yale, von Neumann devoted most of the volume to a detailed comparison between the hardware of the brain (the neurons, axons, synapses, and so forth) and the hardware of the computer (flip-flop circuits, switching speeds, reliability, etc.), taking considerable pains to point out the several-orders-of-magnitude difference between the two in information-processing capability. But there is virtually no mention of the fact that computers and brains, despite their vastly different physical compositions, carry out exactly the same kind of information-processing functions. It's as if one were examining a grandfather clock and a digital watch and were puzzled over the fact that one was made out of wood and brass while the other was formed from plastic and quartz, ignoring the fact that they both performed exactly the same timekeeping function. There are essential differences in the design and construction of the two objects, but functionally they are indistinguishable.

While von Neumann never actually came out and stated that he thought a computer could not duplicate the brain, his writings strongly indicate that he felt that way, and that the computer could never really mimic the brain because it just wasn't made from the right stuff. In other words, when it comes to human-style cognition, hardware counts. In the battle between the philosophers and the scientists over thinking machines, this same argument surfaces again as one of the pillars upon which the "Computers Can't Think" school of thought bases its case. But we're getting ahead of our story, so with the above preliminaries in hand it's time to drag out the scales of justice and listen carefully to the Prosecution and the Defense in an attempt to weed out the few facts from the polemics and hype, and come to some position on the issue of machine cognition. As in all trials, we start with the Prosecution.

TOP-DOWN SYMBOL CRUNCHING

Herbert Simon, winner of the 1978 Nobel Prize for economics, is a soft-spoken, slightly graying man, whose trim figure belies the fact that he is now in his early seventies and still one of the most active practitioners of the "artificial intelligentsia's" arcane art. His Nobel-winning work was for pioneering techniques aimed at understanding behavior in organizations and the planning of industrial activity, originating many of the concepts that we now know under the rubric "management science." Somewhat less well known to the general public is his lifelong interest in the ways of human thought processes, and the possibility of capturing these principles in computational algorithms. Now Simon is not noted as being a man of bombast or hyperbole, so one can imagine the shock when in January 1956 he returned from the holidays to announce to his class at Carnegie-Mellon University that "over Christmas, Allen Newell [his colleague at CMU] and I invented a thinking machine." By this he meant that he and Newell had developed a computer program that displayed behavior they considered to be "thinking." Edward Feigenbaum, now a well-known exponent of the "expert system" school of AI, was a student in that class, and his reaction to Simon's bombshell was what one might expect: "What do you mean by a thinking machine?" What Simon, Newell, and their co-worker, J. C. Shaw of the RAND Corporation, meant by a thinking machine defines what we can now term the *top-down* approach to achieving mechanical thought.

Put crudely, the top-down thesis is that human thought processes take place as a result of rule-based symbol processing in the brain. Thus, just as we can go into a chemistry lab and put atoms of various types of chemicals together according to the Mendeleev Table, so forming more complicated compounds having new properties entirely absent from the individual components, the brain can put together the "atoms" of thought (the symbols) according to various rules, thereby generating the multitude of cognitive states we call thinking. Here we see the problem of correspondence between cognitive states, brain states, and machine states in its purest form. The top-downers blithely forget about brain states altogether, and just assign various machine states to cognitive states, much in the same manner that we earlier assigned ASCII codes to alphanumeric symbols. A set of rules (usually termed a *semantic network* or *conceptual dependency graph*) telling how these machine states can combine with each other is then postulated, and the resulting machine states are "decoded" to give an interpretation of the computation in terms of cognitive concepts. This, in a nutshell, is the strategy of the entire top-down approach to AI.

As a simple illustration of the foregoing ideas, here's a conceptual dependency graph for the idea "John bought a car":

JOHN ⟺ *Atrans* ← MONEY	←	JOHN SOMEONE
SOMEONE ⟺ *Atrans* ← CAR	←	SOMEONE JOHN

In the diagram, ATRANS refers to the transfer of an abstract entity, in this case ownership of the car and the money. Many top-down advocates think that most of our everyday acts can be broken down into a dozen or so primitive actions, like PTRANS for the transfer of a physical object and MTRANS for the transfer of information. The claim (or hope) is that these primitive actions form a language for the representation of meaning in a computer, with the idea being to code each of these actions and associated mental states by certain computational states of the machine, and then put in the rules by which these primitives can interact to form more complex sorts of activities.

The very first working program of this type, the one Simon

announced to his mathematical modeling class in 1956, was called *Logic Theorist* in reference to its ability actually to generate proofs for many of the theorems in Alfred North Whitehead and Bertrand Russell's magnum opus *Principia Mathematica*. Amusingly, Simon and Newell participated in the historic 1956 Dartmouth summer meeting, even producing a working version of *Logic Theorist* for demonstration. But the import of this seminal achievement seems to have been lost on the other participants at the meeting, who more or less ignored what amounted to the first working computer program to have displayed anything approaching real intelligence.

The underlying principle that *Logic Theorist* and its successor, *General Problem Solver,* employed is a form of heuristic reasoning called *means-end analysis.* Basically what this involves is noting that when we have a problem to solve, we always start with (1) a given initial state (data, premises, and so on), (2) a desired terminal set of states (goals), and (3) a set of operators that can transform one state into another. The task then becomes to find a sequence of operators that will transform the initial state into the terminal set. Simon and Newell supplied their programs with two kinds of heuristics:

- Procedures for detecting significant differences between two states
- Rules of thumb about which operators typically reduce differences between various kinds of states

The solution principle is then clear: Detect some difference between the initial state and the terminal set; apply some operator that ordinarily reduces such a difference; if the resulting state doesn't differ from the terminal set, stop; otherwise try the same procedure, but now from the new state.

Example: The Three-Coin Problem.   To see how this kind of analysis works, consider the well-known Three-Coin Problem, in which we have three coins, each of whose initial position can be either heads (H) or tails (T). The goal is to transform the initial configuration into one for which all of the coins are showing either H or T, i.e., the goal states are HHH and TTT. For any given state, there are three possible operators: "turn the first coin over," "turn the second coin over," and "turn the third coin over." A move corresponds to the choice of one of these three operators, and a solution to the problem is a sequence of three moves that will transform the initial state into one of the goal states.

If we designate the three operators as A, B, and C, corresponding to turning over the first, second, or third coin, respectively, then Figure 5.3 shows the sequence of possible moves that can be made in this game. Notice from the diagram that it is not possible to move from the state HTT to the goal state TTT in exactly three moves.

The solving of logical puzzles, the playing of simple games like tick tack toe, and a variety of other heuristic search activities typify what we term *automatic* formal systems. These are formal systems that work by themselves in the sense that in their normal mode of operation, they automatically manipulate the formal symbols of the system according to the system's rules. All of the Simon and Newell work on top-down computer cognition can be classified under the heading of such automatic formal systems. Unfortunately, several years' worth of experimenting with automatic formal systems has led to the sad conclusion (one of many, of course) that, rather than demonstrating that human thought is really just formal symbol manipulation in disguise, what the Simon and Newell exercises show is that game playing, theorem proving, and the like can be done well without anything even approaching the full spectrum of human intelligence. In short, programs like *Logic Theorist* can produce intelligent-looking results in a very restricted domain, but once out of that domain there's a Grand Canyon–sized chasm separating them from what anyone would even charitably call thinking.

As an amusing indicator of the nature of the gap that remains to be bridged between rule-based symbol-manipulation programs and everyday thinking, some years ago an effort was made to produce a Russian ↔ English translation program that could take a text in one language and produce at least a rough translation into the other, the goal being to relieve a human translator of the drudgery of doping out the gist of the text so that his time could be more profitably spent polishing the machine version for final consumption. The basic idea was to program a large vocabulary and the grammar from each language into the machine, give it a few rules and idioms, and then turn it on. The immensity of the task was quickly brought out when the simple phrase "out of sight, out of mind" was translated back as "blind and insane!"

The nature of the difficulty was identified by the well-known

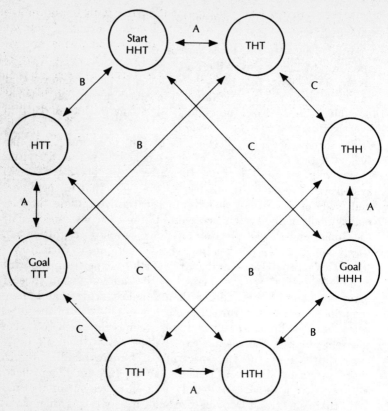

FIGURE 5.3. *The possible moves in the Three-Coin Game*

logician and philosopher Yehoshua Bar-Hillel's claim that a
computer would never be able to distinguish between the phrases
"the pen is in the box" and "the box is in the pen," where "pen"
in the second case would immediately be understood by any
human to refer to a baby's playpen. To make such a distinction,
Bar-Hillel asserted, the computer would need not only a dictionary and grammar, but a universal encyclopedia containing a
vast amount of knowledge about the world, the kind of knowledge that we humans take for granted and routinely acquire as
we stumble through life. Somehow this knowledge must be given
to a machine if it's to act like a thinking agent, at least from a
human's perspective.

The natural-language-processing problem illustrates in the

starkest possible terms the major difficulty with top-down symbol processors: They just ain't got no common sense! There's no way that programs of the Simon and Newell stripe are ever going to "think" in the way humans do until a way is found to code knowledge of the world into the formal symbols that the computer operates with. In retrospect, it's easy to see that when we perceive intelligently, we never perceive an object, but rather a *function* and a *context*. So if I show you a key, you never think of it as just a machined piece of metal; rather you see it as an object that performs the function of unlocking something, perhaps a door, a safe, a car, or whatever the context suggests. It's this kind of knowledge that a computer needs if it's going to think top-down style.

The past decade or two has seen a number of disparate attempts to deal with the common sense acquisition problem for top-down AI. Let's take a glimpse at a couple of the more prominent efforts.

MICROWORLDS

A procrustean approach to giving computers common sense about the world is simply to fence off most of the outside world and let the computer have access only to a very severely restricted universe whose features, idiosyncrasies, folkways, and mores can be spelled out in painstaking detail and then given to the computer in some sort of digestible form. For example, Monopoly is a microworld in which the aspiring real-estate tycoons never have to worry about contingencies like fires, wars, deadbeat tenants, civil action suits, and the zillions of other annoyances that plague owners of pieces of real-world real estate. Board games like chess, go, and checkers are other microworlds of this type.

Probably the best-known microworld program is *SHRDLU,* a block world put together by Terry Winograd in the early 1970s. This universe consists of a few imaginary blocks of various sizes and shapes, strewn about on a flat surface. Figure 5.4 shows *SHRDLU*'s world. The blocks may be colored and cast shadows, but they never have any other physical properties beyond their geometric shapes and dimensions. *SHRDLU* knows all there is to know about this microscopic universe, and is able to converse in a seemingly intelligent fashion when queried about the world

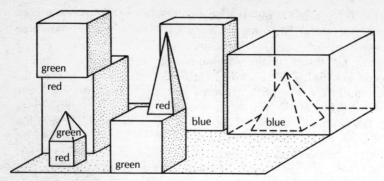

FIGURE 5.4. *The world according to SHRDLU*

or asked to perform certain acts such as placing one block atop another, or picking up a block and moving it to a different location.

Despite what appear to be intelligent dialogues between *SHRDLU* and its inquisitors, the program has a variety of fatal deficiencies as a cognitive entity: (1) *SHRDLU* never initiates any actions but only reacts to queries put to it; (2) the program has no motivational goals whatsoever, other than the goals introduced by inquiries from the outside; (3) the main problems of perception and action involve capturing the interface between symbolic cognition and real objects. But *SHRDLU*'s "world" is *already* symbolic, so it doesn't address this interface at all. But these difficulties pale by comparison with the real problem concerning microworlds in general: They are capable of performing only because their domain is so stripped down that there is nothing left that could require even the slightest glimmer of understanding or real perception. Perhaps the strongest testimony to the inadequacies of the microworlds as a viable approach to computer cognition comes from Winograd himself, now a professor at Stanford, who states:

> The idea is that language and thought can be modeled by such things as formal logic. But I think that that is grossly oversimplified. What people actually do has very little in common with formal logic, and what's missing is the social dimension. Once you take into account what you are using a word for, what part it plays in discourse, there is no boundary to the meaning of that word.

Consequently, microworlds don't appear to be the answer to the commonsense problem. Let's examine another line of attack.

FRAMES

Frames and *scripts* are predicated upon the belief that few situations we encounter in daily life are really new. Technically, frames describe *static* situations, while scripts characterize a *dynamic* set of actions appropriate to a given set of circumstances. Most circumstances that we're called upon to deal with have enough in common with other situations that we can distill the principal features, analyze them, and store them for future retrieval and use. Thus, a frame acts something like one of those IQ tests you encountered as a kid, where some sort of scenario is created with a number of blanks left open to be filled in appropriately to demonstrate your understanding of the story. Although the frame idea appears to have originated with Marvin Minsky as an outgrowth of work on computer vision and language, the high priest of "frameology" is Roger Schank of Yale, a somewhat controversial character in AI circles. What Schank's work demonstrates is that thinking and learning are not just passive processes of filing and retrieving information. The mind learns to build models and structures that can be continually modified and updated as new knowledge becomes available, and that dynamic knowledge base is used to plug the gaps in real-life scenarios as they unfold.

As an illustration of a typical frame, here is a template for a stock market report:

[Because of/Despite] *Current newspaper headline,* the market [staged a broad advance/dropped sharply/rallied/rebounded/crept upward/drifted lower] in [heavy/active/moderate/light] trading with [advances/declines] leading [declines/advances] by a margin of — to —.

Another typical Schankian example is the restaurant script, which has slots for entry conditions like "customer is hungry" and "table is set," and slots for exit conditions such as "customer has less money," "kitchen has less food," and "waiter has more money." Of course, visiting a restaurant is something we do in stages: sit down at the table, read the menu, place the order, eat the food, pay the bill, and so forth, so we divide the script into scenes. Thus, there would be an entering scene, an eating scene, a paying scene, and so on. To allow for different types of restaurants, the script would be divided into tracks. So if we're told that Alex went to McDonald's, the fast-food track containing the necessary variations for the way one proceeds in such a place would be loaded. As Schank says, "We wouldn't place our order over a microphone at Maxim's in Paris, nor would we ask for a wine list at a diner." The fact that we would be surprised at any of these things is evidence that we have some kind of knowledge structures containing information about what usually happens in a given set of circumstances.

One of the acid tests for the ability of a program to "understand" the situation in a given frame is for the program to be able to answer questions about the situation, especially questions whose answers are not directly given by the specification of the frame. For instance, in the restaurant situation we might have the scene:

> The waitress brought the hamburger to John, but it was burned to a crisp, so he got up and stormed out.

Now we can ask, "Did John pay for the sandwich?" On the basis of everyday knowledge about such situations, even a small child would have no difficulty in realizing that John didn't pay. But nowhere is this rather evident fact explicitly stated. Rather it has to be deduced from the facts that are given, together with the background knowledge built into this particular track of the restaurant script.

Of course, if a machine had scripts alone, it wouldn't be able to deal with novelty; it would understand only the prototypical situations that had been programmed into the scripts. Consequently, Schank and others, like Robert Wilensky at Berkeley, have been busy developing programs that would know about people's goals and desires, and how they might go about formulating plans to achieve them. One such program was tested on the story:

> John wanted money. He got a gun and walked into a liquor store. He told the owner he wanted money. The owner gave John the money and John left.

Nowhere does the story make mention of robbery, nor does it explicitly state that the gun was used to threaten the liquor store owner. Nevertheless, the program was able to use its store-house of knowledge about goals and plans in order to infer these facts.

Neither of these approaches—microworlds and frames—has turned out to be a panacea for the ills that plague the top-down approach to intelligent machines; nevertheless, Simon, Newell, Schank, & Co. continue to press on with their hopes of finally achieving the triumph of their rule-based, "symbol-crunching" style of AI. At this juncture, it's well to pause to consider just what the implications of their total and complete success would imply for our basic question: "Can an appropriately pro-grammed computer display strong AI, human?" In order for top-down results to justify an unqualified yes, there would have to be some indication of how the internal states of the machine match up to human cognitive states, as well as to the internal states of the brain, when they're both performing the same sort of task. This means that, at some level, the top-down program states will have to make contact with actual brain states; other-wise, the best that even a perfect top-down program could aspire to would be weak AI, sim-human. So far the top-downers have displayed no such points of contact, and as far as I can see no interest in establishing such a bridge. So while it may be true that a top-down approach can shed some light on some aspects of human thought, it appears unlikely at present that further pounding away at such programs is going to get us any closer to a resolution of the basic question. Consequently, let's move to the other end of the telescope for a look at bottom-up attacks on the matter of thoughts and machines.

BOTTOM-UP EMERGENCE

Herbert Simon is on record with the claim that "everything of interest in cognition happens above the 100-millisecond level—the time it takes you to recognize your mother." This claim com-pactly summarizes one of the principal axioms of faith in the top-down school of AI: that what's going on down at the level of the individual neurons in the brain has no direct bearing on cog-nition, and that somehow we can "skim off" the rules of thought from the higher level of symbol processing and semantic net-working, and just ignore what's happening down below at the

level of the microscopic processing elements. In response to
Simon's 100-millisecond assertion, Douglas Hofstadter has written, "I cannot imagine a remark about AI with which I could
more vehemently disagree." Hofstadter holds to just the opposite view: Everything that is important about cognition goes on
below the magic 100-millisecond level. He is one of the leaders of
the "new wave" school of AI theorists, working on the question
of how intelligent behavior can possibly emerge out of a jumble
of primitive processing elements existing at a subcognitive level.

The basic tenet of the bottom-uppers is that if we're ever to
understand how the brain does its thing, we're going to have to
start at the level of primitive processors functionally equivalent
to neurons, and develop theories of how cognitive states like
your mother, a 747 jet, a migraine headache, and all the other
things the top-downers attach symbolic significance to can possibly come about as the result of connections and interactions between such simple processing elements.

A good analogy for understanding the bottom-up philosophy
is provided by those old-style message boards seen even today in
places like Times Square, where the news of the day and other
types of information are shown by a sequence of flashing lights
on a rectangular board. At the level of the individual lights,
there is no message: All that any of the bulbs can do is blink on
and off. However, by standing above the level of the individual
lamps themselves, we can see a properly timed sequence of such
flashing lamps as communicating the results of the World Series, a report on the state of the stock market, the outcome of an
election, or an announcement of the end of the world. The same
hardware serves for an infinite variety of symbolic messages,
but to recognize that there *is* a message it's necessary, as Hofstadter puts it, to "jump out of the system" somehow. In some
poorly understood way, the system at the level of the light bulbs
would have to have some measure of self-awareness, or self-reference, at a higher level.

The message board example shows that whatever computation
is going on is happening not at the level of the symbolic meaning
(the message), but rather at the much lower level of the flashing
lamps. The computational rules are buried in the program that
tells each lamp when to switch on or off, not in a set of instructions for manipulating the thoughts composing the message.
This is the fundamental difference between the top-downers and

the bottom-uppers: For top-down AI, thoughts and ideas themselves are passive computational entities capable of being pushed around by the rules of a formal system; for bottom-uppers, cognition involves active symbols arising from a collective of computational elements at the subcognitive level—i.e., thinking is an emergent epiphenomenon. So for bottom-up AI, subcognition at the bottom drives cognition at the top, and the brain as hardware is simply a substrate in which active symbols can interact. It's worth noting that this view of thinking requires some kind of hardware in which the active symbols can interact, but there is no absolute requirement that this material substrate be physically the same as a human brain. All that's required is that the playing field in which the symbols frolic have the same computational power as a human brain; i.e., the substrate must be functionally equivalent to a human brain but may differ greatly from it in its actual physical composition.

The key element in the bottom-up program is to identify the bridge between the "meaningless" computations at the subcognitive level and the "meaningful" active symbols. One line of attack has been to try to understand how we do anagrams. How is it that given the word "weird," we can immediately see that its letters can be rearranged to form the words "wired" and "wider," but that no other arrangement leads to a proper word? Surely it's not by trying out all $5 \times 4 \times 3 \times 2 \times 1 = 120$ possible arrangements of the five letters. It seems inconceivable that the brain does anagrams on such a brute-force, straight-line computational basis. Rather, we somehow use our knowledge of what letter combinations tend to go together, form various groups of letters, and let them float around in a sort of "alphabet soup" in our heads, randomly bumping into each other and forming new combinations. Those combinations that look promising are kept, while others dissolve and drop back into the soup where they can link up with another group. Eventually, certain combinations click into place and a new word is formed. Hofstadter and his group at the University of Indiana have developed a program called *Jumbo* to test out various theories of how this subcognitive computation (forming of letter combinations) results in the emergence of active symbols (meaningful English words). The actual strategy implemented in *Jumbo* provides an instructive glimpse into the entire bottom-up program for the creation of mechanized thought.

The workings of *Jumbo* are based upon two analogies: the way in which a living cell conducts its chemical business, and the fashion in which human friendships and romances are formed. Let's first consider the cell. The interior of a cell (its *cytoplasm*) is filled with different kinds of molecules floating around in the cytoplasmic soup. The work of the cell is carried out by enzymes, each of which has a very specific function that it's designed to carry out. On each enzyme there are one or two *active sites,* which will allow only a particular type of molecule to attach itself. The enzyme randomly wanders about in the cytoplasm until it encounters the right type of molecule, which is then attached to the active site. If the enzyme's function is to join two molecules (anabolic), then when the two active sites are filled, the enzyme goes into action and joins the two molecules, whereupon it releases the new compound into the cellular broth. Other types of enzymes function to split compounds (catabolic reactions), or perform more complicated functions like rearrangements and regroupings. *Jumbo* makes metaphoric use of this kind of cellular operation by regarding the molecules in the soup as being the letters of the given word, allowing letters to affiliate randomly with others to form syllables, which in turn can be joined by different sorts of enzymes to form larger groupings ultimately resulting in proper words. In *Jumbo* terminology, these grouping operators are called *codelets,* and there are different types for performing a variety of functions such as combining consonants into clusters, consonants and vowels into syllable fragments, syllables into wordlike objects, and so on. Figure 5.5 is a schematic diagram for such an enzymatic codelet whose purpose is to embody the almost universal rule that, in English, the letter Q is always followed by a U. This codelet floats around in the alphabetic soup until it encounters a Q-shaped character and a U-shaped one, each of which is captured in the appropriate half of the codelet. Once both halves are filled, the codelet joins them into the pair QU, thereby emptying its two slots and making them available again to capture more characters. But once various combinations are formed by this sort of random interaction, how does the program decide whether or not a particular syllable fragment, say, is promising as a step toward formation of an actual word? This is where the analogy with human romance comes into play.

While it may be true that the course of true love never runs smooth, that course invariably follows a path along which recog-

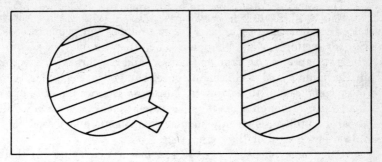

FIGURE 5.5. *An enzymatic codelet*

nizable landmarks appear in a time-honored sequence. First, there is the initial contact. For boy and girl to meet, they must come together at the same place at the same time, in order that Nature may take its course. Following contact, sparks will start to fly if there's mutual interest, and the parties will begin to explore a potential relationship by entering the next phase—dating. During the dating, each party will keep open the option of pursuing simultaneous relationships with others, as well as the possibility of breaking off the current budding romance. After this period of exploration, the parties may decide to strengthen the relationship by making it more exclusive. At this stage, while it's not impossible for the link to be broken by either internal stress or external attractions, there is a deeper commitment and it takes much stronger provocations to break it than in the earlier phases. Following this courtship period, the relationship may be even further strengthened into an engagement, which may then be formalized socially by a marriage. Of course, depending upon social conventions, religious convictions, and the like, even such a strong bond as the marriage may ultimately be dissolved by divorce, with the partners then being sent back into the "social soup" to begin the process anew. Hence, every romance has to go through a sequence of increasingly tough filters, although these steps may proceed in parallel and out of phase as several independent relationships are being explored.

Jumbo makes use of the progressively stronger hierarchies of bonding seen in the paths of love and friendship, in order to decide which of the many random bonds formed between sequences of letters at one level should be taken seriously as candidates for consideration at the next level of combination. Consequently, along with the codelets for acting upon clusters of letters, the program is provided with criteria for assessing which letter combinations seem more likely to appear in proper words. So, for example, *ee* is "happier" (i.e., more stable) as a vowel cluster than *ii; nk* is happier than *kn,* and clusters with a vowel in the middle and a consonant on either end are happier than those composed of three vowels. Thus *senk* is more likely to arise than *kniis,* although when it's discovered that *senk* is not a word, it can be broken down into smaller components and tossed back into a lower level of the soup.

At the same time the "active" enzymes are carrying out their joining and breaking functions, another set of "passive" enzymes called *musing* codelets are imagining what would happen if, say, one syllable was swapped for another. Without actually making the change in the real cytoplasm, the musing codelets consider alternative hypotheses, and explore many paths at once, trying out various kinds of possibilities. But what is it that finally determines when all this random groping, shuffling, combining, and probing finally stops? At what stage does *Jumbo* say enough, and settle upon its best candidates for wordhood?

The program's stopping rule is based upon the notion of *entropy,* the technical term for the measure of randomness, or disorder, present in the cytoplasmic soup. Initially, there are just a lot of individual letters randomly floating around in the soup and the entropy is high; later some structure starts to emerge as individual letters begin to combine with others to form consonant and vowel clusters, as well as short syllables, and the entropy goes down; still later, syllables combine into larger groupings and the entropy is further decreased. While all this is going on, the enzymes are also acting to perform their specific functions, with some enzyme operators decreasing entropy by combining consonants into clusters, consonants and vowels into full syllables and the like, while others, like those that interchange syllables within words, leave the entropy unchanged. Finally, the actions of enzymes that break bonds established by earlier joining enzymes result in actually raising the entropy level of the soup. Roughly speaking, the entropy level can be thought of as a kind of "temperature" of the cytoplasmic soup, and when the enzymes can no longer act to reduce the temperature, *Jumbo* stops and the clusters that remain in the soup are taken as its best effort at forming meaningful words from the initial alphabetic hodgepodge.

A program to do anagrams may appear trivial to some, frivolous to many, and far removed from the general idea of thinking for almost everyone; nevertheless, *Jumbo* captures in particularly transparent form the central ingredient in the bottom-uppers' cognitive paradigm: Whatever intelligence the program displays has not been directly programmed by specifying rules for passive symbol manipulation. Rather, the cognitive behavior emerges as a statistical property of many small things designed to interact with one another that *have* been built directly into the program. Consequently, in contrast to the underlying principle of top-down AI, there are no overall "rules of thought" deterministically governing the manipulation of symbols, and there is no central controller or manipulator and no central program; only a vast number of individual "collectives" whose actions trigger the actions of other collectives resulting in new, more complex patterns of organization. In short, there's no body doing the thinking, only a collection of somebodies.

Hofstadter & Co. have employed the same "statistical emergence" principle in another program aimed at identifying letterforms (how do we recognize that the symbols A, *A*, a, and *a* are all instances of the same letter?), as well as in a program for doing analogies (ABC is to ABD as PQR is to ??). Perhaps not surprisingly, these ideas have not caught the fancy of the mainstream AI community, with its historical bias dominated by rule followers of the Simon-Newell-Schank persuasion, and expert-system peddlers of the Feigenbaum school, who appear to be totally uninterested in any type of AI not marketable in corporate boardrooms or on Wall Street. Certainly Hofstadter's harshest and most vitriolic critic has been Simon's colleague-in-arms Allen Newell, who complained that one of Hofstadter's papers was

". . . somewhat polemical and diffuse with an abundance of strong opinion and argumentation from general conceptual considerations and the absence of concrete scientific data or theory to build on. There is an abundance of attacks on the general opinions of others, with a corresponding promotion of the general opinions of self.

A disciple of the Schank school, Richard Granger of the University of California, Irvine, says:

His [Hofstadter's] AI work is far from the mainstream. He's a loner. His opinions are one man's view. . . . You have to understand that Hofstadter has recognition because he won the Pulitzer Prize. He's a good writer. He's a smart, very clever person. But that doesn't mean that he's right about AI.

To this kind of criticism, Hofstadter replies that

AI people seem trapped in their already-formed modes of thought and their preconceptions. They tend to eschew the whole question of what consciousness means. They avoid the questions of philosophy of mind.

Another researcher following an evolutionary, bottom-up approach to AI is Douglas Lenat of Stanford. In his doctoral work, Lenat developed a program called *Automated Mathematician* (*AM*) whose goal was to learn mathematical facts and prove them—all on its own. Lenat's basic idea was to combine the frame idea with evolutionary adaptation in a program that would learn about the world of mathematical truth on its own. Initially, the program started with a collection of frames with slots like "Definitions," "Examples," and so forth. At the outset most of these slots were empty, so Lenat provided the program with about 250 heuristic rules of thumb that would suggest which slot to work on next, where *AM* should look for new relationships between concepts, and the like. Furthermore, Lenat provided a valuation scale by which each concept's frame would keep track of how each of its slots was doing by recording things like the origin of a concept, and *AM*'s evaluation of its worth relative to the other frames. In this way, the valuation scheme would act like natural selection by identifying the most interesting concepts, dropping those of little "survival value."

The results of Lenat's work surprised even the developer. Within a few minutes of turning on the machine, Lenat saw that *AM* had discovered the concept of number. Soon after, it discovered the rules of arithmetic and the idea of prime numbers. From those building blocks of mathematics, the Fundamental Theorem of Arithmetic (every number can be decomposed into a product of primes in a unique way) was but a small step. Regrettably, after an hour or so of such set-theoretic bliss *AM* ran out of steam, and began looking into such weird and self-contradictory notions as numbers that are both even and odd. Upon examining the situation, Lenat found the difficulty resided in the heuristics that he had originally programmed in to get *AM* off and running. These heuristics dealt primarily with concepts in set theory, and as soon as the program started departing from this well-plowed turf, the heuristics became increasingly useless.

Learning from his experience with *AM*, Lenat began to develop a new program, *Eurisko*. The basic difference between *Eurisko* and *AM* was that *Eurisko* could modify not only its concepts, but also its heuristics—both by the process of natural selection. Lenat's idea was to represent each heuristic by a frame of its own. In this way, "mutations" in the heuristics could also take place one slot at a time. The reader will recognize this procedure as strikingly reminiscent of the way Nature works in altering an organism's DNA by point mutations. The results from *Eurisko* have been most publicly visible in the sound thrashing it gave all human competitors in the national championship of the space-war game *Traveller,* where the program designed space fleets of optimal size, power, flexibility, and so forth. Lenat's work has been heralded by AI guru Marvin Minsky as "a whole new field of knowledge." Currently, Lenat is engaged in trying to utilize the *Eurisko* principle to code up nothing less than the whole field of human knowledge. He estimates that this project, one of the most ambitious ever undertaken in the AI world, will take at least ten years.

Right or wrong, the bottom-up movement is by now most definitely *not* the work of a Lone Ranger with a couple of Tontos riding the desolate plains of the University of Indiana's Computer Center. Variations upon the basic bottom-up theme are popping up daily in many corners of the AI forest, and converts are joining the fold in ever-increasing numbers. One of the most prominent supporters is that doyen of the AI world Marvin Minsky, who predicts that "Hofstadter is one of whom, fifty years from now, they'll say he was on the right track." Minsky's own vision of thought, which he terms the "Society of Mind," is admirably captured in the Disney film *Tron,* in which the hero, a hacker named Flynn, spends most of the film trapped inside a computer, prisoner inside a system that he himself constructed. The film shows the inside of a computer as a community of programs, each portrayed by an actor having a history, a personality, and, most important, a function within a complex political organization. As the story unfolds, the Master Control Program has assumed dictatorial powers, and repressive police programs are employed to bring the other programs under central control. Eventually, with Flynn's help, full-scale warfare breaks out within the society and . . . well, I won't spoil the ending for those who haven't seen the film. Rent it out at your local video shop! Anyway, even this cursory view of Minsky's "society," in which intelligence emerges from the interactions of conflicting, competing parts in a fragmented mind, shows its radical differences in perspective from the rule-following, central actor paradigm beloved by the top-downers. While Hofstadter and Minsky emphasize software in their bottom-up theorizing, let's not forget the hardware side of the house—the new connectionists.

The human brain is composed of about 100 billion neurons linked together in an array of connecting axons and synapses of a bewildering degree of complexity. Inspired by the organization of this "wetware," a group of computer scientists, psychologists, and engineers have banded together to explore the hypothesis that what produces thinking is the establishment, strength, and reciprocal feedback of interneural connections, not computation in the sense of manipulation of formal symbols. In short, thinking "emerges" from the process of neural connections' forming and reforming. Interestingly enough, this thesis is not a new one: In the late 1950s, Frank Rosenblatt of Cornell produced an artificial neural net (the *perceptron*) capable of learning and identifying a variety of letterforms. The basic structure of a perceptron is shown in Figure 5.6, where we can clearly see the threefold character of the machine: a lower level of individual sensory input units wired into a higher-level array of associators (formal neurons or processors), which in turn produce the perceived output of the device.

Unfortunately, a bit later Minsky and his MIT colleague Seymour Papert showed mathematically that such a simple-minded perceptron could never display the kinds of properties that we would associate with genuine thinking, such as recognizing the difference between the letters C and T. The prestige of Minsky, Papert, and MIT, coupled with a serious misperception of just exactly what they had actually proved, created the totally erroneous view that perceptronlike devices were a cognitive dead end, resulting in a two-decade-long hiatus in development of bottom-up AI in general, and connection machines in particular. This whole sequence of events is especially ironic since, as already noted, Minsky is one of the staunchest supporters of bottom-up work. Fortunately, the emergence of a new generation of

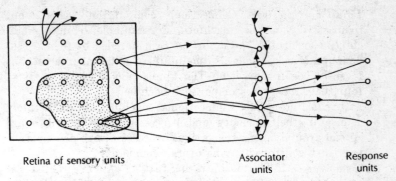

Retina of sensory units   Associator   Response
                                          units   units

FIGURE 5.6. *Diagram of a Perceptron*

AI workers, as well as the general shift in emphasis from serial
to parallel processing in avant-garde computer architectures,
has led to a major revival of interest in connectionism as the
royal road to machine intelligence. Let's have a quick look at the
main planks of the connectionist platform.

The basic idea underlying connectionism is that a densely
linked network of simple, neuronlike processors can behave in
consistent ways to find certain outputs when presented with certain inputs, and that the best way to get the right outputs is not
to specify a rule for calculating them, but rather to let the system find the right answer by trying out different connections in
the network until it settles on those that yield the correct response. Thus, just as *Jumbo* engages in a directed, but still random, assembling of word fragments into trial words and then
settles on real words by lowering the "temperature" of the linguistic cytoplasm, a connection program proceeds in exactly the
same manner to identify other types of patterns such as faces,
geographic features, and letterforms. The bottom-up attitude of
the connectionists sets their program off from traditional AI in
several ways:

• *Hardware counts:* It's simply not possible to separate the message from the medium; high-level symbolic processing cannot
  be abstracted from the hardware. Note that this does not
  imply that all such processing, and hence thinking, must be
  carried out in a medium like the human brain—only that hardware constraints matter when it comes to consideration of the
  cognitive powers of such processing objects.

- *Parallel architectures:* Connectionist computation is carried out in massively parallel machines. For example, Thinking Machines Corporation, a Cambridge firm founded by former students of Minsky, who also maintains a paternal interest, has recently began marketing The Connection Machine, a sixty-four thousand-processor parallel machine.
- *Distributed processing:* Connectionist machines are deliberately diffuse in their memory and processing; the activities are spread around among the various processors with no single supervisory controller having overall command.
- *Unprogrammed:* The most striking feature of connectionist machines is the relative lack of specific instructions. Rather, there are a few general instructions, with the network finding its own solutions by settling down into stable states instead of following detailed, prespecified algorithms.

Currently there are several connectionist programs under way, all employing the foregoing principles but in quite different ways. Amusingly, one of the most active efforts is at that bastion of top-down AI, Carnegie-Mellon, where Geoffrey Hinton and his colleagues are building the *Boltzmann machine,* which is a hardware implementation of Hofstadter's "minimal temperature" notion, predicting the behavior of the overall system from the statistical behavior of its parts. The Hinton group has managed to get the machine to learn a pattern of outputs by varying the strengths of the machine inputs. Another effort utilizing the same ideas, but in a nonprobabilistic manner, is that of Dave Rumelhart at the University of California, San Diego, who makes each processing unit take on a range of input values instead of being just "on" or "off." The sum of the input values then determines the processor's output. Rumelhart has deliberately constructed his processing elements to resemble neurons, the signals being blurred and weighted according to which neuron has transmitted them. The overall result is a system that "relaxes" slowly into a stable state that cannot be changed by small, random input variations. In a quite different direction, Igor Aleksander of Imperial College in London has designed a system that uses random samplings of an image to teach a connectionist array of memory chips to respond to particular patterns of inputs. One of the more striking aspects of Aleksander's work is that after enough inputs of "your mother's face," a prototype of your mother's face becomes stored in the connections of the machine, and thereafter the machine is able to "recognize" your mother when her face again appears at the input. All of these activities speak strongly for the importance of the connectionist program as a major paradigm in the thinking machine derby. To get a better feel for how the connectionist idea works, let's look at a vastly oversimplified version of a Boltzmann machine at work.

Consider the simple Boltzmann machine shown in Figure 5.7. This machine consists of three computing elements labeled X, Y, and Z, together with three lines of connection between them denoted by W_1, W_2, and W_3. What distinguishes a connection machine like this from a conventional computer of the type discussed earlier is that the connecting pathways linking the individual computing elements are variable rather than fixed. This means that each connecting link has a weight associated with it, and this weight determines the nature of the signal that can be passed from one computing element to another. Thus, in a connection machine, it's not just the program that dictates what the output will be, but also the pattern of weights attached to the links. The fact that the weights themselves are not fixed, but can be modified during the course of the computation, enables such a machine to display the capability for learning. Let's see how all this works on our simple machine in Figure 5.7.

The individual elements of the machine can be thought of as

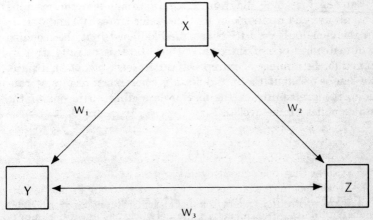

FIGURE 5.7. *A simple Boltzmann machine*

neurons in the brain, which at any moment may either be firing, thereby outputting $+1$, or not firing, thus giving the output 0. Suppose the weights attached to the links are $W_1 = -2$, $W_2 = -1$, and $W_3 = +2$. These weights are applied to all firing signals transmitted along the link. So, for instance, if X and Y are both firing, X will receive a -2 input from Y, and Y will receive a -2 input from X. By convention, an element will fire if and only if the sum of the signals it receives from other elements is positive. To illustrate the way this machine works, let's construct a state diagram showing how the machine will behave under all circumstances.

After a small amount of calculation with Figure 5.7 using the above weights, it's quickly seen that the machine state diagram is as follows:

110
    ↘
       001   ↔   010
    ↗
111

000
    ↘
100   →   000
    ↗
101

011   →   011

From the foregoing diagram, it's easy to see that the machine will always end up in one of the two stable states 000 and 011, or in the cycle 001 ↔ 010. Since the likelihood that the machine will end up in one of these three final states is directly proportional to the number of initial states that lead to that final state, we can say that if the initial state is chosen completely at random, the probability of the machine's ending up in one of the stable states or the cycle is

$$P(000) = \tfrac{3}{8}$$
$$P(011) = \tfrac{1}{8}$$
$$P(\text{cycle}) = \tfrac{1}{2}$$

Unfortunately, this example is a little too small to see the phenomenon of learning, although the network has one very important property that can be seen: It will always go from a state of high energy to one of lower energy. Here a high-energy state is one in which the sum of the weights between pairs of active elements is a large negative quantity, while a low-energy state is one in which this sum is positive. Thus we can think of the energy consumed by the network as being the force needed to operate the negatively weighted connections. It then follows that the states that a Boltzmann machine seeks out are those of minimal energy.

The basic principle of such a machine is that it learns to relate certain input patterns to particular outputs. Inputs and outputs are represented by fixing certain sets of elements by forcing them to fire or not fire, regardless of the weights on the connecting links. Thus, we could require that our elements X and Y fire, thereby representing the input 11. We could then run a series of experiments whose goal would be to teach the machine to output 0 whenever the input quantities were the same (00 or 11), and output 1 whenever they differed (01 or 10). This learning would be carried out by having a feedback mechanism by which the machine itself successively changes the weights W_1, W_2, and W_3 from experiment to experiment. The basic idea is that once the inputs and outputs have been fixed, the network is run in the presence of random noise until it achieves a minimal-energy state. At this point the weights attached to the connections between active elements are increased by a certain amount. Then the inputs, but not the outputs, are again fixed and the process is repeated, except that now when a minimal-energy state is achieved, the weights attached to connections between active elements are decreased by the same amount as was previously added. The result of this process is that if the second set of inputs sent the machine into the "right" internal state, causing it to produce the "right" output, all the weights will have returned to their previous value. But if the output is the "wrong" one, some of the weights will have been permanently changed. By this procedure, the machine will eventually come to a situation in which most of the stable states are those that relate the learned inputs to their corresponding outputs. Moreover, the connections will then arrange themselves so that the network will be able to recognize a wide range of similar, but not identical, input patterns.

Objections to the connectionist view of cognition come in two flavors: theoretical and practical. On the side of theory, the biggest difficulty is that connectionism offers no clear-cut procedure for getting from the low-level energy states to high-level symbolic processing; i.e., there's no prescription for bridging the gap between computation at the hardware level and actual cognition at the level of the software. Critics readily agree that when you turn a connection machine on, something is likely to emerge. But it's not likely to be thinking. The practical objection is that intelligent thinking can never be done in a connectionist network because you could never build a machine with enough connections. Connectionists reply that beyond some minimal level of connectivity, it may be possible to substitute faster switching speeds for more connections.

Connectionism is a very young line of research, and it should clearly be regarded with some measure of reservation. Nonetheless, there is definitely something appealing about the idea of a relatively unprogrammed machine that settles into the creation and recognition of prototypes and patterns. Somehow this strikes me as at least as plausible a model of thinking as a formal, rule-based, highly specified machine. But in either case, what unites the top-downers and the bottom-uppers is the conviction that it is indeed possible for machines to think; they are divided only on the way the thinking is done and the way it can be represented in a medium differing from the human brain. The Prosecution now rests its case: Yes, machines can think! The time has come to allow the Defense to parade its army of philosophers and scientists to the stand in an attempt to convince you that the views of the Prosecution are hopelessly and optimistically misguided. It's to these arguments that we now turn.

PHILOSOPHERS AGAINST: THEY'LL NEVER THINK!

Philosophers have for centuries made a questionable living out of debating issues involving the cognitive capacity of man, and the manner in which various facets of this capacity differ in other forms of life. So perhaps it should come as no surprise that the most virulent arguments heard against the idea of a thinking machine come from the philosophers, as we have already noted in connection with John Searle's Chinese Room experiment. It looks as if the main philosophical arguments opposing the concept of a computer's ever having a real thought come in three primary colors: *phenomenological* arguments based upon the belief that the totality of human understanding cannot be mechanized, *logical* arguments revolving around the limitations imposed by Gödel's theorems, and *antibehavioristic* arguments founded upon the notion that behavioral observation alone is not enough to conclude the presence of genuine cognitive states. Let me now examine each of these philosophical mainstreams in turn.

PHENOMENOLOGY

Moses Hall on the Berkeley campus of the University of California is a little castlelike structure on the other side of the Campanile from the massive, fortresslike concrete blockhouse of Evans Hall, the redoubt of the Berkeley Computer Science and Mathematics departments. This polar positioning is more than just geographic, as over the years Moses has become the command center for a devoted band of loyalists claiming that computers will never think like humans. You see, Moses houses the Berkeley Philosophy Department, and within these hallowed halls walks not only John Searle, he of the infamous Chinese Room, but also Hubert Dreyfus, the philosophical bane of the entire AI community.

Dreyfus is a small, wiry redhead with tortoiseshell glasses, a fondness for plaid western shirts, and a burning zeal for the existential philosophy of the inscrutable German philosopher Martin Heidegger, who promoted the view that a rigorous explanation of the mind would forever be blocked by the impossibility of ever devising a formal representation of the whole of human experience. Dreyfus agrees and, since such a formalization lies at the heart of mainstream AI, he concludes that the development of a program displaying strong AI, human, is a fool's errand; such a research effort is doomed to failure from the very beginning. The core of Dreyfus's claim is that many things central to human thought, like judgment, perception, and understanding, aren't just a matter of following rules. The mind operates against a background of human practices, and it is this shared social background that cannot be formalized.

In his argument against formalization, Dreyfus is joined by

his brother Stuart, also a professor at Berkeley but in the Department of Industrial Engineering and Operations Research, who is responsible for introducing Hubert to the claims and aspirations of AI. As noted earlier, one of the hotbeds of early AI work was the RAND Corporation in Santa Monica, California, where Stuart was employed as an applied mathematician prior to his move to Berkeley in 1967. While at RAND, he observed the work being done by Simon, Newell, Shaw, and others on understanding, problem solving, and chess playing. While a formalist himself at the time, Dreyfus began having reservations about the scientific content of what was being done in the name of AI. During the same period, Hubert, an instructor in philosophy at MIT, was hearing all sorts of wild claims from the students of Minsky and others in the AI Lab that the philosophers were out of date—the traditional problems of philosophy, like perception, understanding, consciousness, and mind, were now being solved at Technology Square on the other side of campus. If this was indeed true, then the philosophers Dreyfus most admired—Heidegger, Merleau-Ponty, and Husserl—must be wrong, since one of the pillars upon which their ideas rest is the notion that these most human of qualities cannot, even in principle, be formalized. Hubert wrote to Stuart at RAND telling him of his MIT experiences, stating that if his philosophers were right, then the AI work being done at RAND was barking up the wrong tree. At this juncture, fate intervened in the form of Paul Armer, now at Stanford but then head of RAND's Computer Science Division, who had already realized that much of the RAND AI work was addressing deep philosophical issues, and who felt that it would be useful to mix a philosopher or two in with the AI crowd. As a result, and at Stuart's suggestion, Armer hired Hubert as a consultant to RAND for the summer of 1964. Little did Armer realize the tempest to be unleashed by that seemingly innocent summer consultancy.

The output of that summer exercise was a paper, "Alchemy and Artificial Intelligence," in which Dreyfus compared the research program of AI to the attempts by medieval alchemists to transmute lead into gold. The paper burst like a bombshell in the AI community, being roundly and soundly denounced both as bad philosophy and as a vicious, inaccurate attack on AI and the motives of the AI researchers. In fact, the paper aroused such strong emotional reactions that for several months the question of whether or not to issue it as an official RAND report was debated within the upper echelons of the corporation. Eventually the matter was settled by an appeal to the principle that just because some people didn't like the conclusions reached by another scholar, that was no reason to suppress publication of the work (at RAND, anyway) if it contained no genuine logical errors or factual mistakes. With the piece's publication as a RAND Working Paper (the lowest category in the RAND publication hierarchy) and the consequent implied corporate imprimatur, the battle was joined between the AI community and the philosophers. Ironically, the long-suppressed paper turned out to be one of the biggest sellers in RAND publication history, no mean feat in a list that has included such influential publications as Herman Kahn's *On Thermonuclear War,* Charles Hitch's work on the economics of defense, and pioneering technical monographs on game theory, computer science, and linear and dynamic programming. Responding to this groundswell of popular support, Dreyfus later expanded the paper into the provocative book *What Computers Can't Do,* exposing his phenomenologically based objections against AI to a wider, public audience, and has recently updated and expanded his arguments in the volume *Mind over Machine,* coauthored with Stuart, who somewhere along the line converted to the existentialist persuasion. It's of psychological, if not intellectual, interest to have a closer look at the style and content of the arguments that could so uniformly incense the entire artificial intelligentsia.

The distilled essence of the Dreyfus position can be expressed in the following syllogism:

I. The AI community claims that thinking is the manipulation of formal symbols according to rules.

II. Phenomenology claims that knowing, understanding, perceiving, and the like involve more than just following rules.

III. Phenomenology is correct.

THEREFORE

No amount of AI, however clever, will ever duplicate human thinking.

It goes almost without saying that Dreyfus's detractors question every one of the premises on this list.

One of the favorite arguments of the brothers Dreyfus involves the way in which one acquires expertise in the performance of some task like playing chess or driving a car. In the Dreyfus scheme of things, gaining expertise at driving a car (or anything else) involves a successive passage through five identifiable stages:

- *Novice:* At this lowest skill level, context-free rules for good driving are acquired. Thus, one learns at what speed to shift gears and at what distance it's safe to follow another car at a given speed. Such rules ignore context-sensitive features such as traffic density and weather conditions.
- *Advanced beginner:* Through practical on-the-road experience, the novice driver learns to recognize concrete situations that cannot be described by an instructor in objective, context-free terms. For instance, the advanced beginning driver learns to use engine sounds as well as the context-free speed as a guide for when to shift gears, and learns to distinguish the erratic behavior of a drunk driver from the impatient actions of an aggressive driver in a hurry.
- *Competence:* The competent driver begins to superimpose an overall driving strategy upon the general rule-following behavior of the novice and the advanced beginner. He or she is no longer merely following rules that permit safe and courteous operation of the car, but drives with a goal in mind. To achieve this goal, the competent driver may now follow more closely than normal, drive faster than is allowed, or in other ways depart from the fixed rules learned earlier.
- *Proficiency:* At the previous levels, all decisions were made on the basis of deliberative, conscious choices. But the proficient driver goes one step further and makes decisions on the basis of a feel for the situation. There is no deliberation; things just happen. So, for example, the proficient driver when attempting to change lanes on a busy freeway may instinctively realize that there's another car coming up on the blind side and delay making a move. This instinctive reaction may arise out of experience in similar situations in the past and memories of them, although it may appear as an unexplainable "lucky guess" to an outside observer. Somehow there is a spontaneous understanding or "seeing" of a plan or strategy.
- *Expert:* An expert driver no longer sees driving as a sequence

of problems to solve, nor does he or she worry about the future and devise plans. He simply becomes one with his car, and experiences himself as just driving rather than as driving a car. Thus, an expert driver has an intuitive understanding of what to do in a given setting. He doesn't solve problems and he doesn't make decisions; he just does what normally works.

The moral of this fable in five parts is that there is more to intelligence and expertise than mere calculative rationality. Expertise doesn't necessarily involve inference; the expert sees what to do *without* applying rules. This is the essence of the Dreyfus argument against the possibility of a rule-based program's ever achieving anything that even remotely approximates genuine human intelligence.

The AI community welcomed this line of argument with about the same level of enthusiasm as Stalin welcomed Trotsky. When Dreyfus was invited to make a keynote address at a general computer conference some years ago, the redoubtable Allen Newell complained to the meeting's organizers that "that kind of platform gives him [Dreyfus] an authority and credibility he's simply not entitled to." Perhaps the most extensive critique of the Dreyfus position was put forth by Seymour Papert, who wrote a long reply to Dreyfus titled "The Artificial Intelligence of Hubert Dreyfus." In this lengthy document—which, interestingly enough, was solicited by Dreyfus's RAND sponsor, Paul Armer—Papert accuses Dreyfus of devoting much of his argument to nothing more than gossip, with most of the remainder composed of statements made by others that Dreyfus felt fit his strongly held preconceptions. Other mainline AI types, like Schank and Feigenbaum, weighed in with comments to the effect that "everything is impossible until you do it" (Schank), and that "every time you confront him [Dreyfus] with one more intelligent program, he says, 'I never said a computer couldn't do that'" (Feigenbaum). To my mind, the most reasonable criticism comes from one of the Young Turks in the field, Robert Wilensky, who states that

certainly there are some things that are formalizable, and some things that resist it more and more. But where do you draw the line? And can you continue pushing the line further and further? Those are the interesting questions, and my real objection to Dreyfus is, why say at this stage that it's going to fail?

In response to his critics, Dreyfus asserts that

> I would be willing to bet that in twenty years it will be settled—
> that people will either be clearly on the right track or that no one
> will be interested any more. And my real hunch is that in twenty
> years people won't be trying—that the wrong-headedness of their
> approach will be as obvious as the wrong-headedness of alchemy.

But the arguments of Dreyfus are not the only philosophical
weapons arrayed against strong AI, human. Let's move from
soft existentialism to hard mathematics and examine John
Lucas's appeal to Gödel's theorems as the basis for discrediting
the very idea of a thinking machine.

MATHEMATICS AND LOGIC

Earlier we saw that Gödel's theorems show both that any reasonably rich formal system is incomplete and that the consistency of
such a system cannot be proved within the system itself. Furthermore, in Turing's work we saw that formal systems and machines are equivalent in what they can do. Ergo, computers are
subject to the same limitations that Gödel imposed on any formal system. Thus, machines are inherently limited in what they
can do and, in particular, there are statements that the mind
knows to be true but that the machine cannot prove. Interestingly enough, Turing anticipated this kind of objection to AI in
his classic 1950 paper on thinking machines, replying that people
may well be subject to similar limitations. The British philosopher John Lucas wasn't convinced by Turing's response, and
wrote a paper in 1961 titled "Minds, Machines, and Gödel," in
which he attempted to strengthen the Gödelian argument against
the view that the mind is a machine or, in Marvin Minsky's wonderfully colorful term, a "meat machine."

The heart of the Lucas argument takes the following course.
By standing outside the incomplete, consistent formal system,
we can see some unprovable statement to be true. But the machine cannot prove this fact; hence, a human can beat every machine, since such a true but unprovable statement exists for
every machine. Furthermore, if the human mind were nothing
more than a formal system, by Gödel's other theorem the mind
could not prove its own consistency. But humans do proclaim
their own consistency. Consequently, the mind must be more
than a machine. Since Lucas's notorious paper appeared in 1961,

long before any computers were being programmed to display behavior that looked much like human thinking, most of the controversy surrounding his arguments was confined to the philosophical community, and provided solid testimony to support Ludwig Boltzmann's observation that "there is much that is appropriate and correct in the writings of these philosophers. Their remarks, when they denounce other philosophers are appropriate and correct. But when it comes to their own contributions, they are usually not so."

As with virtually all philosophical debates, the arguments against Lucas hinge upon the precise meaning he gives to terms like *machine,* as well as the hidden assumptions underpinning his conclusions. For example, Paul Benacerraf points out that Lucas has too limited a view of machines, since any machine that could reprogram itself in the face of a changing environment would be exempt from the Gödel argument. Furthermore, it is also noted that Lucas *assumes* that mind is consistent. In fact, this is far from obvious, as the following paradox constructed by C. H. Whitley shows.

Consider the sentence "Lucas cannot consistently assert this sentence." Lucas cannot assert the truth of this sentence even though he can clearly see that it's true. Why? Because if Lucas could assert it, then that fact would undermine his assumed consistency. Thus, either there is something that Lucas can see to be true but can't assert, or he is inconsistent. Consequently, Whitley claims that Lucas holds too high a regard for humans, since even if there is an unprovable statement that a specific machine cannot assert, humans can't always do it either.

Other arguments countering Lucas claim that he errs in his application of Gödel's results. For instance, the Incompleteness Theorem shows that a machine M cannot prove the Gödel sentence of M from *its* axioms and according to *its* rules of inference. But neither can mind. Furthermore, Lucas doesn't show that he can find a flaw in any machine, but only in any machine that the mechanist can construct. In this same connection, it's well to bear in mind Gödel's own view that there could exist a machine whose abilities equaled human mathematical intuition, but whose program we could never understand. Nonetheless, we would be able to set up conditions leading to the existence of such a machine, e.g., by evolution. Thus, machines too complex to design could nevertheless exist.

To my mind, the most intriguing arguments rely upon turning

Lucas on his head. Rather than suggesting that our self-knowledge proves we are better than machines, one could equally well use the fact that formal systems cannot know themselves to claim that human self-knowledge isn't possible. In other words, if I am a Turing machine, then my very nature forbids me to know everything there is to know about myself. And thus does the mathematically based argument against thinking machines turn out to be as inconclusive as the phenomenological arguments of Dreyfus. So let's go back to the Chinese Room as a last attempt to salvage some philosophical grist for the antimechanist's mill.

ANTIBEHAVIORISM

We have already met John Searle, Dreyfus's colleague at Berkeley, in connection with the Chinese Room argument. Searle is a short, tanned, solid-looking man who speaks with a booming voice, conveying the impression of someone born to the exercise of power. He is also a philosopher of language of some repute, and a virulent opponent of the syntactically based Chomskian school of linguistics. In 1984 he was invited to give the Reith Lectures on the BBC, an annual series in which the speaker is charged with introducing a general audience to some of the leading intellectual issues of the day. He used this opportunity to sharpen and extend the arguments given earlier in his Chinese Room paper about the nature of mind and its possible connection to digital computers. Searle's main assertions are: (1) no computer program is, by itself, sufficient to give a system a mind; (2) the way the brain functions to cause mind cannot be solely by virtue of running a computer program; (3) anything else that causes minds would have to have causal powers at least equivalent to those of the brain; (4) for any artifact that we might build that had mental states equivalent to human mental states, the implementation of a computer program would not by itself be sufficient. Rather the artifact would have to have powers equivalent to the powers of the brain. To support these contentions, Searle offers the following chain of reasoning:

- Brains cause minds.
- Minds have mental content; specifically, they have semantic content.
- Syntax alone is not sufficient for semantics.

• Computer programs are entirely defined by their formal, or
  syntactic, structure.

Searle uses the Chinese Room to illustrate what he claims is the
commonsense, obviously unassailable nature of his position. By
now we recognize that nothing in philosophy is "obvious" and,
as would be expected, the outcries against Searle from within
the AI community are loud and long.

One of the most persistent rebuttals to the Chinese Room
is the claim that while the man inside the room may not understand Chinese, the *entire system* consisting of the man, the
flashcards, the dictionary-rulebook, and so on does display understanding. Searle attempts to counter this "systems" reply by
the ingenious device of internalizing the whole system; i.e., he
suggests placing the entire system within the brain of the man
by having him memorize the rulebook and the flashcards and
doing away with the irrelevant physical confines of the room itself. In this way, the whole system is inside the man but, Searle
argues, the man still doesn't *understand* a word of Chinese. Others base their objections on Searle's claim that "people who accept it [the Turing Test] miss the distinction between simulation
and duplication." For example, the philosopher Richard Rorty
argues that Searle's insistence on the difference between the Chinese Room simulation and real human thought is equivalent to
an orthodox Catholic's argument that the Eucharist conducted
by a "demythologizing Tillichian theologian" or even an Angli-
can does not transform the wafer into the Body of Christ.

In Searle's defense, it has been pointed out that the kind of
behavioral tests for understanding exemplified by the Imitation
Game rest on shaky empirical grounds. For example, in testing
the linguistic understanding of symbol-manipulating chimpanzees, Eric Lenneberg discovered that while the chimpanzees
could successfully manipulate the symbols, high school students
manipulating the same symbols (and with fewer errors) thought
they were solving puzzles, and were unable to translate a single
one of their completed "sentences" into English. Thus, just because a machine (the chimp) can successfully manipulate symbols, there is no necessity for believing it understands the
language.

But not all philosophers subscribe to Searle's pessimistic view
of the adequacy of the Turing criterion. A particularly eloquent

defender of the Turing Test is Daniel Dennett, who stresses the extreme generality of the test and the amount of world knowledge that would be needed to pass it. Dennett concludes that without sufficient world knowledge it would be impossible to pass the test, and that if a machine passes the test, it's safe to *assume* that the machine has the requisite world knowledge. Thus, any computer passing a full-blown Turing Test will be, in every interesting sense, thinking. Dennett claims that Searle underestimates the power of such a machine, and that it could, in principle, actually learn to understand Chinese. In his reply, Searle concedes this as a possibility, arguing that whether this can actually be done is an empirical question. He then concludes by restating his main point, that merely being an instance of a formal system is not enough to prove that a machine is really thinking.

After all the smoke clears away, it seems that the main contribution of *l'affaire Searle* to the thinking machine debate was to provide some very much needed sharpening of the issues involved in the Turing Test, as well as offering a lightning rod upon which many conflicting views of the debate could be focused and put on record. We now rest the case for the Defense: Machines will never think! But before listening to the closing arguments, a bit of supplemental testimony from two friends of the court will be of value in illuminating some additional facets of the overall question.

THE MORALIST AND THE MYSTIC

In the promotional material being ground out by today's go-go expert-system propagandists, a favored place is always reserved for programs claiming to relieve the medical community of the burden of deciding whether a trickle of blood at the lip and a pain in the gut should be attributed to bleeding gums and excess stomach acid or to a slug from a Colt .45. The hope underlying all such programs is that by describing a bunch of symptoms, and then correlating the symptoms with lots of rules of thumb painfully extracted from the minds of human clinical diagnosticians, it will be possible to aid physicians significantly in the performance of their daily chores. Leaving aside the dubious merits of such an argument, it will shed some epistemological light on the overall question of machine intelligence if we spend a moment examining some of the historical accidents lending a mild aura of credibility to the idea of a "mechanical doctor."

THE MORALIST

Around twenty years ago, there was a brief flurry of interest in the notion of creating computer programs that could act the role of a psychotherapist in conversing with mentally ill patients. Joseph Weizenbaum, at the time a young engineer at MIT, had previously been exposed to the idea of computer-aided therapy through an association with Kenneth Colby, a psychiatrist at Stanford who had grown unhappy with conventional one-on-one therapy and was looking toward the computer as a new tool for probing the human psyche. Having been sensitized to the issues of computer therapy, Weizenbaum put these ideas together with his other interests in man-machine interactions to create *ELIZA*, a program that could simulate, under limited circumstances, a Rogerian-style therapist. A brief excerpt from *ELIZA*'s most publicized conversation gives the basic idea:

PATIENT: Men are all alike.

DOCTOR: IN WHAT WAY?

They're always bugging us about something or other.
CAN YOU THINK OF A SPECIFIC EXAMPLE?

Well, my boyfriend made me come here.
YOUR BOYFRIEND MADE YOU COME HERE?

He says I'm depressed much of the time.
I AM SORRY TO HEAR YOU ARE DEPRESSED.

It's true. I am unhappy.
DO YOU THINK COMING HERE WILL HELP YOU NOT TO BE UNHAPPY?

At first glance, this looks like a pretty impressive piece of work. However, a little digging behind the scenes soon exposes the whole "conversation" as an illusion created from a handful of programming mirrors, with the program being designed to respond to particular kinds of clues in prescribed ways. For instance, to words like "father," the program offers back, "TELL ME MORE ABOUT YOUR FAMILY"; statements involving "always" or "all" result in requests for examples; after statements that don't fit into a predefined pattern, the program goes fishing with

"DOES THAT HAVE ANYTHING TO DO WITH . . . ?" mentioning some earlier topic that has been seen to be of significance to the patient because it was prefaced with "my."

Weizenbaum thought of *ELIZA* as nothing more than a challenging programming exercise. Imagine his astonishment when he found people taking the program in dead seriousness. His own secretary insisted on closing the door to the computer terminal room as she unburdened herself to the program, and people were calling him at all hours frantically pleading for just a little time with *ELIZA* to get themselves straightened out. One internationally known Russian computer scientist, interacting with a companion program at Stanford, began unloading a whole plethora of fears about himself, his family, his career, and so forth before an audience of embarrassed onlookers. If a person as knowledgeable about the inner workings of the program as he was could be enticed into making such intimate disclosures to the machine, Weizenbaum felt there was indeed cause to take seriously the moral implications of AI and the potential cost in human terms of such widespread acceptance of the view that human beings were basically just complicated machines. The result of his deliberations on these moral issues was his book *Computer Power and Human Reason,* which appeared in the spring of 1976, ten years after *ELIZA.*

Just like Dreyfus's book, Weizenbaum's was greeted with outrage and polemical attacks by the AI community. The book advances the thesis that the information-processing view of man is one aspect of a twentieth-century trend toward thinking of human beings as means, rather than as ends, and toward considering contemporary social and human problems as being largely amenable to quick-fix, technologically based solutions. Weizenbaum forcefully asserts that the empirical evidence conclusively demonstrates the falsity of the information-processing model of humans and, even more important, that such a view is just plain morally wrong. The critique ends with a call to the computer science community not to promote a vision of human beings that acts to dehumanize them further. He uses his belief that "the computer . . . enslaves the mind that has no other metaphors and few other resources to call upon" as support for his central point: that thinking of people as programmed machines will influence the decisions we make about how we treat them in today's technically oriented world. Finally, he emphasizes that there are domains where computers ought not to intrude, whether or not it's feasible for them to do so. The above psychiatric situation is a prime illustration of the kind of domain Weizenbaum has in mind, one in which interpersonal respect, human understanding, and empathy are required.

Reviewers of the book found much to occupy their typewriters with in projecting their own views of the technology-versus-humanity conflict onto Weizenbaum's *cri de coeur.* John McCarthy noted that if something shouldn't be done, then it shouldn't be done at all—by man or machine. He then went on to compare Weizenbaum's arguments to the objections posed by the Renaissance Church to dissecting the human body because it was the temple of the soul. McCarthy further noted that "when moralizing is both vehement and vague, it invites authoritarian abuse either by existing authority or by new political movements." One of the sharpest criticisms came from Weizenbaum's former associate Kenneth Colby, whose further work on computer-aided psychiatry was the object of a particularly strong attack in the book. Colby wrote:

> Over the past four centuries the scientific community has come to mistrust suppressions of inquiry, not only because they protect the status quo but because so often the finger-wagging moralist has turned out himself to be morally confused, piously self-serving, and irresponsibly blind to the consequences of his own oppressive actions.

This is the kind of sound advice that in my opinion should flash on the screen before the appearance of every politician and TV evangelist. In light of recent history, even Jim Bakker and Jimmy Swaggart might now agree!

Finally, a number of criticisms were directed not so much at the book itself as at Weizenbaum's personal motivations for writing it. Some argued that Weizenbaum could no longer do science and, tenure or not, at a competitive place like MIT the pressure is always on to produce. Hence, the turn away from doing science to becoming the conscience of the AI community. These complaints again underscore the sociopsychological factor at work in the genesis of what's taken to constitute scientific "truth."

Outside the AI world, but still within the rather limited confines of the general scientific community, the reception to Weizenbaum's brand of moralizing went down much more smoothly. Writing in *Datamation,* one of the leading computer periodicals, the noted computer programming author Daniel McCracken affirmed Weizenbaum's view that there were basic differences between men and computers, differences that would never disappear. The British computer scientist N. S. Sutherland, writing in *The Times Literary Supplement*, said that "he [Weizenbaum] raises important issues that are too often ignored. He repeatedly and correctly insists that computers lack wisdom, but if computers are put to ill use, it is because we, not they, lack wisdom."

But what does all this have to do with the basic question of whether machines can think? By taking a moral stand against AI, Weizenbaum has introduced an ingenious argument against the idea of strong AI, human. Rather than basing his arguments against machine cognition on technical and epistemological grounds of the sort put forth by Searle, Lucas, and Dreyfus, Weizenbaum advances the novel contention that even if strong AI, human, is technically feasible, it is *morally* impossible! Of course, he along with everyone else in the game agrees that we are presently very far from anything that even smells like strong AI, human, but then adds the moral imperative that whether we're close or far away is irrelevant, since the very attempt to achieve genuine machine intelligence itself acts to undermine our sense of humanity.

The only way Weizenbaum's misgivings can be other than a moot point is if the research needed to decide the matter is allowed to run its full course, not prematurely terminated as Weizenbaum suggests because it *might* lead to harmful consequences. In fact, a rigorous adherence to the Weizenbaum dictum would lead to the absurd situation in which research in virtually every area of science would be stopped dead in its tracks since any discovery could, in principle, lead to "dehumanization" by showing that something we once thought of as the preserve of humans alone is not so special after all. Rather than pursue this essentially nonscientific line of thought, let's turn our attention to another kind of visionary, but this time one whose ideas on mind and machine come down on the Prosecution's side of the case.

THE MYSTIC

Rudy Rucker is a professional logician as well as the well-known author of several popular books on mathematics, relativity, and geometry, in addition to a number of offbeat science fiction novels traversing some of the same territory. If one can judge from dust jacket photos, Rucker, with his almost shoulder-length hair and studded leather jacket, looks as if he'd be as much at home on a motorcycle on his way to a rock concert as in front of a blackboard selling the Law of the Excluded Middle to a roomful of heavy-lidded undergraduates. Having never met him, I can't say if this is true, but one thing *is* certain: When it comes to matters of machines, minds, and souls, Rucker is a mystic of the first rank.

In his book *Infinity and the Mind,* a semipopular account of various logical paradoxes as well as some of the content and implications of Gödel's work, Rucker devotes considerable space to the question of whether mathematical logic can shed any light on the matter of souls for robots. For all practical purposes, the matter of machine souls and consciousness is indistinguishable from what we have been terming strong AI, human, so Rucker's thoughts on the possibility of "machine dreams" are of some interest.

According to Rucker, there are three possible views on the question of human and robot souls:

• *Mechanism:* Neither people nor robots are anything but machines, so there is no reason why humanlike machines cannot exist.
• *Humanism:* Human beings have souls but robots do not; therefore, no robot can ever be quite like a person.
• *Mysticism:* Everything participates in the Absolute, so it should be possible for humanlike machines to exist.

The pro-AI forces have already argued the first view for us, while the philosophers have waxed long and eloquent over the second. Rucker supports AI, but with the bizarre twist of appealing to the mystical and mysterious Absolute. Let's take a longer look at how one could possibly take such a notion seriously while, at the same time, adhering to the austere rigors of mathematical logic.

The heart of Rucker's argument for mysticism is, first of all, to observe that the individual person consists of three separate parts: (*a*) the *hardware,* composed of the physical body and brain; (*b*) the *software,* comprising memories, skills, and, in general, behavior; (*c*) the *consciousness,* representing the sense of self or personal identity—in short, the soul. The key element in Rucker's position is to now note that it's possible to replace or change any part of either the hardware or the software while leaving (c) unaffected. We're all familiar with changes in the physical body, like artificial hearts, replacement limbs, and false teeth, and no one would entertain for a moment the idea that such changes in any way touch the soul. Rucker makes the not entirely trivial extrapolation that, in principle, an artificial brain could also replace the original and still leave the soul unaffected. For the sake of argument, we'll let this point pass. Alterations in (b), such as forgetting past experiences, learning new skills, or more drastic changes like brainwashing of prisoners of war, also happen without our ever feeling that the person's essential identity has been modified. So what remains for part (c), the soul? According to Rucker, only the primal feeling of existence: Descartes's *sum,* "I am!" This is the only thought that ties us to what we were in the past, or what we may become in the future. This observation is then used to conclude that mere existence means to have consciousness. From here it's smooth sailing and but a small step to the mystic's claim that everything participates in the Absolute, where the Absolute is identified with existence, and hence there is no logical obstacle to machines' having souls (consciousness) just like human beings. Thus, Rucker states that where the classical pro-AI materialist would argue that "men are no better than machines," the mystic replies by claiming that it's just the other way around, that "machines can be as good as men."

We have now run our course and, to some degree, come full circle from formal systems as thinking machines, through philosophical and moral objections to the very notion, and on to robot souls and the universal Absolute. Finally the time has come to listen to the closing arguments and summarize the competing positions, before retiring to the jury room to ponder the verdict.

SUMMARY ARGUMENTS

Our odyssey through the labyrinths of psychology, computer science, mathematics, and philosophy started with the deceptively simple query "Can machines think?" Along the way, we elaborated and sharpened the question of interest to the claim that "an appropriately programmed computer can possess states that are functionally equivalent to the cognitive states of a human brain," and more succinctly expressed this assertion as "strong AI, human." The following diagram compactly summarizes the overall situation:

COMPUTERS		MINDS
Machine states	:	Brain states
+	⟺	⇕
Programs	:	Cognitive states

Briefly, the pro-AI forces (top-down or bottom-up) claim that it's possible to fill in the double arrows with convincing scientific arguments, backed up with actual working computer programs; the anti-AI community says, no way! Tables 5.1 and 5.2 show in abbreviated form the kinds of arguments presented by both sides to support their respective positions. After inspecting the tables, let's step into the jury room and come to some sort of judgment on the question of thinking machines.

YES, COMPUTERS CAN THINK!

PROMOTER	RESEARCH PROGRAM
(top-down school)	
Turing, Dennett	Imitation Game
Simon and Newell	rule-based symbol manipulation
Schank, Wilensky	script following, frames
(bottom-up school)	
Hofstadter, Lenat	subcognitive modules
Minsky	"Society of Mind"
Hinton	Boltzmann machine, statistical mechanics
Rumelhart	deterministic neuronal network
(mystical school)	
Gödel	evolution of "incomprehensible" machines
Rucker	universal participation in the Absolute

TABLE 5.1. *Summary arguments for the Prosecution*

NO, COMPUTERS CANNOT THINK!

PROMOTER	ARGUMENT
Searle	Chinese Room
the Dreyfuses	phenomenology
Lucas	Gödel's theorems
Weizenbaum	immorality

TABLE 5.2. *Summary arguments for the Defense*

BRINGING IN THE VERDICT

The evidence having been submitted, the arguments heard, and the pros and cons weighed, I vote for conviction and cast my ballot with the Prosecution in favor of the possibility of strong AI, human. My reasons? Well, as Sherlock Holmes so wisely noted in *The Adventure of the Beryl Coronet,* "When you have eliminated the impossible, whatever remains, *however improbable,* must be the truth." Basically, I tried to take the arguments summarized in Tables 5.1 and 5.2 and eliminate as many contenders as I could on the grounds of, if not impossibility, then at least implausibility, or in some cases what seemed like pure sophistry or just plain irrelevance. Let's examine the arguments against the philosophers first, in increasing order of difficulty.

To my mind, Weizenbaum's argument from morality can be dismissed at the outset as fundamentally irrelevant to the question to be decided, namely, whether it is in the realm of *possibility* for a machine to think. While I accept the position that scientists bear some measure of responsibility for keeping tabs on the possible social consequences of their work, and even for making these potential consequences known to a wider audience, the possible dehumanizing effect of a genuine thinking machine seems to me to have no bearing whatsoever on the possibility of actually constructing such a device. In fact, if anything I think such potentially negative social and psychological consequences serve as added motivation for pushing on with the research needed to settle the matter. Either strong AI, human, is possible or it's not; if it isn't, then Weizenbaum has raised a moot point; if it is, then it's important to know the nature and degree of that machine intelligence, as this is precisely the sort of knowledge needed to decide just exactly what *kind* of machine we really are. So, all things considered, I think it's safe to eliminate Weizenbaum's case from the competition.

It seems almost as easy to drop the Dreyfus argument from the list of contenders. The core of the Dreyfuses' claim is the phenomenological assertion that many crucial aspects of human thinking like judgment, understanding, and perception cannot be formalized. To support their case, the Dreyfuses present what amounts to anecdotal evidence involving such pursuits as the acquisition of skills and expertise in activities like chess playing, driving, poetry writing, and so forth. There are many things I don't like about this line of reasoning, but the most important is the *ex cathedra*–like pronouncement: Phenomenology says! On what grounds, other than faith, can one swallow the conclusions of the phenomenological philosophers? The whole edifice of the Dreyfus case rests on what amounts to the religious claim that Husserl, Heidegger, & Co. are right. But to my eye, the Dreyfuses put forth anything but a knockdown argument supporting this crucial assumption. Furthermore, I think it's important to note that they are primarily arguing against the top-down AI programs of the Simon and Newell type. Thus, even if through some unforeseeable set of circumstances their phenomenological thesis could be proven correct, I fail to see how this fact would begin to touch the program of the bottom-up school. Putting these observations together, I think it's also safe to scratch the Dreyfuses from the race.

Unlike Weizenbaum's position or the Dreyfuses', Lucas's appeal to Gödel has the surface ring of something you can really get your teeth into: tangible, to the point, and mathematically airtight. But Gödel's results, like all high-precision tools, apply to a very definite and severely restricted set of circumstances, and it seems to me that Lucas has stretched these conditions beyond the breaking point in his efforts to invoke Gödel as an argument against thinking machines. I have already outlined what I see as many convincing objections raised against Lucas's use of Gödel's theorems, so it's not necessary to repeat them here other than to note that Gödel himself didn't appear to see his work as any kind of obstacle to the existence of intelligent machines. And what's good enough for Gödel is certainly good enough for me! Thus does Lucas, too, fall by the wayside.

Oddly enough, I find Searle's argument based on the first-person perspective of the Chinese Room to be the most compelling, and it's with some trepidation that I finally cast it aside along with the others. The two axioms underpinning Searle's claims are (1) brains cause mental states, and (2) no amount of syntax alone can ever generate semantics; i.e., no amount of form will ever produce content or meaning. Personally, I have reservations about the first point and completely disagree with the second. To begin with, when Searle uses the word "brain" he means the kind of human brain that each of us has sitting up there between our ears. He later goes on to say that any program that satisfied the conditions for strong AI, human, would have to have the causal powers of just exactly this kind of brain. While I definitely subscribe to the view that hardware counts, I don't see any compelling reason why those mysterious "causal powers" couldn't be present in a machine, too. As Daniel Dennett has put it, strong AI presupposes that "it ain't the meat, it's the motion," while Searle believes that "it's the meat," and, moreover, only the human brain is the right kind of "meat." Without something more substantial to support this contention, I'm afraid it's unacceptable to me. In fairness to Searle, he has stated that whether any entity other than the human brain could have the right "causal powers" is really an empirical question. So let's grant this "meat versus motion" point to Searle and consider his second pillar, semantics from syntax.

The real heart of the Searle case is that no amount of formal symbol processing will ever enable a system to "understand" what the symbols actually "mean." Referring back to the Times Square message board and its flashing lights, the claim is that no matter how long and hard the board works at switching those lights on and off, it will never "know" whether it's telling you about tomorrow's weather in New York or today's coup in Gabon. All it knows or ever will know is the flashing of lights according to specified rules, i.e., pure syntax. I completely disagree with this claim, at least as Searle states it. The crux of our disagreement is very simple: Searle fails to note that the "syntax $\not\Rightarrow$ semantics" conclusion may fail when one observes the syntax at a higher level. Thus, at the level of the flashing lights there is indeed only form; however, if the message board could somehow jump outside this level and look at itself (as we do), then new possibilities would appear, among them the emergence of content from form. In order to make this kind of level jump, the system must possess some concept of *self-reference*. While it's true that message boards are not known to contain internal models of themselves encapsulating such "self-seeing" abilities, other kinds of systems do display such a capacity.

The canonical example of such a self-referential system is a living cell, in which the information coded in the cellular DNA has both syntactic and semantic content, both of which are used in the cell's metabolic and reproductive cycles. The point here is that the chemical sequence on the DNA strand can, when seen at one level, look like pure syntax, while at another level the *identical* chemical sequence can be interpreted, and thereby acquire semantic content from what originally appeared to be pure syntactic form. It seems unlikely that this dual-level property of the cell has always been present, having most likely evolved over the millennia under evolutionary pressures. So I see no reason why the same kind of evolutionary "emergence" could not happen with machines. In fact, this is exactly the sort of thing that Gödel seems to have had in mind when he spoke of the possibility of our being able to set up the conditions for the coming into existence of a machine whose program we could not understand. Such a machine, too complex to understand, could nevertheless evolve and be *empirically discoverable*. So I'm afraid that I'll have to also reject Searle's claim, and with it the last, best hope for a convincing philosophical case against strong AI, human. Now, since I'm dreaming in print anyway, allow me to take a moment and comment upon the various AI schools and give a somewhat prejudicial assessment of what I see as the plausibility of their respective research programs.

As with all religions, the only thing all the AI faithful can agree upon is the answer to the basic existential question: Is strong AI, human, theoretically feasible? All are in accord that the answer is definitely yes, and that we are far away from having reached this computational state of grace. In fact, the barriers separating the various believers are their different manners of achieving salvation, that is, the philosophies they employ in writing what they hope will be the first genuinely cognitive program. Not to get too exercised at the outset, let's tackle the mystics first.

The mystical school is easy. Its research program consists

merely of trying to show that strong AI, human, is not logically impossible. In this limited but essential task, it seems to me the mystics succeed. Unfortunately, their line of attack is a "program without programs," so to speak, and as a result we're left with the same feelings of dissatisfaction that come over us when we encounter an indirect proof in mathematics—the kind of proof where you assume something is true, then use that assumption to derive a logical contradiction, thereby refuting the original assumption. The most famous example is Euclid's proof of the infinitude of prime numbers (positive integers divisible only by themselves and by 1). Euclid assumed there were only a finite number of primes, and then showed this assumption leads to a logical contradiction; hence, there are an infinite number of primes. While the argument is logically beyond reproach, many mathematicians (myself included) would have been happier with a *constructive* proof, in which a prescription was given for actually cranking out the primes one after the other, together with an argument showing this algorithm would never stop. Regrettably, it can be shown that no such recipe for primes exists, so in this case perhaps Euclid is as far as we can go. But when it comes to AI, the proof is in the program, and the mystics offer no programs. So let's turn to the two main contenders in the AI race, top-down and bottom-up.

At the grand scientific Academy on the island of Laputa, Gulliver encountered a wonderous architect who "had contrived a new method for building houses, by beginning at the roof, and working downward to the foundations." As far as I can tell, the ingenious methods of this architect seem to have been passed on to his intellectual inheritors, the top-down AIers. Somehow the idea of programming meaning into a set of symbols and then letting those symbols interact according to specified rules, thus creating a semantic network of some kind, just doesn't have the right feel to me. It escapes me as to why the given high-level rules for symbol interaction need bear any natural relationship to whatever rules the brain might actually be using—if indeed there are any such rules, in the high-level, top-down sense of that term. In fact, it's manifestly clear in numerous psychological experiments involving chess players, list memorizers, and the like that the way the top-downers have programmed their computers to perform these tasks bears little resemblance to the way humans carry out the same activities.

In addition, there's the not so minor matter of human evolution. Presumably, whatever cognitive capacity the human brain possesses was acquired somewhere along the line of development from an earlier, protohuman, reptilian sort of brain. In other words, the ability to represent the world symbolically and to operate mentally with those symbols arose as an *emergent* property out of whatever hardware happened to be available at the time. So it seems reasonable to me to take as a working hypothesis that there might be something special about that particular type of hardware, and whatever that something special may be, it cannot be omitted if you're in the business of trying to duplicate with another kind of hardware how humans actually think. Now before you start thinking that this contradicts my earlier objection to Searle when he claims that probably only the human brain is the right kind of hardware, let me hasten to note that I firmly believe that duplication of human cognitive processes in a machine is a feasible task. What I don't buy is the top-down idea that hardware doesn't matter. In this regard, I'm totally in sympathy with the bottom-up position that hardware is important, but that there's no reason to think that a brain made out of organic neurons is the only kind of hardware that can have, in Searle's phrase, the right "causal powers." I have yet to see any convincing evidence to indicate that whatever the "something special" is that brought about the emergence of human mental states, it couldn't be functionally implemented in silicon instead of "neuron stuff." Which brings me to consideration of the final school, the bottom-uppers.

By now it should be patently clear that I reserve my real sympathies for the thesis and program of the practitioners of bottom-up AI. A crucial factor underlying my generally favorable view of the bottom-up approach goes all the way back to one of the foundational issues upon which the whole thinking-machine debate rests: the distinction between a model (duplication) and a simulation. John Searle has attached great significance in his argument to the contention that a simulation is not a duplication, and that a machine cannot duplicate human thinking but at best only simulate it. I've already dealt with what I see as the fallacies in Searle's line of reasoning, but I do agree with him on the point that simulation does not equal duplication. Since confusion on this point is rampant in the AI literature on thinking machines, this is a good moment to elaborate upon the distinction, especially as it occupied such a central place in the formulation of my views on the importance of bottom-up AI.

Suppose we have two sorts of objects, let's say a Boeing 767 jet and a second object that someone claims is a "duplicate" or a "model" of the 767. Just what would this mean? What would it take to be a model of a 767? Well, it means just what any ten-year-old kid interested in model airplanes thinks it means, namely that there's a direct correspondence between the external stimuli, internal states, and behavior of the 767 and the inputs, internal states, and outputs of the model. However, the correspondence need not necessarily be either one to one or onto, so there may be some external stimuli, states, and/or behaviors of the 767 that are not represented in the model. So, for example, when you go to Seattle and look at a model of a 767 in the wind tunnel, the seats, movie screens, beverage carts, and all the other paraphernalia forming many of the internal states of the real 767 are not present in the model, for the very good reason that they are irrelevant to the model's purposes, i.e., testing the aerodynamic properties of the real plane. Nevertheless, the external stimuli, states, and behaviors of the model are in direct relationship to a subset of the inputs, states, and behaviors of the real plane. Such a correspondence generates a *modeling relationship* between the real 767 and the object in the wind tunnel. Observe that the model is *simpler* than the real thing it models, in the sense that the model has fewer states. This property is characteristic of modeling relationships: Models are always simpler than what they model. Now what about a simulation?

In my study at home I have a brand-X laser printer whose operating instructions assure me that by suitable fiddling I can make it "emulate," i.e., simulate, a different type of printer, a Hewlett-Packard LaserJet Plus. What does it mean to say my brand-X machine can simulate another machine? Well, it means simply that the inputs and states of the HP machine can be coded into the states of my machine, and those states of my machine can be decoded into the appropriate outputs that would be generated by an actual HP printer. Note that in order for such an encoding/decoding dictionary to be set up, my machine must be more complicated than the HP machine in a very definite sense. Specifically, in order for the inputs and states of the HP machine to be encoded into the states of my "simulator," it must be the case that my machine has more states than the HP printer when both are regarded as abstract machines. Thus, the simulator (my printer) must be more complicated than the object being simulated (the HP printer). This situation is completely general: A simulation is always more complicated than the system it simulates.

The foregoing brief, informal discussion of models and simulations can be formalized in precise, mathematical terms, providing criteria that are, in principle, testable and that we could use to distinguish a program that *models* human cognitive processes from one that merely *simulates* them. In this context, it's interesting to note that a simulation of the brain would necessarily involve a system having more states than the brain itself possesses. However, the brain, with its 100 billion or so neurons, has at least $2^{10^{11}}$ possible states, a number that commands some respect in any company, exceeding the number of protons in the known universe (10^{78}) by a factor of about $2^{100 \text{ billion}}$, a number so large it's difficult even to write it down in words. Thus, we can confidently predict that there will be no simulations of the human brain in the short, intermediate, or very long-term future. Models of the brain are another matter, and it's fortunate that what strong AI, human, needs is models, not simulations.

On balance, it seems to me that the thinking-machine debate is really a battle between philosophers, regardless of the fact that some of them may be masquerading as psychologists, computer scientists, mathematicians, or programmers. And, as it should be in all stories involving philosophers, the debate ends up in complete chaos. My gut feeling is that a genuine machine intelligence will be with us within the next decade or two, but I'll have to confess that that opinion is based as much upon wishing, hoping, and wondering as upon hard facts and philosophical arguments. But I can conclude this excursion into the world of brains, minds, and machines with one opinion that is clear and definite: However the matter of strong AI, human, is finally resolved, the outcome will radically change our view of ourselves and our perception of the place we occupy in the cosmic order of things.

Speaking of the cosmic order of things, the time has come to move our consideration of the uniqueness of human beings away from the literally mundane considerations of biochemical structure, behavior, language, and mind and into the Milky Way Galaxy itself, for a look at the likelihood that there are other intelligent beings out there like us.

6

WHERE ARE THEY?

===

CLAIM:

THERE EXIST INTELLIGENT BEINGS IN OUR GALAXY WITH WHOM WE CAN COMMUNICATE

THE FERMI PARADOX AND PROJECT OZMA

In a conversation with Edward Teller, Emil Konopinski, and Herbert York at a physicists' lunch in the summer of 1950 at the Los Alamos labs, Enrico Fermi responded to someone's claim that extraterrestrial intelligences, or ETIs (be they individual entities, group intelligences, civilizations, or whatever), exist in our galaxy with the now-famous remark "Then where are they?" As one might expect of a comment from Fermi, this commonsense question contains deep, even profound, scientific and philosophical implications that have deservedly received much scholarly attention during the intervening decades, not counting the snowstorm of pseudoscience pulp cranked out by Erich von Däniken and others in the "UFOs are here" genre. The pillar upon which almost all arguments for the existence of ETI rests is the Principle of Mediocrity, asserting that on a cosmic scale there's nothing special about either the Earth or human beings. Consequently, Fermi's question leads to the paradox that if we're nothing special, then intelligent life should have developed in millions of solar systems. Yet we've never seen a single shred of hard evidence to support the existence of ETI, the von Dänikens of the world notwithstanding. On the other hand, if ETIs don't exist, then we are indeed something special, in gross violation of the Principle of Mediocrity. Either of these conclusions is mind-boggling in its implications, and steps on tender toes and egos across the entire landscape of science. But as usual in science, the questions, theories, and armchair philosophy vastly outweigh the experimental evidence needed to assess them, and it's been only rather recently that we have finally started to acquire the real data that many hopefully expect to lead to a definitive resolution of the paradox. That story begins in 1960 with a twenty-nine-year-old astronomer named Frank Drake, and the then rather new field of radio astronomy.

Sometime early in the morning of April 11, 1960, the 26-meter radiotelescope of the National Radio Astronomy Observatory in Green Bank, West Virginia, was turned to the constellation of Cetus the Whale, and Frank Drake initiated Project Ozma, named after the princess of L. Frank Baum's mythical land of Oz. Drake was listening for signals from assumed intelligent beings inhabiting a planetary system surrounding the star Tau Ceti. Thus began the experimental phase to answer the corollary of Fermi's question and one of mankind's oldest puzzles: Are we alone in the universe? Tau Ceti had been chosen as a target since it's not too unlike our own sun in type and age, in addition to being "only" about 11 light-years away, a veritable next-door neighbor on the astronomical scale of things. Drake, now a silver-haired dean at the University of California, Santa Cruz, recalls that when Tau Ceti disappeared over the horizon on that first night of listening, the telescope was then turned to Epsilon Eridani, the second target star in the experiment. To everyone's great astonishment, pulses at the metronomic rate of eight per minute immediately began to pour forth from loudspeakers in the room connected to the telescope. The next day when Epsilon Eridani was again visible, the pulses mysteriously disappeared, only to reappear some days later. The second appearance, however, was also noted on a secondary antenna specifically installed to screen Earth-based "false alarms," thus ruling out an extraterrestrial origin for the pulses. Through a variety of unofficial back-channel sources, Drake later learned that the pulses were due to experimental military radars being tested at the time in the relatively "clean" radio environment of the remote outback of West Virginia. After about two hundred hours of observing Tau Ceti and Epsilon Eridani, no legitimate signals of extraterrestrial intelligence had been recorded, and since the telescope was needed for other tasks, Project Ozma was brought to a close with two definite conclusions: (1) the experimental search for ETI (SETI) was a task well within the realm of modern technology, and (2) SETI can be dangerous to the health of radio astronomers, with false-alarm-induced heart attacks a continuing occupational hazard!

Project Ozma had actually been sparked off by a 1959 proposal made by Philip Morrison of MIT and Giuseppe Cocconi of CERN in a note to the influential British journal *Nature,* in which they argued that on physical grounds the most natural place to look for an ETI signal would be at the radio frequency of 1420 megahertz (MHz), the frequency at which ordinary hydrogen, the most abundant element in the universe, naturally radiates in the cosmic void. By happenstance, it turns out that the background noise in outer space is very low at this frequency, making the Morrison-Cocconi "waterhole" a likely place to look for an ETI signal, at least if it's sent by radio. Drake immediately picked up on the proposal and saw it as a good way both to test the newly installed 26-meter radio telescope, and at the same time call attention to the fact that SETI had now moved from the realm of philosophical speculation to that of experimental science.

So in the decade from Fermi's "where are they?" to Frank Drake's Ozma, the SETI battle lines had been drawn with the major theoretical and experimental boundaries clearly defined: What theoretical arguments can we give from astrophysics, planetary science, biology, cognitive science, anthropology, linguistics, and philosophy to begin to resolve Fermi's paradox, and what kinds of engineering, physics, and computing resources can we bring to bear on Drake's problem of actually detecting an ETI signal? These are the scientific issues that have dominated the SETI landscape for the past couple of decades and around which the SETI arguments, pro and con, revolve to this day.

THEORETICAL ETI: THE DRAKE EQUATION

Undaunted by the failure of Project Ozma to find a needle in the cosmic haystack, Drake convened a small workshop shortly after the conclusion of the search to examine the entire question of ETI and to plot a course for future scientific work on the matter. As a starting point for the discussions, Drake followed the commonly accepted reductionistic path for scientific investigations of the unknown, decomposing the ETI question into a collection of individually digestible pieces involving the physical, biological, psychological, and sociological conditions that would have to be met for ETI to exist. The recombination of these factors led to what is now termed the *Drake equation,* which has subsequently served as the starting point for almost all theoretical speculations about ETI. An understanding of this equation is of prime importance for grasping the way in which science has attacked the ETI question, both theoretically and experimentally, so let's take a more detailed look at Drake's pioneering idea.

In developing the equation expressing the number of communicating ETI civilizations existing in our galaxy, Drake made the not unreasonable assumption that for us to be able to contact such a civilization several conditions would have to be fulfilled. These conditions can be conveniently grouped into the following categories:

- *Astrophysical and geophysical:* An ETI would need to have a suitable physical environment for development, probably on a planet orbiting a star that, like Goldilocks's porridge, is not too hot and not too cold and, furthermore, not too unstable.
- *Biological and psychological:* It must be the case that life, as we know it, should readily arise wherever conditions are suitable (the Principle of Plentitude). Moreover, for the existence of ETI we need the additional requirement that evolutionary pressures force intelligence to emerge.

• *Sociocultural:* Intelligent life must further develop into a technologically based civilization that not only persists for a sufficiently long period of time but also has the desire to engage in interstellar communication.

Clearly, satisfaction of all the above desiderata is a tall order, and the Drake equation was developed to try to give some kind of quantitative measure of how many such planetary civilizations might currently exist in our own Milky Way Galaxy. Now let's look at the individual terms that by common consensus today constitute this basic expression.

The elements forming the Drake equation are:

R^* = the rate at which stars are formed in our galaxy per year

f_p = the fraction of stars, once formed, that will have a planetary system

n_e = the number of planets in each planetary system that will have an environment suitable for life

f_l = the probability that life will develop on a suitable planet

f_i = the probability that life will evolve to an intelligent state

f_c = the probability that intelligent life will develop a culture capable of communication over interstellar distances

L = the time (in years) that such a culture will spend actually trying to communicate

Under the dubious (but simplifying) hypothesis that each of the foregoing factors is independent of the others, an estimate for the number N of advanced communicating civilizations in our galaxy can then be made by just multiplying each of the factors together. This yields the celebrated Drake equation for N as:

$$N = \underbrace{R^* \times f_p \times n_e}_{physical} \times \underbrace{f_l \times f_i}_{biological} \times \underbrace{f_c \times L}_{cultural}$$

Thus we see that to utilize the Drake equation effectively to estimate the likelihood of ETI in our galaxy requires a spectrum of expertise that would make even a Leonardo blanch, representing in my view one of the great multidisciplinary problems of all time.

The heart of the ETI debate then comes down to the development of scientifically defensible estimates for N. We know that N is no less than one; some argue that N is very much larger than one, while others claim that N is either very large or very small. To complete the possibilities, there are those who hold to the position that N is neither large nor small. To make sense out of these mutually contradictory positions, it's useful to take a longer look at the individual pieces making up the Drake mosaic.

SLICES OF THE ETI PIE

Since the various terms in the Drake equation have been the subject of numerous book-length treatments through the years, I'll content myself here with giving only a highly condensed account of some of the more important factors that need be taken into consideration when attempting to assign actual numerical estimates (guesses) to the various terms.

R^*, THE GALACTIC RATE OF STAR FORMATION

Of all the terms in the Drake equation, this one is perhaps the best understood. Theoretical and observational astrophysics over the past few decades has succeeded in creating a picture of stellar formation involving the gravitational coalescence of stars out of interstellar galactic clouds of hydrogen, helium, ammonia, methane, water vapor, and dust grains. As a corollary of this work, we also have a rather detailed picture of the life histories of stars of various masses. It turns out that something on the order of ten stars per year are formed in the galaxy, but only a small fraction of these are suitable candidates to support ETI.

For a particular star to generate an environment suitable for ETI, a number of factors need to be considered. Two of the most important are: Will the stellar environment be conducive to the formation of a planetary system containing Earth-like planets with liquid water, and will the star be too short-lived for life to emerge and move along its evolutionary path to intelligence? Current theory predicts that stars much more massive than about 1.4 solar masses pass through their life cycles far too quickly for living systems to emerge, while stars that are too old would not generate conditions conducive to life, since they will have formed at a time before there was a sufficient abundance of the heavy elements (iron, sulfur, calcium, and so on) currently thought necessary for living organisms. This is because these elements form as the by-products of supernovas, the dramatic explosions of stars in their death throes. Fortunately, such constraints eliminate only about 1 percent of the stars from consideration; unfortunately, there are other constraints as well.

Theoretical and observational evidence strongly suggests that when the stellar cloud coalesces into a proto-star, the general pattern is for the cloud to split into two more or less equal pieces, thereby forming what's termed a *binary system* consisting of two stars orbiting each other. Numerous calculations show that the continually shifting gravitational stresses and strains of binary systems, not to mention the extreme temperature fluctuations, create a physical environment very unlikely to support a stable planetary system, let alone a planetary system with a stable habitable Earth-like zone. It appears that at least half of the stars that are not too massive and not too old belong to such binary systems and hence must be excluded from consideration as an abode of life. Put all these factors together with others, including the inappropriateness of stars that are too small, as well as stars that occupy regions too near the center of the galaxy where exotic events that would be fatal to most conceivable life forms regularly occur, and the quantity R^*, which started in the region of ten stars per year, is dramatically reduced, perhaps by a factor of several thousand. So what we need is not just the crude rate of star formation, but the rate of formation of stars with the "right stuff." In astrophysical terminology, these turn out to be what are called *G-type* stars like our own sun. Consequently, when estimating R^* what we're really looking for is the annual rate of formation of single G-type stars. We'll give specific values later, but for now the important point is that the vast majority of stars make pretty inhospitable homes for the kinds of organisms that we would recognize as being alive.

f_p, THE FRACTION OF STARS HAVING A PLANETARY SYSTEM

In the process of stellar formation, a cloud of interstellar gases begins to contract due to gravitational attraction, changing from

a slowly revolving amorphous blob into a rapidly spinning, pan-
cake-shaped gaseous disk. Since the rate of spin is too great for
the disk to remain stable, one of two things normally occurs:
Either the disk flies apart into a few (usually two) more or less
equal pieces, each of which then spins at a much slower rate, or
the disk throws off a small fraction (1 to 2 percent) of its mass
at a distance sufficiently far from the center of rotation that the
small mass has a great enough lever arm to slow down the spin
of the central disk. The reader will recognize this as the astrophysical equivalent of spinning ice skaters who suddenly throw
out their arms to slow their rate of spin. The first case corresponds to the formation of a binary (or multiple) star system of
the sort discussed above; the second represents the currently
held view as to how planetary systems are formed. It should be
noted, however, that these two processes may not be mutually
exclusive, since calculations indicate that a habitable planetary
system may form if the two stars of a binary system are far
enough apart, say over 20 AU (1 AU equals the average distance
between the Earth and the Sun). But conventional astronomical
wisdom dictates that planetary systems and multiples are like oil
and water: They usually don't mix.

Our own solar system is an example of the second kind of ro-
tation-slowing process, in which about 1 percent of the original
spinning mass was thrown off in the form of the planets (most of
it in Jupiter and Saturn). During this process, though, about 99
percent of the angular momentum of the spinning cloud was
transferred to the planets (again almost all to Jupiter and Saturn), leaving the central Sun with only a modest rate of spin,
low enough to preserve its stability. Since our solar system is the
only one of which we have direct observational evidence, the
question of interest for estimating f_p becomes: How typical is
our own solar system? In other words, if a star does not form as
part of a multiple system, is formation of a planetary system to
be expected?

One line of attack on the planetary question is just to appeal
to the Principle of Mediocrity and say that since our corner of
the universe is nothing special, it's likely to be the case that
planetary systems are common. Clearly, this is more of a philosophical or a religious argument than a scientific one, so to move
beyond it we have two alternatives: direct observational evidence
for extrasolar planetary systems, or stronger theoretical evi-

dence to show how the formation of planetary systems fits into the normal process of star formation.

The difficulty with direct observation of a planet surrounding a nearby star is graphically described by imagining a birthday cake with a single candle placed next to the beacon atop the Eiffel Tower, and then trying to see the candle being blown out by looking at it from the Postal Tower in London. In short, the minuscule amount of light reflected by even a Jupiter-sized planet is totally buried in the more than billion-times-greater luminosity of the parent star. Thus at the moment the only feasible method of obtaining empirical evidence for planetary systems involves searching for small irregularities in the motion of the star due to the gravitational effects of its hypothetical invisible companions. The best candidate for such indirect detection of a planet appears to be the star 36 Ursae Majoris A, where wobbles in the star's orbit have been attributed to a Jupiter-sized planetary companion. However, these observations have been questioned on various grounds, and at the present time all that can be definitely said about observations of extrasolar planets is summed up in a remark by David Black to the 1984 International Astronomical Union Conference on SETI, noting that "there is currently no observational evidence for the existence of any planetary system other than our own." At the time it was expected that the Hubble Space Telescope would provide the experimental muscle needed to resolve the matter, but the tragic *Challenger* accident delayed the planned 1986 launch of the telescope, leaving the experimental situation pretty much unchanged.

On the theoretical side, numerous computer simulations of the coalescence of the interstellar gas clouds have rather strongly suggested the likelihood of planetary systems' emerging over a wide range of initial conditions. Figure 6.2 is a simulation by Stephen Dole showing the kinds of planetary systems that emerge out of a homogeneous condensing stellar cloud of the same mass as our solar system, when various quantities of condensation nuclei are injected into the cloud to provide inhomogeneities needed to get the condensation process started. By way of comparison, Figure 6.1 shows our solar system with planetary distances from the Sun measured in astronomical units (AU), while the planetary masses are given relative to the mass of the Earth, taken to be one. Figure 6.2 shows that a vari-

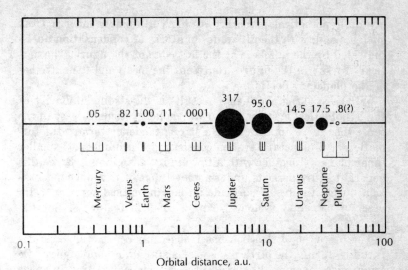

FIGURE 6.1. *The solar system*

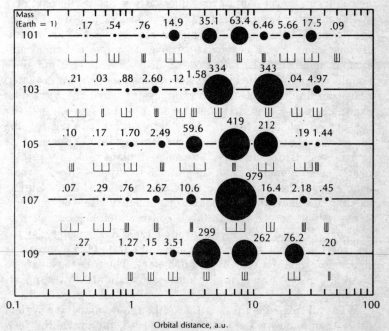

Orbital distance, a.u.

FIGURE 6.2. *Hypothetical planetary systems from computer simulations*

ety of hypothetical planetary systems ultimately emerge from such a cloud, with the different quantities of condensation nuclei indicated by the numbers at the left edge of the figure. The vertical "forks" in the figures represent the mean and the extremes of the planetary orbits.

What's striking about these results is the strong similarity of the hypothetical systems to our own solar system, at least in the sense that there appears to be a strong tendency toward the formation of a planetary system consisting of a number of smaller inner planets, together with a few outer "gas giants." Since this general picture persists under a wide range of random condensation nuclei, the results provide strong theoretical support to the case for planetary systems' being a common feature of Sun-like stars.

The preceding discussion has focused upon planetary systems forming during the birth process of a star. For completeness, we might also consider the possibility of a planet existing in space independent of a star. On physical grounds, it's hard to imagine how such an object could arise unless it was originally part of a stellar planetary system and was then somehow pulled out of the gravitational attraction of its parent star by some kind of cataclysmic event, e.g., a nearby supernova, or maybe a cosmic collision of some sort. In any case, it doesn't really matter since simple thermal equilibrium considerations make such an isolated planet an unlikely place to find life, even if such an object does exist. The problem is that in order for a planet to avoid getting too hot or too cold for life to survive, it's necessary for the planet to radiate back into space the same amount of energy that it receives. Unfortunately, an isolated planet doesn't receive nearly enough energy from the outside to support life, so whatever energy there is must come from internal sources. Simple calculations show that for bodies of planetary size, the temperature gradient needed to maintain a constant 300°K (equivalent to 27°C or 80°F) at the surface is around 1,000°K/kilometer, far too hot for the planet to survive in the solid state (the comparable figure for Earth is only 10°K/kilometer). Thus it seems safe to eliminate such "wandering" planets as candidates for supporting life.

What all this adds up to is that although no planetary system other than our own has ever been observed, the prevailing feeling is that such systems are rather common around single stars, and that the Hubble Space Telescope will soon confirm this prejudice. If so, the value of f_p may soon become the best-understood element in the Drake equation.

n_e, THE NUMBER OF PLANETS HAVING AN ENVIRONMENT SUITABLE FOR LIFE

In the SETI community it's generally accepted that for a planet to be a home for life there must be a plentiful supply of liquid water. In an extremely interesting set of computer simulations, Michael Hart showed in 1978 that if the Earth's orbit had been only 5 percent closer to the Sun, the primordial water vapor outgassed from volcanoes in the Earth's early history would not have condensed to form the oceans, but would have remained in the gaseous state instead. In turn this would have prevented the removal of carbon dioxide, resulting in a runaway "greenhouse effect" of the sort that is now believed to have turned Venus into a planetary version of most people's vision of hell, with surface temperatures hot enough to melt lead (over 800°F) and a permanent cloud cover of sulfuric acid. On the other hand, had our planetary orbit been even 1 percent greater, then the lowered radiation from the "youthful" Sun, coupled with the reduced greenhouse effect, would have left the Earth covered with massive glaciers. Since the surface albedo (reflectivity) of ice is greater than that of water or land, as more and more ice formed, more and more of the Sun's radiation would have been reflected back into space, the result being that the glaciers would never melt. So it appears from Hart's calculations that the early Earth sailed a very narrow path between the Scylla of a Venusian hell and the Charybdis of a Martian deep freeze.

The range of orbits around a star within which a planet can avoid both the greenhouse and glacier effects is termed the *continuously habitable zone* (CHZ), and varies from star to star depending upon its mass. Larger stars have a bigger CHZ, but also burn their fuel much faster, with the result that the CHZ is not stable for the billions of years seemingly needed for evolution to work its magic and transform the cellular slime molds into Einsteins and Leonardos.

Besides the CHZ, planetary size can play a significant role in determining how suitable the planet is for life. For example, planets much larger than the Earth will outgas more material, thereby enhancing the greenhouse effect. Calculations show that if the Earth had been even 10 percent greater in mass, this effect would have prevailed and there would have been no orbit in which the Earth could have traveled and still retained liquid oceans. At the other end of the scale, a planet can also be too small to retain an atmosphere that will be effective in blocking out the solar ultraviolet radiation that is fatal to most forms of life. In fact, Hart also showed that if the Earth's radius had been even 6 percent smaller, this would have been exactly our fate, as then the Earth's gravitational field would not have been strong enough to retain the ozone molecules needed to screen out the damaging rays.

In direct contradiction to the Principle of Mediocrity, it has also been argued that as planets go Earth is not at all typical. The problem is that the Earth and Moon are much more like a "double planet" system than a primary planet and a satellite. For example, the Moon is far larger compared with the Earth than any other satellite of a major planet in the solar system. The large Moon has affected the Earth in many significant ways, e.g., large ocean tides influencing the evolution of crustaceans and amphibians, as well as the appearance of tidewater zones, which could have helped life emerge on land. And the large Moon is not the only thing that's strange about the Earth.

Another anomaly is the Earth's very strong magnetic field. This field is much larger in proportion to the mass and angular momentum of the rotating Earth than that of any other planet. This magnetic field is vital for maintaining the ozone layer protecting life from deadly ultraviolet radiation. Moreover, the Earth also has a very active, molten core. This core is responsible for all volcanoes and mountain ranges, and for the separation of continents, which, in turn, has isolated gene pools, thereby speeding up evolution.

Recent studies claim that all these unusual characteristic of the Earth could well be attributed to the presence of our extraordinarily large Moon. It has been conjectured that the Moon may have been "captured" in an rare encounter in which it passed near the Earth. Since the overwhelming majority of such encounters result in either the complete destruction or the merging of the two colliding bodies, or a simple flyby, such double planets as the Earth and Moon are probably very rare. Thus, if it can be shown that the presence of a large moon in a double planet configuration is necessary for the emergence of life, the term n_e may indeed be vanishingly small.

Putting all these factors together suggests that just as with stars, finding a planet with all the "right stuff" for life may involve an extensive search, and that the quantity n_e may very well turn out to be extremely small.

f_l, THE PROBABILITY THAT LIFE WILL DEVELOP ON A HABITABLE PLANET

Since the considerations given above concerning stars and planets strongly bias our search for life to those regions of the galaxy bearing a strong similarity to our own, when it comes to thinking about the likelihood of life's emerging the most natural approach is to consider how likely it was for life to emerge here on Earth. Here we give only the briefest sketch of this complex issue, referring the reader back to Chapter Two for the gory and glorious details.

There are five basic steps through which life as we know it today emerged on Earth:

Chemical

1. Small organic molecules had to form from the Earth's original material.
2. These small molecules somehow had to combine into the long chains (polymers) required for life.
3. In some fashion the polymers had to form isolated, self-reproducing systems.

Biological

4. Cells and multicellular organisms had to form from the self-reproducing systems.
5. Evolution had to act to produce the multitude of plant and animal species that we call life.

As noted, the first three steps on the list are what are normally termed processes of *chemical evolution,* while the last two are activities associated with *biological evolution.* Let's briefly consider how much we can say we really understand about each of these stepping-stones to life.

According to conventional wisdom, all life on Earth is formed out of a few organic compounds that had to have been created from materials present at the time of the Earth's formation. These compounds, primarily amino acids, mononucleotides, and sugars, are commonly thought to have formed out of simple elements in ready supply on the early Earth such as water, ammonia, hydrogen, and methane, with energy inputs for combining these quantities coming from lightning, volcanic heating, and ultraviolet rays. A famous experiment by Stanley Miller in 1953 showed that if an electrical discharge was passed through a bottle full of these gases, after a week or so many organic compounds would form, including amino acids. (For a diagrammatic representation of Miller's experiment, see page 72.)

An important aspect of these experiments is that they won't work at all if there is even a small amount of oxygen present. In fact, if you try a Miller-type experiment using the present composition of the Earth's atmosphere, all that results is plain old everyday smog. Thus it's crucial for the "primordial soup" theorists that the atmosphere of the early Earth be highly reducing (i.e., deficient in oxygen). We'll return to this point later as it plays a significant role in the question of intelligence.

Subsequent Miller-type experiments by other investigators using variations of the quantities and types of gases thought to have been present in the primordial atmosphere showed similar results, leading to the conclusion that natural formation of the building blocks of life out of inorganic matter seems a good bet. Hence, the first of the five steps to life appears to be one that is relatively easy to negotiate in any early Earth-type atmosphere.

The linking-up of the simple organic molecules into the long polymer chains needed for life poses a bit of a problem. The simple molecules of the type formed in a Miller experiment are very unstable, and can easily be broken apart by the same energy sources that created them. Thus, to survive long enough even to begin to contribute to a polymer chain, these molecules have to be protected from solar ultraviolet rays, which were very much stronger on the early Earth as there was no ozone layer at that time to protect them. The obvious solution is for these molecules to have remained in the sea, where they could easily be protected from dissociation by lying just a few meters below the surface. Unfortunately, when a polymer chain of such molecules comes into the presence of water, there is a strong tendency for the water to break the chain apart, giving us back the original primitive molecules.

The foregoing picture leaves us in somewhat of a quandary: It seems that on the one hand, the sea was necessary to protect the organic compounds from ultraviolet rays, while on the other hand, the seawater acted as a strong deterrent to the formation of the polymers needed for life. It's as if you were driving your car, and every time you stepped on the gas with one foot, you hit the brake with the other. Are there any plausible ways out of this dilemma?

If polymerization was to take place, somehow the organic molecules had to be isolated from water. Or failing this, we must at least propose a mechanism by which the concentration of these molecules could have been sharply increased in the vicinity of the ocean. Several such possible mechanisms have been suggested:

1. Evaporation of water in tidal pools
2. Partial freezing in which the water is removed as crystals
3. Volcanic heating to drive off the water
4. Attachment of the molecules to the surface of clays

Each of these processes is common and has been successfully tested in laboratory experiments showing that polymer chains of up to two hundred amino acids can be produced. Thus, although this step is not quite as well understood as the way in which the primitive molecules could have arisen, there still seems to be no reason why polymerization could not also have come about by fairly straightforward and common physical processes.

The last step in chemical evolution—self-reproduction—is by far the least well understood process in the entire pathway to life. We have already covered this step in excruciating detail in our treatment of the origin of life in Chapter Two, so for now it suffices to keep in mind that there are many contending routes by which all this could have occurred, none of them especially convincing. So for the moment we just have to say that the processes of reproduction and replication remains a weak link in the chain leading to life.

Once we pass into the realm of biological evolution, things begin to get easier again. Oparin's coacervate idea, developed in Chapter Two, deals nicely with the question of how self-reproducing polymer chains could have formed themselves into a cell, while the well-known processes of natural selection and neo-Darwinian evolution provide a tested mechanism by which the many species of plants and animals found today could have arisen over the millennia. But this is not to say that even here there are not still serious questions of detail awaiting answers. For instance, life uses only twenty different sorts of amino acids, while Miller experiments produce far more. Why did life neglect the others? Similarly, sugars and amino acids come in two different "flavors": right-handed and left-handed. Miller experiments produce approximately equal quantities of both, and it's reasonable to assume that the primordial soup contained similar proportions of each type. Nevertheless, the amino acids used in living forms are exclusively left-handed, while all sugars used are right-handed. The only explanation we can currently offer for this puzzling fact is that by chance the left-handed amino acids and the right-handed sugars "took off" first, and their mirror-image competitors were excluded by natural selection. Perhaps. Yet the question remains open, as do a variety of others pertaining to the *exact* manner in which life came to assume its current form on Earth.

One final point: On Earth we find two molecular types, one good for action (the amino acids), one good for replication and reproduction (the nucleic acids). These two molecular types compose the metabolic and genetic components of every living cell. When it comes to the question of ETI, we can naturally ask whether or not it would be possible to have a system in which one molecular type does both. Essential life activity involves preservation of the genetic information, and it's not easy to copy a three-dimensional object. On Earth the process is carried out by translation from the four-letter language of the nucleic acids to the twenty-letter alphabet of the proteins. A corollary of the earlier questions is whether or not there are alternative alphabetical schemes that would do the job equally well, or even better. Current computer experiments with "artificial life," coupled with advances in the informational theory of living organisms, may offer some clues on this question of obvious relevance for ETI. But at present we can say little more.

Taken *in toto,* the foregoing considerations lead to the following conclusions: Many of the building blocks of life are almost certain to form wherever the raw materials exist and there is a sufficient supply of free energy. Further, many kinds of natural processes will lead to the polymer chains needed for catalytic activity and preservation of the genetic information. Provided some kind of mechanism appears to get the process of self-reproduction going, natural selection will then take over and invariably lead to a proliferation of life forms. Thus for assessing the quantity f_l, all seems to hinge upon the likelihood of replication and reproduction emerging as a natural adjunct to the formation of polymer chains. At present this is completely *terra incognita,* with opinions ranging from "It's a one-in-a-zillion fluke that happened only here on Earth" to "It's inevitable wherever life of any kind forms."

f_i, THE PROBABILITY OF THE EMERGENCE OF INTELLIGENCE

If life is going to get into the interstellar communication game, it's clear that some sort of toolmaking is going to be required. This implies intelligence of a kind. While it's still far from clear how necessary intelligence is for biological survival, it's possible to identify several steps that must be taken for the development of a level of intelligence high enough to create the technology needed for communication outside its own environment. These steps include:

• Development of an atmosphere containing free oxygen
• Movement of life from the sea to land
• Emergence of hands and eyes
• Use of tools
• Appearance of social structures

Between $2\frac{1}{2}$ and $3\frac{1}{2}$ billion years ago, microscopic plankton and blue-green algae formed and thus began the process of transforming the assumed reducing atmosphere of the early Earth into one containing large amounts of free oxygen through the process of photosynthesis. Most organisms alive at that time perished in what was for them a highly poisonous oxygen-rich atmosphere. But those that were able to adapt found themselves in a position to take advantage of the increased energy available in chemical reactions involving oxygen. Such organisms could make more efficient use of the available food, and as a result were able to start on the road to the development of the kind of highly energy-intensive brain that governs what we now see as intelligent behavior.

Photosynthesis proceeds by plants' taking in carbon dioxide and combining it with energy from sunlight, giving off free oxygen in the process. A crucial benefit to life forms from this atmospheric oxygen is that some of it forms into the molecule ozone, which acts as an effective shield against the deadly ultraviolet rays from the Sun. Once this shield was in place, it was finally safe for life to leave the sea and begin to establish itself on land. While there's considerable evidence to indicate that intelligence can emerge in sea-dwelling life forms (e.g., the cetaceans), it's difficult to imagine how the kind of technology needed for interstellar communication could develop in an aquatic environment.

The old saying that a picture is worth a thousand words amply underscores the fact that our visual system is capable of taking in an enormous amount of information at a glance. The development of a sophisticated visual system would appear to give a definite selective advantage in the survival game—but only if a correspondingly sophisticated brain developed to process the visual input. It appears that simple brains just ignore most of this input, thus ending up with the crumbs from Nature's banquet table instead of the caviar. Hence, the emergence of eyes acts selectively to promote a larger, more capable brain. Ditto for the appearance of hands, which need a complex brain to make most effective use of their inherent manipulative capability.

Hands and a complex brain make the use of tools possible. Tools, in turn, enable us to extend the capabilities of the body in a variety of ways, all of which contribute to freeing their possessor from some of the pressures of biological evolution. As a simple example, it's not necessary to be able to run very fast to catch a meal if you know how to throw a rock, and you don't have to develop massive jaws for tearing apart your catch if you can cut the meat up into bite-sized pieces.

The argument has been made many times that groups of thinking animals can coordinate and plan their hunting and defensive activities far more successfully than individuals acting on their own. While a social structure does not in itself lead to higher intellect (it hasn't in bees and ants), it appears likely that its adoption in animals of larger brains generally does contribute a supplementary evolutionary shove toward further brain development.

Each of the foregoing factors contributed its share to the development of intelligent life as we know it here on Earth. In our quest for the kind of ETI that we could expect to be able to communicate with, it's plausible to assume that many, if not most, of the items on our list will also appear on theirs, at least if they want to talk with us. But on the other hand, who can really say that what we have observed here on Earth is in any way typical of the galaxy as a whole. Thus the likelihood of primitive life forms' developing intelligence remains one of the big question marks in the Drake equation.

f_c, THE PROBABILITY OF THE EMERGENCE OF A COMMUNICATING ETI CULTURE

The development of a social structure and the capacity for language dramatically affect the path of evolution. Prior to these changes, evolution is primarily at the level of the individual, as information is just passed along in the genetic shuffle. However, as soon as a social order and language enter the scene, evolution then proceeds to act more on the society as a whole than upon the individual. This fact allows the knowledge of one generation to be passed along to the next, contributing to the development of specialized skills that can be used for the entire group. This process is vaguely analogous to the development of multicelled organisms from single-celled predecessors, and shows the advantages of centralization and specialization in the evolutionary setting. So it seems likely that once intelligent life forms, at least some of the species will develop a social order and a technology to go along with it. The major question that remains is: Will they want to communicate with the stars?

Who can ever know the desires of another? The emergence of a technologically based civilization capable of interstellar communication by no means implies that it will wish to contact us. The argument has been made that part of the age-old fascination of humans with ETI has been our puzzlement when we look at the stars and wonder, "What's out there?" But suppose the Earth were just a bit closer to the Sun and, as a result, were continually shrouded in clouds. Would we then have any real interest in ETI? The point here is just to emphasize the fact that possessing an ability and possessing the desire to use it are very different matters, and that the galaxy may well be teeming with ETIs that are happily going their way like the proverbial three monkeys: blind, deaf, and dumb, spending their time in contemplation of eternal philosophical truths and deep mathematical abstractions, with no interest whatsoever in talking with us.

In this same direction, there is also the argument that perhaps ETIs won't *be able* to talk with us even if they want to. Perhaps their science is of such a different nature than ours that there's no basis for a meaningful exchange of information. Or maybe their mathematics is based upon nonnumerical quantities that make it impossible for us to understand what they're doing. We'll return to this line of argument later, but for now we raise it just as a possibility that, if true, would reduce the factor f_c to a negligible level.

L, THE LIFETIME OF A COMMUNICATING CIVILIZATION

Assume there is a technical civilization out there that is valiantly trying to communicate. How long will it be able to persevere in its efforts? This question forms the biggest uncertainty of all in the Drake equation, and this uncertainty is unlikely to be reducible by any sort of terrestrial experiments. To see why, it's helpful to examine the many reasons a communicating civilization may cease its attempts. Some of the most obvious are:

nuclear war	genetic deterioration
overpopulation	overstabilization
exhaustion of resources	loss of interest
pollution	

We have been so bombarded by the media and the doomsayers about the dangers of the items in the first column of the list that by now most of us probably wish to just bury our heads in the sand, ostrichlike, and wait for these possibilities to disappear. Since I have nothing to say on these matters beyond what's already been well chronicled, I'll indulge the reader's desire and move on to the far less familiar possibilities in the second column.

The advent of modern technology and medicine has for the first time opened up the possibility of circumventing Nature's way of weeding out misfits by natural selection. Modern medicine now allows not only the fittest to survive, but almost everyone else, too. In the past, weak, sickly, or genetically defective people had a tendency to disappear from the gene pool early in the game. But no more. Just as those with bad eyesight no longer have to worry about mistaking a hungry saber-toothed tiger for a housecat, those with various genetic defects such as Down's syndrome, sickle-cell anemia, and hemophilia can now survive and even pass these defects along in the gene pool.

Gene-splicing technology has now reached the level where some of the deleterious effects of the foregoing genetic-programming "bugs" can at least theoretically be weeded out of the system. However, these techniques are themselves not without their dark side, opening up the possibility of creating entirely new types of human beings according to any desired set of specifications. Who is to say what kinds of humans should be produced? In Aldous Huxley's *Brave New World,* the kind of static society that emerged from such genetic engineering turned out to be one totally lacking in creativity, a situation good for the government but questionable as a basis upon which SETI is likely to be continued. So genetic deterioration is a very real threat to those societies lucky enough to survive annihilation by nuclear war as well as the other apocalyptic "horsemen" on our list.

Population expansion, excessive energy consumption, and the like cannot continue forever. One possibility for the stabilization of such processes would be for the nations of the world to agree to halt the growth of their economies—the no-growth option. But there are dangers in the no-growth strategy, as the forcible elimination of economic growth may also result in an overly static society in which scientific progress and intellectual curiosity have been destroyed. Economic growth and the growth of scientific knowledge have traditionally gone hand in hand, and suppression of one could cause the elimination of the other as well. The no-growth option would almost certainly result in the cessation of any sort of SETI or space exploration, perhaps even creating a very xenophobic society that has slid back into a primitive, pretechnological life-style from which there can be no hope of recovery.

Finally we come to the possibility that a communicating society will just get tired of trying to make contact and give up. There must be some limit to how long a civilization will try to communicate, and the probability that it will continue to send signals or even listen for thousands or millions of years with no return signal is surely zero. Of course, the communicating phase may recur in cycles, with periods of intense interest followed by loss of interest for long periods, after which communication starts up again. But it's hard to argue that the communicating periods taken in total would necessarily be longer than the periods of silence unless some results were obtained. So from the standpoint of communication, indifference and malaise are just as serious threats as any of the other, more cataclysmic possibilities on our list.

Here it has been possible to touch upon only a few of the more important matters surrounding each term in the Drake equation. For detailed accounts, I refer the reader to the excellent volumes cited in "To Dig Deeper" and now pass on to the matter of assigning actual numbers to the terms in a valiant, but probably foolhardy, attempt to divine on theoretical grounds how likely it is that N is greater than the magic *número uno*.

ANTHROPOMORPHISMS, CHAUVINISMS, AND ETI NUMEROLOGY

Before attaching some numerical "guesstimates" to the terms of the Drake equation, let's discuss for a moment a few of the blatant prejudices built in to the remarks given above regarding these terms. All of these biases trace their origin to one root cause: our interest in ETIs of the type that we could reasonably expect to be able not only to recognize, but also enter into some sort of sensible communication with. A good example of the kind of ETI that we're *not* talking about here is provided in Stanislaw Lem's classic novella *Solaris,* in which the central role is played by a sentient ocean that has been under study for years by scientists who are able to recognize that the ocean is intelligent, but who are totally unable to enter into any kind of meaningful dialogue or interaction with it. Another example of this kind is given in Fred Hoyle's classic *The Black Cloud,* involving an intelligent entity composed of a cloud of interstellar particles. While the heavens may indeed be composed of more things than are dreamed of in our philosophies, those philosophies are exactly what determine the kinds of entities that we can and want to interact with, and hence induce the anthropomorphic slant to our consideration of the Drake equation. But to be explicit about it, let's look briefly at a few of the more important "humanistic" prejudices introduced into the equation.

- *Carbon chauvinism:* A sine qua non of the kind of ETI we're interested in is that it be a life form capable of reproduction. This means that there must be some chemical structure that contains the genetic information to be passed along to progeny. For any reasonably complex life form, the amount of information to be passed along amounts to millions of bits, thus requiring the kinds of long polymer chains we have considered above. According to the known laws of chemistry, there are only two elements capable of forming the kind of long chain needed: carbon and silicon. Terrestrial life is based upon carbon for the simple reason that, at normal Earth temperatures, silicon is not capable of forming these chains. Only at temperatures below −200°C do the chemical properties of silicon allow it to link up into chains of sufficient length to store the needed genetic information. Thus silicon-based life forms may well exist, but only on planets whose oceans are filled with liquid nitrogen! Unfortunately, at such temperatures chemical reactions proceed extremely slowly (that's why we put things into refrigerators to slow down their decomposition), and it seems unlikely that any such silicon-based organism would possess a metabolic rate fast enough to generate a technological base sufficiently advanced to enter into interstellar communication. Hence our anthropomorphic bias toward carbon.

- *Star-type chauvinism:* To be consistent with other anthropomorphic assumptions about the origin of life and the time scale for evolution, it's necessary to assume that a communicating type of ETI will be found on a planet orbiting a G-type star like our own sun. An entertaining and scientifically plausible alternative is provided by the *cheela*, the main actors in Robert L. Forward's novel *The Dragon's Egg,* which are microscopic beings inhabiting the surface of a neutron star. The story describes how in such an environment beings live out their lives on a time scale millions of times faster than ours, and indicates how it might still be possible for meaningful communication to take place. However, scientifically plausible speculations and the way a prudent man would bet are two different things, so we prefer to look at G-type stars until there are compelling reasons to do otherwise.

- *Planetary bias:* Our discussion of the origin of life assumes that it arose on the surface of a planet through natural chemical processes. In other words, it was not imported from interstellar space, and it did not come about as a bolt out of the blue from "elsewhere." Beginning with the Swedish chemist Arrhenius and continuing to the present day with the works of Hoyle, Wickramasinghe, Crick, and others, fanciful proposals have been made that life forms originated elsewhere and were somehow transported to Earth. These are basically untestable and therefore irrefutable hypotheses; nevertheless, strong arguments can be mustered against them on purely physical grounds. So an application of Ockham's razor leads us to planetary chauvinism in the absence of firm contraindications.

The above collection of chauvinisms could be greatly extended, but I think this short list gives the general idea, namely that there is a tremendous amount of subjectivity involved in assessing the terms in the Drake equation and, as a result, any numerical estimates that emerge have to be taken with several shakers full of salt. Now let's finally turn to the process of putting some numbers into the equation in an attempt at least to get a feel for the range of possibilities for the quantity N, the number of communicating ETIs in our galaxy.

Beginning with perhaps the first widely circulated popular account of the SETI question, the still-influential volume *Intelligent Life in the Universe* by the well-known Russian astrophysicist I. S. Shklovskii and the *Cosmos* man, Carl Sagan, a number of authors have thrown their hats into the ring and taken a stab at estimating N numerically. Table 6.1 gives a fairly representative account of these efforts, where H represents the number associated with an optimistic scenario in which everything works out to favor ETI, M denotes a conservative estimate representing the best guess on the basis of current scientific knowledge, and L is the pessimistic, Murphy's Law scenario in which Nature has stacked the deck against ETI.

What kind of sense can we make of an estimate of N that ranges all the way from N = nil ("we're alone") to N = at least 100 million ("the galaxy is crawling with communicating ETI")? Or, put another way, does the Drake equation in any way help us in deciding whether or not it's a good scientific bet to invest our time, money, and energy in looking for signs of intelligent extraterrrestrial life? Some have argued that our high levels of ignorance about most of the terms in the equation make it totally useless as a tool for studying the ETI question;

TERM	SHKLOVSKII AND SAGAN (1966)			HART (1980)			ROOD AND TREFIL (1982)		
	H	M	L	H	M	L	H	M	L
$R*$	—*	10	—*	50	20	10	0.15	0.05	0.005
f_p	—*	1	—*	0.5	0.2	0.025	0.30	0.10	nil
n_e	—*	1	—*	1	0.1	0.001	0.20	0.05	nil
f_l	—*	1	—*	1	0.1	10^{-20}	0.50	0.01	nil
f_i	—*	0.10	—*	1	0.5	0.1	0.10	0.50	nil
f_c	—*	0.10	—*	1	0.5	0.1	1	0.25	nil
L	$> 10^8$	10^7	100	10^6	10^4	100	10^6	10^4	100
N	$> 10^8$	10^6	100	25×10^6	100	nil	4500	$\sim 10^{-3}$	nil

*No upper or lower estimates given.

TABLE 6.1. *Estimates for* N *using the Drake equation*

others point out that even if the numbers are only guesses, attempting to pin down numerical values for the various terms helps us at least identify and focus our efforts on those components of N that we know least about.

Since the job of the statistician is to attempt to provide estimates for various quantities on the basis of incomplete or "noisy" measurements, it's of interest to consider what standard methods of probability and statistics have to say about estimates of N generated from the highly uncertain guesses for its components displayed in Table 6.1. There are two points worthy of note in this connection: First of all, the argument is often advanced that you can't make any statements about how likely something is on the basis of just one observation. If this were indeed true, then approaches to estimating N from the Drake equation would definitely be in trouble, since we have only a single example upon which to estimate all the biological and cultural terms. Fortunately it's not true that a single observation gives no useful information. In fact, any statistician will tell you that a single measurement is all you need in order to estimate the *average,* or *mean,* of a collection of data. Consequently, in the absence of additional data, the best estimate you can make of what the population of data is like is to guess that its average is just equal to that single value you have measured. The reader will recognize this fact as the statistical muscle underpinning our earlier Principle of Mediocrity: What's happening here on Earth is nothing special; as galactic civilizations go, we're very ordinary and typical.

Extending the above argument using far more sophisticated statistical tools, Peter Sturrock has calculated the statistical spread in the value of N using estimates of the component quantities similar to those of Table 6.1. His conclusion is that with 70 percent confidence we can say that N is between 10,000 and 100 million, while with 95 percent assurance we can fix N between 100 and 10 billion. With such enormous levels of uncertainty, we're not really helped very much by the Drake equation itself in estimating N. But the analysis carried out by Sturrock shows that around 80 percent of the dispersion comes from the high level of uncertainty in the quantity L, the lifetime of a communicating civilization, and that almost half of the remaining spread is attributable to the term f_c, the likelihood that a communicating technical civilization will emerge. Thus, even on the basis of a single observation, it's still possible to employ standard statistical methodology to squeeze useful information out of the Drake equation.

A second statistical point to consider is that when we use the term "probability" in regard to the Drake equation, we're not using it in the same sense as when you say that the "probability" of the toss of a fair coin resulting in heads is $\frac{1}{2}$. In this more conventional usage, the value Prob (Heads) = $\frac{1}{2}$ is derived by repeating the experiment many times and then observing that in the long run, the event heads comes up half the time. This is the so-called *relative frequency* approach to estimating the probability of an event. With the exception of the astrophysical terms, we have only a single experiment upon which to base our estimates of the terms in the Drake equation. Thus, when we speak of the "probability" of the emergence of life, or the "probability" of the development of a communicating civilization, we are clearly using a different kind of probability than in the coin-tossing situation. Probabilists and statisticians call this kind of probability a *subjective* probability, since its numerical value is determined not by repeated experiments, but rather by the experience, judgment, and gut feeling of the investigator. While such estimates are less precise than conventional probabilities calculated by the relative frequency approach, they are not totally arbitrary either, since various internal-consistency conditions relating different estimates have to be obeyed. These subjective estimates are bound to improve as we carry out further laboratory experiments into the origin of life, as well as into our linguistic and cognitive capabilities, and as we continue to pursue investigations into the ways and means by which our collection of terrestrial cultures can avoid destroying themselves.

Freeman Dyson is a slender, dark-haired, youthful-looking man of average height, with a long hawklike nose and the intense, penetrating look of someone dedicated to his work. In pursuit of that work he has become one of America's premier theoretical physicists, as well as a thinker deeply concerned about the long-term prospects of a world in which there is enough nuclear weaponry to provide the explosive power of a ball of dynamite six feet in diameter for every man, woman, and child alive on the planet today. From his intellectual redoubt at the Institute for Advanced Study in Princeton, New Jersey, Dyson has through the years tossed out regular tidbits of fact and speculation creating major waves in the small pond of SETI.

At a joint U.S.-U.S.S.R. meeting to discuss SETI held in 1971 at the Byurakan Astrophysical Observatory in Soviet Armenia, Dyson made the characteristically provocative remark, "To hell with philosophy. I came here to learn about observations and instruments and I hope we will soon begin to discuss these concrete questions." Thus did he succinctly highlight the point that despite the utility of the Drake equation as a theoretical basis for many fascinating speculations about ETI, in the final analysis it is not armchair speculation but nuts-and-bolts experimentation that will ultimately settle the issue of whether $N = 1$ or $N > 1$. As an amusing aside, Dyson notes in a later account of the Byurakan meeting how he almost succeeded in creating a minor diplomatic incident when he wrote what he thought was the Russian translation of the English word "philosophy" on the blackboard as part of the opening sentence of his call to arms. It seems that at least in 1971, the Russian word *filosofiya* was used in a very specific sense to denote the kind of Marxist "philosophy" forming the basis for the Soviet political ideology, and not philosophy in the more general sense understood in the West. Fortunately Dyson consulted a Russian friend on the translation of the remainder of his opening sentence before it went on the board, thereby averting an awkward moment, but also showing the delicacy needed to communicate even with earthly intelligences. Anyway, taking our cue from Dyson, let's now turn away from theoretical SETI and pay some attention to the problem that started the SETI ball rolling in the first place: listening for radio signals from ETI.

EXPERIMENTAL SETI:
HOW SHOULD WE LISTEN?

Imagine the following situation: You're an American who has, for reasons unclear even to yourself, taken up residence in a small Central European country. On balance, you're not too sorry to have left behind most of the dubious delights of American culture: neighborhood junk-food emporiums, carbon-copy shopping malls, and the clownish preoccupations with cars and cholesterol, "relationships" and real estate. Nevertheless not all of your cultural baggage has been discarded, and your heart still beats a little faster when the shadows start to lengthen and the football stadiums from Stanford to Yale begin to fill. Unfortunately, you don't live within broadcast range of the U.S. military TV stations in Europe, so it's not possible to tune in to your favorite distraction nor will you again experience the bittersweet pleasures of those regular autumn meetings with your bookie. However, your spirits immediately perk up when a friend calls from America with the welcome news that one of the cable TV companies is putting up a new satellite that will regularly transmit all the football feeds from every network, major and minor, directly to a variety of sister stations all over Europe. To tune in to this bonanza, all you need do is crank up your parabolic antenna and settle back for an autumn's worth of the life you always aspired to—a steady dose of American football without having to actually be there.

Unfortunately, your friend is not exactly the technical type, leaving you in the dark as to how, when, and where to point your antenna to start harvesting this bounty of flying footballs and petulant player strikes. So what kinds of difficulties do you have to overcome? First of all, there's no information about how strong the signal from the satellite will be, so you don't know how sensitive your antenna must be. To be on the safe side, you buy the biggest dish your landlord will allow on the roof. Next, you have no information about the frequency (station) on which the satellite will be broadcasting, so to cover all bets you buy a receiver that will scan all channels. Moreover, the cable company's signal may not be perfectly pure, so you need to have a fairly broad-band receiver allowing you to pick up Channel 4 even if the signal that's coming to you is really Channel 4.2 or Channel 3.8, say. Furthermore, there's no information about the satellite's orbit or broadcasting schedule, so you have to play guessing games with the company as to exactly where in the sky and when you should point your antenna to try to pirate the signal. Finally, even should you manage to surmount all of these hurdles and actually tune in to the broadcast, you'll find that the cable company engineers are no dopes, and that before visions of the Rose Bowl will appear on your screen you'll be faced with the problem of trying to decode the signal. Now let's add a bit more spice to this stew by recalling that your friend (like most of mine), well meaning as he is, blows at least as much smoke as fire, so there may not even *be* a satellite! So now what do you think of your chances of catching Notre Dame doing battle against USC on the tube this fall?

This sad little story is but a pale imitation of the difficulties facing the experimental seekers of ETI. As it's commonly expressed in SETI circles, it's like a blind man in a dark room looking for a black cat—a cat that might not even be there! To get some feel for the real magnitude of the problem, let's take a little harder look at the three most important factors in a radio search for ETI:

FREQUENCY

The Earth's environment is filled with all sorts of radio noise coming from sources ranging from TV stations and military radars to various geophysical activities going on beneath and upon the planet's surface. This noise tends to drown out the reception of a certain band of frequencies from outer space. But outer space itself is far from quiet, containing its own brand of radio noise stemming from cosmic events, not to mention the constant background radiation from the original Big Bang. Figure 6.3 shows how these two kinds of radio noise combine at the Earth's

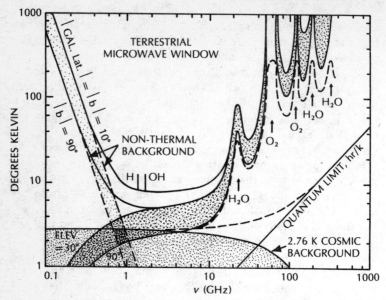

FIGURE 6.3. *Profile of radio noise heard on Earth*

surface to screen out effectively a wide spectrum of radio frequencies.

As noted earlier, every molecule acts as a miniature radio transmitter radiating at its own characteristic frequency. In the figure, the frequencies for interstellar hydrogen (H) and the hydroxyl radical (OH) are marked, clearly showing their favored positions near the rather flat minimum on the thermal noise curve. It was for this reason that Morrison and Cocconi proposed looking for ETI signals at a frequency near 1420 MHz. The region between hydrogen and the hydroxyl radical has, for obvious reasons, come to be termed the *waterhole* in SETI circles, reflecting not only the chemical composition of water (H_2O), but also the metaphorical interpretation of a waterhole as a meeting place for all sorts of "animals."

At present most radio searches are being conducted in or near the waterhole frequencies, although there are occasional proposals to look at other frequencies when seeking special types of signals. But so far there have been no convincing reasons offered to depart from the basic arguments laid down by Morrison and Cocconi, and it seems reasonable to suppose that the majority of Earth-based searches will continue to stay in this region. We'll return to this point in more detail later in the chapter.

SENSITIVITY

On the basis of economy, it's a good bet that the kind of signal an ETI would transmit will come in at least two parts: a beacon to attract our attention, as well as a second signal containing the information to be conveyed. These two types of signals have vastly different frequency requirements. To attract attention over the largest possible distance, all the power in the transmitter needs to be channeled into a single wavelength forming a beacon that will stand out against the cosmic background. On the other hand, when you're transmitting information, the wider the range of frequencies you can transmit over, the more information you can send. This is why a low-fidelity AM radio station transmitting over a bandwidth of only 5000 Hz (hertz) can't match the fidelity of an FM station using a bandwidth of around 100,000 Hz, not to mention a TV station operating on a bandwidth of 6 MHz.

Since we'll have to see the beacon before getting the message, it seems likely that our receivers will need to have very fine resolution, even down to 1 Hz. To see why, just imagine a beacon broadcasting exactly at the waterhole frequency of 1420 MHz, and suppose we had a receiver whose resolution was such that we could distinguish signals separated in frequency by no less than 100 MHz. In other words, we could hear signals at 1300 MHz, 1400 MHz, 1500 MHz, and so on, but could not distinguish anything in between. Such a receiver would pass right over the magic frequency of the beacon, leading us down the garden path of no ETIs, when in reality they're just waiting with bated breath to get in touch with us. Unfortunately, it takes a lot longer to search from 1400 MHz to 1500 MHz if you do it in steps of 1 Hz than if you do it in one fell swoop, so most current search strategies try to make some sort of compromise between high resolution and search time.

SEARCH DIRECTION

In Project Ozma, Drake & Co. pointed their telescope at Tau Ceti and Epsilon Eridani primarily on the basis of the G-star chauvinism discussed earlier. Regrettably, the Principle of Mediocrity comes into play here with the unhappy fact that the universe is just filled with G-type stars. In fact, in almost any direction you look there's an abundance of stars of "our type," so this requirement doesn't really restrict the search space to any appreciable degree. About all that can be said in this regard is that all things being equal (as they never are), it's probably a good idea to stay away from the galactic center as there are all sorts of events going on there, none of them conducive to living the good life—or any life at all.

One interesting variant on the direction theme has been proposed by Michael Papagiannis, a seemingly tireless astronomer at Boston University who is also president of the International Astronomical Union's Special Commission 51 on Bioastronomy (as SETI is euphemistically termed in polite scientific circles). Papagiannis suggested that if ETIs existed and were in a colonizing mood, the most likely place for them to take up abode in our solar system would be in the asteroid belt between Mars and Jupiter, since there they would find plenty of the raw materials needed to sustain an exploratory colony. Consequently, his idea is to focus some attention on looking for signs of ETI in our own solar system, as well as searching the stars. Seeking ETI in the asteroid belt may indeed be a stroke of divine inspiration. But at the moment it appears that most telescopes are not focused in this direction.

Before considering some of the actual searches that have been conducted so far, it's worth pausing to emphasize that we have been talking here only about searches in the radio-frequency part of the electromagnetic spectrum (10,000 Hz to 1000 MHz). Some have advocated searches at other wavelengths, primarily in the infrared 100,000 to 400 million MHz. The initial suggestion along these lines came in a short note to *Science* in 1960 by Freeman Dyson, who noted that a truly advanced civilization would surely have developed the technology needed to harness the entire energy output of its parent star. His suggestion was that such a civilization would dismantle all the planets of its solar system, using the matter to form a shell surrounding the central star in order to prevent the escape of enormous amounts of solar energy into outer space. Such a sphere would capture all the solar radiation for use by the ETI civilization, and a by-product of such capture would be that the sphere would radiate strongly in the infrared part of the spectrum.

A civilization capable of the kind of engineering magic needed to construct such a *Dyson sphere* is termed a Type II civilization in the classification scheme of Nikolai Kardeshev, a Russian SETI expert. According to this scheme, a Type I civilization is one at a level of development similar to our own, capable of utilizing most of the energy of its own planet, while a Type III can command the energy of an entire galaxy. By Dyson's arguments, we should tune our telescopes to the infrared part of the spectrum to see signs of a Type II civilization. Of course, eavesdropping on the radiated energy of ETI and listening for deliberate signals are quite different matters, requiring correspondingly distinct observing strategies. So at present there's not too much attention being devoted to looking for Dyson spheres. Other proposals are even more fanciful, involving signals sent by beams of neutrinos, tachyons (faster-than-light particles), and other mechanisms that at present are more properly left to the speculations of science fiction than to the realm of science fact. Now let's take a look at some of the searches since Ozma in an attempt to understand how we might recognize an ETI signal if we saw one.

WHAT ARE WE LISTENING FOR?—THE SYNTAX AND SEMANTICS OF SETI

Suppose you're a radio astronomer interested in SETI, and you manage to talk your bosses into giving you a little bit of the telescope's "dead time" to indulge your curiosity. You decide to do a "conventional wisdom" search at the waterhole frequency of 1420 MHz, and turn your telescope to one of the likely G-type stars in our galaxy, such as Tau Ceti. What exactly would you see, and how could you tell if your record really did include a signal from an advanced Tau Cetacean civilization?

To address this question, at the historic 1971 Byurakan meeting I. S. Shklovskii presented the diagram shown in Figure 6.4, which represents an artificially generated record of the type obtained by a radiotelescope. The graph is formed by superimposing random noise upon eighteen weak signals. The places where the signals occur are indicated by the small windows marked along the top of the diagram. This example should easily convince you that it's not possible to detect the existence of an ensemble of signals just by eyeballing the usual radiotelescope record. But by making use of statistical techniques of cross-correlation of the two independent spectra of the signal and the noise, the pattern of Figure 6.5 is obtained. The rather marked peak at reference delay 0 indicates the presence of a nonrandom component in the original record, i.e., a signal. This is one of the

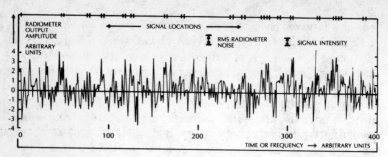

FIGURE 6.4. *An artificial radiotelescope record with eighteen signals*

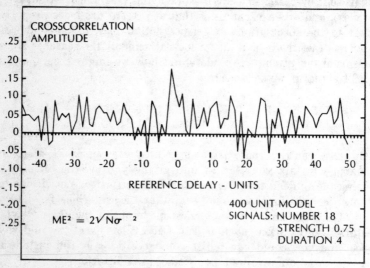

FIGURE 6.5. *The cross-correlation record of the signals and the noise*

standard procedures for recognizing that a real signal is embedded within an otherwise noisy record.

Another way of checking for the presence of a signal, especially if it's of the beacon variety, is just to point the telescope directly at the star and measure the received energy, then point it slightly away from the star and see if the energy from the comparison point differs in any significant manner. Figure 6.6 displays an experiment of this type carried out on Tau Ceti by Gerritt Verschuur at the waterhole wavelength of 21 cm (i.e., frequency 1420 MHz). In this case, it's clear just by inspection that there is no real difference in the received patterns from the

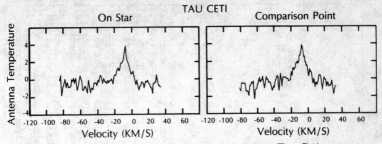

FIGURE 6.6. *The telescope record from Tau Ceti*

star and from the comparison point, which is about 20 minutes of arc away.

One of the most intriguing signals ever recorded is displayed in Figure 6.7. This is the famous "WOW" signal recorded in 1977 by the Ohio State University SETI project headed by John Kraus and Robert Dixon. The strength of the signals received in each of the fifty channels of observation are recorded on the left side of the figure, while the right side just indicates where in the sky the telescope was pointing. Note that mostly the received energy can be represented by a small number, usually 1 or 2. However, the WOW signal was so strong that it was necessary to go beyond the integers and use letters through Q to represent its magnitude. Regrettably, the signal was never seen again, despite repeated efforts to reacquire it by many investigators around the world over the last decade. So for now it's necessary to relegate the WOW signal to the ever-growing category of heart attack–inducing SETI anomalies.

Over the past several years, Jill Tarter of the NASA Ames Research Center and the University of California, Berkeley, has become the unofficial keeper of the books for ETI radio searches, having at last count (in 1987) recorded forty-five such efforts beginning with Project Ozma. So far there have been no successes, although all theoretical arguments conclude that we should expect to search for centuries before having a betting man's odds of actually finding an ETI signal, even if it is out there. (In this regard, Frank Drake sets a search horizon of five thousand years as an off-the-cuff estimate.) Nevertheless activity has never been more feverish in this area, with NASA currently in the process of putting together a decade-long SETI radio effort that should start around 1990. The NASA SETI program is divided into two parts: an *all-sky survey,* which will look at the

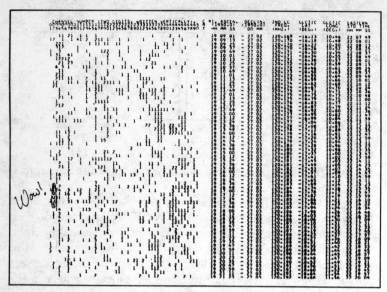

FIGURE 6.7. *The Ohio State "WOW" signal*

entire sky over a wide frequency range but with rather low sensitivity, and a *targeted search,* which will take a narrow-bandwidth look at about eight hundred stars over a restricted range of frequencies bordering the waterhole. Figure 6.8 displays the section of the cosmic haystack that the NASA SETI program will look at.

Oddly enough, the Russians, who were very positive about radio searches for ETI in 1971 at Byurakan, seem to have discontinued all efforts along these lines. Rumor has it that Shklovskii, who was head of the Astrophysics Division of the Soviet Academy of Sciences, apparently became disillusioned about the prospects for either the existence or the detection of ETI (it's not clear which), with the result that virtually all Russian radio search activity seems to have stopped. However, Shklovskii's death in 1985 may reopen the possibility of the Russians' rejoining the hunt.

But the NASA SETI program is far from the only search being contemplated over the next few years. Paul Horowitz of the Harvard-Smithsonian Project has made use of the explosive developments in microelectronics to develop an 8.4-million-channel narrowband spectrum analyzer, enabling his Sentinel Project to complete the equivalent of a hundred thousand *years* of

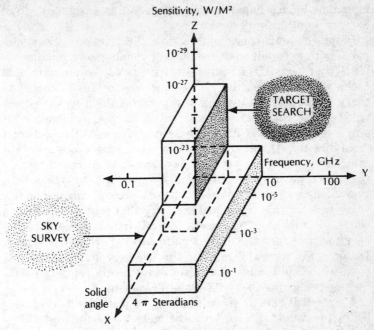

FIGURE 6.8. *The NASA SETI program parameters*

OZMA listening in one *minute!* Despite this incredible technological advance, in five years of listening, Horowitz jokes, "we've discovered the Sun twice." We still need years of searching to cover just the NASA parameter set, showing how truly immense the galaxy really is, and what a microscopic sliver of the haystack the searches so far have actually looked at.

For those cost-conscious SETI consumers, it's noteworthy that the cost of developing Horowitz's piece of equipment was a paltry $95,000, while the operating budget for the project itself is an anemic $20,000 per year. Both sums, incidentally, have been provided by the Planetary Society, a nonprofit SETI organization formed by Carl Sagan and partially sustained by movie mogul Steven Spielberg, who apparently is willing to invest at least a bit of his *E.T.* proceeds into the search for real ETs. At present, the Sentinel Project, along with the Ohio State SETI program which has been going since 1973, are the only totally dedicated SETI radio searches, not devoting telescope time to any other purpose. All the others involve either piggybacking SETI on the search for other astronomical phenomena, or grabbing time on telescopes in the odd moments when they're not engaged in other work.

Despite the almost negligible costs of SETI, especially as compared with multibillion-dollar particle colliders or a trillion-dollar SDI system, SETI enthusiasts have faced a continual uphill battle to obtain even the microscopic level of funds needed to carry on the search. A poignant example of this problem is provided by the Ohio State project, which has been run on a cost-only basis for almost fifteen years now. None of the personnel from directors Dixon and Kraus on down have taken a cent of salary for their time, yet they have somehow managed to keep the search alive by dint of Yankee ingenuity and sheer hard work. But even this noble effort almost came unglued a few years back when a different university that owned the land on which their telescope sits wanted to sell it to a property developer for transformation into a golf course! Fortunately Kraus, Dixon, & Co. were able to avoid this close encounter of the golf course kind, but only after a sustained media campaign reinforced strong appeals by the scientific community.

A somewhat similar situation occurred in 1981 when NASA's SETI program was axed from the budget by an extraordinary legislative amendment proposed by that eternally vigilant guardian of the public purse, Senator William Proxmire. Unfortunately Congress passed the amendment, thereby mobilizing the U.S. SETI community in a lobbying effort to get the money reinstated. Enter the Planetary Society's biggest gun, Carl Sagan, who had had earlier interactions with Proxmire and thought of him as being a reasonable man despite his public image to the contrary. So off to Washington went Sagan, who listened to Proxmire's argument, which, in essence, came down to the standard $N = 1$ argument that if ETI existed, we would have seen it by now. Sagan's rejoinder was to point out the enormous importance of the factor L, the lifetime of technical civilizations, in the Drake equation, and the crucial importance of SETI for discovering whether or not there have been other civilizations that have avoided self-destruction. It turned out that Proxmire had never even heard this line of argument before, and after reflecting upon the whole issue decided to drop his objection. The final battle came when the issue had to be taken up on the floor of Congress. But by good fortune the film *E.T.* had just been released at that time, and was grossing more box-office receipts per day than the entire amount NASA was seeking to look for E.T.'s real-life siblings. Given the glacial pace of congressional deliberations, nothing happened for weeks and months. But finally, on September 30, 1981, the last day before the new fiscal year, Congress passed interim funding to keep the country solvent and in the process also passed the budget for the independent agencies—including NASA. Thus was American SETI saved from legislative extinction.

Most of the foregoing SETI stories deal with the syntactic aspects of radio searches, i.e., how we might recognize a signal if we saw one. But what about the *semantic* component? Imagine we actually had a live ETI transmission in hand. What would it be likely to say? What sorts of messages might be contained in a collection of pulses of the type that most searchers think will compose an information signal? No one really has any idea, of course, about what an ETI might think is important enough to try to transmit across the galaxy. So most studies of the semantics of SETI naturally tend to focus upon the kind of message that *we* might want to send to them (another good example of the irresistible anthropomorphic bias of most SETI work).

On the northern coast of Puerto Rico, near the town of Arecibo, there is a natural dish-shaped hole in the rock over 1,000 feet wide. Inside this bowl sits the world's largest radio-telescope. So large, in fact, is this dish that its collecting area exceeds that of all the optical and microwave telescopes ever built in the history of man. Put another way, it would take around 4 billion bottles of beer to fill the Arecibo bowl. In 1974 modifications were made to this telescope enabling it to transmit a radio beam of unprecedented power, up to 20 terawatts (1 terawatt = 1 trillion watts) for a short interval. As an inaugural test of these changes, it was decided to use the Arecibo dish to transmit a signal to the edge of our galaxy informing potential listeners that "we are here!" This pathbreaking signal, composed of a sequence of 1,679 binary 1's and 0's, was transmitted in a little under three minutes on November 16, 1974, at a frequency of 2380 MHz with a bandwidth of 10 Hz. Note that this is not the waterhole frequency but is still in the low part of the cosmic thermal noise curve of Figure 6.3. The actual sequence transmitted is shown in Table 6.2. What kind of informational content about we earthlings could be contained in such a sequence of pulses?

The logic underlying the message is to assume that any receiv-

```
000000101010100000000000000101000001010
000001010010100010001000100101011001010101
010101010100100100000000000000000000000
00000000001010000010100000000000000000000
11010000000000000011000000011010000000000
00000000101010000000000000000011110
0000000000000000000000000000000110000
11100001000011000100000000000000010
0001101000110001000110101111011:::1
011111011111100000000000000000000000
01000000000000000110000000000000000000
00000000000001000000000000000011111100
0000000011111000000000000000000000000
00110000110000111000100000010000001000
000000100001101000011000110010101011
110111110111101111000000000000000000
000000000100000010000000000000000000
00110000000000010000010000001000000000
11111100000111111000000000000000000110
0000000000000000011000001110000000000
000001100000001100010000100000000000
0110010000000000000010000000001000000
0000110001000110000001000001000000010
00000100000100000001100000000010001000
01011100000000010000000000000100000000
1000000100000000100000010000001000000
00000011000000000001000100000001000000
0101110100010110100000000110010001111
1111011100000110000110111000000001
10000011011001000000001000000111111100
10000001010000010000000000000110011000
00000000000000000000000000000000110001
00011100000000000011101010001010101
01001110000001100111010000000000000
0101000000000000011111:1000000000000
0001111111100000000000001100000011111
000000000100000000001100000001100
0000000101100000010011000000110011100
001000101000001000000100010001001001
001000100000000100010100000000000000
01000010000100000000000001000000000000
000000000001000101000000000011:1001
1110100111000
```

TABLE 6.2. *The 1974 Arecibo transmission*

ing ETI will soon recognize that the number of pulses is the product of the two prime factors 23 and 73; i.e., the unique expansion of 1,679 into prime factors is $1{,}679 = 23 \times 73$. (Recall: An integer is prime if it is divisible only by itself and by 1.) Since every integer can be written as the product of primes in exactly one way, the fact that 1,679 has only two prime factors suggests that the signal is actually a code for the construction of a two-dimensional picture. By breaking up the message into seventy-three rows of twenty-three characters each, arranging each row one under the other, and letting 0 stand for a blank with 1 being a dark space, a clever ETI would arrive at the picture shown on the left in Figure 6.9, with its interpretation given to the right.

Starting at the top, the first part of the message is a counting lesson that describes the number system that is to be used. The numbers 1 through 10 are written across the top in binary notation. Notice that each number has a "number label" associated with it, both to indicate that it *is* a number and to show from which direction it is to be read. The numbers 8, 9, and 10 are deliberately written on two separate lines to show how numbers

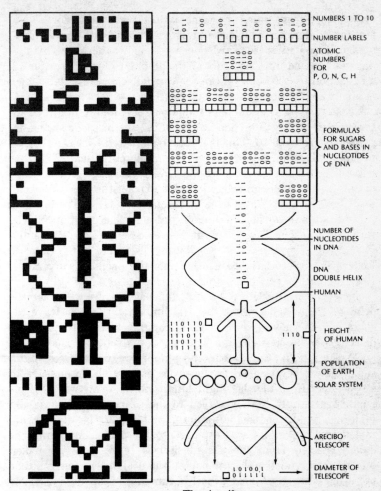

FIGURE 6.9. *The Arecibo message*

too large to be specified on a single line will be written later. The rest of the message deals with various physical, chemical, and biological features of life on Earth. In fact, these parts of the message are exactly what we would like to know about ETI in order to fill in some of the blanks in the Drake equation. The message concludes with a description of the telescope that sent it, shown as centered on the third planet with the number across the bottom indicating that the telescope is 2,430 wavelength units (1,000 feet) wide, where the natural wavelength for ETI to assume is that on which the signal was sent. All very simple, logical, and straightforward—once you know the key. This signal was beamed into the heart of the globular cluster Messier 13, a collection of 300,000 stars in the constellation Hercules, which is about 25,000 light-years from Earth.

Those readers who dabble in amateur cryptology might be interested in trying their hand at decoding the message string depicted in Figure 6.10. This is a message created by Frank Drake to show what we might receive from a hypothetical ETI. The principles are the same as for the Arecibo message, but don't be discouraged if you can't decode it—to extract the message would probably require a whole team of specialists. (Hint: The message consists of a total of 551 binary pulses.) The solution can be found in "To Dig Deeper." But just in case ETI isn't listening in M13, or is perhaps on a trip visiting elsewhere in the Galactic Federation, there have been other efforts to send a souvenir of Earth to the stars.

Sometime in 1989, the *Pioneer 10* space probe will pass the orbit of Pluto and become the first human artifact to leave the solar system, moving on a heading roughly toward the star Aldebaran in the constellation Taurus. Shortly before the launch on March 2, 1972, Carl Sagan and Frank Drake suggested that a small plaque be attached to the probe as a symbolic message to any wandering ETI who might bump into it on its way to Taurus. Surprisingly NASA agreed to the proposal, and a six- by nine-inch gold anodized aluminum plate with the display depicted in Figure 6.11 was attached to the probe. As one might have expected from the use of figures representing a naked human male and female, as soon as the design was made public the lunatic fringe started a mail-in campaign accusing NASA of trafficking in space pornography. One can only wonder whether E.T., Jabba the Hutt, or the Blob would find the figures one bit erotic! In any case, the chances are nil that *Pioneer 10* will ever enter another solar system, so the whole exercise was far more symbolic than real anyway.

Heartened by the generally positive response to the *Pioneer 10* plaque, and never one to miss a chance to promote SETI in the public arena, Carl Sagan saw the launch of the *Voyager 1* and *2* probes in 1977 as another opportunity to spread human cheer and goodwill outside the solar system. Since there was much

```
11110000101001000011001000000001000001010 0
10000011001011001111000001100001101000000
00100000100001000010001010100001000000000
00000000001000100000000000101100000000000 0
00000001000111011010110101000000000000000
00001001000011101010101000000000101010101
00000000011101010101110101100000001000000
00000000000010000000000000010001001111110 00
00111010000010110000011100000001000000000
10000000010000000111110000001011000101110
10000000011001011111010111110001001111100 1
00000000000011111000000010110001111110000 0
10000011000001100001000011000000011000101
001000111110010111 1
```

FIGURE 6.10. *A message from a hypothetical alien*

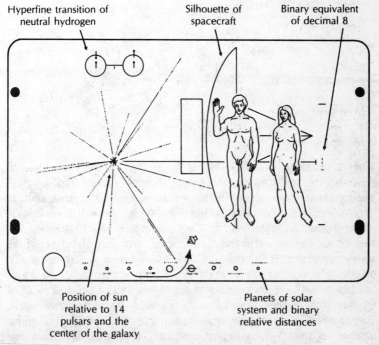

FIGURE 6.11. *The plaque on* Pioneer 10

more time to prepare the message than for the *Pioneer* effort, the *Voyager* communiques could be far more complex and imaginative than just a simple plaque. As a result, both of the *Voyager* probes carried a special kind of videodisk upon which was encoded much of our scientific information, as well as a medley of earthly sights and sounds, sort of an interstellar version of "The Earth's Greatest Hits." Table 6.3 gives a list of the contents.

It's not without interest to note that Sagan seems to have had a change of heart sometime after these exercises, since in a 1983 article in *Science* he argues rather strongly in favor of the current SETI programs of listening instead of sending, basing his case on the following points:

- *New kid on the block:* Since we're just entering the SETI game, few ET civilizations could be more backward than we are. Hence, we should be listening, not sending.
- *Poor mouthing:* Civilizations considerably more advanced than we are would have vastly greater energy resources and more sophisticated technologies that they could use for transmission.
- *Barbarism:* Two-way conversations that may take centuries have not yet entered into our long-term planning processes, which mostly don't extend beyond the next election or war.
- *Hide-and-seek:* By sending, we might "give our position away" to an unscrupulous ETI who might want to plunder our planetary resources or use us for slaves or food.
- *Village idiot:* It's not clear that we have anything interesting to say.

Most of these points are debatable at best, with the exception of "hide-and-seek," which is not even debatable since we gave away our position years ago with the escape of our TV transmissions of *I Love Lucy, Dallas,* and *Mork & Mindy,* as well as military radar signal leakage outside the ionosphere. Nevertheless, the fact remains that at present no one is worrying about sending, all the programs being devoted to various forms of listening. My own guess is that it's a matter of pure economics. It's hard enough to talk the Proxmires of the world out of the few million dollars that NASA spends on SETI each year. Imagine what they'd say if you told them you were going to spend the money and merely send information, not try to receive it. I rest my case.

PICTURES (in sequence)

calibration circle
solar location map
mathematical definitions
physical unit definitions
solar sys. parameters (2)
the sun
solar spectrum
Mercury
Mars
Jupiter
Earth
Egypt, Red Sea, Sinai
  Pen., Nile (from orbit)
chemical definitions
DNA structure
DNA structure magnified
cells and cell division
anatomy (8)
human sex organs (drawing)
conception diagram
conception photo
fertilized ovum
fetus diagram

fetus
diag. of male and female
birth
nursing mother
father and daughter (Malaysia)
group of children
diagram of family ages
family portrait
continental drift diagram
structure of earth
Heron Island (Australia)
seashore
Snake River, Grand Tetons
sand dunes
Monument Valley
leaf
fallen leaves
sequoia
snowflake
tree with daffodils
flying insect, flowers
vertebrate evolution diag.
seashell (Xancidae)

dolphins
school of fish
tree toad
crocodile
eagle
S. African waterhole
Jane Goodall, chimps
sketch of bushman
bushmen hunters
Guatemalan man
Balinese dancer
Andean girls
Thai craftsman
elephant
Turkish man with beard
  and glasses
old man with dog and
  flowers
mountain climber
Cathy Rigby
Olympic sprinters
schoolroom
children with globe

cotton harvest
grape picker
supermarket
diver with fish
fishing boat, nets
cooking fish
Chinese dinner
licking, eating, drinking
Great Wall of China
African house construction
Amish construction scene
African house
New England house
modern house (Cloudcroft)
house interior with
  artist and fire
Taj Mahal
English city (Oxford)
Boston
UN building (day)
UN building (night)
Sydney Opera House
artisan with drill

factory interior
museum
X-ray of hand
woman with microscope
Pakistan street scene
India rush-hour traffic
modern highway (Ithaca)
Golden Gate Bridge
train
airplane in flight
airport (Toronto)
Antarctic expedition
radio telescope
  (Westerbork)
radio telescope (Arecibo)
book page (Newton's System
  of the World)
astronaut in space
Titan Centaur launch
sunset with birds
string quartet
violin with score

GREETINGS IN MANY TONGUES (alphabetically)

Akkadian
Amoy (Min dial.)
Arabic
Aramaic
Armenian
Bengali
Burmese
Cantonese
Czech
Dutch

English
French
German
Greek
Gujarati (India)
Hebrew
Hindi
Hittite
Hungarian

Ila (Zambia)
Indonesian
Italian
Japanese
Kannada (India)
Kechua (Peru)
Korean
Latin
Luganda (Uganda)

Mandarin
Marathi (India)
Nepali
Nguni (SE Africa)
Nyanja (Malawi)
Oriya (India)
Persian
Polish
Portuguese

Punjabi
Rajasthani
Romanian
Russian
Serbian
Sinhalese (Sri Lanka)
Sotho (Lesotho)
Spanish
Sumerian

Swedish
Telugu (India)
Thai
Turkish
Ukranian
Urdu
Vietnamese
Welsh
Wu (Shanghai dial.)

SOUNDS OF EARTH (in sequence)

whales
planets (audio
  analog of
  orbital velocity)
volcanoes
mud pots
rain

surf
cricket frogs
birds
hyena
elephant
chimpanzee
wild dog

footsteps and
  heartbeats
laughter
fire
tools
dogs (domestic)
herding sheep

blacksmith shop
sawing
tractor
riveter
Morse code
ships
horse and cart

horse and carriage
train whistle
tractor
truck
auto gears
Saturn 5 rocket
liftoff

kiss
baby
life signs:
  EEG, EKG
pulsar

MUSIC (in sequence)

Bach: Brandenburg Concerto #2, 1st m.
Java: court gamelan— "Kinds of Flowers"
Senegal: percussion
Zaire: "Pygmy girls" Initiation song
Australia: horn and totem song
Mexico: mariachi— "El Cascabel"
Chuck Berry: "Johnny B. Goode"
New Guinea: men's house
Japan: shakuhachi (flute)—
  "Depicting the Cranes in Their Nest"

Bach: Partita #3 for violin
Mozart: "Queen of the Night"
  (from "The Magic Flute")
Georgia (USSR): folk chorus— "Chakrulo"
Peru: pan pipes
Louis Armstrong: "Melancholy Blues"
Azerbaijan: two flutes
Stravinsky: "Rite of Spring" conclusion
Bach: Prelude and Fugue #1 in C Major
Beethoven: Symphony #5, 1st m

Bulgaria: shepherdess song—
  "Izlel Delyo hajdutin"
Navajo: night chant
English 15th cent.: "The Fairie Round"
Melanesia: pan pipes
Peru: woman's wedding song
China: ch'in (zither)— "Flowing Streams"
India: raga— "Jaat Kahan Ho"
Blind Willie Johnson: "Dark Was the Night"
Beethoven: String Quartet #13, "Cavatina"

TABLE 6.3. *The contents of the* Voyager *disk*

* * *

Having now considered the main theoretical and experimental underpinnings of the ETI question, it's time to let the ideologists of the $N = 1$ and the $N > 1$ schools have their day in court. But before entering the courtroom and listening to their respective arguments, let's try to summarize the various subdivisions of the problem by listing ten possible answers given by astronomer John Ball to the original Fermi question: "Why are we unaware of ETI?"

1. *There is no ETI.* Either Earth is unique or ours is the first civilization in the galaxy to reach this level of development.
2. *ETI exists, but it's very primitive.* It doesn't know we're here, but it might like to know.
3. *ETI exists and is at about our level of development.* It suspects we might be here and it might like to talk with us (the Mirror Hypothesis).
4. *ETI exists and it knows we're here.* It would like to talk with us if it could just attract our attention.
5. *ETI exists and knows we're here, but it doesn't care.* We pose no threat and we have nothing that it wants.
6. *ETI exists and we are of some interest to it.* A few ETI scientists are discreetly studying us.
7. *ETI exists and we are of considerable interest to it.* It is studying us in some detail, but inconspicuously.
8. *ETI exists and it occasionally dabbles in our affairs.* We are of considerable interest to ETI and it wants to interact with us directly (the UFO Hypothesis).
9. *ETI exists and is experimenting with us.* We are laboratory animals for it (the Petri Dish Hypothesis).
10. *God exists.* A supernatural being who is omnipotent and omniscient exists (i.e., God is identical with ETI).

All of these views except the first imply the existence of ETI, although Cases 2 through 9 are not mutually exclusive. Cases 1 to 4 are the popular views, with 2, 3, and 4 representing the dominant view of the SETI scientific community. Cases 6 and 7 are commonly termed the Zoo Hypothesis, for obvious reasons. Beginning with Case 8, one leaves the realm of science and enters into the domain of religion and philosophy, Case 10 being the popular nonscientific position.

Since what we're interested in here is science, let's lump Cases

3 through 7 under the general label $N > 1$, while the other side of the trial, $N = 1$, will be associated with Case 1. Case 2 involves ETIs so primitive or profoundly alien that no communication is yet possible, so I also lump this case in with the $N = 1$ side of the house. Now, having completed the preliminaries, let's listen to the Prosecution arguments for why we are not alone in the galaxy.

$N > 1$: ETI EXISTS!

The early 1970s were a particularly cordial period in U.S.-Soviet relations, when even the chronically overbooked Moscow "gourmet restaurant" Aragvi was always ready to accommodate a "famous visiting American professor" by mysteriously conjuring up a table without benefit of the traditional *na leva* gratuity, an opportunity I myself was always ready to exploit during a 1972 tour of duty at the Control Sciences Institute of the USSR Academy of Sciences. From September 5 to 11, 1971, during this all-too-brief golden age of détente, the Byurakan Observatory near the Armenian capital of Yerevan hosted what is still one of the most extraordinary SETI meetings ever convened. This Soviet-American gathering at the foot of Mount Ararat, which we mentioned briefly earlier, had as its unofficial agenda a detailed analysis of each of the terms in the Drake equation, together with a consideration of the various experimental attacks upon ETI as outlined above. While the experimental state of ETI research has improved by several orders of magnitude since this historic event, a reading of the transcript shows that the theoretical speculations are still as fresh and timely as the day they were proposed over fifteen years ago (a good indicator of the level of hard data versus soft speculation in the theoretical ETI business).

After a week of "Armenian breakfasts," an indispensable ingredient of which is a shot or two of the fiery local cognac, the theoretical underpinnings of the entire $N > 1$ school of SETI thought were laid down, mostly by an American contingent nicknamed the Cornell group by the distinguished historian William McNeill, himself a meeting participant. This constellation of SETI devotees consisted of Carl Sagan, Philip Morrison, Frank Drake, and Thomas Gold, all of whom were or had recently been

on the faculty of Cornell University at the time of the meeting.
The essence of the position put forth at Byurakan was that by
inserting their best scientific estimates, subjective probabilities,
and just plain hunches into the Drake equation and turning the
crank, the outcome would be a number N far greater than 1.
Since this line of argument has already been addressed in some
detail above, let me try to summarize the core of the $N > 1$
thesis using the following chain of reasoning:

I

Every shred of genuine scientific fact points to the conclusion
that the Earth and our solar system are perfectly ordinary and
typical in every possible way (the Principle of Mediocrity).

II

Since life, intelligence, technology, and all the rest have
developed here on Earth, in the absence of further information
we must assume that these conditions are typical
elsewhere as well.

THEREFORE

ETI exists elsewhere in our galaxy; i.e., $N > 1$.

As a corollary to the claim that N is greater than 1, it's of
interest to remark upon the final resolution of the meeting,
which, incidentally, serves as an exemplary model for East-West
scientific cooperation and goodwill. It states in part that "the
Conference participants . . . agreed that the promise of contacts
with such extraterrestrial civilizations is sufficiently high to justify initiating a variety of well-formulated search programs."
Thus was the manifesto of the $N > 1$ enthusiasts laid down, and
thus it stands to this day: The likelihood of N being larger than
1 is sufficiently great to justify the costs of actively searching.
In following the literature over the years since Byurakan, it's
intriguing to see just what kinds of proposals for "searching"
the Byurakan declaration has generated.

While the majority of scientists concerned with SETI have
understandably concerned themselves with the sorts of radio
searches considered earlier, there have been the usual extremists
at both ends of the scientific spectrum who interpreted the word
"search" in the literal sense, and focused their energy, calcula-
tors, and typewriters on the matter of *direct* contact. These vi-

sionaries fall into two totally distinct groups: the UFOers and the space travelers. Since there has never yet been an unambiguously documented case of an extraterrestrial visit to Earth, I won't open this can of worms here, leaving those who feel a deep psychological need to believe in direct extraterrestrial intervention in our puny affairs free to do so. Instead, in what follows I'll settle for considering some of the less contentious scientific arguments for space travel as a means of contact.

First of all, the various Apollo, Viking, Voyager, and Pioneer programs, as well as similar Soviet ventures to Venus and more recently to Mars, leave little doubt that space travel of at least a limited sort is well within the realm of our current technology and purse. The problem for SETI is that by now it's generally conceded that there are no intelligent life forms on any of the planets of our solar system, implying that if we want to meet ETI face to face, we're going to have to wind up our big toys and set off into the interstellar void. Just how technically and economically feasible is it to mount an expedition to visit even one of the "nearby" stars?

To get some feel for the magnitude of the problem, think about the distance involved in traveling to the Moon. The Moon is about 240,000 miles from Earth and represents the greatest distance that man has yet ventured beyond Earth. If we imagine this distance as being the same as walking across your living room, then on the same scale a trip to the nearest star is equivalent to going to the Moon. In other words, such a trip represents about 100 million trips to the Moon! And this is only to the nearest star, Alpha Centauri, which, unfortunately, is not very interesting from an ETI standpoint. To get to Ozma's candidates, Tau Ceti and Epsilon Eridani, each around 11 light-years away, would involve about 300 million such trips. These are not distances to be taken lightly (no pun intended). So distance alone imposes severe restrictions on what can be done about going out to find ETI.

But let's suppose, as some have done, that it's feasible to construct a ship using some sort of superduper fusion or antimatter drive that will allow a ship to travel at $0.1\,c$, one-tenth the speed of light. Studies have shown that no new physical principles are involved in making such a vision a reality, although the engineering hurdles are enormous. At such a velocity, you might be able to visit one of the stars shown in Figure 6.12 within your

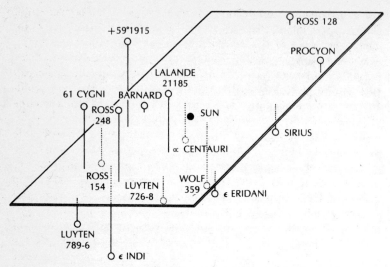

FIGURE 6.12. *The fifteen nearest stars to the Earth*

lifetime. Techniques of suspended animation could extend this limit, but probably not by much without introduction of major new, and totally unpredictable, biological and physical principles. So on physical grounds alone the prospects for a generation of Captain Kirks venturing where no man has gone before appear rather bleak, at least if that venturing extends much beyond a few light-years from Earth. But let's imagine that all the physics and engineering works out and you're determined to check up on doings at Wolf 359 or Procyon. How much would it cost to indulge your curiosity?

The question of economics ultimately comes down to how much energy you need in order to transport yourself and your belongings to a nearby star. Some back-of-the-envelope calculations are sobering. Let's suppose that E represents the amount of energy needed to maintain what you might think of as the "good life." Assuming as above that you can travel at $0.1\,c$ and that you require 10 tons of mass in your spacecraft per passenger (about the same ratio as for a large jet passenger plane), the amount of energy needed for our prototypical 100-year trip is about 2 million $\times$ E. To pin down a value for E, let's consider the annual energy consumption in the United States. In 1979 this figure was 10^{20} joules, leading to a value for E of 4×10^{13} joules (1 joule = the amount of energy needed to raise the temperature of 1 cubic centimeter of water by about $\frac{1}{4}$°C). Putting these figures together, we come to the sad conclusion that the minimum energy needed for one passenger to Tau Ceti is around 8×10^{19} joules. Thus, for a 100-passenger colony this represents an amount of energy sufficient to sustain the entire American population, the most profligate in history, for a period of *several hundred* years. The bottom line of this elementary calculation is that unless there is some development that makes energy literally cost-free, no society will ever be able to underwrite the cost of sending you on your journey to the stars.

But just to keep the fiction alive, let's suppose such a free energy source was miraculously discovered and we set sail in search of ETI. What might we find? Would ETI be a cute little wide-eyed charmer like E.T., or would it be more like the nightmarish creature of *Alien,* or perhaps neither? And what kind of social order might ETI have developed to enable it to survive the various perils of postindustrial life outlined earlier? These are some of the issues that ETI theoreticians of the $N > 1$ persuasion have fun speculating about when their *real* day's work is done. It's impossible not to be sucked in by this kind of speculation, so let's carry the $N > 1$ claim to ridiculous extremes and consider a few of the more sober, or at least scientifically defensible, possibilities for the form of ETI.

THE SHAPE OF ETIS TO COME

Most of the reasons put forth by the SETI community as justifications for trying to make contact with ETI are of a somewhat lofty and depressingly sober character, viz., renewed hope in knowing that another civilization managed to survive our current market basket of nuclear, ecological, and psychological crises, membership in a galactic federation, technological wonders such as free energy, teleportation, and immortality. Such noble aims are just the ticket for scientists, sages, and congressmen, but as for myself these supposed benefits of SETI are only moderately interesting, and my own interest in ETI is distinctly more visceral—I want to know what it looks like! Direct contact is the best way to scratch this itch, although radio searches *might* supply the information if the message is one of a pictorial nature like those considered earlier, or if the message gives in-

structions about how to construct a living ETI here on Earth.
But in the absence of any contact, direct or otherwise, we have
to fall back upon our own biological knowledge to speculate on
what might step out of that first UFO to become an IFO. There
are two diametrically opposed views on this matter.

The first line of reasoning about the form of ETI is to argue
by appeal to the process of *convergent evolution*. On Earth, whenever Nature has had a problem to solve, whether it be the optimal design for a sense organ to process visible light or an
efficient way to tear food apart into bite-sized pieces, there has
been a tendency for the problem to be solved in a very similar
manner across a wide variety of species. Movement in water is a
good example. In the course of Earth's history, there have been
three species that made their living by swimming rapidly in
coastal waters preying on small fish to fill their stomachs. These
three species are the tuna (a fish), the dolphin (a mammal), and
the ichthyosaur (an extinct reptile). These three animals have
very little to do with each other, biochemically or phylogeneti-
cally. But if we examine their physical forms, we find they all
look about the same—like a living torpedo. This is a good example of how several different evolutionary paths may "converge"
to the same ecological position. It just happens to be very efficient to have a torpedo-shaped body if you need to swim fast to
catch your dinner. The convergent evolution school applies this
same general principle to speculating about ETI. We can sum
up the convergent evolution thesis in the following diagram:

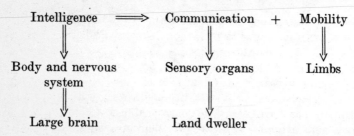

Life on earth has evolved two distinct types of symmetry, bilateral and radial, and it's no accident that the most successful
life forms have bilateral symmetry. As we've noted, it appears
very likely that life evolved in the oceans, and in such a watery
medium an organism with a streamlined body has a distinct com-

petitive advantage when it comes to catching prey or escaping from predators. On the other hand most life forms with radial symmetry lead rather starfishlike sedentary lives and have simple nervous systems.

It further appears that a necessary precondition for the development of a complex nervous system is an active, mobile, predatory life-style. In such a predatory life form with a complex nervous system, the central controlling brain should be close to the primary sense organs so that the connecting nerve paths are short and the animal's response time is correspondingly fast. Such an animal must also have its sensing and grasping organs at the front of the body near the mouth and, if it must smell its food before eating, the organ for this sense must be located above its mouth. Thus bilateralism and the presence of large ganglia of nerves near the front of the body and close to the primary sense organs are essential characteristics of intelligent creatures in the convergent evolution scheme of things.

On Earth, birds, fish, and mammals all conform to the above requirements. However, birds are not likely to develop high intelligence because they must be light and have a large surface area in order to fly. Thus they can't afford the weight of a large brain and the heart needed to supply such a brain with enough blood to keep it going. Life in the water doesn't suffer from these drawbacks, as witness, for example, the whales, which are the largest creatures ever to exist on this planet. However, it can be argued that life in the water is too easy to afford the kinds of challenges to survival necessary to stimulate development of a complex nervous system. That is, challenges of the type leading to higher brain functions are usually associated with three things: the use of tools, the development of language, and the formation into social groups. Only land-based mammals fulfill all of these conditions.

As a final exemplar of convergent evolution, there is the development of jointed legs, which seem to be the best solution for moving over different types of terrain. But a large number of legs make for difficulties in coordination and slowness of movement, while an odd number would create an awkward imbalance. Thus, the swift runner would probably have only a small number of legs in pairs, of which one or two pairs would possibly have been modified to act as arms for the manipulation of tools.

Putting all these considerations together, one comes up with an ETI whose physical form would be remarkably humanoid; in fact, remarkably like the kinds of forms reported by people who are abducted by the occupants of UFOs, or maybe like the benevolent aliens depicted in the film *Close Encounters of the Third Kind.* In all cases these ETIs look just about like you and me, usually with the exception of a more pronounced egg-shaped skull, presumably an indication of the far more advanced state of their cerebral development.

Frankly I find this sort of anthropomorphic argument totally unimaginative and quite boring. What a delicious cosmic joke it would be to spend a few zillion dollars to look in on happenings around Barnard's Star, and find a planet where everyone drove Fords, ate at McDonald's, and watched *The Cosby Show!* But for alternatives it's not sufficient to accept the standard arguments that there may be many different pathways to creatures functionally equivalent to, but physically unlike, ourselves. As one might suspect in matters of the imagination, in a search for alternatives we have to leave the mainline scientific community behind and turn to the science fiction writers and philosophers for some mind-bending, yet physically feasible, candidates.

One of the great depictions of an alien life form in fiction is given in Donald Moffitt's novel *The Jupiter Theft,* which tells of the plight of the Cygnans, a race of creatures that evolved on the satellite of a gas giant planet orbiting a binary star system. One of the members of the stellar pair collapses into a black hole, but the Cygnans have sufficient warning that a small colony escapes in five 30-mile-long ships, the interiors of which are primarily huge, open, artificial forests where the Cygnans live alongside the small arboreal animals they catch for food. The story tells of how these creatures enter our solar system and begin dismantling Jupiter as a material source of energy, and the feeble attempts of humans to do something about it. Figure 6.13 is an artist's interpretation of how the Cygnans look. The Cygnan is about 1.5 meters tall, with six limbs that can be used as either arms or legs, and a long, three-petaled tail that folds to conceal the sexual organs. The slender, tubular body is built on a cartilaginous skeleton, with the brain located between the upper pair of limbs at the top of the spinal cord. The three eyes are placed on stalks in an equilateral triangle around a broad, flexible mouth. The Cygnan has a harsh, rasping plate in the mouth, and a spiked, tubular tongue. It has a well-integrated nervous

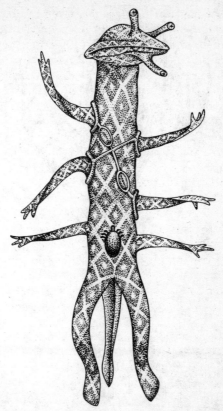

FIGURE 6.13. *A Cygnan from* The Jupiter Theft

system, with much faster synaptic reflexes than those of a human being. Cygnan speech is musical, consisting of chords produced by multiple larynxes, and depends on absolute pitch. The language is incredibly rich and varied; it has more than a million phonemes, and each word is made up of several phonemes. But if you think the Cygnans are still too humanoid, let's look at another possibility.

On Earth, the only intelligent entities that differ radically from the kind of bilateral humanoid considered earlier are colonies of social insects like bees and termites. Arguments have been made that ETI life forms might also adopt this kind of bottom-up mode of group intelligence. A sci-fi example is depicted in Figure 6.14.

This picture shows the Cryer, a creature from Joseph Green's

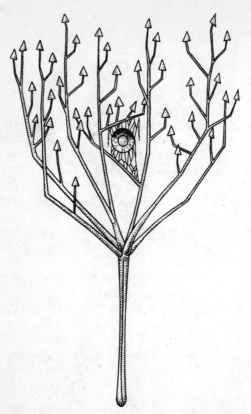

FIGURE 6.14. *The Cryer from* Conscience Interplanetary

book *Conscience Interplanetary.* The Cryer is an independently functioning unit of a planet-wide silicon-based plant intelligence inhabiting the planet Crystal, which has an atmosphere of 18 percent oxygen, the rest being nitrogen and hydrogen. Life on Crystal is based on silicon, with a high percentage of metallic elements. The Cryer resembles a two-meter-high bush with a crystal-and-metal trunk and branches, with small, sharp glass leaves. The trunk contains silicon memory units, powered by a low-voltage solar storage battery and connected by fine silver wires. About six feet up the Cryer's trunk is an organic air-vibration speaker membrane created for it by the planet-wide entity to enable it to speak with human beings. It is a broad, saucer-shaped leaf held in place by stretched wires to provide a vibrating diaphragm. A magnetic field generated in silver wire coils hanging on either side of the speaker causes it to vibrate to produce sound.

The planet-wide intelligence consists of thousands of smaller units like the Cryer, connected by an underground nervous system of fine silver wire. Each unit has a specialized function, some storing electricity generated by sunlight, some extracting silver for constructing the nervous system, some providing memory storage, and some acting as sensor units. The overall intelligence is able to perceive temperature, motion, position, electrical potential, and vibrations through its member units.

Cygnans and Cryers give only a small taste of the kinds of ETIs that may be out there, if one is to believe the science fiction writers' union. I bring them up here only to show that the argument from convergent evolution, while scientifically defensible on the grounds that it's happened at least once, is far from the last word on the matter of the kind of physical form ETI might assume. As to how ETI might *act,* we have already spent a chapter examining the degree to which human actions are biologically determined, coming to no definitive answers. So when it comes to ETI's actions, I think discretion is the better part of valor. Consequently I'll now return to the courtroom and listen to the claims that there is no ETI and that the above science-fiction possibilities are just that—fiction.

ETI? THERE'S NO SUCH THING: $N = 1$

Alfred Adler was one of the giants of modern psychoanalytic thought, a onetime associate of Freud's and the originator of the notion that compensatory mechanisms are often developed to combat what is now termed an inferiority complex. In 1974 another Alfred Adler, a man who in my opinion could do with a few sessions on the couch himself, published an absolutely hilarious article in *The Atlantic* in which he takes the Byurakan meeting as a vehicle for expounding an evidently deep-seated sense of resentment against what he terms "the modern technologist." After soundly denouncing most of the Byurakan *speculations* (which had continually been advertised as such by the speakers themselves) as "lunatic assertions" and "intellectual pollutants," Adler goes on to note that "the human qualities most displayed by the conferees were . . . cupidity, inanity, and triviality." At this point the article moves into high gear with its main message: the nature of the technological mind as seen by Professor Adler. According to this vision, "The modern technologist is a gifted, highly trained, opportunistic, humorless, and unimaginative ass." A couple of sentences later we learn that "none of his fatuous pseudo-science is science; all of it is empty of intellectual content, inflated with self-importance, and held accountable for nothing." Does this go for *all* modern technologists? My last employer definitely, but to condemn all modern technologists seems a bit much even to my cynical eye. And whom does the ever cheerful Adler choose as the focal point of his vivid invectives? None other than Johnny Carson's SETI consultant, Carl Sagan, who even in 1974 was already becoming the lightning rod for discharge of the petty resentments and jealousies of a host of less visible (and less talented) scientists, disaffected writers, and academics.

This whole Adler business would hardly even be worthy of mention if it were not for the fact that the article displays in a particularly blatant manner the sophomoric attitude toward science and scientists held in certain quarters of the academic community. But what's more important for our purposes, it serves as a symbolic opening salvo fired against the initial euphoria emerging from the Byurakan sessions regarding the likelihood of contacting ETI. While it's hard to imagine anyone taking Adler's arguments as anything other than dark grumblings and light entertainment, by the mid-1970s a backlash against the $N > 1$ claims was definitely in the air, with the resulting fallout threatening for a while to destroy even the small and tenuous foothold that SETI had carved out for itself in the remote foothills of the mountainous terrain of mainline science.

Arguments claiming $N = 1$ tend to come packaged in one of two quite distinct wrappers: factorization and observation.

- *Factorization:* Here we find all arguments centering upon one or more of the terms in the Drake equation. All that's needed to show that N is negligible is to demonstrate conclusively that one term of the Drake equation is close enough to zero that, for all practical purposes, it *is* zero. This is the goal of the factorization artists—to produce a knockdown argument showing just this, focusing upon one of the astrophysical, bio-

logical, psychological, or sociocultural terms in the equation.
• *Observation:* The observers use a quite different line of reasoning, the classical *reductio ad absurdum,* which goes as follows: Suppose ETI does exist. What observable consequences would be likely to follow from this assumption? Do we actually observe any of these consequences? If not, then it's highly likely that $N = 1$.

Let's look at each class of claims in turn.

In 1975 the disaffection with the prevailing $N > 1$ attitude toward ETI foreshadowed in Adler's outburst was given scientific form by Michael Hart, a young astronomer at the National Center for Atmospheric Research in Boulder, Colorado. Hart, now at Anne Arundel College in Maryland, and probably the only practicing astronomer who also holds a degree in law, took a Talmudic view of the ETI question, zeroing in on the one hard, incontrovertible fact surrounding the whole issue: *There are no intelligent beings from outer space on Earth right now.* His pathbreaking paper, titled "An Explanation for the Absence of Extraterrestrials on Earth," offers a detailed consideration of this empirical observation, termed Fact A in the paper, concluding that the most reasonable explanation for Fact A is that there are no other advanced civilizations in our galaxy. It's illuminating to consider Hart's analysis of Fact A in more detail.

In good, logical, legal fashion, Hart divides the possible explanations for Fact A into five categories:

• *Physical:* Some kind of physical, biological, astronomical, or engineering difficulty makes space travel unfeasible.
• *Sociological:* ETIs have not arrived because they have chosen not to. This includes all explanations involving lack of interest, motivation, or organization, as well as political obstacles.
• *Temporal:* ETIs have arisen so recently that they haven't had time to get here yet, even though they want to visit us.
• *Historical:* ETI has been here in the past, but is not here now.
• *Uniqueness:* There are no other civilizations in our galaxy. If there were, Hart says, they would have colonized the solar system a long time ago, and we would not be asking, "Where are they?"

Hart dismisses the physical explanations by asserting that the usual arguments against space travel involving time of travel and energy requirements are vastly overstated. It's interesting to note here that according to his calculations, the energy needed to accelerate a ship to 0.1 c and decelerate it requires that the ship carry about nine times its own weight in fuel. This calculation should be compared with the much later, and far more pessimistic, estimates of Drake considered earlier under what, in my opinion, are far more realistic assumptions. Hart here also dismisses other possible physical hazards, e.g., the danger of collision with meteorites (traveling at 0.2 c, a 4-ounce rock will impact a ship with the force of a 40-kiloton bomb, twice the force of the atomic blast that leveled Hiroshima), cosmic rays, and so forth.

As to sociological explanations for Fact A, Hart has the uniform argument that no sociological explanation will suffice unless it can be shown that the same argument will apply to every race in the galaxy at all times. So if you think that ETI is not here because it blew itself up in a nuclear *Götterdämmerung,* then you must show that every ETI that ever existed also blew itself up. Hart claims that this argument is universal, and can be used to counter any of the sociological explanations presented for Fact A.

To address temporal explanations, it's necessary to estimate how long it would take ETI to reach us in a wave of colonization. Hart calculates that with a ship velocity of 0.1 c, such an expansion wave would move across the entire galaxy in about 2 million years. But the age of our galaxy is on the order of 10 *billion* years, so to accept the temporal explanation it's necessary to assume that it took 5,000 time units (1 time unit = 2 million years) for the first civilization to emerge that had the inclination to explore the galaxy, but that the second such species (i.e., mankind) arose less than 1 time unit later. Hart concludes that while the temporal explanation is theoretically feasible, it should be considered highly unlikely.

There are several versions of the historical explanation, the most common being that ETI was here rather recently (less than five thousand years ago) but didn't hang around. The weakness in this explanation is that it doesn't explain why Earth was not visited earlier. On the one hand, if ETI could have visited us earlier, then we need a sociological explanation for why it didn't; on the other hand, if ETI visited us as soon as it was able and this was only within the last five thousand years (only one four-

hundredth of a time unit), then this requires an even more remarkable coincidence than that mentioned earlier in connection
with the temporal explanation. Another version of the historical
explanation is that the Earth was visited long ago, say over 50
million years in the past. The problem here is that one again
needs a sociological explanation to show why, in all the intervening years, no other ETI ever came to Earth and stayed.

The sum total of Hart's arguments comes down to a collection
of reasons why the final four alternatives are even less likely
than the uniqueness explanation for Fact A. Thus follows
Hart's assertion that $N = 1$, ushering in a period of critical
scrutiny of the whole SETI enterprise from every point on the
scientific compass.

Within a year of the appearance of Hart's claims, counterarguments were put forth by Laurence Cox, who appealed to the
principle that in order for any civilization to enter the colonization game, it would be necessary for it to stabilize its population.
If not, then even at a population growth rate equaling our own,
the population of such an ETI society would quickly outstrip
all the colonizable planets in the galaxy. Under the hypothesis
that the society could solve the population explosion problem,
Cox calculates that the temporal explanation is the most likely
way to account for Fact A, and that the ETIs just haven't yet
had time to reach us.

An intriguing variant of the Hart argument was put forth in
1980 by another of the Young Turks in the anti-ETI camp,
Frank Tipler, a mathematical physicist at Tulane with a penchant for contentious views on modern cosmology. Tipler
launched a broadside against what he called the semireligious
overtones of the entire SETI program by arguing that any civilization much more advanced than ours would surely be able to
construct the types of self-reproducing machines that were mentioned in our discussions of artificial life in Chapter Two. The
idea of sending probes to search the galaxy for signs of emerging technological life was first put forth in the 1960s by Stanford astronomer Ronald Bracewell, who suggested that any
advanced civilization would surely choose this cost-effective way
of exploration in preference to direct travel. Such devices,
termed *von Neumann probes,* would represent an extremely cheap
way of exploring outer space, being able to cover the whole gal-

axy for a few billion dollars. *Star Trek: The Motion Picture* was based on the use of such machines, which are really nothing more than *very* souped-up versions of the kinds of primitive probes that we have already sent to the Moon, Mars, and Venus, as well as to other bodies in our own solar system. In his article "Extraterrestrial Intelligent Beings Do Not Exist," Tipler strongly argues the same point as Hart, that an expanding wave of colonization by such probes would fill the galaxy in a time much shorter than the current lifetime of the galaxy. Yet we don't see even the faintest sign of such a von Neumann probe; hence, they don't exist, and neither does ETI.

As a fascinating commentary on the sociological ways of modern science, Tipler later published what he claims is the clearest evidence for a "save-the-world, semireligious motivation" underlying mainstream, establishment SETI. The "evidence" he presents involves the treatment that pro-ETI reviewers gave his critical paper. It seems that a shortened version of the full paper was submitted to the prestigious journal *Science,* whose editor sent it to Carl Sagan for review. Apparently Sagan rejected the paper for what Tipler saw as at least superficially valid and relevant reasons. Tipler proceeded to revise the paper to answer (to his satisfaction, at least) the objections raised by Sagan, and then submitted the revised paper to the well-respected astrophysical journal *Icarus.* As fate would have it, the editors of *Icarus* also sent the paper to Sagan for refereeing, with the result that it was again rejected with a referee's report *identical* to that earlier submitted to the editors of *Science.* While it's difficult to know the precise circumstances surrounding this particular case, the general phenomenon is well known to any denizen of the academic deep, as few pagan rites quite equal the atavistic, troglodytic satisfaction of setting your colleague's ego on its ass—the equivalent in academic circles of what on the gridiron is known as The Sack. As Tipler tells it, "Had Sagan rejected the paper with a claim that my changes were inadequate, or asked someone else to referee the paper (and reply to my changes), I would have, of course, disagreed with the rejection, but I would have felt the rejection was based on scientific grounds. As it is, I feel as if I have become involved in a theological debate." Tipler also recounts similar adverse remarks from Philip Morrison, who commented on how imprudent it would be to abandon radio searches, a matter not even mentioned in Tipler's article.

Of course, Sagan and Morrison represent the bastions of SETI and the scientific establishment, and one has to keep in mind that these attempts to keep Tipler out of print were taking place in the period when SETI advocates were having their funding problems with Congress. At that time, the last thing the pro-ETI community wanted was for some young upstart from Tulane to place a loaded gun in William Proxmire's hands by publishing a difficult-to-rebut argument in a well-respected and widely circulated American scientific journal. Consequently, Tipler's article finally appeared in a British publication, *The Quarterly Journal of the Royal Astronomical Society,* an eminently reputable journal but hardly coffee-table reading matter for congressmen or their aides.

The moral of this strange little tale is only that scientists are no more selfless than anyone else when it comes to recognizing what side their bread is buttered on. And when matters come down to the eternal struggle between the ideology of science and the realities of economics, as Damon Runyon once remarked, "The race is not always to the swift, nor the battle to the strong—but that's the way to bet." In modern science, as in the rest of modern life, ideals and ideologies are pretty feeble competitors when they come into conflict with the pocketbook.

Brandon Carter of the Meudon Observatory in Paris has advanced a different observation-type of argument, based on anthropic considerations showing why the search for ETI looks to be a bad bet. To outline his reasoning, consider the following quantities:

t_E = the time needed for evolution to produce an intelligent species on Earth

t_0 = the length of time that evolution can proceed on Earth

t_{ms} = the time during which the Sun remains in a state of temperature and size amenable to life, i.e., the time the Sun remains on the "main sequence"

t_{av} = the average length of time needed to evolve an intelligent species on an Earth-like planet

The Principle of Mediocrity says that $t_E \approx t_{av}$, and that t_{av} is either much smaller or much greater than t_{ms}. But both of these expectations are contradicted by the fact that we observe $t_E \approx$

t_{ms}, which comes from assuming that the actual time needed to evolve an intelligent ETI species on Earth is about the same time needed to evolve such an intelligent entity on a planet *like* Earth. Furthermore, if there are many improbable steps on the road to developing intelligent life, then we would expect to see t_{av} much, much greater than t_{ms}. Consequently, the observation that $t_E \approx t_{ms}$ is hard to justify beforehand, since a priori we would have expected to find $t_E \approx t_{av}$. The implication that follows is that t_E is either much greater or much less than t_{ms}.

At this point Carter invokes the so-called Weak Anthropic Principle, which for our purposes can be expressed by saying that whatever we observe is biased by the presence of conditions needed to ensure that we, as observers, exist to make the observation. We will give a much more detailed discussion of this principle in the next chapter, but for now it suffices to note that there is only a certain kind of universe that we could possibly see: the kind in which the conditions are such as to allow our own existence as astrophysicists who make the observations. Using this kind of reasoning, Carter then argues that t_E might not come close to equaling t_{av}. The fact that we observe $t_E \approx t_{ms}$ strongly implies that t_{av} is much larger than t_E, and that the observed numerical coincidence $t_E \approx t_{ms}$ is due to the Weak Anthropic Principle self-selection effect. Thus, Carter concludes that t_{av} is much larger than t_E, which itself is about equal to t_{ms}. Hence, the existence of ETI is highly unlikely since most Earth-like planets will be destroyed by their star's leaving the main sequence long before intelligent life has a fighting chance to emerge.

To cap off his claim, Carter derives a simple formula for how long life on Earth can continue to evolve. Carter's formula predicts that the biosphere on Earth can continue for at most another 450 million years. This is a very short time indeed, implying that the evolutionary window on Earth is already 90 percent of the way closed. Putting all his arguments together, Carter concludes that there are no ETIs in the galaxy, and probably not anywhere else in the universe either.

The cases made by Hart, Carter, and Tipler for $N = 1$ are representative of what we've termed above the observation category of counter-ETI claims. Now let's look at some of the factorization-based claims that one or more of the terms in the Drake equation must be negligible.

* * *

One of the most compelling cases of the factorization type has been put forward by the philosopher Nicholas Rescher against the likelihood of our being able to communicate with ETI, even if it does exist. The standard pro-ETI argument for our being able to engage in meaningful communication is that while ETI's social, political, and cultural systems may be radically different from our own, its science is quite likely to be very nearly the same. Rescher asks: Why should this necessarily follow? The usual argument is the following anthropomorphic chain:

1. Common problems constrain common solutions.
2. ETI civilizations have in common with us the problem of cognitive accommodation to a shared world.
3. Natural science as we know it is our solution to this problem.
4. Therefore, natural science is likely to be ETI's solution, too.

Rescher notes that the obvious difficulty with this reasoning is that ETI's problems and ours are *not* the same, since the two civilizations are literally worlds apart and have significantly different environments and resources. To suppose a common problem is to beg the question. Let's consider for a moment what it would take for ETI's science to be the same as ours.

For ETI's science to be functionally equivalent to ours and hence form the basis for a meaningful exchange of information, the following conditions would have to be the same:

• *Formulation:* The mathematics ETI uses must be the same as ours. But there's no reason why this must be so. For example, it may use some kind of nonnumerical arithmetic.
• *Orientation:* It must be interested in the same sorts of problems that we are. But this may not be the case either, as ETI may devote all its efforts to the social sciences, or might never develop electromagnetic theory if its physical environment doesn't suggest it; e.g., ETI may live in on a murky world of Stygian gloom where the main sensory inputs come from sound rather than light.
• *Conceptualization:* It must have the same cognitive perspective on Nature as we do. For instance, it's not that seventeenth-century biologists had something different to say about genes, DNA, and the process of inheritance than we do; they had *nothing* to say about these matters.

Putting all these factors together, whatever ETI's science is, it's going to be geared to ETI's sensors, cultural heritage (which determines what's interesting), and environmental niche (which determines what's pragmatically useful). Sameness of the object of contemplation does not guarantee sameness of the ideas about it; e.g., primitive people regarded the Sun as a god, while we think of the same object as a giant thermonuclear reactor. Note that there is no argument here against the principle of the uniformity of Nature. The problem is that it's different thought-worlds that are at issue in the elaboration of science. Thus does Rescher conclude that the quantity f_c in the Drake equation must be vanishingly small.

Michael Hart has contributed to the factorization school of argument as well, as noted earlier in discussion of his calculations showing that the continuously habitable zone of a star is depressingly narrow, suggesting that the term n_e, the number of planets that are suitable for life, is small. In addition, Hart has argued that the term f_l, involving the probability that life will emerge, is also negligibly small. This line of reasoning is worth examining in detail, especially as it occurs in several other biologically based attacks on the Drake equation.

We have already seen from the Miller-type experiments that many of the basic chemical building blocks of life can be formed by natural chemical reactions in the primordial soup. However, in order to have f_l be large, it's necessary to display a commonly occurring mechanism by which these raw materials can form themselves into self-replicating molecules of DNA. This is one of the main points of attack on the Drake equation.

All earthly organisms have DNA strands that consist of a chain of millions of individual nucleotides arranged in very specific ways. Any other ordering is usually useless, and may even be fatal. This is why most mutations are deleterious and are quickly weeded out of the gene pool. Hart supposes, for the sake of argument, that on the early Earth there existed a "genesis DNA," which, when seeded into the primordial broth, acted as a template around which other such strands formed, thereby giving the initial push needed to start evolution on its merry way. Suppose that this prototype replicator needed only a sequence of six hundred nucleotides in order to work, rather than the millions required by modern DNA. Further, imagine that only one hundred of the six hundred positions on the strand have to be occupied by a particular nucleotide element (A, G, C, or T), with the remaining five hundred positions capable of being occupied by any of the four bases. Then by random assembly of the 4 nucleotide bases, there are 4^{100} possible chains of 100 units. After a few more calculations, Hart comes to the conclusion that the chances of such a strand of genesis DNA spontaneously forming are less than 1 in 10^{32}. This number is inconceivably small, implying that f_l is essentially zero.

But if $f_l \approx 0$, what are we doing here? If the likelihood of a particular event is negligible, how can we account for our own existence? This Fact B obviously requires an explanation, particularly as it flies in the face of the cherished Principle of Mediocrity. Hart's ingenious answer to this dilemma is to point out that according to modern cosmological thought, the universe is not finite, but infinite! Therefore, even though the chances of success of any single experiment are negligibly small, there will be an infinite number of experiments, thus assuring many successes; in fact, a gambler's dream come true—an infinite number of winners. Consequently, if we accept this line of reasoning, there are an infinite number of planets where life has formed, but these planets are so sparsely sprinkled throughout the universe that our chances of ever meeting with an ETI from one of them are essentially zero.

Physicists and astronomers like Hart are not the only ones to have made this kind of calculation. The eminent evolutionary biologists Ernst Mayr and George Gaylord Simpson have put forth similar, but less quantitative, discussions of ETI using exactly the same chain of reasoning. Mayr points to the convoluted combination of seemingly chance circumstances that led from the primordial slime to modern technological humanoids, noting that the ancient Greek, Chinese, and Mayan civilizations were created by individuals essentially indistinguishable from us anatomically, yet they never developed a technological society. Simpson makes what amount to the same kinds of arguments, with both biologists concluding that money spent on SETI is betting in a game whose odds are the most adverse in history.

A common thread running through all these biological and sociological objections to the Drake equation is the assumption that each slot on the nascent DNA strand has to be filled *independently*. To illustrate, imagine you were set the task of constructing a necklace consisting of one hundred beads, each of which is to be one of four colors: red, blue, green, or yellow. Further, suppose that for aesthetic reasons the beads have to be arranged in a very particular order, so that each of the hundred positions must be occupied by a prespecified color. Now imagine you start dipping your hand into a barrel full of beads of the various colors and begin to assemble the necklace by placing the first bead you get in position 1, the second in position 2, and so on. What are your chances of getting the necklace put together properly in one hundred trips to the barrel? Under the assumption that each of your turns at the barrel has an equally likely chance of turning up any of the 4 colors, then the chances of selecting 100 colors in exactly the right order are 1 in 4^{100}, just the odds used by Hart in his analysis of genesis DNA. In short, even if all the people on Earth spent all their time trying, the odds are infinitesimally small that the necklace would ever be completed. But complex systems in Nature just are not put together like necklaces. Let's see why.

The assembly scheme coming out of the independence assumption implies that after a necklace one hundred beads long has been assembled, we examine it bead by bead to see if every position is occupied by the right color. If not, then the entire necklace is torn apart and we start over from scratch. Nature works in quite a different manner. On our trial necklace, even though not all the positions are occupied by the right color, many will be. In fact, there's a 25 percent chance that any particular location on the necklace will have a bead of the correct color. So if the necklace as a whole isn't perfect, we keep the part of it that contains the right color in the right place and throw away only those parts of it that don't match up. What we might have after the first round of such an experiment is something like the string seen in Figure 6.15, where X represents a proper match and O denotes a color mismatch.

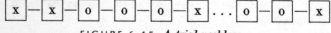

FIGURE 6.15. *A trial necklace*

For the next trial, all the pieces corresponding to X's will be kept and only those necklace fragments having mismatches will be filled from the barrel. It's easy to see that with this "ratcheting effect" of keeping the subsystems that somehow "work," the entire necklace can be assembled in rather short order.

The above ratcheting principle forms the basis for Herbert

Simon's Watchmaker Parable, illustrating the way in which complex systems can be formed out of individual subsystems. We briefly looked at this parable in Chapter Four, the idea being simply that it's far quicker to form a complex system of one hundred pieces from ten subsystems of ten pieces each than it is to try to assemble a single system of one hundred components. Computer experiments using this idea for the assembly of genesis DNA have been made by the chemists Manfred Eigen and Peter Schuster, as well as by the biologist Richard Dawkins, all of whom come to the conclusion that formation of genesis DNA using a dependent and directed, rather than an independent and random, assembly of nucleotides from the primitive components coming out of a Miller-type experiment is perfectly feasible within a geological time frame.

The fly in the ointment is that in order for the ratcheting principle to work, it's necessary for the assembler of the necklace to know what the necklace is supposed to look like. There has to be a target design that all this shuffling-about of beads is aiming at. Otherwise, there's no way of telling whether a particular fragment should be kept or discarded. This is all very reminiscent of Douglas Hofstadter's *Jumbo* computer program described in the last chapter, which tries to do anagrams by a directed assembly of individual letters. In that case, there are very well understood and definite targets—recognizable words of the English language. But if Nature is engaged in trying various combinations of nucleotides to find a string that will self-replicate, how can she decide whether a particular fragment is or is not part of such a string before the entire string is assembled? Unless a way out of this dilemma is found, one is thrown back to the case considered by Hart, Mayr, and Simpson. At present no one has any clear-cut idea of how to break through this crucial bottleneck in the Drake equation, which serves as our cue to conclude the arguments for the Defense and move on to final summaries.

SUMMARY ARGUMENTS

Let's first recall the precise question to be settled:

Is N, the number of civilizations *within* our galaxy with which we are capable of *communicating*, greater than one?

Note that I have highlighted the crucial point that our concern is only with ETI civilizations within our own Milky Way Galaxy, and even then only with those ETIs with whom we can exchange meaningful information. Thus, the question as stated must be answered negatively if the nearest ETI is in Andromeda, or if we encounter a clearly living but totally incomprehensible ETI like Lem's sentient ocean in *Solaris.*

The various subgroups of the $N > 1$ position are displayed in Table 6.4, along with representative members of the different groups and a brief indication of the arguments they employ to defend their positions. I hasten to point out that in some cases I have made use of a bit of literary license to assign certain individuals to particular groups, since their writings are not absolutely explicit as to precisely where they stand regarding the magnitude of N. Nevertheless, on the basis of what they have written I feel the assignments of Table 6.4 are close enough for government work, and certainly acceptable for our purposes here. Following the Prosecution's summary, Table 6.5 displays the various counterclaims offered by the Defense.

So there it is: the usual collection of eminent scientists stridently arguing mutually exclusive positions. After the smoke clears, the situation seems ultimately to turn upon an act of

$N > 1$: E T I E X I S T S !

PROMOTER	ARGUMENT
(N is large)	
Sagan, Morrison	Principle of Mediocrity
(N is small or large)	
Dyson	comets or Dyson spheres
Papagiannis	asteroid belt
(N is moderate)	
Drake	travel/colonization too expensive
(agnostic)	
Rood	Drake equation
Bracewell	von Neumann probes

TABLE 6.4. *Summary arguments for the Prosecution*

$N = 1$: ETI DOES NOT EXIST!

PROMOTER	ARGUMENT
Hart	no colonization; f_e small
Tipler	absence of von Neumann probes
Mayr, Simpson	f_l, f_i, L small
Trefil	no colonization
Carter	Anthropic Principle
Rescher	otherworldly science

TABLE 6.5. *Summary arguments for the Defense*

faith—just as Frank Tipler claimed. In fact, the situation is strikingly similar to the famed psychologist Carl Jung's analysis of alchemy: "When facts are few, speculations are most likely to represent individual psychologies." If you're a believer in the Principle of Mediocrity, then it's inconceivable that N could be small; on the other hand, if you have faith in the irresistible urge of all living things to seek out new worlds, then you have to feel that $N = 1$ and we are that *número uno*. As to my own brand of spiritual firewater, read on.

BRINGING IN THE VERDICT

To the ETI question as stated, I vote for acquittal, thus supporting the Defense argument that $N = 1$. Oddly enough, while most of the Defense arguments center about the Fermi Paradox and the issue of colonization, it is not this line of reasoning that leads me to side with the Defense. Nor is it a rejection of the Principle of Mediocrity forming the heart of the Prosecution's case. Rather, my view is that ETI may very well and probably does exist, even somewhere in the Milky Way. However, what I find difficult to swallow is the implicit corollary of the Principle of Mediocrity that if ETI is around, we will be able not only to recognize it, but even to enter into some sort of meaningful dialogue. In this regard I find the arguments put on the table by Rescher difficult to rebut. And in view of these arguments I think the issue is not that ETI science may be more advanced than ours. Rather the issue is that the likelihood is essentially zero that they will be doing our sort of science at all. So there may well be intelligent extraterrestrial civilizations out there, but the chances are negligible that we'll ever contact one doing "our kind" of science. Thus it's at the communication level that I draw the line, and since communication is an integral part of the judge's charge to the jury, I have little recourse other than to conclude that $N = 1$. Hence, my vote for the Defense.

While it's not part of my argument for $N = 1$, what can we expect if SETI is actually successful and a signal from the deep is received? The conventional wisdom of the pro-ETI crowd always emphasizes how the receipt of such a signal will profoundly change our concept of ourselves. Just what could this actually mean? At one end of the scale, if the signal shows that the entire universe is run by a band of angelic swans from 61 Cygni who control every aspect of our lives, then such a discovery would indeed have profound implications for our notion of self. If, on the other hand, the signal shows that there is a "second Earth" out there where ETIs worry about stock market crashes, go on vacations to "Hawaii," and play baseball, then the message would probably result in a vast, almost unbelievable disappointment, but would surely not influence our self-concept in the slightest. So just what are the advantages of detecting a signal, other than of course satisfying our curiosity?

The benefits of a message from the stars will ultimately depend upon whether the ETI civilization is sufficiently close to ours for a meaningful transmission of useful information. If the civilization is totally alien, then there will really be nothing to learn from the signal since we will have nothing at all in common with it. After all, what could we have to say to members of a civilized, technological species inhabiting the surface of a neutron star, living out their lives in a fraction of a second (by our clocks)? Robert L. Forward thought there was something to say in *The Dragon's Egg,* but interesting as his arguments are, I'm skeptical. In fact, a message from such an alien may not even be decipherable, in much the same way that the mysterious Voynich manuscript here on Earth has defied all attempts to ferret out its meaning.

If the ETIs are close enough to us for some sort of meaningful exchange of information to take place, unless their message was specifically tailored to a culture like ours at our point of development, their science would probably be no more comprehensible to us than the wiring diagram of an IBM PC is to an aboriginal tribesman. And when it comes to political, cultural, and ethical information, the signal may suggest practices or systems that we would find immoral or just plain unworkable, e.g., rationing of children or abolition of money.

To conclude on a somewhat sober note, I find the supposed benefits of SETI to be vastly oversold by the pro-ETI enthusiasts, principally because I think that even if ETIs are out there, we'll either never know it or we'll never get any real benefit from it, simply because they are truly and fundamentally *alien.* Consequently, when dividing up the research-dollar pie, it seems to me a poor investment to put too much faith, hope, or money into SETI on the grounds that the expected return from a message will enable us to amortize the investment easily, even if the signal comes hundreds or thousands of years from now. On the other hand, a few million dollars a year is petty cash from the NSF and NASA budgets, so why not spend a little money now and then on looking for the pot of gold at the rainbow's end? After all, curiosity is a wondrous thing, and it's hard not to wonder when you look up at a clear night's sky, "Where are they?" We'll never know if we don't look. And looking and hoping are what science is all about.

7

HOW REAL IS THE
''REAL WORLD''?

===

CLAIM

THERE EXISTS NO OBJECTIVE REALITY
INDEPENDENT OF AN OBSERVER

BUILDING THE STAGE

In his comedy *As You Like It,* Shakespeare makes the well-known statement that "all the world's a stage, and all the men and women merely players." This Shakespearean remark conjures up the commonsense, everyday view of physical reality: The universe of material objects—chairs, cars, trees, atoms—exists independently of us, just as the theater and its stage exist independently of the actors and their audience. This image of an impersonal, aloof cosmos was engraved onto the scientific consciousness by the authority of Newton and his idea of events that unfold in an arena of absolute space and time. Since this idea forms the framework upon which our story in this chapter is draped, let's quickly review the essential components of the world according to Newton.

The essence of the "objectivist" position, nowadays termed *naïve realism,* is that the world consists of a collection of independently existing "things" that are simply "out there" whether we observe them or not. To be more specific, we can identify the principal components of this ontology as follows:

- There exist identifiable things that possess intrinsic attributes.
- It is not necessary for these things actually to be observed in order for them to exist.
- We as observer/participants are part of this reality, but think of it as being independent of us and as existing both before and after ourselves.
- The observers have predetermined roles to act out within the framework of this reality.

Einstein himself pithily summarized the core of this taken-for-granted reality when he remarked to Pascual Jordan, "Do you really think that the Moon exists only when you look at it?" In his view, it was the job of science to go beyond mere surface appearances and to describe and understand the nature of this objective, independent-of-human-affairs, rock-bottom kind of physical reality.

The Newtonian picture is by now so deeply ingrained in our ways of thinking about life, the world, and the universe that it's hard to imagine anyone doubting it. And indeed few did until early in this century, when the relativity and quantum theorists recognized that Shakespeare and Newton had always been living behind the facade of a Potemkin village, at least when it came to dealing with the very small, the very large, and the very fast. But the world at large, including most practicing physicists, was happy to accept the tacit assumption that these non-Newtonian effects really count only in the microworld of the atom or the macroworld of distant galaxies. And it's been only in the past decade or so that the reality crisis of the physicists has spilled over into the realm of everyday life, with accounts in both the popular and the New Age press unveiling for the general public such seemingly romantic notions as observer-created realities and the intertwining of modern physics and Eastern mysticism—with the blessings of at least some renegade physicists, no less! To get a glimpse of what's involved in this wholesale revamping of our concepts of physical reality, there's no better place to start than with the familiar parlor game of twenty questions.

A common form of the twenty-questions game involves a group of people who send one of their number out of the room to act as the questioner. The group then decides upon a target word and the banished party is asked to return. It is then the task of the questioner to identify the target word using at most twenty questions, such as "Is it alive?" or "Is it liquid?" The winner of the game is that questioner who identifies the target word using the smallest number of questions, under the stringent condition of having only one chance at actually guessing what the word is.

The physicist J. A. Wheeler likes to tell of the time he played an interesting variant of the game following a dinner party at the home of physicist Lothar Nordheim. According to Wheeler, he was sent from the room for what seemed an inordinate length of time. Returning to the room, he saw a smile on everyone's face—a sure sign that some sort of mischief was afoot. He then started his questioning with the customary sweeping queries: "Is it animal?" No. "Is it mineral?" No. "Is it alive?" No. But as the questioning went on, Wheeler noted that the answers were slower and slower in coming, with the person being questioned thinking for a long time before responding with a simple yes or no. Finally Wheeler felt he had narrowed the possibilities down to the point where he was ready to take the plunge. "Is the word 'cloud'?" he asked. At which point everyone broke out laughing and told him he was correct. It seemed that while he'd been out of the room the others had agreed that they would not select *any* word, but rather would let some word emerge as a consequence of Wheeler's questioning. The agreement was that the parties being questioned could respond with either a yes or a no, the only constraint being that whichever response they gave, they would have to have a definite word in mind that would be consistent with all the preceding responses. So the game was at least as difficult for the others as it was for Wheeler!

The point Wheeler makes when recounting his twenty-questions story is that the game serves as a metaphor for two competing versions of what constitutes physical reality. Let's call them *objective* and *contextual reality.* Objective reality corresponds to the standard form of the game in which the word is preselected. This is just our old friend Newtonian reality again. The things (words) of this world exist and have real properties independent of human observers or measuring devices. Wheeler's game corresponds to a contextual reality, and involves a world that is literally created by the way in which it is probed by the observer. Just as there was no definite word but only potential words when Wheeler (the observer) entered the room, no stage is out there waiting for us to step forward and read our lines either. This situation calls to mind Gertrude Stein's withering assessment of Oakland: "There's no 'there' there." Actually, there are only potential "theres," and the stage of reality is constructed in real time as we proceed to act out our roles as observer/participants.

So is Wheeler's word really there or isn't it? Is there an honest-to-god objective reality underlying the surface appearance of things? Or is it necessary to introduce some kind of observer as the creator/constructor of what we think of as being "real"? Shakespeare, Newton, and my barber say yes, the world really is "there"; the modern quantum physicist tells us maybe not. To see why, as well as to understand the many senses in which Wheeler's word and our world might not really be out there at all, we must set out on an all-too-brief tour of a few prominent landmarks in the wonderfully weird world of the quantum.

GHOSTS IN THE ATOM

Newton's world is a world of particles and forces. One might think of it as a world composed of little billiard balls, each characterized at any given moment by three attributes: a mass, a position in space, and a speed of movement in some spatial direction (technically, a velocity). The mass is what we call a *static attribute,* since its value doesn't change during the course of time. The position and velocity are examples of *dynamic attributes.* Everything that happens in Newton's world happens as a result of these little balls flying around, colliding, combining, and breaking apart according to forces acting upon them from the outside. The formula for these interactions has been enshrined in the physicist's lexicon as Newton's Second Law, and is expressed in the form $a = F/m$, i.e., the acceleration of a particle (the rate of change of its velocity) equals the force imposed upon it divided by the particle's mass. As to the nature of these mysterious imposed forces, Newton, cagey as ever, evaded the issue entirely with his classic disclaimer *hypotheses non fingo* (I make no hypotheses).

In this Newtonian kind of universe, everything is unbelievably tidy and orderly. As soon as the imposed forces are specified, together with the initial position and velocity of each particle, events unfold with metronomic regularity upon a preexisting stage of space and time. Implicit in this rosy clockwork world is the assumption that the attributes of the particles are present at each moment, quite independently of whether or not there is a voyeur on the scene taking a quick peek at them with some kind of measuring device. The unchallenged success of this Newtonian picture in predicting phenomena of concern in the eighteenth and nineteenth centuries, coupled with the close agreement between the billiard ball metaphor and everyday common sense, led to a kind of "soft brainwashing" of both the scientific community and the general public. The prevalent belief of those times was that

Newton's universe = The real universe

The first cracks in the Newtonian facade came with the Special Theory of Relativity, in which Einstein showed that the playing field of space and time could not be as clear cut as Newton thought. In fact, the Special Theory showed that the only kind of reality consistent with observational evidence was one in which space *and* time were not considered as separate entities at all, but as a single indivisible unit—spacetime. Furthermore, the Special Theory asserted that the separation between two given events observed in the new playing field of spacetime might be seen as positive by one observer, negative by another. In short, the two different observers could be seeing two quite different "realities," making it impossible for them to agree upon the answer to even such a seemingly simple question as which of the two events preceded the other.

With the introduction of the idea that there is no such thing as an objective, observer-independent event, at least insofar as describing its location in space and time, Einstein showed that there was something fishy about the kind of reality that Newton had in mind. However, as has often pointed out in the past, Einstein's work was in many ways the last gasp of the Newtonian world, as neither the Special nor the General Theory of Relativity had much to say about the material objects themselves. On matters pertaining to the static and dynamic attributes of Newton's particles—e.g., mass, electric charge, velocity, spin—relativity theory is silent or, more accurately, tacitly accepts the Newtonian precepts hook, line, and sinker. Instead Einstein's theories focus upon the other half of the Newtonian doublet, the unexplained forces (particularly gravity), in effect centering attention on the nature of the playing field on which the particles act out their predetermined Newtonian destinies. Especially in the General Theory, which is "nothing more" than a general theory of gravitation, Einstein showed that the playing field itself is in some way created by the particles, which are then told how to move by the topography of the terrain they generate. So rather than having an independent reality of its own, the playing field exists in a kind of symbiosis with the players. This is a queer enough notion in its own right, at high variance with human perceptions generated and sharpened by the events and vicissitudes of everyday life. But when it comes to weirdness, you ain't seen nothin' yet.

At about the same time Einstein was slaving away at the Swiss Patent Office in Bern putting the finishing touches on the Special Theory, Max Planck in Berlin was working the other side of the Newtonian street with his discovery of the quantized nature of the radiation given off by a hot object. This work showed that some of the basic quantities of physics, like energy and angular momentum, come in minimum-sized "chunks." In particular, Planck demonstrated that light of any energy comes in such chunks whose size depends upon the frequency of the light, i.e., its color. The implications of this work drove the last nail into the coffin of Newtonian reality, serving as the impetus for what today J. A. Wheeler terms *recognition physics:* the study of why there are such things as time and space and dimensionality at all. Just as it's impossible to say you've really visited America without seeing the Statue of Liberty, the Grand Canyon, and the Golden Gate Bridge, it's equally impossible to talk about the "reality of reality" without visiting a few of the sights in the land ruled by the iron hand of the quantum. So let's start the tour.

* * *

To understand the profound implications of quantum theory for describing the way the world really is, there's no better place to start than with three versions of the traditional double-slit experiment. The experimental setup includes a projector, which produces three different types of material objects upon command: bullets, water waves, and electrons. For any given run of the experiment, only one of these types is produced. Whichever type of object is chosen, the device projects it toward a screen containing two slits (or gaps), either or both of which may be open. Behind the screen sits a line of detectors capable of registering the appearance or absence of the projected objects after their passage through the screen. Now let's run a few experiments.

First of all, suppose the projector is set to produce a stream of bullets. Figure 7.1 shows the results of three such experiments: with slit 1 open, with slit 2 open, and with both slits open. We'll call the number of bullets reaching the detectors in each case P_1, P_2, and P_{12}, respectively. In the figure, the bullets passing through slit 1 are shown as white-centered circles, while those passing through slit 2 are depicted as solid black circles. It's important to notice here that the number of bullets reaching each detector when both slits are open is just the sum of the numbers obtained when only one or the other of the slits is open. This is exactly the result we would expect to obtain from the classical view of bullets as individual particles going about their appointed rounds. Now let's change the projector setting from bullets to water waves and see what happens.

In Figure 7.2 the projector sends water waves instead of bullets through the slits (which we now might envision as gaps in a jetty) and on to the line of detectors. In this case the detectors can be thought of as floating buoys whose bobbing up and down measures the height (energy level) of the waves passing beneath them. The symbols I_1, I_2, and I_{12} denote the situations in which gaps 1, 2, and 1 and 2 are open, respectively.

A crucial point to notice here is that when either gap 1 or gap 2 is open, the detection pattern is similar to that seen when projecting bullets. But when both gaps are open, the patterns diverge dramatically. This divergence is the result of the phenomenon of wave interference in which two waves can interact to form a new composite wave, either by reinforcing each other through constructive interference, or by neutralizing each other by means of destructive interference at places where peaks

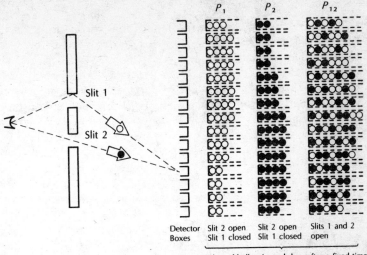

FIGURE 7.1. *The double-slit experiment with bullets*

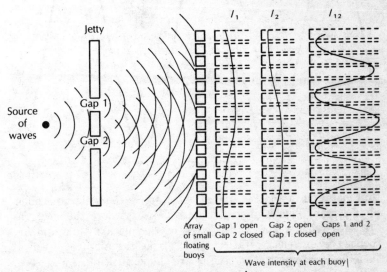

FIGURE 7.2. *The double-slit experiment with water waves*

in one wave encounter troughs in the other. The basic idea, which is important for our later discussions, is depicted in Figure 7.3. Now let's again flip the projector switch and this time produce electrons instead of water waves.

In this experiment we can regard the slits as two holes in a

(a)

(b)

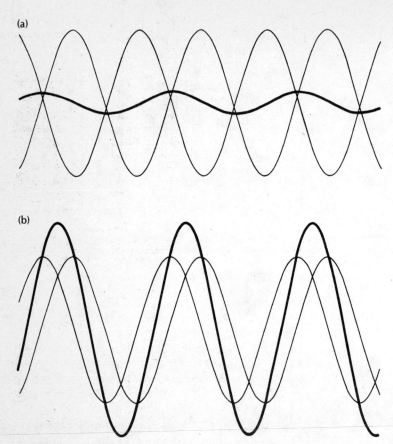

FIGURE 7.3. (a) *Destructive and* (b) *constructive wave interference*

thin metal plate and the line of detectors as elements on a phosphor screen. This is just like your TV set, where the electrons
are shot from a gun at the back of the picture tube and are
focused by an electrostatic lens to strike the picture screen at
just the right spots. The results are shown in Figure 7.4.

When we open only slit 1 or slit 2, we get the same pattern
seen in Figure 7.1 (the bullets). In Figure 7.4, the open-centered
circles represent electrons passing through slit 1, and the solid
black circles those that pass through slit 2. The surprise occurs
when we open both slits. Column P_{12} shows the same interference pattern we saw in the experiment using water waves, hence

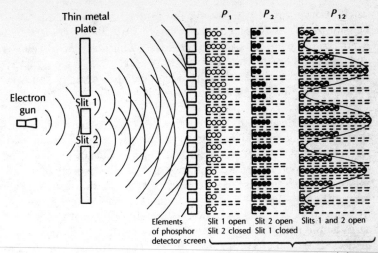

FIGURE 7.4. *The double-slit experiment with electrons*

it, too, necessarily involves some kind of wave motion with interference effects. In this case, however, P_{12} is *not* the sum of columns P_1 and P_2, so we can't say which of the two slits any particular electron went through. This fundamental lack of knowledge is indicated in the figure by representing the electrons as half black and half white. It's crucial to note here that the electrons still arrive at the phosphor screen as individual particles, i.e., as "bullets." It's just that their pattern of arrival makes it look like they collectively obey some sort of wavelike law of motion, making it impossible to assign a given slit to a given electron. Thus we arrive at:

THE MYSTERY OF THE QUANTUM WORLD

How can electrons possess the attributes of both particles and waves, yet behave like neither?

With the discovery that the fundamental particles of Newton's world—the constituents of the atom—have such surprising and contradictory behavior, the last thread of support for classical Newtonian reality was cut away, confronting physicists with the task of describing and explaining a wondrously bizarre new world. Comparatively speaking, the description part of this double-sided chore turned out to be easy. But the explanation of what the description *means* divides the community of physicists and philosophers to this day. So let's start with the easy part, and work our way into the conundrums of present-day thinking as to the true nature of quantum reality.

From our discussions in Chapter One, the reader will recall that for a theoretical scientist, to *describe* some phenomenon means to construct a mathematical representation or model of the phenomenon that takes into account all the aspects of interest about it. In the case of quantum objects like the electron, this means that we need to find some kind of mathematical structure that encompasses static attributes like charge and mass, as well as dynamic attributes like position, momentum (mass times velocity), direction of spin, and so forth. Furthermore, our mathematical structure must reflect the strange behavior discussed above, in which the electron shows the characteristics of both a particle and a wave without actually being either. The solution to this Quantum Description Problem is a tall order, yet one that was filled rather rapidly in no less than three different ways by Werner Heisenberg, Erwin Schrödinger, and Paul Dirac about sixty years ago. As it turned out, these seemingly different mathematical descriptions all ended up being mathematically equivalent. So I'll content myself here with just a brief sketch of one of the solutions (Schrödinger's), since even today it forms the main weapon in the working physicist's mathematical arsenal for dealing with quantum phenomena.

The heart of Schrödinger's scheme is to represent the "state" of a quantum entity like an electron at any moment by a mathematical gadget displaying wavelike behavior. What this means is that such an object can show the type of interference phenomena associated with waves when it interacts with other such objects. As a key component in his solution to the Quantum Description Problem, Schrödinger derived an equation that tells us how the state of the object changes at each point in space over the course of time. In this solution the state somehow encapsulates all the dynamic attributes that the object can possess. So to calculate the chances of any particular value of any one of these attributes turning up at any particular moment if a measurement is actually made, Schrödinger argued that we must perform some additional mathematical operations on the state to extract the

desired likelihoods. While the technical details are out of bounds
here, the basic idea is rather simple to describe.

To fix a specific situation, suppose we have an atom existing in
an excited energy state. Such an atom gives up energy by throwing off an electron, just as in the Planck experiments discussed
earlier. Quantum mechanics represents this event as a wave
function that spreads out from the atom in an ever-widening
spherical wavefront. This is exactly like dropping a stone into a
pond and watching the ripples of water move away from the
stone. The amplitude of this spreading wave function at a point
in space and time gives the probability of finding the electron at
that location at that point in time. Now suppose that the electron
eventually runs into a silver atom in a piece of photographic
film. When it hits the film, the electron gives up its energy and
leaves a black spot on the film. At that precise moment the electron's wave function "collapses" in a way that is reminiscent of
the breaking of a soap bubble. The wave function disappears
from all of space except the region of the struck silver atom.
Since the electron has given up all of its energy to the silver
atom, it has no probability of existing elsewhere. The wave function vanishes or, more properly, becomes a "spike" at the location of the silver atom in space at the time of the collision.
Keeping this concrete situation in mind, let's now turn to the
general situation described by Schrödinger.

Suppose the quantity $W(x, t)$ represents the wavy state of
the particle at time t in a spatial region described by the quantity x. Further, imagine that A represents the attribute we want
to know about, e.g., the particle's position. Schrödinger showed
that each such attribute can be associated with its own characteristic *family* of waveforms. A sample of some of these waveform families is shown in Figure 7.5. The figure, incidentally,
shows why some attributes can assume only quantized values
while others can take on a continuous spectrum of values. Waveform families, like the spherical harmonics, are constrained in
the kinds of vibrations they can display by the geometrical region of their action. Thus, such waveform families can vibrate
only at certain resonant frequencies, while all other frequencies
are physically inaccessible. Unconstrained families like the sine
waves can vibrate at any frequency whatsoever, hence the attribute corresponding to such a family can assume a continuum of
values.

Suppose the waveform family corresponding to the attribute

WAVEFORM-ATTRIBUTE DICTIONARY

Waveform   Attribute

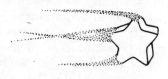

Impulse wave   Position

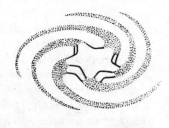

Sine wave   Momentum

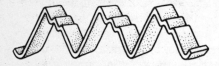

Spherical harmonic   Spin

Piano wave   Unnamed attribute

FIGURE 7.5. *A waveform-attribute dictionary*

A (such as an electron's position) is denoted by $w_1(x, t)$, $w_2(x, t)$, $w_3(x, t)$, ... Part of Schrödinger's genius was to see how to associate each member of such a family with one of the possible values that the attribute A can assume—once it's actually measured. In general, each such family has an infinite number of members; hence, usually an attribute A has the possibility of taking on any one of an infinite number of possible values. Again, whether the number is quantized or not depends solely upon the waveform family associated with A. For example, if A is the position of an electron within a closed box, then at any particular moment the electron might be found at any one of the infinite number of spatial locations within the box. Appealing to general mathematical results, it can be shown that the state $W(x, t)$ may be *uniquely* decomposed in terms of the waveform family associated with the attribute A. This means we can find a set of numbers $c_1, c_2, c_3, \ldots$ such that

$$W(x, t) = c_1w_1(x, t) + c_2w_2(x, t) + c_3w_3(x, t) + \cdots$$

where $\{w_1(x, t), w_2(x, t), \ldots \}$ is the waveform family corresponding to the attribute in question, e.g., position.

A good way of thinking about this decomposition process is to recall the grade-school science experiment in which your teacher shone ordinary white light through a prism and a rainbow came out the other side. In the quantum case, the object's *wave function* $W(x, t)$ corresponds to the white light; the waveform family $\{w_i(x, t)\}$, $i = 1, 2, \ldots$, to the various colors of the rainbow. In this metaphor, each particular attribute of interest about the object of study corresponds to a different prism through which we can view the wave function. Of course each type of prism will break up the wave function into its own particular "rainbow," so we'll get a different waveform family $\{w_i (x, t)\}$ and a different set of numbers $\{c_i\}$ in our decomposition depending upon which attribute (prism) we're using to separate W. Now, how does the above decomposition allow us to get a handle on the spread of values that the attribute A might take on?

Recall that in the decomposition there is a unique number c_i associated with each family member $w_i(x, t)$. Furthermore, by Schrödinger's scheme there is a way to pair up $w_i(x, t)$ with the ith value that the attribute A might conceivably assume when measured. Then Schrödinger's rule for the likelihood of dynamic

attribute A taking on its ith possible value is simplicity itself:
Just square the quantity c_i. That's all there is to it. Just multiply the number c_i by itself and the result will be the probability
that, when measured, the value of the attribute A will be found
to be its ith possible value. (Technically, the number c_i is a complex number, not real. Thus, we should use the complex modulus
rather than c_i^2. For details, see the To Dig Deeper section.) Of
course, the *specific* numerical value seen when we do measure A
will be conditioned by the precise correspondence between the
waveform $w_i(x, t)$ and the set of theoretically possible values
that A can take on. But the underlying idea of how to calculate
the dispersion of possible experimental outcomes is, I think,
clear and straightforward.

Since the idea is so central to all of quantum theory, let's reca-
pitulate the steps in Schrödinger's solution to the quantum description problem.

QUANTUM DESCRIPTION ACCORDING TO SCHRÖDINGER

1. Calculate the wave function $W(x, t)$ for the given experimental situation from the Schrödinger equation.
2. Decide which attribute A you wish to measure.
3. Look up the waveform family $\{w_i(x, t)\}$, $i = 1, 2, \ldots$ corresponding to A in the waveform-attribute dictionary.
4. Decompose the wave function in terms of the appropriate
   waveform family as $W(x, t) = c_1 w_1(x, t) + c_2 w_2(x, t) + c_3 w_3(x, t) + \ldots$
5. Compute the probability that A will assume its ith possible
   value by squaring the number c_i.

Let's pause here for a moment and reflect upon the dramatic
difference between the above prescription for describing a quantum object such as an electron, and Newton's procedure for describing a classical particle like a bullet. For the bullet, the
Newtonian description regards the state as being the *actual* position and momentum of the bullet at any instant of time; for
Schrödinger, the state is the wave function, which measures only
the *likelihood* that the particle (an electron, say) has a certain
position (or momentum) at a given time. Conceptually and otherwise, these are radically different views of the "reality" of the
particles' attributes. In the Newtonian case there's no question

but that the position and momentum are innate attributes of the bullet existing at all times. For the electron, the Schrödinger description is silent on the matter of the innateness of these attributes, and only gives a prescription for how to compute the likelihood of an attribute's taking on a given value when a measurement is actually performed. Note that this is true even though the traditional quantum view outlined above has tacitly reinstated the Newtonian vision of absolute and separate space and time. To incorporate Einsteinian spacetime into a quantum description would take us right up to the forefront of contemporary research on quantum gravity, far beyond where we either can or need to go in an elementary account of this type.

After all is said and done, we come to see that the Newtonian state of the particle (the position and momentum) has the appearance of something substantial and agrees with everyday common sense. On the other hand, the quantum state (the wave function W) appears as a physical fiction, a mere wave of probability, taking on a tangible quality only when a measurement is actually made. Yet it would appear that this mathematical wave is the very thing that's needed in order to build a description that is in harmony with what's actually seen in the laboratory. And there's no sweeping the dirt under the rug either, since the quantum description is the undisputed king of all theories of physical phenomena, having been tested thousands of times in laboratories and research centers around the world and never yet failing to be in accord with what our instruments report. Nevertheless, to physicists of a philosophical inclination and to philosophers of a physical bent, the whole quantum business is shrouded in an aura of mystery when it comes down to what it all really *means*. This cloud of philosophical and physical uncertainty hangs like a mist around the peaks of two sacred mountaintops on the quantum horizon: the Quantum Measurement Problem and the Quantum Interpretation Problem. So as the next stop on our package tour, let's take a longer look at these two so-far-unscaled peaks.

MEASUREMENT TO MEANING

When I first encountered the weirdness of the quantum world as a student too many years ago, one of my first thoughts was

"How could it be like that?" Little did I realize at the time that my futile plea for an explanation of just what was going on had already been answered by the late physicist, educator, and general *bon vivant* Richard P. Feynman when he remarked:

> I think it is safe to say that no one understands quantum mechanics. Do not keep saying to yourself, if you can possibly avoid it, "but how can it be like that?" because you will go down the drain into a blind alley from which nobody has yet escaped. Nobody knows how it can be like that.

I think that I left the pursuit of physics at just about the time I came across this remark.

To my way of thinking then (and now), the Schrödinger solution was no solution at all, just a set of formulas and mathematical tricks for predicting the results of experiments. Not being of a very practical orientation even then, I didn't think this was nearly enough. Somehow I thought physics was going to talk about the world of reality, but what I found was merely a discussion of the world of phenomena. Only later, after abandoning both the world of reality and the world of phenomena for the otherworldly universe of mathematics, did I come to see more clearly that perhaps the two worlds of reality and phenomena could be brought into contact after all. Since the link between them lies in the act of observation, i.e., measurement, the first step in this rapprochement necessarily has to be a deeper understanding of what's so special about the nature of measurement, and why it seems to play such a distinguished role in the consideration of quantum processes.

All solutions of the Quantum Description Problem, Schrödinger's or otherwise, share a common feature: Prior to a measurement, the quantum object is described only by a wavelike quantity specifying the relative likelihood of an attribute's taking on one or another of its potential values when actually measured. As we saw in the Schrödinger scheme, these likelihoods are given by a set of numbers that, taken together, form a probability distribution for the outcome of an observation made on the object. Of course, once the measurement is actually taken all uncertainty fades away, since one of the possible values of the attribute has been singled out by the measuring device. To hammer home the point, suppose the attribute of interest is A and that A can theoretically take on N possible values in a given experimental situation. Let's label these values $v_1, v_2, \ldots, v_N$. Note that each of these values is just a symbol that might be physically displayed as a pointer position on a dial, the number of clicks from a counter, or some other form of output produced by the measuring device once we actually do the experiment. By the Schrödinger procedure discussed above, associated with each such value v_i is another number c_i^2, the likelihood that the experimental outcome will show the value v_i, $i = 1, 2, \ldots, N$. Thus before the measurement is made we have the situation depicted in Table 7.1.

Now suppose the measurement is made and it turns out that the resulting value for the attribute in question is, say, v_2. Then after the measurement the situation is that given in Table 7.2. Thus after the measurement is taken, the a priori set of likelihoods $\{c_1^2, c_2^2, \ldots, c_N^2\}$ has "collapsed" into the degenerate set $\{0, 1, \ldots, 0\}$ in which every element is zero except the second, corresponding to the actual outcome, which now has likelihood equal to one, i.e., complete certainty.

As a particularly simple concrete illustration of the foregoing situation, consider a spinning electron. We can think of this as a basketball spinning on the end of some Harlem Globetrotter's finger, with the finger corresponding to the axis of spin for the ball. Suppose a fixed direction in space is prescribed, and that the attribute we're concerned about is the component of the electron's spin in that direction. The axis along which the electron is spinning then either points in the direction in question or points in the opposite direction. For the sake of definiteness, let's call the attribute value in the first case UP and in the second case call it DOWN; i.e., for this experiment $N = 2$, $v_1 = $ UP, and $v_2 = $ DOWN. Incidentally, this example illustrates the point that the "values" of an attribute don't always have to be thought of as numbers. They just need to be distinguishable labels like UP and DOWN, characterizing different possible outcomes of measurement. If we have no special information about the electron, then its spin axis before measurement is equally likely to be pointing in any direction. Consequently, it's reasonable to assume that the two possible outcomes are equally likely, i.e., $c_1^2 = c_2^2 = \frac{1}{2}$. As soon as we actually measure the electron's spin, we find out in which direction its spin axis is pointing, with the consequence that the a priori likelihood set $\{c_1^2 = \frac{1}{2},$

possible experimental outcomes	v_1 v_2 ... v_N
likelihood of outcome	c_1^2 c_2^2 ... c_N^2

TABLE 7.1. *The situation before making a measurement*

possible experimental outcomes	v_1 v_2 ... v_N
likelihood of outcome	0 1 ... 0

TABLE 7.2. *The situation after the measurement*

$c_2^2 = \frac{1}{2}$} collapses to either $\{0, 1\}$ if the axis points "DOWN" or to $\{1, 0\}$ if it points "UP."

In terms of the above experiment, we are now in a position to state the essential features of the Quantum Measurement Problem. But before doing so, let's pause to clarify one dangling loose end: What *is* a measuring device anyway? In the community of quantum theorists even this seemingly simple question is unsettled. Some say a measuring device is any instrument capable of leaving a permanent record. According to this view, which seems to concur with most people's sense of what's right and proper, things like Geiger counters, meter sticks, and photographic plates all constitute valid measuring devices. But others claim that the only kind of measuring device capable of collapsing the quantum probability set (or, equivalently, the wave function) is consciousness; i.e., the observation has to enter a conscious mind before the magical collapse can occur. And even in this stringent view, it's still unclear whether any conscious mind will do, or whether only the kind of consciousness displayed by *Homo sapiens* will suffice. Can the probabilities be collapsed by your family dog? By the roaches in the kitchen? By an amoeba? No one is saying for sure. So for the moment we'll leave the issue of measuring devices necessarily vague, returning to it with a vengeance in later sections. Now let's get back to a statement of the Measurement Problem itself, which is composed of the two following commonsense queries:

QUANTUM MEASUREMENT PROBLEM

A. *At exactly what point in the measurement of the electron's spin does the probability set "collapse"?*

B. *How does the act of observing the electron's spin collapse the set of likelihoods?*

To see just how strange and puzzling, not to mention philosophically troubling, this Measurement Problem really is, let's take a moment to discuss these points in somewhat more detail.

In everyday life when we think of making a measurement—say, measuring the size of our living room for a new carpet—the moment at which the measurement occurs seems self-evident. Or does it? For example, does the measurement occur at the precise instant when we lay the yardstick down for the last time on the other side of the room? Or does it occur when the result of the measurement impinges on our consciousness? Or did it occur before we ever laid the ruler on the floor, perhaps when we first decided to make the measurement? Common sense would probably argue for the first alternative, but if there's one thing that physicists have learned about the world of the quantum, it's not to trust everyday, macroworld common sense. And when we descend to the level of quantum objects, the situation doesn't get any easier. For instance, at a large experimental particle-physics laboratory like CERN in Geneva, a particular experiment designed to measure an attribute of a quantum object may go on for months. So even here we are faced with the problem of exactly when the measurement of the attribute takes place. Is it when the experiment is planned? When the accelerator is turned on? When the ghostly tracks of the particle are seen in a bubble chamber? The fact is, no one really knows. And until the situation can be resolved, the question of when a quantum object actually acquires its attributes will remain open. And with it will remain open the kind of reality that underlies the surface world of observed phenomena.

Equally troublesome is the second point, the mechanism by which a physical measuring device acts to collapse a metaphysical wave of probabilities. For the sake of concreteness, let's assume that an ordinary meter stick qualifies as a valid "collapsing device." How could it be that such a material device, when used to measure the position of an electron (admittedly this is a *very* finely graduated meter stick), could act upon the quantum wave function, an object composed of pure information with no tangible material reality at all? Or, put the other way around, how could such an ephemeral object as a wave of probability (i.e., information) act to give tangible physical attributes like position or spin to material objects? In the sections that follow we'll explore a number of competing answers that have been offered by the quantum theory community. But at the moment we have other fish to fry.

Since it bears significantly upon the Quantum Measurement Problem, at this point I'd be remiss if I didn't make at least a small gesture of obeisance in the direction of the famous Heisenberg Uncertainty Principle. No account of quantum phenomena, popular or otherwise, can omit this most striking of results, if for no other reason than its enthusiastic application, as well as concomitant misunderstanding, within a wide range of disciplines from modern physics to modern art and a lot in between.

Recall from Schrödinger's solution to the Description Problem that for every attribute A there is a family of waveforms that goes along with A, given by the waveform-attribute dictionary. As with most dictionaries, the converse is also true here: With every family of waveforms, there is a corresponding quantum attribute. Let's call this fact the Dictionary Correspondence Theorem. The "attribute" might not be anything to which we would ordinarily be able to attach physical significance or meaning; nevertheless, in the abstract space of attributes it has full voting rights along with more familiar citizens like position, momentum, and all the other celebrity attributes. This situation is graphically illustrated in Figure 7.5 by the family of "piano waves," which correspond to an as yet unnamed attribute. Perhaps in honor of my amateur pianist neighbor, we could dub this attribute "out of tune." Anyway, the point is that there is a duality between attributes and waveform families. Just as the Supreme Court has decreed "one man, one vote," in the quantum world the laws of Nature are equally strict and dictate that to every attribute there is a waveform family and conversely.

Now suppose we are given a particular waveform family $\{w_i(x, t)\}$, $i = 1, 2, \ldots$, together with its associated attribute A. Mathematical fact tells us that there is something else besides A that we can associate with the given waveform family: another waveform family that is as *unlike* the family $\{w_i(x, t)\}$ as a family can be. Let's call this the waveform family *conjugate* to $\{w_i(x, t)\}$, and denote it by $\{m_i(x, t)\}$, $i = 1, 2, \ldots$ But by the Dictionary Correspondence Theorem, associated with this conjugate waveform family $\{m_i(x, t)\}$ is an attribute V, which is in some sense as "unlike" the attribute A as possible. There is an important mathematical relationship between the two waveform families associated with the conjugate attributes A and V. To describe this relationship, it's necessary to recall our earlier metaphor about prisms and waveforms.

Suppose we have two prisms, one for the attribute associated with the waveforms W and the second for the attribute paired with the family M (here and in what follows, I have abbreviated the family names to bold symbols for ease of writing). Now let's pass an arbitrary waveform family X through the W prism. We will obtain a "W rainbow" consisting of N_W colors. The number N_W is an inverse measure of how closely the family X resembles the prism family W; i.e., if N_W is large, the resemblance is small, and conversely. Similarly, if we pass the waveform family X through the M prism, we obtain an "M rainbow" composed of N_M colors inversely measuring the resemblance of the X family to the M family. The crucial mathematical fact about this situation is that the product $N_W \times N_M$ is always greater than zero. In fact, it can be shown that there is a constant $R > 0$, such that the product $N_W N_M \geq R$. And this constant R is independent of the particular waveforms. As a technical aside, the specific value of R depends upon the particular units used in the problem and is not too important for us. What *is* important is that R is always fixed by those units and is always bounded away from zero. In the jargon of mathematics, the foregoing relationship between N_W and N_M is termed the *Spectral Area Theorem,* and in plain language it merely states the fact that two prisms corresponding to conjugate waveforms (hence, to conjugate attributes) cannot each resolve the same waveform family X to an arbitrarily fine degree of precision. There is some irreducible level of coarseness in the joint resolution of the family X, with the joint uncertainty in the overall resolution being bounded from below by R. The celebrated Heisenberg Uncertainty Principle is a direct consequence of this Spectral Area Theorem, which holds for any two prisms corresponding to conjugate waveform families W and M and any third family X. Let's see why.

The commonly held view of the Heisenberg Uncertainty Principle is that it involves an irreducible disturbance, or uncertainty, introduced into the measurement of one attribute due to the intrusion of the measuring device when making a measurement on a different attribute. To illustrate this misleading idea, suppose we have a ball rolling along a straight line and we want to measure its current position. One way to do this would be to take a fast-frame photo of the ball, thereby "freezing" its position at some instant. But to do this we would have to bounce photons off the ball in order to get the picture, and the photons would necessarily impart some energy to the ball, thereby disturbing its velocity at the moment in question. We might argue that the influence of a photon or two would be insignificant, which is true enough—for ordinary footballs or baseballs. But if the "ball" is an electron or another quantum object, the photon's impact does wild and woolly things to the ball's speed and direction of motion. The upshot of this entire chain of reasoning is that the more accurately we want to measure the ball's position, the more uncertainty we have to be willing to accept in our measurement of its velocity. This, in a nutshell, is the distilled essence of the popular view of Heisenberg uncertainty: We can't simultaneously measure two conjugate attributes with perfect accuracy, nor can they both have well-defined values at the same moment.

Probably on account of this picturesque, easy-to-understand idea that measurement seems necessarily to involve physically intruding upon the object being measured, the idea grew up in the world outside physics that the *cause* of Heisenberg uncertainty can be laid at the doorstep of the measurement act itself. To illustrate the popular view, allow me to quote from a recent popular book purporting to describe the Uncertainty Principle:

> In the subatomic world, the act of measurement changes the system being measured, giving rise to what is known as the Heisenberg Uncertainty Principle. The principle tells us that if we choose to measure one quantity (e.g., the position of an electron), we inevitably alter the system itself and therefore can't be certain about other quantities (e.g., how fast the electron is moving).

This interpretation is just plain wrong. Leaving aside the fact that not every measurement act involves a physical interaction with the system, the common misconception described here holds only when the attributes involved are what we have termed conjugate. So, for example, there's no particular problem (at least in principle) in arbitrarily accurate simultaneous measurements of both a particle's position and its energy, as position and energy are not conjugate attributes. Since it's now evident that the physical act of measurement in and of itself has nothing to do with *causing* the measurement uncertainty noted by Heisenberg, what is the basis of this striking principle of ignorance? From what has gone before, the proverbial perceptive reader will by now be sensitized to the claim that the rock-bottom cause lies in the Spectral Area Theorem. Let me sketch the argument.

Let's suppose we want to measure some attribute A, like position. Using our prism metaphor, we know that the attribute A has its own special prism. From our earlier discussion, we also know there is a waveform family **A** associated with the attribute A. In addition, we automatically obtain free of charge a conjugate attribute V, with its own special prism and its own waveform family **V**. The Spectral Area Theorem tells us that if **X** is any waveform family whatsoever, corresponding to its own attribute X, if we pass the family **X** through the A and V prisms, the "rainbows" emerging must satisfy a relationship that, in effect, says that if there are a lot of colors in one of the rainbows, then there can only be a small number of colors in the other, and conversely. Here it's crucial to note that the number of colors that come out of a prism is an inverse measure of how good a job that prism does in pinning down (i.e., measuring) the values of the attribute X. But since this inverse relationship must hold for *any* waveform family **X**, i.e., for any attribute X, let's just take $X = A$. In this case, by passing the waveform family for A through its own prism we will naturally get a rainbow with only a small number of colors, since that's what the A prism is designed to do when faced with the waveform family **A**. But the Spectral Area Theorem now requires that passing the conjugate waveform **V** through the A prism will give a rainbow with the *maximal* number of colors; i.e., the A prism will not be able to pin down values of the attribute V at all! Clearly the argument is the same if we interchange the roles of A and V, taking $X = V$ instead. The dilemma is that we have only one prism with which to do our resolving, and that prism is terrific at resolving only its design-type of waveform and awful at resolving the waveform conjugate to it. This is the real meaning of the Heisenberg Uncertainty Principle, and it should now be clear why there is at least no theoretical obstacle to simultaneous perfect measurements of two attributes that are not conjugate. Since the

Spectral Area Theorem applies only to conjugate attributes, if the attributes in question are not conjugate there is no Spectral Area Theorem, hence no Heisenberg uncertainty. Having paid our respects to the genius of Heisenberg, let's now move on to the other mountaintop—the Quantum Interpretation Problem.

We found the top of the first quantum mountain littered with all the problems of measurement just considered. These issues all center upon the meaning of observation, and what precisely an act of measurement can do in the way of generating knowledge about the dynamic attributes of a quantum object. By way of contrast, at the top of the second quantum mountain lie scattered a plethora of problems concerning the properties of a quantum object when it's *not* being measured. In short, what we find is the question: To what degree does a quantum object possess any dynamic attributes when it's cavorting about in its birthday suit, blissfully unobserved? Both quantum theory and experimental quantum fact support the position that a quantum object like an electron behaves like a wave when it's not being measured, and that it behaves like a particle when a measurement is made. Thus we can state the main question as:

THE QUANTUM INTERPRETATION PROBLEM

What is the true "nature" of an unmeasured quantum object?

The best way to see what we mean here by the term *nature* is to examine what might be called the orthodox and the reactionary schools of thought on the Interpretation Problem.

Orthodox View

1. The wave function gives a complete description of any *single* quantum object.
2. All quantum objects represented by the same wave function are physically identical.
3. The information an observer lacks about an unmeasured object is simply not there to be known.
4. The observed differences between identical unmeasured objects are due to inherent, i.e., quantum, randomness in the objects.

Reactionary View

1. The wave function gives only a statistical description of an *ensemble* of quantum objects, hence a necessarily incomplete description of any single such object.

2. Quantum objects represented by the same wave function may not be physically identical.
3. The observer's ignorance about the attributes of an unmeasured object is due to the effect of certain "hidden" variables, which quantum theory conceals from view.
4. Objects with the same wave function may show differences upon observation because they were physically different before the measurement.

Those swearing an oath of fealty to the reactionary creed are often called *hidden variables* theorists for the obvious reason that they cling to a classical view of reality. Their credo is that once the properties and values of these hidden variables are known, then all the uncertainty about the values of attributes will fade away, and the quantum object will be seen as no different from a Newtonian particle. The primary motivation for this vision of reality is the desire somehow to avoid placing the measurement process upon a pedestal of special honor among the myriad physical actions that the universe might allow. The key assumption separating these two views of reality is the second point on each list: the contention that there is a one-to-one correspondence between the grass-roots physical reality of dynamic attributes for objects, and the hard-to-get-your-hands-on mathematical reality of wave functions.

Before going into the courtroom, I think it's worth noting that the vast majority of working physicists are neither orthodox nor reactionary, but pragmatic. The typical physicist in the lab just is not bothered by these ontological questions, and regards quantum theory solely as a "machine" for making predictions about the world of phenomena. Thus the mountaintops of the Measurement and Interpretation Problems hold no fascination for him, since they are concerned with the Shangri-La of deep reality, not with the dusty flats of observed phenomena. As long as he can use the quantum machine to describe and predict the results of his experiments, the average physicist is just like the average car owner: He doesn't care what makes the magic work. He just wants to know what levers to pull and what knobs to twist in order to get from A to B. Fruitful as that attitude is in the world of phenomena, it takes us no closer to an understanding of what kinds of miracles underlie the workings of the machine. Ultimately it's at this level that the battle must be fought, and the strategies employed are completely dependent

upon the attitudes the expedition leaders take toward scaling
our twin peaks, Mount Measurement and Mount Interpretation.
Before turning the floor over to the various climbers and an account of their strategies for reaching the summits, let's briefly
review the impressive volume of vocabulary introduced in the
preceding sections. The main items are compactly summarized in
the box below.

TERMS AND CONCEPTS

QUANTUM OBJECT   an object of any size that displays both
 wave and particle behavior in the quantum manner

STATIC ATTRIBUTE   a property of a quantum object that doesn't
 change over time, such as mass, charge, and spin

DYNAMIC ATTRIBUTE   a time-varying property of a quantum
 object, like position, velocity, energy, and spin axis
 orientation

WAVE FUNCTION   a mathematical object displaying wave
 behavior that encapsulates all the attributes of a
 quantum object

WAVEFORM FAMILY   a collection (usually infinite) of waveforms
 sharing certain characteristics that enable them all to be
 associated with one dynamic attribute

MEASUREMENT PROBLEM   the question of how and when the act
 of measurement "collapses" the wave function

INTERPRETATION PROBLEM   determination of the nature of a
 quantum object when it is in its unmeasured state

UNCERTAINTY PRINCIPLE   Heisenberg's assertion that conjugate
 attributes cannot both be simultaneously measured to an
 arbitrary degree of precision

HIDDEN VARIABLES   postulated variables hidden from
 observation whose values, if known, would account for
 measurement uncertainty

With this lexicon at our side, let's give the podium over to the
Prosecution and its parade of witnesses claiming that when it
comes to objective reality independent of an observer, there just
isn't any such thing.

THE ROMANTIC REALITIES

Not far from my old apartment in the center of Vienna, there's one of those raucous, sawdust-on-the-floor, student-hangout type of beer halls advertising 101 or so brands of brew from around the world—all of them wet! On odd occasions, fortunately rare, visitors from North America are intrigued by this place and want to drop in and sample the waters. So despite my best efforts to dissuade them, the dictates of good hospitality demand my entry through the portals of this smoke-filled den, at least for a quick pint or two. To make the best of a bad situation, on these occasions I try to throw my beer-swilling schilling in a productive direction by ordering the Danish poison Carlsberg, telling myself that by doing so I'm at least casting a small vote for science. Why science? you ask. Well, unlike the competition, which tends to spend its promotional budget on the sponsorship of pro football telecasts, auto racing, or some other macho type of activity, Carlsberg invests in quantum theory! More precisely, Carlsberg invested in Niels Bohr, the spiritual father of all quantum theorists, and Niels Bohr used that Carlsberg largesse (the gift of a rather elegant mansion, no less) to house an institute for theoretical physics in Copenhagen that served for decades as the mecca for all quantum theorists. The output from Bohr's institute still serves as orthodoxy in the community of physicists when it comes to the Measurement and Interpretation Problems, so it's fitting that we start our account of what I've labeled the *romantic realities* with a consideration of what's now usually known as the Copenhagen Interpretation.

Before outlining the case from Copenhagen, let me set the terminological stage. All of the Prosecution's witnesses will be presenting realities that are "romantic" in the sense that they come straight from the fantasy novelist's pen—literally incredible. It's the romantic realities that you're reading about when you scan those Sunday supplement accounts of quantum theory as a basis for mysticism, telepathy, parallel worlds, the dialectic, altered states of consciousness, astral projection, meditation, pyramid power, tarot reading, and all the other subdivisions of the occult found at your favorite bookshop. With the imprimatur of such intellectual giants as Bohr, von Neumann, Wigner, Heisenberg, and Schrödinger, who could blame the occultists for appropriating at least the form, if not the content, of the romantics' far-out views of what's really what? Here I'll try to remain within the confines of the Measurement and Interpretation Problems as outlined above, but if the reader notices the narrative occasionally slipping off the track in the direction of the occult, it's only because the romantic realities suggested by the quantum facts are truly so strange that it's sometimes difficult to separate serious science from both hopeful and hopeless speculation. With these disclaimers on the record, it's on to the Little Mermaid and the Tivoli Gardens of Copenhagen for the testimony of our first romantic, Niels Bohr himself.

THE COPENHAGEN INTERPRETATION

There is no deep reality

Bohr's position on reality is simple: There is no deep reality. Just that. No deep reality of any kind whatsoever. The implication of such a claim is that quantum objects in their unmeasured state literally have no dynamic attributes. In contrast to the pragmatists, who might say that the question of the existence of such attributes is literally meaningless, the Copenhagen Interpretation developed by Bohr goes much further. Copenhagenists say that such attributes definitely do not exist. Or, more accurately, whatever attributes objects might possess are *contextual:* They depend upon the measurement situation, so they cannot be ascribed to the object independent of the measuring device and the act of measurement. This claim gives rise to Bohr's famous Complementarity Principle, which states that whether the object displays wave properties or particle properties depends upon the measurement situation and not just on the object itself. In other words, the Heisenberg Uncertainty Principle is an intrinsic property of Nature, and that the observer, the measuring device, and the system to be measured form a whole that cannot be divided. More prosaically, we might express this wave-particle complementarity idea using Bohr's own phrase: "The opposite of a big truth is also a big truth."

Then where do these attributes come from if they don't exist for unmeasured objects? Well, if they're there in the object's measured state, then the only place they can come from, according to Copenhagenists, is out of the measurement act itself. In other words, for a Copenhagenist the dynamic attributes are not a property of either the quantum object or the measuring device taken separately, but are a property of the *joint relationship* between the object and the device. Somehow measurement seems a little like nitroglycerine: Neither nitric acid nor glycerine is explosive on its own, but when you bring them together, BANG! This summarizes the Copenhagen view of attributes, too. Bring an object together with a measuring device and BANG: instant attributes.

There are several drawbacks to the Copenhagen view, not the least of which is that it assigns a privileged role to the measuring instrument. As far as the Measurement Problem is concerned, the Copenhagenists put all the mysteries of the wave function collapse right at the boundary between the quantum object and the measuring device. This leads to the puzzling situation in which two radically different types of systems are forced to interact: a *classical* measuring device and a *quantum* object. So in actuality the Copenhagen view doesn't solve the Measurement Problem at all, but merely sweeps it under the rug into the one place that's inaccessible to all observers—the inside of the measuring device itself. As to the Interpretation Problem, the Copenhagenists are clear: An unmeasured quantum object has no attributes; ergo, there is no deep reality underlying the world of phenomena. In David Mermin's words, the traditional Copenhagen view answers Einstein by saying, "The Moon really *isn't* there if you don't look." It should be noted that more recent detailing of the Copenhagen view by W. Zurek and others softens this conclusion somewhat to *maybe* the Moon really isn't there.

Oddly enough, despite the major drawbacks to the Copenhagen position, to this day it constitutes the conventional wisdom of the physics community. One of the reasons is undoubtedly Bohr's immense prestige, as well as the fact that his institute got its oar in the water first in the reality generation game. But an equally substantive reason is a hard mathematical fact proved by von Neumann that tends to give support to the Copenhagen view. Interestingly enough, although the Copenhagenists latched on to his result as evidence for their case, von Neumann himself leaned toward our next romantic reality, the school of consciousness.

CONSCIOUSNESS-CREATED REALITY

The observer's consciousness creates reality

As a reaction to the classical/quantum schizophrenia of the Copenhagen view, von Neumann argued that both the measuring device and the quantum object should be treated as quantum systems. Pursuing this symmetry, von Neumann produced an elegant mathematical basis for quantum phenomena in his 1932 treatise *Die Mathematische Grundlagen der Quantenmechanik.* In this magisterial work von Neumann showed that if the predictions of quantum mechanics are correct, then the world cannot be made out of ordinary objects possessing innate attributes. In fact, by this result the world cannot even be constructed out of combinations of *unobservable* ordinary objects. This conclusion seems to banish forever any kind of hidden variable theory from the reality game. The nature of this banishment will be considered in far greater detail below. As noted a moment ago, this hard fact was pounced upon by the Copenhagenists as providing mathematical ammunition for their claims. But the world of quantum theorists is just as tricky as the world of quantum theory, so things were not nearly so clear-cut as Copenhagenists might have hoped.

Recall that the Copenhagen position maintained that there was a definite separation between the measuring device and the quantum object being measured, and that the wave function collapse was assumed to occur in some vague neighborhood between the two. Von Neumann wanted to pin down the size of this neighborhood. To everyone's surprise and consternation, when he put the object and the device on the same footing by thinking of them both as quantum objects (under rather idealized circumstances), von Neumann discovered that as far as the final observed results were concerned, he could put the "cut" between the two anywhere he pleased! From the standpoint of the Measurement Problem, this means that the wave function collapse can occur in the system, in the device, or anywhere in between—take your pick. In the example of measuring our living room carpet, this would mean that as far as the final quantum-theoretic description of our living room goes, we could think of the measurement as taking place at the moment we decided to buy the carpet and knew that such a measurement would have to be made, or at the very moment the actual measurement entered our minds, or anywhere in between.

As a consequence of this truly shocking result, von Neumann focused upon the one even slightly questionable (from a rigorous scientist's viewpoint) link in the entire measurement chain: the human mind. Although he never actually said so in print, one can infer from his many parables and remarks on the matter that his "Cut Theorem" forced von Neumann into taking refuge in human consciousness as the final "collapsor" of the wave function. In this last refuge of quantum theorists, von Neumann is joined by his fellow Central Europeans Eugene Wigner and Erwin Schrödinger, who between them cooked up what are probably the most celebrated and colorful thought experiments in the annals of the quantum theory—Schrödinger's Cat and Wigner's Friend—to illustrate graphically the difficulties involved. The point of Schrödinger's experiment is to illustrate the profound weirdness of the wave function as a complete description of a macroscopic object like a cat. The setup for the two thought experiments is shown in Figure 7.6.

The experiment involves a sealed and insulated box (A) containing a radioactive source (B). The source has a 50–50 chance of triggering the Geiger counter (C) during the course of the experiment, thereby activating a mechanism (D) that causes a hammer to smash a flask of prussic acid (E), thereby killing the cat (F). An observer (G) has to open the box in order to collapse the wave function into one of the two possible states (cat = DEAD, cat = ALIVE). A second observer (Wigner's Friend) (H) is then needed to collapse the wave function of the larger system comprising the first observer (G), the cat (F), and the equipment $(A–E)$. The problem here is that now the original observer (G), Wigner's Friend (H), and the apparatus $(A–E)$, plus the cat, constitute a new system, which may itself require an "Acquaintance" to collapse *its* wave function, and so on.

Wigner's interpretation of the foregoing experiment is that quantum theory breaks down when the conscious awareness of the observer is involved. For Wigner his own conscious mind is the basic reality, and the things in the world "out there" are not much more than useful constructions built out of his own past experiences, somehow coded into his consciousness. In this picture of reality, the moment when the information about an observation enters the consciousness of an observer is when the

FIGURE 7.6. *The Schrödinger's Cat and Wigner's Friend Experiment*

mathematical wave function collapses into physical reality. Despite the stature of its supporters, the feeling of most physicists today when they hear this kind of explanation is aptly summed up by Stephen Hawking's remark: "When I hear of Schrödinger's Cat, I reach for my gun." On such an unambiguous note of rejection of consciousness-generated reality from today's premier cosmologist, let's hop across the Atlantic from the old world of Copenhagen, Budapest, and Vienna to the hill country of central Texas for our next romantic contender.

THE AUSTIN INTERPRETATION

Reality is observer-created

Texas may call itself the Lone Star State but Texans have always done things in a big way, so when the agenda item is reality generation no one will be surprised to find that the "lone star" is magically transformed into an entire universe of glowing objects, the centerpiece being nothing less than the meaning of meaning itself. The chief architect of this Texas-sized version of reality is John A. Wheeler, director of the Center for Theoretical Physics at the University of Texas at Austin.

The heart of the Austin Interpretation championed by Wheeler is the idea of a reality created by the observer through exercise of the measurement option. The Austin school believes that we are wrong to think of the past as having a definite existence "out there." The past exists only insofar as it is present in the records we have today. And the very nature of those records is dictated by the measurement choices we exercised in generating them. Thus, if we chose to measure an electron's position yesterday in the lab and recorded the resulting observation, then that electron's position from yesterday exists but its velocity doesn't. Why not? Simply because we chose to measure the position and not the velocity.

Because this very act of *choosing* is always involved in what we measure, Wheeler feels the act of observation is "an elementary act of creation." In actuality, the Austin Interpretation doesn't go quite so far as to claim that these choices dictate the reality of macroworld objects like tennis balls, but rather confines its claims to the microworld of quantum objects like electrons. Nevertheless, Wheeler's message is clear: "No elementary phenomenon is a phenomenon until it is an observed phenomenon." To illustrate the point, Wheeler has introduced an important variation on the classic double-slit experiment discussed earlier. Recall that in the standard situation we first decide which of the slits is to be open, then we turn on the projector and observe the pattern of response at the detectors. In Wheeler's Delayed-Choice Experiment, we wait until *after* the quantum objects have passed the slits before we decide which gaps are to be open.

To illustrate the idea, consider receipt of light on Earth from a distant point source (a quasar) as shown in Figure 7.7. One of the great theoretical predictions of relativity theory was that the gravitational field around massive bodies would act to bend beams of photons as they passed nearby. This is the so-called gravity lens effect, and it works in much the same way that a magnifying glass bends light rays here on Earth. Just by chance, there happens to be a large galaxy standing directly between Earth and the quasar QSO 0957+561. This means that light from the quasar has to pass around the galaxy in order to be collected in Earth-based telescopes. In their observations of

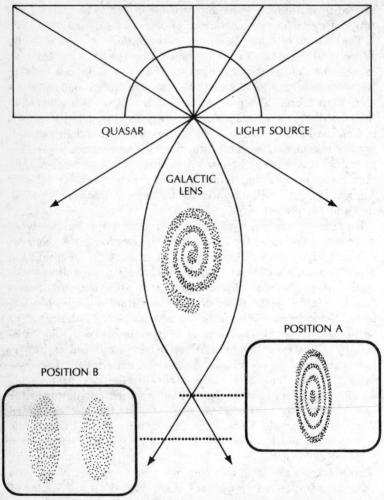

FIGURE 7.7. *The Galactic Delayed-Choice Experiment*

the quasar, astronomers have detected a double image that they attribute to this bending of the quasar's light as it passes around the rim of the galaxy on either side, as shown in the figure.

From the perspective of delayed choice, the interesting question becomes: When we detect a photon from the quasar today on Earth, which side of the galaxy did it pass around on its trip here? Common sense would argue that this question must have been settled billions of years ago when the photon moved past the galaxy, since the quasar is so old that the light started on its way to us even before our sun began to shine. But remember the rule: Never trust earthly common sense when it comes to quantum objects. The Austin crowd says no, we can actually influence what we have a right to *say* about the past by what we choose to measure today. Here's how. First, simply use standard optical means to bring together the two beams that have gone around the two sides of the galaxy. Then allow them to cross. Now exercise the measurement option by deciding whether to put your detector at the intersection point A, or at B where the beams have again separated. This option can be exercised differently for each photon, but only one choice per photon, please! If you choose the first option, interference effects indicating the photon took both paths will be seen; the second choice will show that the photon took only one of the two paths around the galaxy. Thus the Delayed-Choice Experiment seems to show that which path a photon took around the galaxy billions of years ago is dictated by the measurement choice we make here on Earth today. In this way, Wheeler argues, the observer creates reality.

We should hasten to note that the Austin Interpretation champions an *observer*-created reality, not a consciousness-created one. The Austin view, while differing from Copenhagen in significant ways, still accepts some of the crucial aspects of Bohr's position. Most important, the two schools agree that scientists can communicate unambiguously only about the final results of a measurement. For Wheeler, the essence of existence (reality) is meaning, and the essence of meaning is communication defined as the joint product of all the evidence available to those who communicate. In this view meaning rests on action, which means decisions, which in turn force the choice between complementary questions and the distinguishing of answers. Putting all these links together, out pops the Austin Interpretation of reality generation by exercise of the quantum measurement option.

Of course the reliance upon an observer to create reality is also a part of the von Neumann–Wigner consciousness school of romantic realities. However, the Texans are very clear on the point that their brand of reality has no need to invoke the special role of consciousness. They endorse the Copenhagen view that what constitutes a measuring apparatus is any device that records a quantum phenomenon, giving rise to Wheeler's statement "Let's not invoke consciousness as a prerequisite for what in quantum mechanics we call the elementary act of observation."

To summarize the Austin position on the twin problems of measurement and interpretation, the stance is clear on the matter of the nature of unmeasured quantum objects: These objects have no attributes until a measurement is taken; i.e., there is no objective reality without measurement. As for the Measurement Problem, the Austin Interpretation seems to be pretty much in agreement with that of the Copenhagenists: Possibility becomes actuality at the moment the record is made. In attempting to pin down when this moment takes place, the Austin Interpretation invokes the communication postulate, which seems to imply that the wave function collapse occurs when the elementary quantum process is brought to a close by an irreversible act of amplification. This act of communication closes Wheeler's Meaning Circuit of existence, shown in Figure 7.8. Here the quantum aspects of existence appear in the "underground" part of the loop back from meaning to physics. Wheeler's self-referential universe logo, also shown, neatly sums up the Austin view of the universe as a Delayed-Choice Experiment in which the existence of observers who see what's happening gives tangible reality to everything else.

By now it should be evident that problems of language are becoming increasingly severe as we try to bring the testimony of the Prosecution's romantic realists into some semblance of contact with ordinary images of space, time, matter, and all the rest. One of the first to try to come to grips with this language/reality gap is our next witness, Werner Heisenberg, who champions a reality in which what's real is a combination of what may be and what is. Let's swear him in.

THE DUPLEX INTERPRETATION

Reality consists of potential and actuality

We've repeatedly emphasized that the quantum wave function somehow encapsulates all the possible attributes a quantum object can display—once we finally get around to making a measurement. After years of pondering Feynman's forbidden ques-

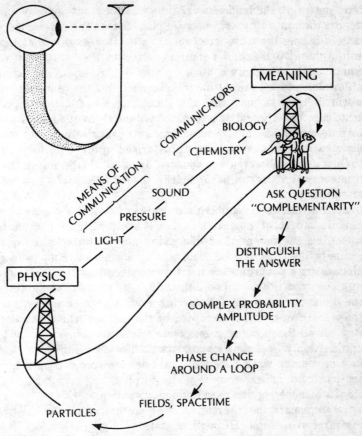

FIGURE 7.8. *Wheeler's Meaning Circuit*

tion, Heisenberg eventually concluded that reality consists of two disjoint worlds: the world of *potential (potentia)* and the world of *actuality,* with the two joined by the act of measurement. How does his vision differ from that put in the record by the previous distinguished witnesses for the Prosecution, who themselves also sanctified the measurement act?

For Heisenberg the only reality, as that term is usually employed in ordinary life, is the world of actuality, i.e., the world of phenomena. But phenomena have to be constructed out of something, don't they? What is that something? In the Duplex Interpretation that something underlying tangible reality is pure potential, the tendency for things to come out one way and not another once they are observed phenomena. Thus Heisenberg is here taking the wave function at face value, saying that this will-'o-the-wisp realm of potential comprises the "stuff" out of which things like knives, forks, plates, tables, chairs, and medium-rare steaks are ultimately formed. So the unmeasured world literally *is* just what the quantum wave function represents it to be—a world of unrealized potential. At the moment of measurement, one of these tendencies is magically granted a more exciting life-style, being transformed into the world of actuality as an observed phenomena. It is at this moment that whatever attributes were implicit in the *potentia* surface as real attributes.

At first glance Heisenberg's universe of potential looks a lot like the world of potential in Las Vegas, where the roulette wheel has the potential of displaying any of thirty-seven outcomes before the croupier decides to give it a spin. But, in fact, Heisenberg's potentials are much less well defined than this. For him even the range of possibilities is not set until you specify the measurement option. Referring again to the casino setting, the *potentia* would be represented by the entire world of possibilities for all the different games and devices the casino offers, like craps, twenty-one, roulette, and keno. The possibilities present in a particular wave function would become specified only once we decide which game we're going to play, i.e., once we've decided upon the specific measurement situation.

To emphasize the "unreality" of quantum objects in the underworld of *potentia,* Heisenberg states:

> In the experiments about atomic events we have to do with things and facts, with phenomena that are just as real as any phenomena in daily life. But the atoms or the elementary particles themselves are not as real; they form a world of potentialities or possibilities rather than one of things or facts. . . . Atoms are not things.

Like the other romantics, Heisenberg disavows any sort of objective, observer-independent reality propping up the world of everyday phenomena. The world of *potentia* cannot really be seen as anything other than a kind of shimmering mirage of dreamlike reality, waiting to be awakened into actuality by the magical Midas touch of measurement. But let's consider for a moment some of the differences between potential and actuality as seen by Heisenberg, and the realities put forth by his Prosecution cohorts.

First of all, the Measurement Problem. Given his association with Bohr and the Copenhagen school, it would be unthinkable to find Heisenberg's position on wave function collapse departing from the orthodox Copenhagen line. Indeed, the Duplex Interpretation, just like the Copenhagen, Austin, and Consciousness interpretations, asserts that there's something very special indeed about making measurements. Furthermore, all indications are that Heisenberg was just as vague as the Copenhagenists as to exactly when this sacred event takes place. However, his position does insist upon the exercise of the measurement option before the possibilities inherent in the wave function can actually be specified, a stance that puts the Duplex and Austin positions into momentary conjunction, an alliance that is soon broken by the Austin insistence on meaning as the core of all existence. The Duplex Interpretation seems to have nothing at all to say on this key point, a fact that also separates it from the advocates of consciousness-created realities.

When it comes to the Interpretation Problem, the Duplex position is unambiguous: Quantum objects have no meaningful existence other than in the world of *potentia,* and they certainly don't possess anything that could be called attributes in the unmeasured state. So again we are faced with testimony arguing that objective reality is a physical fiction brought on by our lack of linguistic sophistication and inability to comprehend what it could possibly be like to live in a world of pure potential. But like all the other romantics heard from so far, Heisenberg proposes a world in which there are two halves separated by the all-embracing and all-consuming act of measurement. It's the task of our last Prosecution witness to convince you that perhaps the Measurement Problem is no problem after all—provided you're ready to entertain the idea of lots of realities instead of none at all.

THE MANY-WORLDS INTERPRETATION

*There is a universe for every possible observation,
each of them equally real*

In 1941 the Argentine writer and poet Jorge Luis Borges published a small volume of fantastic stories, *The Garden of Forking*

Paths. In the title story, the sinologist Stephen Albert tells the protagonist, Hsi P'êng, about the infinite labyrinth of Hsi's ancestor who, according to Albert,

> did not think of time as absolute and uniform. He believed in an infinite series of times, in a dizzily growing, ever spreading network of diverging, converging and parallel times. This web of time—the strands of which approach one another, bifurcate, intersect or ignore each other through the centuries—embraces *every* possibility. We do not exist in most of them. In some you exist and not I, while in others I do, and you do not, and in yet others both of us exist.

This phantasmagorical Borgesian world was brought to the august pages of *Reviews of Modern Physics* sixteen years later when Hugh Everett III, a student of Wheeler's, published his doctoral dissertation, leading to what is now termed the Many-Worlds Interpretation of quantum theory.

Everett started from the same place as von Neumann in that he regarded both the system that was to be measured and the measuring device as quantum objects. But then instead of worrying about *when* the wave function collapses, Everett said, in effect, forget about the collapse. Following this dictate to its logical conclusion, Everett's theory claims that whenever the system and measuring device interact, the new system composed of the two *splits* into as many copies as there are possible outcomes of the measurement. So if the measurement could have yielded one of M possible outcomes, after the interaction Everett says there are now M equally real "worlds." In World 1, the measuring device shows possible outcome number one; in World 2, it shows possible outcome number two; and so on. Consequently, instead of the wave function's collapsing, the quantum system realizes all possible outcomes, and each of them is actually realized in its own separate world. At this juncture, the practical man poses the obvious question: "If there are so many different worlds out there, each of them real, why do I seem to see only one of them (at a time, at least)?"

Everett's answer to the above query is one that will brighten the day of every science fiction writer, mystic, and modern cosmologist: The inhabitants of these worlds live on parallel planes of existence. So the sci-fi writers are on the right track after all, and there is a universe in which the Confederacy did win the

Civil War and another in which UCLA didn't win even one NCAA basketball title. And we have the final authority of the physicists to assure us that these worlds really are out there. But if you're a USC man, don't start making plans yet for emigrating to that UCLA-less heaven. There is a censor in the cosmos who ensures that we humans can occupy only one universe at a time. So we can't see all the others, even though Everett assures us they're out there, and each of them is just as real as the universe we actually experience. Just why we should be confined to a single universe at a time is anybody's guess, and most commentators on the matter (non–science fiction writers, that is) fall back upon that most Chomskian of explanations that we humans just happen to be wired up that way.

Popular expositors of the Many-Worlds Interpretation have obviously fallen in love with the idea of a myriad of universes continually branching away from each other as each observation takes place. Somehow the idea of $10^{100}+$ universes does capture the imagination. In all fairness to Everett, however, it should be noted that he thought of the situation in slightly less romantic terms, in which only the measuring apparatus itself branches into these different possibilities. Of course this in itself is a pretty wild notion, but it pales by comparison with the popular image of branching universes rather than branching meter sticks or Geiger counters. An equivalent, but even less romantic, view is offered by David Deutsch, who thinks of there being a fixed, but infinite, number of universes at all times. In this setup, whenever a measurement is made this infinity of universes just reconfigures itself to account for the possible experimental outcomes. Thus, rather than an ever-increasing temporal sequence of Borgesian parallel worlds, the Deutsch picture is more like that suggested much earlier by Boltzmann in which the worlds all exist simultaneously and always have. They somehow just occupy different "spaces" that are mutually inaccessible to each other.

Whichever way you call it, the Many-Worlds Interpretation, despite its frankly bizarre character, is a favorite with a number of physicists for several reasons. The most important is that it's the only reality that doesn't sanctify the measurement act. As far as Everett's thesis goes, measurement devices and actions exist on the same footing as any other physically realizable activity. Since there is no collapse of the wave function, there is no

Measurement Problem. Beyond any shadow of a doubt, this is the cleanest possible solution to the Measurement Problem: Just banish it from the realm of problems by pulling the rug out from under the essence of the difficulty—the collapse of the wave function. But the price we pay for this solution is our willingness to accept a resolution of the Interpretation Problem that stretches credulity. Instead of asking us to picture unmeasured quantum objects that have no definite existence (as in all the other romantic realities), Everett jumps to the other end of the scale and says not only do they really exist, but there are an uncountably large number of them. It's hard to imagine concluding the testimony for the Prosecution on a more flamboyant note than this.

So we've come to the end of the Prosecution's case, and what a case it's been: The opposite of a truth is a truth; reality comes straight out of consciousness; observers dictate what's real; reality is potential; anything that can happen does happen. No wonder journalists love these romantic realists. Just as in literature, where there are many shades of romanticism, so it is in physics, with the romantic physicists conjuring up many answers to the Quantum Measurement and Interpretation Problems. For future reference, the answers proposed are summarized in Table 7.3.

The Defense is going to have a tough time putting up a battle against such an armada of dazzling realities and intellectual muscle. While their visions of reality have somewhat less flair than those of the Prosecution, and involve a slightly more pedestrian view of the cosmos and more legwork in developing the details, the Defense experts are no slouches themselves when it comes to playing the reality game. So let's now give our attention over to their arguments for why there may be something to the idea of objective reality after all.

THE DOGWORK REALITIES

In his account of his years as Einstein's assistant at the Institute for Advanced Study, Abraham Pais tells of an incident in 1948 when Niels Bohr was visiting the institute. Since he didn't like the large office assigned to him, Einstein generally used the smaller office next door that had originally been allocated to his assistant. Consequently, during his visit Bohr was using Ein-

SCHOOL	WAVE FUNCTION COLLAPSE	UNMEASURED ATTRIBUTES
Copenhagen	by measuring device	don't exist
Consciousness	by conscious mind	don't exist
Austin	from communication	created by meter option
Duplex	from measurement act	only phenomena are real
Many Worlds	there's no collapse	all possibilities are real

TABLE 7.3. *The romantic realists*

stein's large office and pondering once again his decades-long debate with Einstein over what we have termed the Quantum Interpretation Problem. As Bohr paced the room muttering in his inimitable way, the one word that came through clearly was "Einstein . . . Einstein . . ." As he stood for a moment peering out the window with his back to the door, Bohr once again mumbled the magic name, and at that very moment Einstein silently entered the room from next door to get some tobacco from the humidor on his desk. After a moment, Bohr turned around and saw Einstein standing there like a genie conjured up by his magic incantation. Following a few seconds of astonishment, the two old adversaries both broke out laughing at this seeming stroke of synchronicity. In quantum terms, Bohr might have said that only the intervention of his "measurement device" brought Einstein into the room's reality. Einstein, no doubt, would have claimed that he existed all along as a "hidden variable," and that once his "value" was known by Bohr's observation, all the mystery of the situation disappeared. This little vignette illustrates the position of the first and by far most prominent witness for the Defense, Albert Einstein, the universally acknowledged king of post-Newtonian physics. We've already described Einstein's position on the matter of reality as naïve realism. For the sake of completeness, let's briefly summarize its content.

NAÏVE REALISM
Deep reality consists of ordinary objects

Since we've already considered the realist position in some detail, it suffices now to note that the idea of an ordinary object is exactly what the words imply: an object whose attributes really exist whether or not one is observing them. Naturally, in the classical view of attributes, the act of measuring a quantum object is no more sacred than the act of measuring your living room for a carpet. It's just a confirmation of something that existed all along. Thus for the naïve realist the solutions to the problems of measurement and interpretation are quite clear-cut.

First of all, measurement. Since attributes uniquely exist at all times, the wave function description is incomplete. This means there must be hidden variables whose values, when known, in effect collapse the wave function into a single possibility. Consequently, the Measurement Problem is the result of the incompleteness of the quantum description, and disappears as soon as the additional variables are accounted for. As for the Interpretation Problem, the Einstein position is equally clear: All attributes exist at all times, observed or not. So there does indeed exist a single, objective, observer-independent reality. End of story.

In the realist picture, attributes like position and velocity combine in ordinary ways to form new attributes. So, for instance, if two particles collide to form a single new entity, their respective velocities before the collision can be added to determine the velocity of the new particle. All operations of this kind in which attributes are combined involve the Boolean logical operations of <u>AND, OR, NOT</u> and so on. Our next Defense witness claims that quantum objects may have real attributes after all, but the logic of the quantum world is just different from that we normally use.

QUANTUM LOGIC

The quantum world uses a nonstandard type of logic

Shortly after publication of his quantum bible in 1932, von Neumann and Garrett Birkhoff invented a new type of logic that can be used for describing how quantum objects combine their attributes to form new attributes. To explain the basic idea of these *nondistributive lattices,* imagine we have a set of objects that can possess three sorts of attributes. Call these attribute types X, Y, and Z. Using normal rules of logic, we can combine these attribute sets in various ways. For instance, we can form a new set composed of those objects that possess both attribute X

<u>AND</u> attribute Y, called the *intersection* of X and Y and denoted $X \cap Y$. Similarly, the set of those objects having attribute X <u>OR</u> Y is termed the *union* of X and Y, denoted $X \cup Y$. One of the most important laws of normal logic is the so-called *distributive law*, which states that

$$X \text{ } \underline{\text{OR}} \text{ } (Y \text{ } \underline{\text{AND}} \text{ } Z) = (X \text{ } \underline{\text{OR}} \text{ } Y) \text{ } \underline{\text{AND}} \text{ } (X \text{ } \underline{\text{OR}} \text{ } Z)$$
$$X \cup (Y \cap Z) = (X \cup Y) \cap (X \cup Z)$$

To illustrate the foregoing relationships, suppose the attributes in question are the types of polarization that can be displayed by a photon. Polarization is an attribute that is associated with a particular direction in space, and any given photon is either completely polarized in that direction or completely polarized at right angles to that direction. Thus the polarization attribute can take on only one of two values relative to a given direction. Suppose three directions are given: vertical, horizontal, and diagonal, denoted V, H, and D, respectively. Here, as the names imply, the directions H and V are assumed to be at right angles to each other, while D is intermediate between the two. Using the above notation for union and intersection, we could describe those objects polarized in *either* the horizontal or vertical direction as $H \cup V$, while those polarized in *both* directions would be expressed as $H \cap V$ (of course, this set is empty here since photons cannot be completely polarized in two orthogonal directions at once). The distributive law now states that the collection of all those photons polarized both horizontally and diagonally plus those polarized vertically consists of those that are vertically or horizontally polarized plus those that are vertically or diagonally polarized. Simple, ordinary common sense, right? Well, we already know about the value of common sense in the quantum world.

Let's consider the Three-Polarizer Paradox as an example of the failure of the distributive law in the quantum jungle. Recall that a polarizing filter is just a slab of material that passes only a single type of polarized light. A good example is a pair of polarized sunglasses that pass only visible light in a single direction, screening out the light from other directions that is the cause of annoying glare. Suppose we have three such filters, each designed to pass light polarized in one of the above directions H, V, and D. The experimental setup is displayed in Figure 7.9.

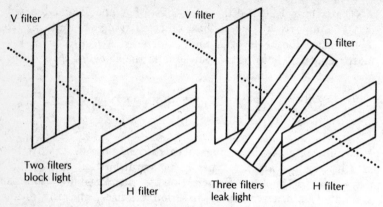

FIGURE 7.9. *The Three-Polarizer Paradox*

When we use only the H and V filters, all light is blocked, reflecting our earlier remark that light cannot be polarized in two directions at once. Now let's throw a joker into the deck and consider the third filter, D.

The D filter passes light polarized diagonally to that passed by either the V or H filter. If we place the D filter either before or after the $H + V$ filter stack, the result is just what we'd expect: No light gets through. However, if we place it in between the H and V filters as shown in the diagram, a miracle occurs. Light gets through the stack. How can this possibly be? According to normal Boolean logic, it can't. The only "logical" explanation emerges when we pass to a non-Boolean kind of reasoning in which the above distributive law is no longer valid. This is the essence of the argument made by the quantum logicians: Things are just different in the quantum realm, including the kind of logic that underlies the combining of attributes of quantum objects. According to the non-Boolean view, the quantum world consists of individual islands on which the ordinary rules of logic apply (the case of individual attributes). But these islands combine their attributes in a way that can be described only by some weird, nonstandard rules applicable solely to the world of the quantum. It's as if you had a group of Pacific islands on each of which the natives speak the same language. But when the islanders get together for their joint annual festival, the only language allowed is an ancestral tongue in which some of the ordinary grammatical rules no longer apply. Assume for the sake of argument that the quantum logic picture is correct. What does it say about our benchmark tests, the Measurement and Interpretation Problems?

A major spokesman for the quantum logicians is David Finkelstein of the Georgia Institute of Technology, who argues that there is nothing strange about the idea of unmeasured quantum objects' possessing definite attributes at all times. What is strange is the way these attributes are combined to form what we see with our measuring instruments (such as the three polarizers). Thus, on the Interpretation Problem the logicians say yes, there is an objective reality consisting of quantum objects having definite attributes. As to the Measurement Problem, quantum logic says nothing about the when, why, or how of the wave function collapse, but speaks only about the properties of the object in the unmeasured state. Thus quantum logic offers us no help whatsoever in scaling this peak of the quantum terrain. On this disappointing note, let's call in our next witness for the Defense.

While von Neumann's proof against hidden variable realities seems to close the books on the naïve realist position, dealing what looks like a mortal blow to the gleam in Einstein's eye, as always with quantum theory things are not what they seem. Our next Defense expert shows us how even the genius of von Neumann may not be beyond reproach, as he manages to do the impossible: construct an interpretation of quantum theory involving only ordinary objects. Let's hear how this seemingly impossible task was carried out.

THE QUANTUM POTENTIAL INTERPRETATION

Reality is undivided wholeness connected by "pilot waves"

In 1951 McCarthyism was running rampant across the American intellectual landscape, the father of the atomic bomb, J. Robert Oppenheimer, being one of its prominent victims. At the Atomic Energy Commission hearings against Oppenheimer, one of the witnesses called was David Bohm, a young physics professor from Princeton, who had been one of Oppenheimer's Ph.D. students. Bohm refused to testify against his old professor, an action that clearly irked the commission. Given the temper of those times, such an act was tantamount to confessing to Communist leanings of one's own, and ways were found to strip

Bohm of his professorial post. Following this clash with the authorities, Bohm left the United States, finally settling in London as professor of physics at Birkbeck College after leaving temporary havens in Brazil and Israel. Having brushed the dust of McCarthyite witch-hunting reality off his boots, Bohm proceeded to the safer and infinitely saner and more rewarding consideration of quantum reality. At this time he began to develop an earlier idea of Louis de Broglie into a mathematically consistent interpretation of quantum theory involving only ordinary objects.

Interestingly enough, prior to his forced withdrawal from teaching at Princeton, Bohm had authored what is still a highly regarded textbook on quantum mechanics that follows the conventional Copenhagen party line. But even though he was serving up this traditional Danish pastry to his students, Bohm was becoming increasingly convinced through conversations with Einstein that both Bohr and von Neumann were wrong—an ordinary-reality interpretation of quantum theory was possible. The chink Bohm identified in von Neumann's armor had to do with an implicit assumption he made about the interaction of quantum objects. Von Neumann assumed that they interact in what he termed "reasonable" ways. The kind of interactions that Bohm had in mind would definitely not be "reasonable" by von Neumann's criteria, as we'll soon see.

The key theoretical idea that Bohm based his approach upon was the notion of a *pilot wave.* This idea had been introduced in the 1920s by de Broglie but quickly laughed out of court by the Copenhagenists in view of what looked to be insurmountable mathematical difficulties. But Bohm showed how to overcome those difficulties, reviving de Broglie's idea of regarding a quantum object as a particle with an associated pilot wave that, in effect, tells it how to move. Let's look at one or two of the details.

In the pilot wave picture, every quantum object is a real particle possessing definite attributes at all times. Associated with each such object is a pilot wave that is also real but undetectable other than through its effects on the particle. This wave is termed the *quantum potential,* and serves the function of "reading" the environment and reporting its findings back to the particle. Let me emphasize here that this is a real wave and should not be confused with the quantum *wave function,* a purely mathematical gadget for making predictions. The particle then acts in accordance with the information provided by its associated pilot wave. As a result, in the Quantum Potential Interpretation a quantum object is not composed of a single "thing," particle or wave, but is both. Notice how objective reality is restored in this picture, as there is no longer the ongoing schizophrenia between the object as wave or particle. At all times it is both, and at all times the particle side of the house possesses all the usual classic attributes. Bohm's genius was to show how this scheme could be made to work. But there's no free lunch in life or in quantum theory either, and to those of a traditional outlook there is a heavy price to be paid for this restoration of objectivity.

The first major complaint against the quantum potential approach is that it invokes a physically unobservable wave to restore objective reality. To most physicists, if you can't measure it, then it doesn't exist, and the advantage of postulating an entity such as the quantum potential is not worth the price of undetectability. When it comes to the trade-off between an objective reality and physical observability, the verdict of the working physicist is that without observability you have nothing, or at least whatever you do have isn't physics. But this objection to the quantum potential pales by comparison with the other main argument against it: the need for faster-than-light signaling.

It's ironic to realize that the quantum potential was developed as a step toward rescuing Einstein's idea of reality from the scrap heap where it was tossed by the work of Bohr and von Neumann. Ironic because the biggest obstacle to the rescue operation is one that Einstein himself created by his Special Theory of Relativity, with its ironclad prohibition against any kind of signals being transmitted at a velocity faster than that of light. Remember that the quantum potential acts something like a radar wave sending out probes into the environment, with the "reflected" signals used by the particle half of the quantum alliance to decide what to do. Thus the quantum potential senses the presence of a measuring apparatus of a certain type and immediately notifies the particle, which then adjusts its behavior to accommodate to the kind of attribute the device is designed to measure. It can be shown that this kind of signaling from the quantum potential back to the particle involves information transfer of *some sort* that must move at superluminal rates—a direct violation of Einstein's cosmic speed limit.

Bohm's partial answer to this difficulty is that the quantum potential is not a wave of matter, just a wave of active information. Its effect depends only on its form, not upon its magnitude; consequently, unlike matter waves such as sound or water whose effect diminishes with distance from the source, the quantum potential can have big effects at long distances. This is the phenomenon of nonlocality, which will occupy our attention shortly. In Bohm's view relativity is a statistical effect, not an absolute one. The superluminal effect is seen only when we look at the correlations between signals at two separated locations; if, however, we look at what's happening in the local neighborhood of either location, the statistical properties of the signals appear to be independent. Therefore no superluminal aspects show up.

In recent years Bohm has become an active spokesman for the school of thought that sees the universe as a giant hologram, arguing that to truly understand and be able to explain quantum processes, we must abandon our traditional modes of reductionistic thinking. Beneath the world of surface phenomena there is an undivided seamless whole, and it is this "underworld" that is the domain of quantum objects. In this realm, every object is connected to every other because of the intertwining of their quantum potentials, ensuring that every quantum object carries a trace of every other object with which it has ever interacted.

With regard to the Measurement and Interpretation Problems, the Quantum Potential Interpretation comes off rather well. It definitively resolves both problems in the way that Einstein would have liked best (aside from the superluminal aspect, which he most definitely didn't like at all). Bohm's theory deals with the Measurement Problem in much the same way as the Many-Worlds Interpretation: He says that the wave function does not collapse because it doesn't represent a complete description of the object. Once the additional variables are provided (the quantum potential), there is no collapse, hence no Measurement Problem. The Interpretation Problem is disposed of in a similarly clean manner: All quantum objects are ordinary objects having all attributes at all times. Consequently, reality is objective and independent of whether or not we happen to be looking. Thus the Quantum Potential Interpretation solves every problem that's interesting about quantum reality—just as long as you can accept "real" entities that are undetectable and superluminal transfer of information. Before closing the book on the Defense case, let's hear from one more witness who has revived an old idea of Wheeler and Feynman's to provide yet another interpretation that restores the objective reality of quantum objects, but this time *with* a wave function collapse.

THE TRANSACTIONAL INTERPRETATION

Reality is a wave function traveling both backward and forward in time

In our earlier story about the excited atom giving off its excess energy to an atom of silver to darken a photographic plate, we noted that prior to the process of absorption, the electron's position is described by a wave function that is created at the moment of its emission from the excited atom. In scientific parlance, this is termed a *retarded wave,* for reasons that will become apparent in a moment. Toward the end of the Second World War, J. A. Wheeler and Richard Feynman proposed what they termed an *absorber theory* of such emission processes in which *advanced waves* are produced in such emission-absorption processes on an equal basis with retarded waves. The idea is that when the retarded wave is absorbed by the atom of silver (at some time in the future), there takes place a cancellation erasing all traces of advanced waves and their effects. In this theory, the silver atom absorber carries out its absorption of the original retarded wave by manufacturing a second retarded wave that is identical in amplitude but exactly out of phase with the retarded wave from the emitting atom. In this way, the two waves cancel and we speak about the original retarded wave as being "absorbed." In the Wheeler-Feynman theory, the silver absorber also makes an advanced wave that "backtracks" the retarded wave, moving backward in time along the path taken by the retarded wave from the emitter. This advanced wave reaches the emitter exactly at the instant of emission. It then continues backward in time, but now is accompanied by the advanced wave from the emitter. Since the two waves are exactly out of phase, they also cancel, removing all "advanced" effects in the process.

When we observe this absorption of energy by the silver atom, we don't have access to these inner mechanisms of Nature. As a result, all we see is that a retarded wave has traveled from the excited atom to the photographic plate. Thus, from an observational standpoint, the absorber theory leads to precisely the same observations as any of the usual descriptions, e.g., the Copenhagen Interpretation. But the conceptual difference is considerable, since now there has been a two-way exchange transferring energy across spacetime from the excited emitter atom to the absorbing silver atom. Recently, this idea has been taken up by John Cramer of the University of Washington, forming the heart of what he calls the Transactional Interpretation.

The key ingredient in Cramer's view is that every quantum event (interaction) involves such an advanced-retarded "handshake" across space and time. This handshake is a sort of two-way contract between the past and the future serving as the vehicle for the transfer of energy, momentum, spin, and so forth. While the details of Cramer's arguments are a bit heavy for our purposes here, the essential point is that the transaction is explicitly nonlocal in the sense that the future is in some way affecting the past, at least insofar as it enforces correlations between quantum events. For example, when we look through our telescope at the light from the star Tau Ceti, which is eleven light-years away, not only have the retarded light waves from Tau Ceti been traveling for eleven years to reach our eyes, but the advanced waves generated by the absorption processes within our eyes have reached eleven years into the past, thereby completing the transaction that permitted Tau Ceti to light up our lives.

The great advantage of the Transactional Interpretation is that it eliminates the observer from the formalism of quantum mechanics. By this process, all of the paradoxes associated with observer-dependent realities such as half-dead/half-alive cats, waves of knowledge, and splitting universes vanish. The drawbacks are that the vanishing act is performed by unobservable phenomena (the advanced waves) transferring information and energy at superluminal velocities. Note also that with the Transactional Interpretation there is still a wave function collapse; in fact, there are *two* collapses: one for the retarded wave, one for its time-reversed counterpart. On the other hand, the naïve-realist requirement that quantum objects have well-defined properties at all times is retained. So like the Quantum Potential Interpretation, the Transactional Interpretation involves real entities that are unobservable and superluminal transfers of information. Swallowing the first is more a matter of taste than experiment. Pursuit of the second takes us deep into the heart of one of the most startling results in modern physics—the Bell Interconnectedness Theorem. But before giving our attention to this friend of the court, let's pause to summarize the testimony for the Defense in Table 7.4.

SCHOOL	WAVE FUNCTION COLLAPSE	UNMEASURED ATTRIBUTES
Naïve Realist	no	always exist
Quantum Logic	??	always exist
Quantum Potential	no	always exist
Transactional	yes	always exist

TABLE 7.4. *The dogwork realists*

THE BELL TOLLS FOR LOCALITY

In his experiments on the paranormal, Duke psychologist Joseph B. Rhine often employed card-matching experiments in which a subject was shown the back of a Zener card whose front might bear a star, cross, circle, square, or wavy lines. The subject was then asked to call out the pattern on the card being shown, with significant deviations from the 20 percent random-guess hit rate deemed evidence for ESP. Suppose we cook up a variant of Rhine's experiment to test for telepathy.

Our experiment will involve two subjects, Alexander and Anastasia, placed on opposite sides of an opaque screen. Instead of showing cards with patterns, the experimenter will show our subjects questions randomly chosen from a fixed set of, say, three possible queries, which are themselves randomly selected from a supply of three-question sets. Further, assume that each subject's question is randomly selected on every trial. To keep things simple, suppose the questions have only a simple yes or no answer. Thus a specific set of questions could be: (1) "Do you believe this experiment genuinely reveals anything about the existence of ESP?" (2) "Do you think the experimenter has any idea of what he's doing?" (3) "Are you doing this just for the money?" Now imagine that the responses by Alexander and

Anastasia repeatedly display this peculiar feature: Whenever they are both shown a question bearing the same number they both give the same answer. After several repetitions of the experiment, always with the same result, the experimenter concludes that the two are definitely in telepathic contact. He then publishes papers in all the right ESP journals, claims a cushy professorship, and is granted time on *The Today Show* to report his astonishing findings to a world eager for scientific confirmation of its deeply held beliefs about such matters. Is there anything amiss here?

Despite public acclaim, fortune, film offers, and a cover appearance on *People* magazine, Alexander and Anastasia are roundly denounced as fakes by impassioned scientists who claim that the whole experiment is a fraud. The scientists point out that the entire circus can be accounted for by the simple assumption that the two are in communication *before* the experiments begin. All they need to do is to agree in advance on what their answers to the particular questions will be. Thereafter the results are foreordained. For example, if they agree that they will answer no to all questions numbered 1, yes to questions numbered 2 and 3, there is no need for communication during the experiment, and the "astonishing" outcome of their agreement whenever the questions are the same is assured.

The less vocal (but more thoughtful) members of the intellectual fringe note that not only are their more excitable colleagues correct, but that the sort of agreement described is the only procedure by which the two fakers can possibly arrange things so that the final outcome is always complete agreement on identically numbered questions. The claim of this less vocal, but far more insightful, group of scientists is that it is not only sufficient but also necessary that Alexander and Anastasia both know the answer they will give to each question in order for them always to concur whenever the question numbers are the same. In addition, the thoughtful scientists have also kept track of the answers given by the two "telepathists" when the question numbers are ignored, i.e., when the pair are not told the number of the question that they are being asked. These investigators find that over a long run of experiments, the complete set of answers given by the two differ exactly half the time, as would be expected if the subjects just randomly answered yes or no. In other words, if we just look at the string of answers from each subject over a long sequence of experiments when they are uninformed as to the question numbers, the number of agreements equals the number of disagreements, on the average. From this fact they conclude that Alexander and Anastasia *must* have been in some type of communication in those trials where they were told the question numbers, trials that led to perfect agreement when the question numbers were the same. Let's see why.

We've already seen that the pair must have had some kind of plan for how they would each answer the questions, for instance, the pattern no-yes-yes stated above. Let's abbreviate this to NYY. Furthermore, if we observe that they always agree when shown identically numbered questions, then our earlier argument demands that each uses the same plan. Our problem is to show that Alexander and Anastasia must have actually *communicated* their respective plans to each other. To show this, first of all assume that there was no communication. Under this assumption, they should agree only half the time—on the average. Now let's compute the average number of hits and misses with the NYY plan. Since there are three questions in each set, the total number of cases is nine. These cases are shown in Table 7.5, together with the answers given by the subjects using the NYY plan.

It's easily verified that any other plan involving two Y's and one N or two N's and one Y will have the same result: five agreements and four disagreements. Of course if the plan is NNN or YYY, then there will be complete agreement. The upshot of this entire business is that with *any* conceivable plan there are more

QUESTION ASKED (ALEXANDER FIRST)	(1, 1)	(1, 2)	(1, 3)	(2, 1)	(2, 2)	(2, 3)	(3, 1)	(3, 2)	(3, 3)
Alexander's answer	N	N	N	Y	Y	Y	Y	Y	Y
Anastasia's answer	N	Y	Y	N	Y	Y	N	Y	Y
Agree/disagree	A	D	D	D	A	A	D	A	A

TABLE 7.5. *Experimental results with the plan NYY*

agreements than disagreements. This directly contradicts the 50-50 split seen when there is no communication. So we conclude that Alexander and Anastasia are indeed telepathic.

While the saga of Alexander and Anastasia may seem rather fanciful, it illustrates perfectly a crucial fact about the logic of measurement processes. And when transferred to the quantum domain, this experiment forms the basis for the Bell Interconnectedness Theorem, a result that some have hailed as the most profound discovery of science. Since Bell's result was motivated by von Neumann's "proof" of the impossibility of hidden variable theories and Bohm's subsequent "disproof," the right place to start our search for the essence of Bell's message is to go back to hidden variables and the notoriously puzzling Einstein-Podolsky-Rosen (EPR) Paradox.

By now it should be evident that Einstein was always profoundly unhappy with the idea of a statistical kind of quantum world underlying the world of surface phenomena, spending the last half of his life fighting an unrelenting guerrilla war against the Copenhagenists and their anti-hidden-variable ontology. The biggest salvo Einstein ever fired in this battle was the paradox he concocted in 1935 with two colleagues, Boris Podolsky and Nathan Rosen, designed to show that the quantum theory as touted by the Copenhagenists provides only an incomplete description of Nature; i.e., there must be a more complete theory involving variables hidden from the view of the current theory. Let's consider a simple version of the EPR experiment due to David Bohm.

Imagine a device that generates pairs of electrons and shoots them off in opposite directions. One of the attributes of an electron is its spin direction, the axis of which can point in one of two directions. Let's call them UP and DOWN. Since the total spin in this two-particle system must be zero in order to conserve angular momentum, when one electron's spin axis points DOWN, the other must be UP, and conversely. EPR argued as follows: Generate such a pair and let them separate so that one remains here on Earth, with the other going to the great spiral galaxy Andromeda 2 million light-years away. Now measure the spin direction of the member of the pair that stayed here on Earth. The Copenhagen view says that before the measurement this electron had no definite spin at all. Rather it was in the state "half UP, half DOWN," just like the sorry state of

Schrödinger's Cat. And similarly for its sibling over in Andromeda. But quantum theory states that as soon as the measurement is made, *voilà:* It's UP or DOWN with no ifs, ands, or buts. Moreover, *at that very moment* the other electron in Andromeda somehow "knows" about the measurement here on Earth and its spin direction is also definitely fixed DOWN or UP, opposite to whatever was seen on Earth. The paradox is now clear: How did that information get from here to Andromeda so fast? By Einstein's own Special Theory of Relativity, it should have taken at least 2 million years for any kind of signal to get to Andromeda. Yet according to the Copenhagen position, the electron's spin is somehow communicated instantly. What's going on here?

Naturally Einstein used the above thought experiment to claim that the quantum description was deficient and that there must be hidden variables whose values, if known, would have given the earthbound electron and its Andromeda counterpart definite and opposite spins at all times. In that case there would be no need for the superluminal transfer of information to Andromeda, because the Andromeda electron wouldn't need it to know in which direction it should be spinning. In this way Einstein tried to restore objective reality to the quantum world by pitting the Copenhagen Interpretation against his own well-tested Special Theory of Relativity. The quantum theorists and Einstein batted the EPR ball back and forth across the net for the better part of thirty years, with no definite resolution of the point. The impasse was finally broken in 1964 with the publication in the first volume of an obscure physics journal of a six-page paper using no more than elementary undergraduate mathematics. That paper was by CERN physicist John Bell, and served as the rallying cry for a whole new era of quantum reality research.

The EPR argument rests upon the assumption that superluminal information transfer is impossible, hence quantum theory cannot be complete. Bell took his cue from Bohm's demonstration that a hidden variable theory can make sense—at least if information gets around faster than light—and discovered that Einstein was wrong: *Any* viable hidden variable theory *must* allow for faster-than-light communication. To dispel a possible misconception right at the outset, this does not imply that you can send a message instantaneously to a friend in An-

dromeda. Remember: Bell's Theorem talks about the world of
deep reality, not the world of phenomena. We'll return in a moment to a more detailed consideration of this point. First let's
look a little more closely at Bell's magnificent achievement.

To see the basic idea underlying Bell's result, let's go back to
our electron pair-generating device and assume that it regularly
shoots out electrons whose spin axes are randomly oriented. This
means that as each pair comes out, the chances are equal that the
spin axes of the pair point in any particular direction in space,
with, of course, one electron's axis pointing opposite to its
twin's. The general situation is shown in the figure below, where
the random spatial direction is p and the generator is denoted by
the symbol $\otimes$.

$$\swarrow \leftarrow \otimes \rightarrow \nearrow \quad {}^{+p}$$
$$-p$$

Now suppose we have two identical spin-detection devices, one
on Earth and one on Andromeda, each with a direction knob allowing the device to detect electrons spinning either in the direction of the knob setting or in the opposite direction, e.g., the
directions $+p$ or $-p$ as above. Suppose at the outset that the
two devices are both set to the same direction—call it d. Since
detection of an electron is a yes/no proposition, let's also agree
that each device records its observations on a tape, writing "1"
if it detects an electron, and "0" if nothing is detected. Since
initially both knobs are placed at the same directional setting, we
would expect the tapes from Earth and Andromeda to agree.
Thus a typical run might produce the following records:

Earth's record: 01000101110011001101

Andromeda's record: 01000101110011001101

Note the crucial point that although each sequence is random
(since the generator fires electrons whose spins are randomly
oriented), there is perfect correlation between the two records.
So although an observer on Earth and one on Andromeda would
see what looks like a purely random sequence of 0's and 1's,
someone with access to *both* records would probably start drawing conclusions about the two streams of electrons being related
in some nontrivial way.

Now let's change the situation and set the knob on Earth at a new direction, say $d + 10°$, leaving the Andromeda device alone. In this situation, some of the electrons detected on Earth will not be detected on Andromeda, and vice versa. Typical records in this case might look like this:

Earth's record: 01*1*001*0*11100110*1*1101

Andromeda's record: 01*0*0011*1*110011001101

Here the three mismatches are indicated in italics. Since there are twenty trials in the run, when the offset between Earth and Andromeda is 10° we have an error rate of 15 percent. Obviously, since the situations here and in Andromeda are perfectly symmetrical, the error rate would have been exactly the same had we left our knob alone and our friends in Andromeda been the ones to make the 10° adjustment. Let's denote this error rate as $E(10°) = 15$ percent.

To continue the experiment, suppose now that the Andromedans decide to twist their knob by 10° in the opposite direction; i.e., their new setting is now $d - 10°$. Thus the relative difference between Earth's and Andromeda's directions is now 20°. Big question: What is the error rate $E(20°)$? This question is easy to answer if we assume that the errors on Earth are *independent* of those in Andromeda; i.e., the errors we see when we use Earth's device as the standard are independent of the errors we see when we use the Andromeda device as the benchmark. What this assumption involves is the claim that whatever's happening on Earth has no bearing on what's going on in Andromeda, and vice versa. Under this working hypothesis, we can easily work out a bound for the new error rate. Since the errors seen on Earth were $E(10°)$, we have to add to this the further errors introduced by setting the Andromeda knob at $d - 10°$. By the symmetry of the situation, this error is also $E(10°)$. So we might first conclude that the new error rate should be twice this amount, i.e., $E(20°) = 2E(10°)$. But wait. When the Andromedans shifted their knob, we lost the standard for Earth's record, and similarly, when we twisted Earth's knob, we destroyed the standard for Andromeda's record. Thus the overall effect is that there will be some cases in which "double errors" cancel each other out. That is, an error will be detected on both Earth and Andromeda, each canceling the other's effect so there appears to be a match instead of a mismatch. Taking this factor into account, we can say only that the error rate of interest at 20° could not be greater than twice the rate at 10° but could possibly be less. Symbolically, we can write this as $E(20°) \leq 2E(10°)$. It should be clear that the particular angles 10° and 20° have no bearing on the argument, which is valid for *any* angle A. Thus we can write $E(2A) \leq 2E(A)$, which is the famous Bell inequality, the basis for Bell's Theorem.

Now let's look at the hypotheses we used in deriving the foregoing result. There are two:

• *Objectivity:* We assumed that the electrons' spin axes really had a definite direction at all times between emission from the generator and their measurements on Earth and in Andromeda. In other words, the electrons are ordinary objects.

• *Locality:* The errors seen on Earth and in Andromeda are completely independent of each other. In short, twisting the direction knob on Earth has no bearing on what's seen in Andromeda, and conversely.

By now you might be saying to yourself, "OK, this all looks perfectly reasonable. What's the point?" The point is that if you actually perform this experiment with real electrons (but without a confederate on Andromeda), you will find that the Bell inequality is violated. In fact, according to the experiments of John Clauser and more recently Alain Aspect, it's violated to such a degree that the possibility of attributing the deviation to experimental error is negligible. In short, Bell's result says that either locality or objectivity (or, perhaps, both) has got to go! Thus, if you want to keep an Einstein-Bohm type of hidden variable reality, then you have to do as Bohm did and sacrifice locality. On the other hand, if you want to retain locality, which most physicists of the Copenhagen stripe insist upon, then there can be no hidden variables to bail out a naïve-realist position. And even if quantum theory ultimately turns out to be false, Bell's result will still hold: objectivity or locality, but not both. So this fact, sometimes called Bell's Interconnectedness Theorem, places severe constraints on any pretender to the reality throne, imposing the strict condition that if you're advocating a hidden variable approach and you haven't explicitly included a place for superluminal connections, then don't bother submitting your paper. Your theory cannot possibly be correct. No wonder some have called this one of the most important results in the history of physics.

* * *

The normal long-distance communication channels in America used to be offered by Ma Bell; in the universe it seems that the long lines have been laid down by Doc Bell. This is a good time to reconsider the question of whether we could use this cosmic Bell System to send a superluminal invitation for cocktails to our colleagues in Andromeda. There's been no small amount of confusion on this point since Bell's Theorem came out of the physicists' closet, an indicator being a flash letter from a California think-tank executive to an Undersecretary for Something at the Pentagon informing him of the result, and suggesting that the ability to send such messages would offer an unjammable command-and-control communication system for submarines. No doubt the second paragraph of the letter was an offer to look deeper into the matter for a small consideration. Unfortunately for that enterprising executive, the prospects appear distinctly bleak for using Doc Bell's channels for submarines or any other kind of human or ETI contact. Let's see why.

In our spinning-electron experiment, we saw that twisting the knob on Earth had a definite effect on the correlation between the record seen on Earth and that seen on Andromeda. Thus we can definitely say that some kind of nonlocal effect was "caused" on Andromeda by our action here on Earth. The problem is that the Andromedans won't notice anything unusual. All that will happen is that they will get a record consisting of a different random sequence of 0's and 1's. Since they don't know what record they would have received if we hadn't twisted our knob, there is no real transmission of information between us. The only way that information could be transmitted would be if the Andromedans had advance knowledge of the settings we were going to be using. But there is still no superluminal method known for transmitting this information. So, since one random sequence looks pretty much like any other, Andromeda will have no way of detecting the *difference* between two sequences attributable to different settings of our spin detector. And it is only in such differences that a message can be coded. Thus, inspection of the output of *their* detection device gives them no information about the input to *our* device, because they don't know the hidden variables (the setting of our device). So it seems that Doc Bell's communication channel is even better than the California executive claimed: You can use the channel to send a signal that's not only unjammable, but so perfectly scrambled that only Nature holds the decoding key!

Despite his appearance here as a friend of the court, as long as we have Bell on the stand it's impossible to resist the temptation of asking for his own position on the cases put forth by the Prosecution and Defense. Given the choice of abandoning either objective reality or locality, Bell casts his vote for retaining objective reality and the Quantum Potential Interpretation of Bohm. Says Bell, "In my opinion, the pilot wave picture undoubtedly shows the best craftsmanship among the pictures we have considered." On this unambiguous note, let's call in one more friend of the court to tell us why information about the earliest moments of the universe has some light to shed on how reality really is.

IN THE BEGINNING, THE VERY BEGINNING

In 1964 two physicists from Bell Labs were trying to calibrate a microwave communication antenna and found their efforts continually thwarted by some sort of background noise that they were unable to account for by any Earth-based interference. The ultimate explanation of that noise resulted in the award of the 1978 Nobel Prize in physics to the two researchers, Robert Wilson and Arno Penzias, for their discovery of "fossil evidence" of nothing less than the moment of creation of the universe itself. This discovery of the so-called microwave background radiation was the final factor in tilting the scales between the Steady-State Theory of the universe, which held that things have always been more or less as they are today, and the Big Bang Theory, claiming that the universe began in an explosion of literally cosmic proportions. By consensus, the Wilson-Penzias noise is the electromagnetic residue of that primordial fireball and, along with the observed expansion of the universe in all directions and the abundance of the light elements—hydrogen, helium, and deuterium—it serves as the major selling point for the Big Bang Theory today.

If the Big Bang Theory is correct, the implication is that at some time in the past, currently estimated at 12 billion years ago (plus or minus a few hundred million), the universe was compressed into a microscopic point of hard-to-believe proportions and properties. What is absolutely astonishing is that physicists now feel that they can give an almost letter-perfect account of what happened after the first 10^{-30} second or so following the universe's birth. Just incredible! This amount of time is so short that no kind of clock imaginable can even come close to measuring it. Yet the large-scale structure of the universe as we see it now is reasonably well explained by current theory after the first 10^{-30} second. Unfortunately for the cosmologist as well as the reality theorist, those first few ticks of the clock were where most of the action took place, since that's the time when what we now call the laws of Nature were laid down and the structure we see today was determined. To dig a little deeper into the fundamental nature of these laws, let's take a closer look at what we do see today when we look at the large-scale structure of the universe, as well as some of the puzzling aspects about it.

Two things you'd soon notice about today's universe if you looked at it through a powerful telescope is that it's extraordinarily homogeneous and isotropic (i.e., it is smooth and looks the same in all directions). Thus, on the large scale the visible matter has a remarkably uniform distribution; it is not organized into "clumps" separated by regions of empty space. Furthermore, this is the picture you would observe regardless of the direction in which you pointed your telescope. Besides the homogeneity and isotropy, after a few calculations you'd immediately notice another peculiarity. The universe's rate of expansion is a bit like Goldilocks's porridge: not too big and not too small, but just right. So right, in fact, that a change of just a fraction of a percent in one direction or the other in the force of gravity would lead to an uninhabitable universe: either one in which stars were born and died much too quickly for our kind of life to evolve, or a universe in which matter could not have coalesced into stars and galaxies at all. In short, the universe is "flat," precariously balanced on a knife edge between an open cosmos of runaway expansion and a closed universe of rapid recollapse. In view of the way things look today, the Big Bang might be compared to a group of blindfolded football players gathered in a huddle. The players are instructed to run away from the center of the huddle in a straight line when they hear the referee's gun go off. The shot is fired and they start running, with the miraculous result that the original huddle expands outward not in a ragged, roughly circular fashion, but into an ever-growing *perfect* circle! This remarkable state of affairs cries out for an explanation, and somehow what was going on in that first 10^{-30} second holds the key. The first clue comes from what some call number mysticism.

NUMEROLOGICAL PHYSICS

Homogeneity, isotropy, and flatness are not the only puzzling coincidences we observe about the way the universe appears today. There are also some oddly disturbing relationships between many of the basic constants that go to make up the so-called laws of physics. In 1923 the British cosmologist Arthur Eddington noticed a curious relationship between the gravitational constant G, Planck's constant h, the speed of light c, and the mass of the proton m_p. When he combined these basic constants of Nature so as to cancel out their respective units of measurement, thus obtaining a pure, dimensionless number, Eddington found the following ratio:

$$\frac{hc}{Gm_p^2} \approx 10^{39}$$

What struck Eddington about this incredibly large number was that, to within a factor of 10 or so, it is exactly the square root of the number of protons in the universe, which is the immense quantity $N_p \approx 10^{78}$ (here the symbol $\approx$ means "approximately equal"). Since there's no a priori reason why the number of protons should bear such an uncanny relationship to the earlier quantity, Eddington felt that he was on the track of a deep, undiscovered principle of Nature, and invented an elaborate theory to account for this type of numerical "coincidence."

Later another eminent British physicist, Paul Dirac, followed up Eddington's ideas and discovered further remarkable relationships of the same sort linking the electric force e between the proton and electron, the gravitational force between the same two particles, the age of the universe t_μ, and the time needed for light to cross an atom. Here are Dirac's relations:

$$\frac{\text{Electric force}}{\text{Gravitational Force}} = \frac{e^2}{Gm_p m_e} \approx 2.3 \times 10^{39}$$

$$\frac{\text{Age of the universe}}{\text{Time for light to cross an atom}} = \frac{t_\mu}{e^2/m_e c^3} \approx 6 \times 10^{39}$$

To have these basic constants combine in such a way as to arrive at virtually the same outrageously large number was just a little heavier dose of coincidence than Dirac was prepared to accept. Thus, he made the bold assertion that the two ratios were in fact identical, leading (after a minor amount of algebra) to the estimate:

$$t_\mu \approx \frac{1}{G} \times \frac{e^4}{m_p m_e^2 c^3} \qquad (*)$$

This relation for the age of the universe is written to show that the only part of it that is not fixed is the term involving the gravitational constant G. The rest of the terms involve masses, the speed of light, and the like, all of which are assumed to be unchanging over the course of time. But since the age of the universe is obviously not time-invariant, Dirac concluded that in the relationship he had discovered, the gravitational coupling constant is steadily decreasing as time goes on, to keep things in balance. Furthermore, from Eddington's relations we see this has the consequence that the number of protons in the universe must be increasing with the square of the age of the universe, implying that matter is continually being created.

At the time Dirac made these claims in the late 1930s, they caused a small stir in the cosmology community. However, later experiments using the Viking lander to measure the orbital period of Mars showed that Dirac's idea of a time-varying G is very unlikely to be correct, since the period was not changing as would be required if G were not constant. So are these "coincidences" just coincidences, or is there still a real explanation lurking in the wings? In 1961 Robert Dicke of Princeton published an argument asserting that Dirac's coincidence was no coincidence, not at least if one accepts what has now come to be termed the Weak Anthropic Principle. Since Dicke's anthropic argument has formed the basis for more than a little controversy in the physics community over the years, it's worth our time to devote a few pages to a deeper consideration of its basis and conclusions.

ANTHROPIC PRINCIPLES

At the most uncontroversial level, anthropic reasoning comes down to the well-accepted principle that when you're engaged in measuring anything, it's necessary to take into account the particular properties of the measuring instrument. When the in-

struments happen to be ourselves as human beings, then the conclusions from our measurements have to respect the peculiar
features giving rise to our situation as observers. And the most
important such features are the physical conditions that appear
to be necessary for our very existence at this time, on the third
planet circling a typical G-type star in the suburbs of the Milky
Way Galaxy. This idea, in effect, is what underlies the so-called
Weak Anthropic Principle (WAP), which can be stated as:

WEAK ANTHROPIC PRINCIPLE: The observed values of all physical
quantities are restricted by the requirement that they be compati-
ble with our existence as observers.

The reader will recognize this kind of reasoning as a middle
ground between the pre-Copernican view of mankind as the center of the universe, and the post-Copernican cosmology, which
denies mankind any special status or position. The Weak Anthropic Principle states, in effect, that while our position may
not be central, it is privileged to some degree.

In his 1961 paper, Dicke employed the WAP as an explanation for Dirac's numerical relations. His argument is instructive. On the basis of well-known principles of nuclear physics,
Dicke calculated that the expression on the right side of relation
(*) on page 479 should very closely approximate the lifetime of
a typical star. So it's not at all surprising, he claimed, that these
same constants will combine to equal roughly the age of the universe. Reason? Since the matter from which we are constructed
must have first been synthesized in the nuclear reactions at the
core of a star, the universe cannot be younger than the lifetime
of a star, or we would not be here to worry about the question.
End of proof. Since it's crucial for understanding the heated
debate between the proponents and opponents of anthropic arguments, note carefully the chain of reasoning in Dicke's argument:

1. Given the existence of mankind, the age of the universe could
   not have a value much different from the one it actually has.
2. Thus, Dirac's relations don't apply to *any* universe, but only
   to the universe that we actually observe today.

This pattern of logic completely reverses the direction of reasoning usually employed in science. Generally we start by specifying the initial situation and the laws of Nature, then predict

the subsequent state of affairs. Anthropic reasoning proceeds in the opposite direction: Start from the final observed state (now), and try to constrain the initial situation by asserting that it could only have been one that would give rise to a universe that's inhabited today by intelligent observers like ourselves. So Dicke's technique is to cite a present condition (our existence) as an explanation for a phenomenon having its source in the past (the age of the universe). Up to this point—the introduction of intelligent observers—most physicists will at least grudgingly accept the tenets of the WAP, even if many of them do believe it smacks more of tautology than principle. But there is a stronger version of the principle, termed (not very imaginatively) the *Strong* Anthropic Principle (SAP), which makes intelligence the key actor in the cosmic drama.

The WAP says nothing about the laws of physics themselves, nor does it comment on the actual values of the fundamental constants like the speed of light or the gravitational coupling constant. It simply tries to explain various observed features of the universe, taking these items as givens. The SAP, on the other hand, tries to use anthropic reasoning to attach actual values to these quantities. An example will illustrate.

Suppose the gravitational constant G were a million times larger than it actually is. Then the lifetime of a star in its life-giving phase would be about a million times less, since the higher gravitational forces would greatly accelerate the burning of its nuclear fuel. But even in such a universe Dicke's argument would still apply. If an observer exists in such a universe, when the age of that universe is around ten thousand years he would see a universe whose mass would be a trillion times smaller than ours. Question: Would life arise in such a vastly accelerated universe? The WAP is totally silent on this issue; the SAP says no, life can exist only if the fundamental constants have values very close to their observed levels.

The foregoing sort of argument leads to the most familiar form of the SAP:

STRONG ANTHROPIC PRINCIPLE: The universe must be nearly as we know it or life would not exist; conversely, if life didn't exist, neither would the universe.

The reader will immediately note that the gap separating the SAP from the classic argument from design invoking a supernatural Creator is no more than a hairsbreadth, omitting only an explicit invocation of a Designer. Finally, for the sake of completeness, let's note the so-called *Final* Anthropic Principle (FAP), which asserts the kind of ultimate fate for intelligence that virtually all traditional religions would endorse, namely, that our descendants will become like gods.

> FINAL ANTHROPIC PRINCIPLE: Once life is created, it will endure forever, become infinitely knowledgeable, and ultimately mold the universe to its will.

If this kind of argument sounds familiar it should, since it's central to the Austin Interpretation of J. A. Wheeler considered earlier. Wheeler argues that for a universe to be real, it must evolve in such a way that observers come into existence. One of the main pillars supporting his contention is what he calls the *Participatory* Anthropic Principle (PAP), asserting that the universe is brought into existence by the collective observations of all intelligent observers who have ever existed or who ever will exist. At about this point, skeptics like Martin Gardner start trotting out principles of their own, like the Completely Ridiculous Anthropic Principle (CRAP), as antidotes to the high-flying assertions of these "anthropic physicists." As a micro-example of a typical academic feud, let's look at a few of the arguments for and against the anthropic principles to see how they may or may not help us come closer to understanding the true nature of Nature.

Heinz Pagels of Rockefeller University was one of the most vocal opponents to the use of anthropic principles in physics prior to his untimely death in a rock-climbing accident in 1988. Pagels claimed that anthropic principles are "the lazy man's approach to science." He cited at least three main deficiencies in the use of such reasoning in the practice of science, arguing that anthropic principles: (1) use the unknown to explain the known; (2) never predict anything and are entirely *post hoc;* (3) are both immune to experimental falsification and untestable. Pagels concluded his indictment of the "anthropicists" by saying that the anthropic principles are "the closest that some atheists can get to God." It's amusing to note that in a popular article on the topic arguing *in favor* of anthropic reasoning, the physicist Tony Rothman used similar words: "When confronted with the order and beauty of the universe and the strange coincidences of nature, it's very tempting to take the leap of faith from science into religion. I am sure many physicists want to. I only wish they would admit it."

As to the claims that the anthropic principles are untestable, unfalsifiable, and *post hoc,* supporters point out that Dicke could have used the WAP to argue against the Steady-State Theory of the universe even before the observations of Wilson and Penzias. The argument is that in the Big Bang Theory the age of the universe happens to be approximately equal to $1/H$, where H is the expansion rate of the universe. However, in the Steady-State Theory, H must by definition be constant and therefore has nothing to do with the age of the universe. Consequently, in the Steady-State Theory there's no reason for $1/H$ to equal the lifetime of a typical star. So the fact that it does is either a gigantic coincidence or an anthropic argument in favor of the Big Bang. The fact that Dicke did not make this claim in no way argues against the inherent possibility of generating a testable prediction on the basis of the WAP. To see another kind of prediction that can be obtained using anthropic arguments, let's return to the topic of the last chapter and quickly reprise Brandon Carter's anthropic-based argument against the existence of ETI.

As we know, those arguing in favor of the likelihood of ETI often invoke the Principle of Mediocrity to buttress their claims. What this comes down to is a special case of the Copernican Principle asserting that there's nothing special about life on Earth. Consequently, since the universe is so vast, and since we are here, the chances are great that "they" are there. Wheeler, for one, counters with the anthropic claim that the universe is vast only because it's several billion years old, and needs to be that old to give rise to one intelligent civilization (ours). Since there's no particular reason why there should be ETIs out there, additional civilizations would be wasteful of the universe's resources. This line of reasoning has been vastly sharpened by Carter, who sparked off the current "anthromania" in a 1974 address to the International Astronomical Union in which he coined the term "anthropic principle." We have already given the essence of Carter's argument against ETI in the last chapter, so there's no need to repeat any more than the basic idea here.

First of all, Carter assumes that there are a number of individually improbable steps on the road to intelligent life. Next he predicts the average time between the emergence of an intelligent species and its death from, say, the burning out of its sun. Finally, he argues (on the basis of the WAP) that intelligent life is exceedingly rare. So we conclude that if ETIs are found with any frequency, Carter's WAP-based prediction is wrong. This constitutes a testable prediction using anthropic arguments: Just find lots of ETIs out there, and Carter's WAP-based argument will be falsified. But we're starting to wander off course from our original goal of looking at the quantum cosmological doings in the first 10^{-30} second of the universe's existence. So let's close this short excursion into anthropic thinking with the following food for thought from Freeman Dyson: "As we look out into the Universe and identify the many accidents of physics and astronomy that have worked together to our benefit, it almost seems as if the Universe must in some sense have known that we were coming."

QUANTUM COSMOLOGY

Since in the Big Bang picture the universe was much smaller than an atom in the very early going, we have to use the concepts of quantum theory to describe what was happening in those first few picopico . . . picoseconds. You might object that it seems to go well beyond the bounds of credulity to imagine that the whole universe could be compressed into a volume far less than that of an atom, since the energy density must have been intolerably large. But remember, according to quantum theory energy and time are conjugate variables, so we can get large amounts of energy into a small volume if the time is short enough. If 10^{-30} second isn't short enough for you, perhaps you'll need to seek your fortune on the astral plane after all. In any case, let's stay in this universe and look at some of the explanations for how the large-scale features of the universe could have emerged out of this "point" of matter-energy.

The two main puzzles surrounding the moment of the Big Bang center on the seemingly highly ordered nature of the flyspeck of matter-energy constituting the initial state, and the extraordinarily delicate balance in the gravitational force that left our universe teetering right on the edge between a runaway expansion and a too-rapid collapse. Let's talk here only about the Initial State Paradox.

The crux of the paradox is that the observed homogeneity and isotropy of the universe today, not to mention the conditions necessary for our existence, are difficult to account for by anything other than a highly ordered initial state. Yet if we choose initial states of the universe randomly, the chances are overwhelmingly high that the state that pops up will be very disordered. The situation here is exactly the same as that faced when we deal out the cards in a game of poker. According to the probability theorists, there are a total of 2,598,960 possible initial hands that could turn up on the deal in a round of five-card draw poker. Let's assume that these possibilities stand for the various possible initial states of the universe. Now let's randomly dip into the deck and select a hand for our universe. According to current theory, the chances of getting a hand that corresponds to an initial state favorable to our type of life, and that is consistent with the kind of large-scale structure we observe, are vastly less than the likelihood of having a royal flush staring you in the face when you pick up your poker hand. And this probability is only 4 in 2,598,960, or a bit better than 1 in a million! So how can we account for the apparently highly unlikely initial state of our universe? Several answers have been proposed.

•*Many-Universes Theory:* This resolution of the paradox is the cosmologist's appeal to Everett's Many-Worlds Interpretation in quantum theory. The Everett theory postulates a different branch of the universe for each possible value of an observable quantity, so what could be more natural than to claim that our universe just happens to be one of the few in which all the conditions and constants came out "just right"? Note here the appeal to the WAP as a self-selection mechanism for choosing a "good" universe for life from the set of possibilities, almost all of them "bad."

Because of its neat disposition of the Initial State Paradox, the Many-Worlds Interpretation is a favorite among cosmologists. In fact, among all the quantum interpretations considered earlier, Everett's is the only one that really gives a consistent and coherent picture of how to deal with the initial state problem. However, opponents argue that it is the very antithesis of

Ockham's Razor, being far too extravagant in dispensing "universes for all occasions" to be taken seriously as a solution to the dilemma.

•*Dissipation:* Adherents to this view claim that the initial state was not so well ordered at all, but that frictional and other dissipative forces smoothed out the initial inhomogeneities. Thus turbulent mixing and recombination of the primordial matter soon led to the kind of regular state we see today. Opponents argue that if we admit disordered initial states, there are always some such states that are so nonuniform that even after billions of years the irregularities would not have been dissipated. Furthermore, as we know from rubbing two rough surfaces together, friction generates heat, and calculations show that the amount of dissipation needed to arrive at today's universe would have generated an amount of heat far in excess of what's observed in the Wilson-Penzias background radiation. So at the moment dissipation isn't seen as a likely solution to the paradox.

•*Inflation:* At present the main scientific opponent to the Many-Universes Theory is the idea that the early universe enjoyed a short inflationary period, which smoothed out the initial state; thereafter, the universe settled into its current expansionary mode. This is a little bit like what happens when you blow up a balloon. Initially the balloon has no air and is just an irregular, crinkly rubber sack. However, as soon as you pump in the first couple of breaths of air, the balloon immediately springs into a smooth, regular shape, which expands uniformly thereafter.

The inflationary model, originally proposed by Ed Tryon in the early 1970s and later developed by Alan Guth at MIT, postulates a repulsive force that operated against gravitational forces to expand the universe to about the size of a basketball during the first 10^{-35} second after the Big Bang. At this point the repulsive primeval force split into the four forces we know about today (gravitational, electromagnetic, weak nuclear, and strong nuclear), and the familiar radiation-dominant force of expansion took over. An important feature of this scenario is that it allows the universe to have come into existence as nothing more than a quantum fluctuation in a total vacuum. The matter needed for Nature to pull off this conjuring trick came, of course, out of Einstein's famous formula $E = mc^2$, which shows the equivalence between matter and the energy contained in the vacuum. In short, everything comes out of nothing!

At present inflationary models seem to have the upper hand in the cosmological derby, although those committed to a more anthropic view note that one can give an anthropic explanation for why the universe is so isotropic that's just as convincing as the one obtained by invoking inflation. For example, Hawking and Collins have argued (via the WAP) that if the universe were not isotropic, then we wouldn't be here to observe it. Carrying this argument one step further, they would claim that the initial state *must* have been special, too. Opponents would (and do) say that while there's nothing wrong with this line of reasoning, it's certainly not necessary, and that an idea like inflation is aesthetically more satisfying. On this note, let's move to the final contender for the solution of the Initial State Paradox.

•*God:*   This is clearly the most straightforward solution of all. Simply invoke a Grand Designer who stirred up Goldilocks's porridge to exactly the right temperature and consistency so that both the initial state and the fundamental constants of Nature came out "just right" for us to be here. This is the familiar argument from design, which has been the mainstay of all non-scientific accounts of the universe from time immemorial and needs no further amplification here.

As a postscript to the quantum cosmology issue and the Initial State Paradox, it's amusing to consider for a moment the *final state* from the anthropic point of view. If we believe in the Final Anthropic Principle, there might not be much to choose between the argument from design and the idea that our successors will ultimately come to be indistinguishable from God. This argument follows from Wheeler's Participatory Anthropic Principle, which requires intelligent life to have a significant effect upon the large-scale properties of the universe. Following up the implications of the FAP, many scientists and philosophers have come to the conclusion that if life evolves in all the many universes in a quantum cosmology, and if life continues to exist in all these many worlds, then all of these universes will approach what the French Jesuit priest and mystic Pierre Teilhard de Chardin called the Omega Point. As noted by the anthropicists Frank Tipler and John Barrow:

At that moment, life will have gained control of all matter and forces not only in a single universe, but in all universes whose existence is logically possible; life will have spread into all spatial regions in all universes which could logically exist, and will have stored an infinite amount of information, including all bits of knowledge which it is possible to know. And this is the end.

And this is the end for us, too, in our account of the anthropic principles and their possible relevance for the problem of reality. Let's now give the floor back to the lawyers for their final arguments.

SUMMARY ARGUMENTS

Both the romantic realists and the dogwork realists have argued extensively and persuasively to convince us of the rightness of their respective causes. Before summarizing the positions, let's again review the issue before the house. Put simply, we have the Prosecution's claim:

There is no such thing as a unique, observer-independent reality.

On the other side of the courtroom, we hear the Defense say, "Maybe not." At least it says there is no irrefutable evidence to conclude that an objective deep reality, independent of observers, does not underlie the world of phenomena. Tables 7.6 and 7.7 summarize the competition.

Before I enter into a justification for my own conclusion on this ultimate question, let me pull an ace from up my sleeve and say that whatever position you care to hold, the experimental data will not refute you. As it turns out, *each* of the above positions is in complete accord with the experimental evidence! So until there's an experimental breakthrough of some kind, the position you hold on the quantum reality issue is more like a religious conviction than a matter of science. All positions are defensible, and your choice becomes as much a matter of aesthetics and a gut feeling for "how it could be that way" as a logical consequence of hard facts. With this extraordinary situation in mind, allow me to close out this all-too-brief tour of life, behavior, cognition, language, machines, and universes with my private prejudices as to the reality of reality.

THERE IS NO OBJECTIVE REALITY!

PROMOTER	ARGUMENT
Bohr (Copenhagen Interpretation)	overall measurement situation
von Neumann, Wigner (Consciousness Interpretation)	consciousness determines reality
Wheeler (Austin Interpretation)	measurement option
Heisenberg (Duplex Interpretation)	*potentia* and actuality
Everett, Deutsch (Many Worlds Interpretation)	every world is a reality

TABLE 7.6. *Summary arguments for the Prosecution*

A SINGLE, OBSERVER-INDEPENDENT REALITY MAY EXIST!

PROMOTER	ARGUMENT
Einstein (naïve realist)	Newtonian reality is real
von Neumann, Finkelstein (quantum logic)	nondistributive logic
Bohm, Bell (quantum potential)	pilot wave theory
Cramer (transactional events)	advanced and retarded waves

TABLE 7.7. *Summary arguments for the Defense*

BRINGING IN THE VERDICT

The paradox of the quantum realm is that although common sense dictates that the universe exists "out there" independent of acts of observation, the universe does not actually seem to exist "out there" independent of acts of observation. One view is that we are insignificant specks playing out totally uneventful roles in a vast cosmic play; the alternate position says that in some way we are not only the players, but the drama's writer, director, and producer, as well as critic and audience, too. It's hard to be of more central importance than that! As I've tried to cut this Gordian knot of conflicting scientific visions of reality, my own oscillations between the arguments of the Prosecution and Defense have come to symbolize for me the essence of the dilemma itself: "How can it possibly be that way?" In the final analysis perhaps we all think of ourselves as romantics at heart, so my personal struggle with the nature of reality comes to a temporary halt with a vote for the Prosecution and its clients, the romantic realists. Specifically, I give the nod to Everett's Many-Worlds Interpretation.

When it comes to sifting the evidence and claims, as I mentioned above the experimental evidence really offers no help. Everything known from the laboratories is perfectly consistent with the MWI and any of the other contending views. So it ultimately comes down to a matter of aesthetics, and to my mind at least, the MWI just has a few more selling points than the competition. To begin with, it has fewer *ad hoc* assumptions, especially about the mysterious measurement act. That the physical act of attaching a Geiger counter, camera, microscope, or meter stick to some system should dramatically affect the basic nature of things is still a difficult notion for me to swallow. The MWI manages a fairly clean resolution of this problem by the simple expedient of denying that there is any problem. Second, the MWI appears to be the only quantum reality that gives a coherent picture of the Initial State Paradox of cosmology. Since the way we see the laws and state of the universe today is conditioned by the character of that initial state, an interpretation that gives some kind of scientifically defensible account, even if it does seem bizarre, looks better to me than the scientific equivalent of a Gallic shrug or an even more outlandish explanation. Finally, there is Bell's Interconnectedness Theorem. Such a result cannot be proved in the MWI for obvious reasons: The proof relies on the fact that while many outcomes of a measurement are possible, only one of them is actually realized; i.e., we need a counterfactual condition to prove the result. In a cosmos where *all* possible outcomes are realized, there is no Bell's Theorem. To my mind, banishing this kind of superluminal connection is a definite plus for the MWI. Of course, any reality in which happenings around Procyon or over in Andromeda affect earthly doings has plenty of nonlocality of its own, even without Bell's result. Nevertheless, I feel more comfortable with this kind of nonlocality than with the Bell type.

Before closing I should say a word or two about the dogwork realists, in particular the quantum potential crowd. When I first learned about quantum mechanics and started pondering the fateful question, I naïvely wondered why it wasn't possible to regard an electron simply as a particle that moved along its appointed Newtonian path in wavelike fashion, with a continual back-and-forth "wavy" type of locomotion like that of a fish or a snake. While my ignorant musings were hopelessly adrift in a technical sense, they don't seem that far away in spirit from what's presented in the quantum potential, or pilot wave, picture. The view of a quantum object as a particle with an associated wave appears to me to be only one step (albeit a gigantic one, conceptually) removed from my early vision. So when it came time to vote in the reality game, I was sorely tempted to cast my ballot for the quantum potential. But when all was said and done, as a romantic at heart I just couldn't resist a romantic reality, and the MWI is far and away the most romantic of them all. So while my mind is with the quantum potential, my heart is with the MWI. And so is my vote.

CONCLUSION

THE BALANCE
SHEET

===

ARE HUMANS REALLY
SOMETHING SPECIAL?

WHERE DO WE STAND?

Physicists and philosophers love principles: the Heisenberg Uncertainty Principle, the Principle of Conservation of Energy, the Principle of Parsimony (Ockham's Razor), Fermat's Principle, and many more. For centuries one of the most inviolable of all such principles was Aristotle's Principle of Continuity, by which Nature passes gradually from the most imperfect forms here on Earth up to the most perfect works of God in heaven. By this reckoning, hell was at the center of the Earth, hence the center of the universe. A natural corollary of this principle is that mankind occupies a central position in the universal scheme of things. Later Copernicus displaced mankind from its unique position in the most dramatic fashion possible. With his Copernican Principle, he argued that no one part of the universe is more privileged than any other. The Principle of Continuity and the Copernican Principle represent the antipodes of the human role in the universe: mankind at the heart of all things versus mankind as an insignificant speck on the cosmic horizon. We are living at one of those rare moments in which the pendulum is swinging through its midpoint, on its way back to the human-centered universe of Aristotle. And our age has its own principle, the Anthropic Principle, asserting man's role as the measure of all things. In one way or another, all of our stories in this book have been accounts of what science has to say about this anthropocentric claim. So as prelude to a summing-up, let's reexamine our multiple foci.

The Big Question serving as the leitmotiv of our journey through the jungles of modern science can be simply stated as:

Is there anything special about human beings?

Each of the stories I've told in traversing the uncharted terrain of science has addressed this Big Question from its own particular vantage point: human biochemical structure, behavioral patterns, cognitive capacities, and so on. Some of the stories relate to the uniqueness of humans here on Earth; others deal with our role in the galaxy, or even the universe at large. And as we pass from one of these venues to another, the precise form of the Big Question varies accordingly. Yet the overall theme has always remained the same: Are we unique in any way that really counts? To come to a verdict on this question, let's briefly revisit each of our topical areas and rephrase the Big Question in terms suitable for illumination by that area's special sort of lamp. In this way, perhaps, a few glimmerings of our "specialness" may emerge from these individual pieces of evidence.

Origin of Life.   In this chapter we considered our material structure, the particular carbon-based biochemical processes by which all known life forms on Earth operate. A version of the Big Question appropriate to this context would be: Is the particular way in which life arose here on Earth a statistical fluke, unlikely to be repeated anywhere ever again? Or is the combination of steps leading to Earth's life forms an almost inevitable outcome given similar environmental conditions?

To be alive, any object must somehow possess the capability for metabolizing raw materials from its environment into products needed to maintain itself. Moreover, the object must also be capable of some kind of self-repair of its metabolic and reproductive machinery, as well as production of copies of itself, perhaps in conjunction with other members of its species. In our consideration of these matters, we saw that on the basis of general theoretical arguments by von Neumann and others, any such object must possess structures that perform certain distinct functional activities: a constructor, a controller, a copier, and so forth. So we could regard the Big Question in this setting as being tantamount to asking: How likely is it that organisms would arise elsewhere that possessed these functional capabilities as part of their physicochemical makeup?

On the basis of the various explanations put forward for the emergence of life here on Earth, my impression is that should the Earth be wiped clean of all life today in some kind of planetary Armageddon, the likelihood of life forms of any kind reemerging in a few billion years would be a bet that not even Lloyd's of London would put on the board. Consequently, origin-of-life considerations suggest to me that there is indeed something special not only about humans, but about life in general, as we see it here on Earth today.

Sociobiology. Moving from biochemical structure, we next examined the degree to which human behavioral activities, especially those of a social nature, somehow distinguish us from the animals. In particular, we were concerned with whether these behavioral traits were primarily determined by genetic programming or, alternately, were principally a product of environmental (read: cultural) considerations. In this instance, a good phrasing of the Big Question might be: Are most human social behavior patterns innate, or are they primarily acquired by means of learning and/or cultural conditioning?

In our examination of this question, the arguments flew fast and furious. Relevant aspects of biology, genetics, and sociology were mixed with politics and ideology in an ever-shifting blend of logic, experiment, and raw emotionalism. While there was considerable evidence to support the claim that many higher animals behave *as if* they are following the dictates of their genes, the gap between these animals and *Homo sapiens* is a large one, and one that the more vocal opponents of the sociobiologists assert will never be bridged.

After all the rhetoric, smoke, and ashes drift away, my feeling is that the sociobiological debate offers the least clear-cut evidence one way or the other on the Big Question. Even at this point, about the best I can offer is the opinion that human behavioral repertoires could very well be special, differing in essential ways from the basically genetic determination of other living things. For me the sociobiological verdict still comes out as nothing more conclusive than a definite maybe.

Language Acquisition. General social behavior is one thing; the specific behavioral trait of spoken language is something else altogether. This area took us into a consideration of whether language capacity is part of the genetic birthright of every human being. Or is language a human skill that's acquired along with a variety of others as part of a general learning capacity? Here our Big Question comes down to asking: Is the human language acquisition capacity a unique product of the way that the human brain and body happen to be put together? Or can it be expected to occur in any sufficiently complex organism capable of general probing, learning, and problem solving in its environment?

Of all the evidence put forth in this book for the uniqueness of humans, in my view the language acquisition case is by far the strongest. The Chomskian assertion that there is a language acquisition device that's part of our genetic makeup seems far and away a more convincing explanation for the observed facts about language acquisition than any of the countertheories offered by either the behavioral or cognitive psychologists. While the neurophysiological evidence for the location, or even existence, of this device may still be far from conclusive, my gut feeling is that the day is not far off when the boundaries of the language device will be precisely determined and Chomsky's position vindicated. Thus my view is that the language acquisition evidence points strongly toward the position that a human being is indeed a pretty queer bird.

Artificial Intelligence. Closely related to the language acquisition problem is the general question: Is it possible, in principle, to construct a machine that displays the same kind of cognitive powers as a human being and, moreover, carries out these cognitive tasks in the same way? When translated into these terms, our Big Question becomes: Is there anything unique about our way of thinking? Or, more specifically, can we duplicate human cognitive processes in a machine?

Some, like Wittgenstein, have argued that there is no distinguishable difference between our language and our thinking. While I'm far from convinced of the validity of this claim, at least in any strong sense, even a weak form of it immediately suggests close connections between the "thinking machine" question and the problem of language acquisition. In our consideration of the AI problem, a lot of philosophical arguments were put forward showing why a machine could never think like you and me. On the other side of the field stand the computer scientists and engineers arguing that the final score should not be tallied when the game has barely begun.

Strangely enough, while I feel that the language evidence points clearly to our special nature, here I find myself siding, at least provisionally, with the computer scientists and engineers. Thus, on the basis of the AI evidence, I might conclude that our cognitive capacities are not so special after all. How can I explain this clear-cut contradiction to my earlier position vis-à-vis language? Basically, I can't. My best effort is to argue that the language problem indicates that humans are special as compared with all other living agents here on Earth. But computers are not living agents (at least not yet), and I find no essential contradiction in thinking that perhaps a genuine thinking machine is yet a possibility. Anyway, I'm afraid I must come up with a negative reading on the Big Question here.

Extraterrestrial Intelligence. Moving away from Earth, our first stop was the Milky Way Galaxy and the question of whether there are other living, intelligent beings out there for us to communicate with. Here we might phrase the Big Question in the form: As living, intelligent, communicating entities, are human beings unique in the galaxy?

Using the Principle of Mediocrity, a corollary of the Copernican Principle, astronomers gave us arguments showing why the galaxy should be teeming with ETIs. On the other hand, we examined a number of biological, physical, and anthropic arguments indicating that the chances for the existence of an ETI are vanishingly small, essentially zero. Unhappily, I find the latter category of pessimistic arguments far more convincing than those of the optimists, leading to the sad conclusion that we probably are alone, at least in the galaxy. And, in fact, if the universe is finite, the same arguments seem to point to the even more disturbing conclusion that we could very likely be alone in the universe as well. So on the strength of these ETI considerations, humans again start looking like something very special indeed.

Quantum Reality. The final stop on our stroll through the wonderland of science was nothing less than the universe of phenomena with its accompanying puzzler: What is the nature of the deep reality underlying observed phenomena? In particular, we examined the role of humans as observer/participants in the creation of the underlying "stuff" from which the world of phenomena is built. Here we could pose the Big Question in the form: Is a human presence necessary to bring reality into existence?

Most of the group we termed the romantic realists gave arguments suggesting that there is no such thing as an objective physical reality, independent of human observers. The opposition, led by Einstein, argued otherwise. On the basis of the actual experimental evidence, we saw that there are no grounds for accepting either side's case as the last word. Nevertheless, a variety of aesthetic considerations make it at least plausible, if not desirable, to lean toward the romantics, thereby thrusting mankind into the role of creator as well as observer and participant.

In one last attempt to bring everything together, Table 8.1 summarizes my overall impressions on the Big Question from each of the foregoing perspectives.

To my eye, the overall conclusion is that *homo sapiens* is a very special creature, at least here on Earth, and maybe in the universe as a whole. While it may not yet be a conclusion to bet your pension on, I think the odds favoring our uniqueness are high enough that my bookie would tell me, "Off the board, doc." Since it would take a volume nearly the size of this one to address adequately the many implications of this conclusion, let me close this brief survey of science and the nature of mankind by mentioning just one of them.

ARE HUMANS SPECIAL?

origin of life	probably
sociobiology	hard to say
language acquisition	very likely
artificial intelligence	maybe not
extraterrestrial intelligence	very probably
quantum reality	arguably yes

TABLE 8.1. *The bottom line*

In his scathing indictment of the modern American university in the recent bestseller *The Closing of the American Mind,* Allan Bloom notes with alarm the gradual transformation of the university from a community of scholars providing a liberal arts education to what one of my colleagues has described as "a trade school for the bewildered." An important count in Bloom's indictment is the disappearance of any systematic study of the great works of literature, philosophy, and the arts from the program of today's undergraduate, an observation that goes hand in hand with the increasing illiteracy rate in the population at large. Bloom, a humanist, sees the problem from the vantage point of the college of liberal arts; many of the same signs also appear in the college of science and engineering. As a longtime habitué of this corner of the campus, I too have noted with alarm an ever-accelerating trend toward more and more specialized courses and programs of the trade-school variety, necessitating elimination of broader perspectives on the domains of science and their many interrelations. My conclusion is that it's not just the concept of a classical liberal arts education that's endangered; it's the very notion of education itself, liberal arts or otherwise.

An important part of Bloom's solution to the problem is a return to the Great Books: Plato, Shakespeare, Tolstoy, & Co. In the same spirit I would advocate a return to the Great Problems to reverse the trend toward fragmentation and incoherence in the sciences. And in my view, the problem areas we've covered in this volume—the origin of life, quantum reality, sociobiology, and all the rest—are certainly prime candidates for inclusion on anybody's list of Great Problems. These problems share the same virtue as the Great Books: They force one to learn about the mutual interrelationships of many things. For example, it's inconceivable to me that anyone could even begin to address the origin-of-life question without a good knowledge of chemistry, molecular biology, evolutionary biology, and, most likely, combinatorics and computer modeling as well. In the same vein, contributing to the AI question requires an understanding of mathematical logic, the theory of computation, cognitive psychology, neurophysiology, computer engineering, and programming languages. Similar remarks could be made for the other topics we've looked at in this volume. The point is not even that the Great Problems are solvable by these means, but rather that there's so much to learn about the overall landscape of science and the different ways of scientific thinking by expanding our horizons and going beyond narrow, traditional, intradisciplinary thinking.

To close on a somewhat somber note, Table 8.1 appears to lead to the verdict that there is truly something special about humans. Nuclear holocaust, cosmic catastrophe, AIDS, and a thousand other demons sit waiting to snuff out this small flicker of intelligence and light in what looks like a vast, empty void. Whatever we humans are and whatever we can be, I think we have a responsibility not to lose it through carelessness and neglect, benign or otherwise. Periodic reflection on the assessments given here may help us keep this imperative in mind. Let's hope so.

TO DIG DEEPER

===

CHAPTER ONE

WORLD VIEWS IN COLLISION

The story of Jocelyn Bell and the discovery of pulsars is surely one of the more exciting episodes in science during the turbulent 1960s. A firsthand account of the work by the lady herself is given in

Wade, N., "Discovery of Pulsars: A Graduate Student's Story," *Science,* 189 (1975), 358–364.

See also the interview with Bell in the volume

Judson, H. *The Search for Solutions.* New York: Holt, Rinehart and Winston, 1980.

The first account of pulsars as rapidly rotating neutron stars appears to have been given by Thomas Gold at a 1968 symposium at the International Centre for Theoretical Physics in Trieste, Italy. The precise citation is

Gold, T. "The Nature of Pulsars: Survey of Present Views," in *Contemporary Physics, Trieste Symposium 1968,* Vol. 1, pp. 477–481. Trieste, Italy: International Centre for Theoretical Physics, 1969.

A detailed, scholarly account of the pros and cons of the entire *l'affaire Velikovsky* is given in the highly enlightening work

Bauer, H. *Beyond Velikovsky.* Urbana, IL: University of Illinois Press, 1984.

This work is notable not only for its thorough investigation of the scientific basis of Velikovsky's claims, but also the detailed discussion of the form and content of the criticism Velikovsky received. All in all, the author concludes that while Velikovsky was very likely wrong from the standpoint of science, it's not possible to *prove* him wrong. Furthermore, the critics were themselves far from beyond reproach, at least insofar as the methods they employed. In this connection, Bauer cites the impassioned criticism by astronomer Carl Sagan, who got so carried away in his denunciation of Velikovsky that he ended up using the unconsciously held belief that science offers certainty and truth, the creed of "scientism." I highly recommend this book as a demonstration of how modern science operates when wearing both its logical and sociological hats.

However, Bauer himself is not beyond using some of the same rhetorical tricks he accuses Velikovsky's acolytes of employing. For a sympathetic, nevertheless critical view of Bauer's book, see

Gardner, M. *The New Age: Notes of a Fringe Watcher,* pp. 65–71. Buffalo, NY: Prometheus, 1988.

For an account of the ideas that got the whole Velikovsky business off and running, see the "source":

Velikovsky, I. *Worlds in Collision.* New York: Doubleday, 1950.

DID YOU SAY SCIENCE?

The common perception of science is as a means for "gadget production"; a collection of facts leading to practical ends. But scientists think of science as a set of methods and conceptual schemes leading to an understanding of natural processes. For an informative, educational, and easily readable discussion of this profound misunderstanding, see

McCain, G., and E. Segal. *The Game of Science,* 4th Edition. Monterey, CA: Brooks/Cole, 1982.

The conventional ideology of science is an amalgam of the views of philosophers, historians, and sociologists about the logic, progress and norms of the scientific process. It is presented in digestible form in

Broad, W., and N. Wade. *Betrayers of the Truth: Fraud and Deceit in Science.* New York: Simon and Schuster, 1982.

The foregoing book is especially notable for its detailed discussion of the "missing link" in the conventional ideology—the human factor. The authors conclude that the very nature of the ideology increases the attractions of fraudulent activity in science, as well as the likelihood that such activity will go undetected. The authors claim that the root cause of the problem is that the system based upon the conventional ideology ends up rewarding not only genuine success, but also the *appearance* of success. As an account of the dark side of science that many in the scientific establishment go to great pains to pooh-pooh, this book is hard to beat.

THE NATURAL PHILOSOPHER'S STONES

A first-rate discussion of all the difficulties associated with the use of induction, as well as details on the various "-isms," see

Chalmers, A. *What Is This Thing Called Science?*, 2nd Edition. Milton Keynes, UK: Open University Press, 1982.

For a gentle introduction to philosophical problems arising in connection with science, the following work is highly recommended:

Kemeny, J. *A Philosopher Looks at Science*. Princeton, NJ: Van Nostrand, 1959.

Wittgenstein once wrote that he thought it would be possible to write a serious work on philosophy that consisted entirely of jokes. His idea was that if you understood the joke, then you would get the philosophical message. John Allen Paulos took this dictum seriously, producing the following extremely entertaining, as well as informative, work from which I shamelessly lifted the little joke in the text on the Problem of Induction:

Paulos, J. *I Think, Therefore I Laugh: An Alternative Approach to Philosophy*. New York: Columbia University Press, 1985.

For a detailed, even relentless, pursuit of the diagram for mathematical modeling displayed in Figure 1.2, see the volume

Rosen, R. *Anticipatory Systems*. Oxford: Pergamon, 1985.

A novel work that attempts to explore the nature of reality as seen from the perspectives of literature, sociology, physics, art, film, and a variety of other fields is

Exploring Reality, D. Cohn-Sherbok and M. Irwin, eds. London: Allen and Unwin, 1987.

The quote by Kalman relating to the instrumentalist view of the world is taken from

Kalman, R. "Identification from Real Data," in *Current Developments in the Interface: Economics, Econometrics, Mathematics*, M. Hazewinkel and A. Rinnooy Kan, eds., pp. 161–196. Dordrecht, Netherlands: Reidel, 1982.

This paper, as well as several others noted in its bibliography, presents a particularly graphic portrayal of an attitude commonly held by many so-called hard scientists: If you can't measure it, it doesn't exist—literally! Happily, as time goes by such unimaginative and increasingly indefensible prejudices are being weeded out of the scientific mindset, to be replaced by far less precise, but vastly more enlightening, perspectives.

RATIONALITY FOR REALISTS

The text discussion of the work of Wittgenstein, Popper, et al. is nothing more than a caricature of their deep, insightful views on the theory of knowledge, language, science, and reality. Two of the best general references to appear in years on the interplay between the ideas of these philosophers and the logical workings of science are

Newton-Smith, W. *The Rationality of Science.* London: Routledge and Kegan Paul, 1981.

Oldroyd, D. *The Arch of Knowledge.* New York: Methuen, 1986.

A wonderful picture of the entire political, psychological and sociological climate in Austro-Hungary leading up to the views of the Vienna Circle and much, much more is offered in the volume

Johnston, W. *The Austrian Mind.* Berkeley, CA: University of California Press, 1972.

Another work purporting to cover somewhat the same territory, and one that has received enormous amounts of (in my opinion) undeserved publicity, is

Janik, A., and S. Toulmin. *Wittgenstein's Vienna.* New York: Simon and Schuster, 1973.

A personal survey taken through the years I've lived in Vienna shows that of seventeen friends who've started reading this paralyzingly dull volume, not one has gotten further than the middle of Chapter 3. Frankly, the only thing I can see that the book has going for it is a catchy title, which undoubtedly accounts for its continuing sale to unsuspecting tourists in the Viennese bookshops. My recommendation: Stick with Johnston unless, of course, you're looking for instant insomnia relief.

In addition to the general philosophical sources cited below, a firsthand account of the discussions of the Vienna Circle and their relationship to the work of Wittgenstein is provided in

Ludwig Wittgenstein and the Vienna Circle: Conversations Recorded by Friedrich Waismann, B. McGuiness, ed. Oxford: Basil Blackwell, 1979.

A good, short biography of Wittgenstein is

Pears, D. *Wittgenstein.* Glasgow: William Collins and Sons, 1971.

Amusingly, given his later stance, Popper was initially attracted to Marxism in his youth and spent some time working as a laborer. He later renounced these leftist views, and has subsequently placed great emphasis upon the importance of democratic principles. A good sample of the various philosophical and social ideas of Popper can be found in the collection

A Pocket Popper, D. Miller, ed. London: Fontana, 1983.

For a personal account by Popper himself of the evolution of his views, see his autobiography:

Popper, K. *Unended Quest: An Intellectual Autobiography.* Glasgow: William Collins and Sons, 1976.

Lakatos died in 1974 at the relatively young age of fifty-two. As a result, much of his work was published posthumously. For a summary of this work and its significance, see

Essays in Memory of Imré Lakatos, R. Cohen, et al., eds. Dordrecht, Netherlands: Reidel, 1976.

Feyerabend, P. "Imré Lakatos." *British Journal for the Philosophy of Science,* 26 (1975), 1–18.

Lakatos's own statement of his idea of a Scientific Research Program is presented in his classic essays

> Lakatos, I. *The Methodology of Scientific Research Programmes.* Cambridge: Cambridge University Press, 1978.
>
> Lakatos, I. *Proofs and Refutations.* Cambridge: Cambridge University Press, 1976.

In considering the matter of public debate, it's of interest to note Feyerabend's description of his experiences as a young student attending the well-known Alpbach symposia: "I met outstanding scholars, artists, politicians and I owe my academic career to the friendly help of some of them. I also began suspecting that what counts in a public debate are not arguments but certain ways of presenting one's case. To test the suspicion I intervened in the debates defending absurd views with great assurance. I was consumed by fear—after all, I was just a student surrounded by bigshots—but having once attended an acting school I proved the case to my own satisfaction." By his own admission, Feyerabend not only comprehended a useful social truth, but also used it to lay the basis for his later intellectual eccentricities, some of which are recounted in his famous work

> Feyerabend, P. *Against Method: Outlines of an Anarchistic Theory of Knowledge.* London: New Left Press, 1975.

As an aside, the Dadaist movement promoted a somewhat sacrilegeous, irreverent attitude toward art, with nothing to be taken seriously. It is exactly this kind of attitude that Feyerabend advocates for the philosophy of science. When portrayed in this light, perhaps his ideas aren't so outlandish, after all. Unfortunately for Feyerabend and the rest of the "sociology of knowledge" theorists, it's difficult to point to even a single form of a physical relation that was determined by the social order or structure in which it was formed.

BUDDY, CAN YOU PARADIGM?

The story about Julian Bigelow, von Neumann, and "nobody's" dog, as well as much background information about Thomas Kuhn, is presented in the immensely entertaining history of the geniuses and eccentrics of the Princeton Institute for Advanced Study:

> Regis, E. *Who Got Einstein's Office?* Reading, MA: Addison-Wesley, 1987.

Ironically, Kuhn's pathbreaking work on paradigms in science appeared in *The International Encyclopedia of Unified Science,* a series of books from the University of Chicago Press that was an outgrowth of the logical positivist movement led by Rudolf Carnap when it moved to America during World War II. The precise citation is

> Kuhn, T. *The Structure of Scientific Revolutions,* 2nd Edition. Chicago: University of Chicago Press, 1970.

This edition contains a lengthy postscript by Kuhn in which he addresses many of the critical remarks leveled at the ideas in the original edition of 1962.

Those readers looking for a somewhat gentler introduction to the paradigm notion, without having to wade through the customary dense prose of historians and philosophers, should see

Briggs, J., and D. Peat. *The Looking Glass Universe.* New York: Cornerstone Library, 1984.

This little masterpiece of scientific exposition addresses not only some of the basic epistemological issues raised by Popper, Kuhn, and others, but also treats a variety of the more speculative and exciting paradigms in the scientific world today. Among the items considered are Prigogine's theory of far-from-equilibrium systems, Bohm's ideas on language and quantum theory, and Sheldrake's theory of morphogenetic fields in developmental biology. All in all, one of the best volumes available to give the general reader a glimpse into some of the edges of today's frontiers of science—and thought!

The heart of Shapere's continuing critique of Kuhn's position is found in his review of the first edition of Kuhn's book, which appeared as

Shapere, D. "The Paradigm Concept." *Science,* 172 (1971), 706–709.

PHILOSOPHICALLY SPEAKING

For an outstanding reference on the ways replication and induction are carried out in scientific practice, see

Collins, H. *Changing Order.* London: Sage Publications, 1985.

This volume is particularly important for its in-depth consideration of how the Problem of Induction is solved in a sociological, or practical, sense in everyday science. The author details the mechanics of this procedure by considering three case studies in physics, engineering, and psychology: the detection of gravitational radiation, the construction of an infrared laser, and the emotional life of plants. Throughout, the author presents a lucid, enlightening and entertaining summary of the interactions between the philosophical difficulties of induction and the practical means by which science goes about dealing with them. Highly recommended.

A TALE OF TWO SUICIDES

An account of Boltzmann's suicide set against the general social and intellectual climate of turn-of-the-century Vienna is given in the Johnston book cited earlier. Kammerer's life and tragic death are recounted with great detail and sympathy in

Koestler, A. *The Case of the Midwife Toad.* London: Hutchinson, 1971.

A less detailed account of the Kammerer case, told within the general context of fraud in science, is given in the Broad and Wade volume cited earlier.

The original form of Merton's norms can be found in his classic work

Merton, R. K. *The Sociology of Science.* Chicago: University of Chicago Press, 1973.

Other excellent accounts of the sociology of science easily accessible to the general reader are

Richards, S. *Philosophy and Sociology of Science: An Introduction,* 2nd Edition. Oxford: Blackwell, 1987.

Ziman, J. *An Introduction to Science Studies: Philosophical and Social Aspects of Science and Technology.* Cambridge: Cambridge University Press, 1984.

While the above treatments look at the practice of science from the sociological standpoint, an alternate approach is to look upon the whole enterprise from the perspective of an anthropologist. In this view we regard scientists as if they were members of some strange, hitherto unknown tribe, who spend their days practicing arcane and mystical rites. The job is to understand the structure, language, customs and so forth of this "tribe" by using the commonly accepted concepts, methods, and procedures of cultural anthropology researchers. A fascinating account of an experiment of just this type involving biological work at the famed Salk Institute is given in

Latour, B., and S. Woolgar. *Laboratory Life: The Construction of Scientific Facts,* 2nd Edition. Princeton, NJ: Princeton University Press, 1986.

A short account of the Summerlin incident is given in the Broad and Wade book noted above. For all the gory details, the interested reader should consult

Hixson, J. *The Patchwork Mouse.* New York: Doubleday, 1976.

ON THE FRINGE OR AT THE CUTTING EDGE?

Two classic treatments of monkey business masquerading as science are the volumes

Gardner, M. *Fads and Fallacies in the Name of Science.* New York: Dover, 1957.

Gardner, M. *Science: Good, Bad and Bogus.* Buffalo, NY: Prometheus Books, 1981.

However, my own tastes lean more toward the outstanding discussion given in

Radner, D., and M. Radner. *Science and Unreason.* Belmont, CA: Wadsworth, 1982,

from which the list of pseudoscience "fingerprints" in the text is taken.

THE PULPIT AND THE LAB

The story of Mrs. Fernandez and her "trial by prayer" is recounted in

Raup, D. *The Nemesis Affair.* New York: Norton, 1986.

This book gives a participant's account of one of the most heated of today's scientific controversies, the problem of what happened to the dinosaurs. However, the author uses this issue as a vehicle to discuss much more general and far-reaching questions about belief systems in science and the role they play in shaping what a particular community comes to think of as "good work." Thus, the book serves as an admirable attempt to explain the evolution of a paradigm crisis as it unfolds in real time.

On the matter of belief systems in science, Raup thinks most scientists would claim that science involves the use of experiments to test hypotheses and careful scholarship with no prior commitment to a particular answer. Also, he feels they would argue that religion is not science because it involves no experiments, tests no hypotheses, and is committed beforehand to a set of beliefs. Raup says that these scientists' claims contain a lot of bunk. In other words, the ideal of science as broadcast far and wide by the PR division of the scientific establishment, and the actual practice of science as carried on down at the lab bench, bear little if any resemblance to one another. Just as I've always suspected! This little confession by Raup calls to mind the remark by Austin when informed that Gödel had shown that there were truths of arithmetic that could not be derived from the Peano axioms. Austin remarked, "Whoever thought otherwise?"

The interplay of observations, laws, theories, and models, not only in science but also in religion, is covered nicely in

Barbour, I. *Myths, Models, and Paradigms: A Comparative Study in Science and Religion.* New York: Harper and Row, 1974.

This book is to be recommended not only for its comparative analysis of the scientific and religious enterprises, but also for much worthwhile background information about the role of myths in the process of reality generation. Of special note is Barbour's discussion of the uses of models in religion, where he offers the following competing visions of the relationship between God and man:

$$
God = \begin{cases}
\text{monarchical—king and kingdom} \\
\text{deistic—clockmaker and clock} \\
\text{dialogic—one person and another person} \\
\text{agent—agent and his actions} \\
\text{social process—individual and community}
\end{cases}
$$

Another volume covering some of the same ground, but with a slightly more biased stance, is

Hummel, C. *The Galileo Connection: Resolving Conflicts Between Science and the Bible.* Downer's Grove, IL: InterVarsity Press, 1986.

An excellent volume giving not only an overview of the quasi-religious character of much of science, but also a general audience introduction to a spectrum of questions, problems, and proposed solutions in science is

Stableford, B. *The Mysteries of Modern Science.* London: Routledge and Kegan Paul, 1977.

INTO THE COURTROOM OF BELIEFS

The quote from Bauer is taken from his book on Velikovsky cited above.

CHAPTER TWO

GENERAL REFERENCES

Since Oparin, the origin of life has been a topic of continuing fascination for scientists and the lay public alike. In recent years there have been several excellent treatments for the general reader. Two that were instrumental in shaping my own view of the field are

Scott, A. *The Creation of Life: Past, Future, Alien.* Oxford: Blackwell, 1986.

Shapiro, R. *Origins: A Skeptic's Guide to the Creation of Life on Earth.* New York: Summit, 1986.

The Scott book is notable for an excellent account of the biochemical aspects of life, as well as for a truly first-rate collection of diagrams and figures illustrating some of the trickier points in the way life works. The Shapiro volume, while not illustrated, is highly recommended as an account of the competing positions from a skeptical, but not hostile, point of view.

A slightly more technical presentation of the "facts of life" is the following textbook account aimed at university undergraduates:

Day, W. *Genesis on Planet Earth: The Search for Life's Beginnings,* 2nd Edition. New Haven, CT: Yale University Press, 1984.

OUT OF THE FIRE AND INTO THE SOUP

By today's standards the details of Oparin's program for the origin of life, if not the direction, seem hopelessly adrift. But the importance of his work for a scientifically based attack on the problem cannot be overemphasized. To illustrate the type of "scientific" view as opposed to religious dogma held prior to Oparin, one need only recount the "spontaneous generation" ideas of the Flemish chemist and physician Jan Baptista van Helmont, who gave the recipe: "Dirty undergarments encrusted in wheat; twenty-one days is the critical period. The mice that jump out are neither weanlings nor sucklings, but fully formed." While Pasteur stamped paid to this ridiculous idea in the nineteenth century, it was not until the work of Oparin that a serious scientific attack on the origins question was mounted. As an entertaining aside on the spontaneous generation hypothesis, despite Pasteur's work the theory didn't finally expire until its last bastion, the British scientist Henry C. Bastian, died. Regrettably, this seems part of the typical life cycle of discredited theories. In any case, the original work of Oparin can be found in the following English version:

Oparin, A. *Origin of Life,* S. Morgulis, trans. New York: Macmillan, 1938. New York: Dover reprint, 1965.

The independent proposal of Haldane, which gave rise to the term "primordial soup," is found in the essay

Haldane, J.B.S. "The Origin of Life," in *On Being the Right Size and Other Essays.* Oxford: Oxford University Press, 1985.

An account of Oparin's political activity during the Lysenko period is given in the Shapiro book noted earlier. See also the definitive account of the whole deplorable Lysenko affair given in

Medvedev, Z. *The Rise and Fall of T. D. Lysenko.* New York: Columbia University Press, 1969.

Miller's personal account of the circumstances surrounding his classic experiment appears in

Miller, S. "The First Laboratory Synthesis of Organic Compounds Under Primitive Conditions," in *The Heritage of Copernicus,* J. Neyman, ed., pp. 228–241. Cambridge, MA: MIT Press, 1974.

In connection with Miller's experimental parameters, it's worth taking note of the fact that the initial run of the experiment produced nothing of interest. Only when Miller interchanged the order of the spark discharge and the condensing chamber did measurable amounts of any kind of amino acids emerge.

This is a point worth pondering in regard to the problem of unacceptable investigator interference with prebiotic simulations; evidently, the problem was there from the very outset.

As an indication of the faith that Cyril Ponnamperuma places in Nature's ability to generate amino acids from simple chemicals, as a side activity he is chairman of the council of the Dambala Institute, a center devoted to exploitation of the dambala plant (a kind of winged bean) as a protein source to solve the Third World hunger problem. Ponnamperuma states, however, that this is only an interim solution, his ultimate goal being to generate proteins directly from primitive elements in the atmosphere (carbon, nitrogen, hydrogen, and so on). He thinks we could make up to 20 percent of our food that way, the principal limitation being the energy required for the synthesis. As to where he stands on the origin of life on Earth, his statement "If I can demonstrate a replicating molecule, I'll die a happy man" says it all. For a more complete account of his ideas on the food problem, as well as on prebiotic synthesis, see

"Seeds of Life: An Interview with Cyril Ponnamperuma." *Omni,* 1980.

Additional references to Ponnamperuma's work are given later under Chapter Six, devoted to the existence of extraterrestrial intelligence.

A CRASH COURSE ON HOW LIFE LIVES
Simple and entertaining general accounts of the mechanisms of life include the Scott book cited earlier, as well as

Hofstadter, D. "The Genetic Code: Arbitrary?" in *Metamagical Themas,* pp. 671–699. New York: Basic, 1985.

Rosenfield, I., E. Ziff, and B. Van Loon. *DNA for Beginners.* London: Writers and Readers Publishing, 1983.

A somewhat more technical presentation of the facts is

Rose, S. *The Chemistry of Life,* 2nd Edition. London: Penguin, 1979.

Those not convinced that a self-reproducing factory is possible are urged to read the prologue of the book

Hogan, J. P. *Code of the Lifemaker.* New York: Ballantine, 1983.

POTHOLES ON THE ROAD TO LIFE
The "junk" DNA problem has recently been attacked by computer simulation experiments run in Material Mode by Loomis and Gilpin at UC, San Diego. They speculated that much of the excess DNA is just there by chance. Using a simulation program embodying various rules for DNA replication, they found that a single gene will blossom into a genome containing many genes, some of which are members of multigene families, and all of which are embedded in a very large proportion of dispensable sequences. Hence, they conclude that: (1) most of the DNA in eukaryotic genomes does nothing at all, and (2) large quantities of dispensable sequences will accumulate in the genome before it will stabilize. An account of their work is found in

Loomis, W., and M. Gilpin. "Multigene Families and Vestigial Sequences." *Proceedings of the National Academy of Sciences USA,* 83 (1986), 2143.

A summary of these experiments is
Lewin, R. "Computer Genome Is Full of Junk DNA." *Science*, 232 (1986), 577–578.

The WEES simulator idea is discussed in great detail in the paper
Lahav, N. "The Synthesis of Primitive 'Living' Forms: Definitions, Goals, Strategies and Evolution Synthesizers." *Origins of Life*, 16 (1985–86), 129–149.

MONSTERS, HYPERCYCLES, AND NAKED GENIES
A detailed description of the background, experimental setup, and results of Spiegelman's pioneering experiment is given in
Spiegelman, S. "An *in Vitro* Analysis of a Replicating Molecule." *American Scientist*, 55 (1967), 3–68.

The complementary experiment by Eigen is presented in
Eigen, M., W. Gardiner, P. Schuster, and R. Winkler-Oswatitch. "The Origin of Genetic Information." *Scientific American*, 244 (1981), 88–118.

A fairly complete description of Orgel's ideas about creating template-directed RNA without benefit of enzymes is presented in
Orgel, L. "The Origin of Life and the Evolution of Macromolecules." *Folia Biologica*, 29 (1983), 65–77.

The Gilbert scenario for the origin of life out of self-catalytic RNA is outlined in
Gilbert, W. "The RNA World." *Nature*, 319 (1986), 618.

On the puzzle of junk DNA, Gilbert's view is that it arises as the relic of the old intron-exon structure left imprinted on the DNA from the RNA molecules that originally encoded proteins.

A complete expository and technical account of much of the "hypercycle" theory underlying the ideas of Eigen is found in
Eigen, M., and P. Schuster. *The Hypercycle: A Principle of Natural Self-Organization.* Berlin: Springer, 1979.

See also the 1981 *Scientific American* article cited earlier.
The computer experiments of Niessert are reported in
Niessert, U. "How Many Genes to Start With? A Computer Simulation About the Origin of Life." *Origins of Life*, 17 (1987), 155–169.
Niessert, U., D. Harnasch, and C. Bresch. "Origin of Life Between Scylla and Charybdis." *Journal of Molecular Evolution*, 17 (1981), 348–353.

A spectrum of theories beyond those discussed in the text have also been offered to explain why nucleotides came first. Perhaps the most intriguing is the hydrated electron theory of John Scott, who argues that in a primordial atmosphere short on ozone, the ultraviolet radiation would strip electrons away from water molecules. Such electrons would immediately be surrounded by four additional water molecules, forming what is termed a *hydrated electron*. Before being absorbed into another water molecule, such a hydrated electron could do a lot of destructive damage to chemical compounds nearby, especially those having a net positive charge to which the electron would be attracted. The essence of Scott's idea is that the net negative charge of most nucleotides would offer a protective barrier that would give them a preferential survival rate in such an environment, their positively charged competitors being wiped out by the energetic hydrated electrons. Scott presents the details of this case for a general audience in

Scott, J. "Selection in the Soup." *The Sciences,* Nov.–Dec. 1983, pp. 36–42.

For additional details on the role of the hydrated electron, see also the Scott book noted under General References.

THE CHICKEN'S STORY

For an introductory discussion of Oparin's coacervates and Fox's proteinoid ideas, as well as some personal views expressed by Fox about his critics, see the Shapiro book cited under General References. Additional material on the proteinoids can be found in

Fox, S. *The Emergence of Life.* New York: Basic, 1988.

Fox, S. "New Missing Links." *The Sciences,* January 1980, pp. 18–21.

Fox, S. "The Proteinoid Theory of the Origin of Life and Competing Ideas." *American Biology Teacher,* 36 (1974), 161–172.

Recent studies indicate that the problems of conducting useful chemical syntheses in the high-temperature environment of the hydrothermal vents on the ocean's floor seem to be insurmountable. For a discussion of the reasons why, see

Miller, S. L., and J. L. Bada. "Submarine Hot Springs and the Origin of Life." *Nature,* 334 (1988), 609–611.

LIFE: A TWICE-TOLD TALE

It has been argued that the transition from prokaryotic cells to eukaryotic was the biggest single advance in the whole course of evolution. The current theory is that it occurred by prokaryotes gobbling up bacteria having useful properties; so useful, in fact, that the prokaryotes decided not to let them go. Lynn Margulis has offered virtually incontrovertible evidence that not only did the mitochondria arise in this fashion, but also the cellular flagellum and the centriole (the device that separates the chromosomes at cell division). Her view that the hosts and their invaders evolved into a mutually beneficial symbiotic relationship leading to the eukaryotic cells is detailed in

Margulis, L. *Origin of Eukaryotic Cells.* New Haven, CT: Yale University Press, 1970.

Margulis, L. *Symbiosis in Cell Evolution.* San Francisco: Freeman, 1981.

Shapiro's proteins-first scheme is given in greater detail in his book noted earlier. It's interesting to note Leslie Orgel's response to Shapiro's idea. Orgel commented that he wasn't too enthusiastic about speculations that didn't carry some good experimental evidence along with them—the typical response of experimentalists everywhere to the unbridled enthusiasms of theoreticians. Shapiro concedes the point, however, and then goes on to suggest several lines of experimental attack on the question of whether proteins could, in principle, come first.

The second genetic code discussed in the text is written into the structure of the enzymes that couple the transfer RNA with its corresponding amino acid. These enzymes, or *synthetases,* are the real translators between the language of the proteins and the language of the nucleotides. Recent work suggests that this code may be much older and more deterministic than the classic genetic code considered in the text, and it may be less redundant as well. The original results indicating the possible presence of such a second code are in

Hou, Y.-M., and P. Schimmel. "A Simple Structural Feature Is a Major Determinant of the Identity of a Transfer RNA." *Nature,* 333 (1988), 140–145.

For a less technical summary of the work and its possible implications, see

de Duve, C. "The Second Genetic Code." *Nature,* 333 (1988), 117.

"DNA Loses Its Monopoly on Genetic Code." *New Scientist,* May 19, 1988, p. 34.

The short version of Dyson's "toy model" for the emergence of a system of metabolizers is presented in his book

Dyson, F. *Origins of Life.* Cambridge: Cambridge University Press, 1985.

If your interests in the origins question are of the one-nighter variety, this little masterpiece is the book for you. It offers, in my view, the best possible introduction to the overall origin-of-life problem, in many ways serving the same purpose as Schrödinger's classic *What Is Life?* in presenting a modern physicist's view of life. The main difference is that Schrödinger centered his attention upon the process of replication, while Dyson focuses on metabolism. It's interesting to note that Schrödinger's volume served to direct attention to problems that soon led to the major breakthroughs underpinning much of modern molecular biology. Perhaps the experimental gaps noted by Dyson will act in the same manner to stimulate a renaissance in the area of cellular metabolism rather than replication. For a more technical account of Dyson's ideas, see

Dyson, F. "A Model for the Origin of Life." *Journal of Molecular Evolution,* 18 (1982), 344–350.

ASHES TO ASHES, LIFE FROM DUST

The initial suggestion that life might have been based upon silicon in the form of clays rather than carbon appears to have come from J. D. Bernal, although he gave them only the secondary role of helping to gather the chemicals needed to synthesize carbon-based proteins and/or nucleotides.

For an introductory and highly entertaining presentation of Cairns-Smith's "seven clues to the origin of life," see his scientific detective story:

Cairns-Smith, A. G. *Seven Clues to the Origin of Life.* Cambridge: Cambridge University Press, 1985.

The technical arguments supporting the conclusions drawn in the above volume are given in

Cairns-Smith, A. G. *Genetic Takeover.* Cambridge: Cambridge University Press, 1982.

IT CAME FROM OUTER SPACE

It seems that scientists are never happy about the way they're portrayed by other scientists, especially in books for a general readership. By all accounts Crick wasn't too pleased with Watson's description of him in *The Double Helix,* ostensibly because he didn't like the idea of personal publicity. As he tells it, however, he later changed his mind about the book (including dropping his idea of a libel suit) because he thought it did a better job than he'd anticipated in showing the general reader how a certain type of scientific research is done. Later, *both* Watson and Crick came under fire from Erwin Chargaff of Columbia, one of the pioneers of molecular biology, who dismissed them with the withering remark, "In our day that such pygmies throw such giant shadows only shows how late in the day it has become." There's just nothing like public visibility and a Nobel Prize to incite your colleagues' ire, especially if you're young, brash, and seemingly lucky. But as one of my teachers once said, "I'd rather be lucky than good."

In explaining the directed panspermia theory, Crick claims that the benevolent aliens would probably send yeast or bacteria as the initial seeds of life, since these organisms can survive very harsh environments and can live without oxygen. The book was later criticized by the paleontologist Nils Eldredge (of "punctuated evolution" fame) as being an attack on religion. Crick later argued that he had nothing against religion, only against beliefs that he feels don't correspond to the facts, e.g., antiscientific views, dogmatic fundamentalist views, irrational views. To see for yourself what he had in mind, "the source" is

Crick, F. *Life Itself.* New York: Simon and Schuster, 1981.

The popular books outlining the wild visions of Hoyle and Wickramasinghe are

Hoyle, F., and N. C. Wickramasinghe. *Diseases from Space.* New York: Harper and Row, 1979.

Hoyle, F., and N. C. Wickramasinghe. *Lifecloud.* New York: Harper and Row, 1978.

In all fairness to the H&W comet theory, there is some real scientific evidence showing that the fundamental idea (*not* that proposed by H&W) may be sound. For a discussion of what needs to be done to settle the matter, see

Bada, J., M. Zhao, and N. Lee. "Did Extraterrestrial Impactors Supply the Organics Necessary for the Origin of Terrestrial Life? Amino Acid Evidence in Cretaceous-Tertiary Boundary Sediment." *Origins of Life,* 16 (1985–86), 185.

Hobbs, R., and J. Hollis. "Probing the Presently Tenuous Links Between Comets and the Origin of Life." *Origins of Life,* 12 (1982), 125–132.

A very readable and fairly devastating critique of the technical basis of the H&W Version I theory is given in the Shapiro book noted above. Version II has not been the object of any kind of scientific critique for obvious reasons.

AND GOD CREATED... FROM FISH TO GISH

Like many popular legends, often perpetuated by the filmmakers, the story of the Scopes Trial as portrayed both in the play and in the movie *Inherit the*

Wind is at considerable variance with what actually happened. Contrary to popular belief, Scopes was not a persecuted biology teacher, but rather a physical education instructor who was substituting for the regular instructor on the day in question. More important, Scopes was an enthusiastic participant in the incident, which had been cooked up by local power brokers as a means of getting the town of Dayton on the map, as well as to test the constitutionality of the law in the courtroom. For a fuller account of the real facts surrounding this bit of Americana, see

Gould, S. J. "A Visit to Dayton," in *Hen's Teeth and Horse's Toes,* Chap. 20. New York: Norton, 1983.

The creationist controversy has been extensively treated in so many places that it's impossible to do anything other than offer a brief sampler here. For the position of the creationists themselves, a basic source is

Morris, H. *Scientific Creationism.* San Diego: CLP Publishers, 1974.

Scientific arguments against the idea of creationism are presented in the following works, of which the next-to-last is especially recommended for its complete text of Judge Overton's opinion in the Arkansas case:

But Is It Science?, M. Ruse, ed. Buffalo, NY: Prometheus, 1988.

Gurin, J. "The Creationist Revival." *The Sciences,* April 1981, pp. 16–19.

Jukes, T. "Quackery in the Classroom: The Aspirations of the Creationists." *Journal of Social and Biological Structures,* 7 (1984), 193–205.

Kitcher, P. *Abusing Science.* Cambridge, MA: MIT Press, 1982.

Science and Creationism, A. Montagu, ed. Oxford: Oxford University Press, 1984.

Scientists Confront Creationism, L. Godfrey, ed. New York: Norton, 1983.

While the creationist position is clearly nonscientific, the scientists are not beyond reproach in this matter either. Several of the articles in the foregoing compendiums make arguments not so much for science as against the creationists, on a variety of social, psychological, and political grounds. For example, an article by A. Kehoe in the Godfrey collection begins by giving a nice overview of the history of the creationists creed, as well as the strategies they have employed to try to get their ideas institutionalized in the school system. The article then departs entirely from any sort of "scientific" critique and becomes an emotional plea for anyone who values the principles upon which the United States is based to resist the claims of the creationists, since it's just plain un-American for any group to have its personal doctrinal beliefs legislated. This article makes it evident that what is involved here is not a scientific controversy, but rather a political one. Of course, this has been pretty obvious almost from the moment the Dayton sheriff put the cuffs on John Scopes, but it's depressing to see that the controversy hasn't really progressed beyond this level, even in the so-called scientific literature. To my mind, this isn't a very compelling example of "scientists" confronting creationism. The Ruse volume is notable for its emphasis on the philosophical, as opposed to political, aspects of the debate, as well as the editor's firsthand account of the Arkansas trial as a participant.

A particularly detailed discussion of the many problems with the classic Primordial Soup Theory is given in

Thaxton, C., W. Bradley, and R. Olsen. *The Mystery of Life's Origins.* New York: Philosophical Library, 1984.

In addition to geological, thermodynamic, and chemical evidence against most of the soup theories, this book also presents an excellent account of the difference between operations science and origins science. Interestingly, the authors ultimately end up supporting an off-Earth position on the origins question, but at least their arguments are cogent and well presented, if somewhat biased against the conventional wisdom.

THE LOGIC OF LIFE

For more information on von Neumann's ideas about self-reproducing machines, see

Essays on Cellular Automata, A. Burks, ed. Urbana, IL: University of Illinois Press, 1970.

von Neumann, J. "The General and Logical Theory of Automata," in *John von Neumann—Collected Works,* Vol. 5, pp. 288–328. New York: Macmillan, 1961–63.

Interestingly, the idea of a machine that could make copies of itself and be "harvested," much as plants are today, was considered not long after von Neumann's original work. A popular account of the economic possibilities is given in

Moore, E. F. "Artificial Living Plants." *Scientific American,* 195 (1956), 118–126.

Moore concludes that such "plants" would have an enormous advantage if we could solve the design problems, since then we could free agriculture from its dependence upon the natural characteristics of plants and produce any crop instead of just those that Nature happens to supply. He ends by noting that he thinks creation of such an artificial plant would be more easily attainable than human flight to another planet!

There is by now an extensive literature on the Life game detailing the enormous complexity that can emerge from the very simple rules defining what happens at each cell. A good introductory presentation, complete with computer programs, is

Poundstone, W. *The Recursive Universe.* New York: Morrow, 1985.

For those interested the details of Conway's proof of a self-reproducing Life pattern, perhaps the most accessible account is that in

Berlekamp, E., J. Conway, and R. Guy. *Winning Ways for Your Mathematical Plays.* Volume II. London: Academic Press, 1982.

A natural extension of Conway's version of Life is to consider playing it in three dimensions. So instead of the infinite checkerboard, we use an infinite "egg crate" in which the cells are cubes instead of squares. In many ways this is a more appropriate version of the game for studying real life, which unfolds in our three-dimensional space rather than in Conway's planar world. Interestingly enough, this idea had been suggested as early as 1976 by science fiction writer Piers Anthony in his book *Ox*. Recently, Carter Bays of the University of South Carolina has explored a wide variety of such three-dimensional versions of Life. A good introductory account of the difficulties and possibilities, together with further information, is found in

Dewdney, A. K. *The Armchair Universe*, pp. 149–159. New York: Freeman, 1988.

Life is by no means the simplest or the most complex cellular automaton imaginable. In fact, studies of the far simpler one-dimensional automata whose "universe" consists only of cells on an infinite *line*, rather than a plane, have shown equally complex behavior. A good summary of what can happen is given in the collection.

Wolfram, S. *Theory and Applications of Cellular Automata* Singapore: World Scientific, 1986.

The paper by Langton outlining how cellular automata could be used to represent the functional activities of living entities is

Langton, C. "Studying Artificial Life with Cellular Automata." *Physica D*, 22D (1986), 120–149.

Further information on the whole circle of artificial life questions, as well as an account of some fascinating experiments, can be found in

Dawkins, R. *The Blind Watchmaker*. Essex, UK: Longman, 1986.
Artificial Life, C. Langton, ed. Reading, MA: Addison-Wesley, 1988.

The following articles represent a selection of material outlining the nature of computer viruses, as well as some of the difficulties they can cause and what might be done to create "antiviral" remedies:

Denning, P. "Computer Viruses." *American Scientist*, 76 (1988), 236–238.
Dewdney, A. K. "A Core War Bestiary of Viruses, Worms and Other Threats to Computer Memories." *Scientific American*, 252 (1985), 14–23.
Reid, B. "Reflections on Some Recent Widespread Computer Break-ins." *Communications of the Association for Computing Machinery*, 30 (1987), 103–105.
Witten, I. "Computer (In)security: Infiltrating Open Systems." *Abacus*, 4 (1987), 7–25.

CHAPTER THREE

GENERAL REFERENCES

The bible of sociobiology, whose publication sparked off the furor over how we act, is

Wilson, E. O. *Sociobiology: The New Synthesis*. Cambridge, MA: Harvard University Press, 1975.

A good textbook discussion of the principles of sociobiology by one of its foremost advocates is

Barash, D. *Sociobiology and Behavior*. New York: Elsevier, 1977.

Three books that are must reading for anyone seriously interested in pursuing the many and varied threads composing the sociobiology "debate" are

Kitcher, P. *Vaulting Ambition*. Cambridge, MA: MIT Press, 1985.

Ruse, M. *Sociobiology: Sense or Nonsense?* Dordrecht, Netherlands: Reidel, 1979.

The Sociobiology Debate, A. Caplan, ed. New York: Harper and Row, 1978.

The Caplan volume is a compendium of most of the important papers by the warring factions in the debate, including Hamilton's original articles on inclusive fitness and kin selection, the Boston Group's notorious letter to *The New York Review of Books*, and Wilson's extended response in *BioScience*, as well as much, much more. These papers are indispensable reading if you want a clear view of what created the debate and why it has taken the form that it has. The book by Michael Ruse is an excellent nonpartisan account by a philosopher of science assessing the pros and cons of the debate circa 1977. For reasons that are hard to fathom, Ruse has been labeled a sociobiologist by later commentators, especially those of the "anti" camp, probably on the grounds that "if you're not with us, then you're against us." In any case, I find his account to be a quite impartial, illuminating, thoughtful, and well-written discussion of all sides of the issue, both scientific and philosophical. Finally, there is the book by Kitcher. This is another attempt by a philosopher of science to deal with the whole sociobiology business from the perspective of ten years afterward. Some reviews have labeled the book a definitive treatment of the topic, one that will sound the death knell for sociobiology and close out the debate once and for all. Reading these sorts of kudos, I wanted to be enthusiastic when I first picked up the book, but my expectation of reading an objective assessment of the facts and theories was dealt a blow when I looked at the publisher's dust-jacket blurb and found glowing testimonials by none other than Richard Lewontin and Stephen Jay Gould—hardly uninvolved or detached observers of the sociobiological scene. After digesting the material, my view is that a death blow to sociobiology this book is not. Frankly I think Kitcher, unlike Ruse, has failed to maintain an appropriate arm's-length distance from his topic—a dangerous oversight in a philosophical assessment. Nonetheless, if you can overlook the author's somewhat pompous literary style, there's a lot of valuable material here and a number of arguments that must be given serious consideration.

NATURE/NURTURE: SENSE OR NONSENSE?

An excellent description of Milgram's experimental setup and results is found in the book

Koestler, A. *Janus*. New York: Random House, 1978.

Koestler notes the important variation of the experiment in which Milgram allowed the subjects to inflict any level of shock they wished as a punishment for a wrong answer, rather than being compelled to use a level determined by the Leader. In this case, 38 out of the 40 subjects refused to go beyond a level of 150 volts, the level at which the pupil made his first loud cry, with the average shock administered a measly 54 volts. Milgram's own account of these experiments can be found in his 1974 book *Obedience to Authority*.

NEO-NEO-DARWINISM AND SOCIOBIOLOGY
Library shelves groan under the weight of books expounding the Darwinian and neo-Darwinian theories of evolution, so let me content myself here with the following short, well-written and easily accessible sources:

Arthur, W. *Theories of Life.* London: Penguin, 1987.

Ayala, F. "The Mechanisms of Evolution." *Scientific American,* 239 (September 1978), 56–69.

Smith, J. Maynard. *Problems of Biology.* Oxford: Oxford University Press, 1986.

The emergence of sociobiology as an interdisciplinary amalgam of ethology, population ecology and evolutionary genetics is traced in

Barlow, G. "The Development of Sociobiology: A Biologist's Perspective," in *Sociobiology: Beyond Nature/Nurture?,* G. Barlow and J. Silverberg, eds., pp. 3–24. Boulder, CO: Westview Press, 1980.

The distinction between the Central Dogma of Molecular Biology and what I've termed here the Central Dogma of Social and Behavioral Biology can be made more explicit by the following diagram:

GENETIC INHERITANCE

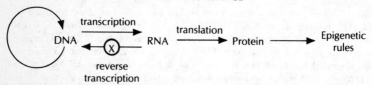

CULTURAL INHERITANCE

The dogmas of genetic and cultural inheritance

Here the prohibition against information flow from the proteins to the genotype is indicated by the x in reverse transcription for genetic inheritance. On the other hand, while DNA can replicate itself, its cultural equivalent, the semantic network, cannot. This diagram also shows the connection between the epigenetic rules of Wilson and Lumsden and the processes of genetic and cultural inheritance.

ANIMAL ANTICS
A thorough discussion of game theory in the evolutionary context by the master himself is given in

Smith, J. Maynard. *Evolution and the Theory of Games.* Cambridge: Cambridge University Press, 1982.

An introductory, textbook-level account is found in Chapter Six of
  Casti, J. *Alternate Realities: Mathematical Models of Nature and Man.* New York: Wiley, 1989.

Full details of the experiments by Riechert on ESS for grassland spiders are reported in
  Riechert, S. "Spider Fights as a Test of Evolutionary Game Theory." *American Scientist,* 74 (1986), 604–610.

The discussion of the parental investment question, together with a much more detailed (and entertaining) treatment of the "arms race" between males and females, is given in
  Dawkins, R. *The Selfish Gene.* Oxford: Oxford University Press, 1976.

A nice account of the sex determination procedure in Hymenoptera, as well as an easily readable discussion of associated matters like altruism and inclusive fitness is found in
  Smith, J. Maynard. "The Evolution of Behavior." *Scientific American,* 239 (September 1978), 176–192.

For the source papers in which Hamilton introduced the notion of inclusive fitness, see the Caplan book noted earlier.

For a discussion of why there might be no more to be learned about people from observing animals than by reading Aesop's fables, see the article
  Simon, M. "Sociobiology: The Aesop's Fables of Science." *The Sciences,* 18 (1978), 18–21.

THE STRANGE CASE OF ALTRUISM
The original paper by Trivers that outlined the case for reciprocal altruism is
  Trivers, R. "The Evolution of Reciprocal Altruism." *Quarterly Review of Biology,* 46 (1971), 35–39, 45–47

THE GENETIC IMPERATIVE
Like Darwin, who devoted only a few words in his epic works to the special problems of human evolution, in *Sociobiology* Wilson addresses the matter of human sociobiology only in the book's last chapter, and in a purely speculative mode. However, also like Darwin, Wilson had clearly been thinking long and hard about the implications of his work for *Homo sapiens,* as shown in his full-length treatment of the matter in
  Wilson, E. O. *On Human Nature.* Cambridge, MA: Harvard University Press, 1978.

For Wilson's personal statement about many of the ideas put forth in this work, see the interview
  "Genetic Destiny," *Omni,* 1978.

The coevolutionary circuit of Lumsden and Wilson is completely described in all its painstaking mathematical and sociobiological detail in
  Lumsden, C., and E. O. Wilson. *Genes, Mind, and Culture.* Cambridge, MA: Harvard University Press, 1981.

See also the updated treatment given in the paper
  Lumsden, C., and E. O. Wilson. "The Relation Between Biological and
  Cultural Evolution." *Journal of Social and Biological Structures,* 8 (1985),
  343–359.

Stephen Jay Gould has eloquently put forward a case for the flexibility of
the human brain as the principal reason why it's not necessary to invoke a
genetic explanation for human behavior. A popular account of his argument is
found in
  Gould, S. J. "Biological Potentiality vs. Biological Determinism," in *Ever
  Since Darwin,* pp. 251–259. New York: Norton, 1977.

GETTING INTO HER GENES

As a good example of the kind of direct attacks in the literature claiming that
sociobiology is sexist, see
  Alpher, J., J. Beckwith, and L. Miller. "Sociobiology Is a Political Issue,"
  in *The Sociobiology Debate,* cited above.

Excellent popular accounts expounding Wilson's views of the biological ori-
gin of religion and morals are
  Masters, R. "Sociobiology: Science or Myth?" *Journal of Social and Biological
  Structures,* 2 (1979), 245–252.
  Wilson, E. O. "Human Decency Is Animal." *New York Times Magazine,* Oc-
  tober 12, 1975.

Somewhat more detailed discussions are given in the books
  Flanagan, O. *The Science of the Mind.* Cambridge, MA: MIT Press, 1984.
  Schwartz, B. *The Battle for Human Nature.* New York: Norton, 1986.
  von Schilcher, F., and N. Tennant. *Philosophy, Evolution and Human Nature.*
  London: Routledge and Kegan Paul, 1984.

Each of these books is quite remarkable in its own way, giving a critical picture of sociobiology from a particular vantage point. Flanagan speaks as a
philosopher of science, emphasizing the critical arguments against Wilson's
claims for the origin of morality and normative principles out of biological
necessity. Schwartz focuses his attention on the doctrine of self-interest as it
arises in Adam Smith's economics, the evolutionary biology of Darwin, and
Skinner's behavioristic views in psychology. Within this setting, sociobiology
is treated primarily as an attempt to show that the behavior of animals (including humans) serves reproductive fitness, by applying the notion of economic self-interest to social behavior. Finally, the von Schilcher and Tennant
book is a critical analysis of modern evolutionary theory, assessing its philosophical consequences in relation to morality, knowledge, consciousness, and
language with special attention to problems of cultural evolution. Taken together, these volumes provide extremely good coverage of sociobiology as seen
from the philosopher's point of view.

CANT VS. KANT

For an easily digestible recounting of the story of social Darwinism in America, see
  Morris, R. *Evolution and Human Nature.* New York: Putnam, 1983.

A good journalistic report from the sociobiology front written at the time the accusations were flying hot and heavy is

Wade, N. "Sociobiology: Troubled Birth for New Discipline." *Science,* 191 (March 19, 1976), 1151–1155.

Apparently, the first that Wilson knew of the Boston Group's attack was when he received a phone call from science journalist Boyce Rensberger asking for his reaction. Wilson, of course, was dumbfounded at the fact that *The New York Times* had a copy of an attack that had been prepared by colleagues whom he regarded as friends and who, moreover, occupied offices within a few hundred meters of his own. For an excellent discussion of the circumstances surrounding the infamous *New York Review of Books* letter, as well as a compact summary of the claims and counterclaims, see

Currier, R. "Sociobiology: The New Heresy." *Human Behavior,* Nov. 1976, 16–22.

Another good source for what the debate between Wilson and his colleagues is all about is

Ruse, M. "Sociobiology: Sound Science or Muddled Metaphysics?" *Proceedings of the 1976 Philosophy of Science Association Meeting,* F. Suppe and P. Asquith, eds., pp. 48–73. East Lansing, MI: Philosophy of Science Association, 1977,

The original letter of the Boston Group to *The New York Review of Books* is reprinted in the Caplan compendium noted earlier. A more extensive version of this critique can be found in

Allen, E., et al. "Sociobiology: Another Biological Determinism." *BioScience,* 26 (1976), 182–186.

Wilson's responses to the two attacks are reported in

Wilson, E. O. "Academic Vigilantism and the Political Significance of Sociobiology." *BioScience,* 26 (1976), 183–190.

Wilson, E. O. "Letter to the Editor," *New York Review of Books,* 22 (1975), No. 20, 60–61.

On the other side of the Atlantic, Lewontin's British comrades in arms were also not hesitant to chip in with their own two cents' (or pence) worth of criticism of Wilson, as well as of their countryman Richard Dawkins. A couple of representative samples are

Midgley, M. "Gene-Juggling." *Philosophy,* 54 (October 1979).

Rose, S. "Pre-Copernican Sociobiology?" *New Scientist,* October 5, 1978, 45–46.

In his inimitable style, Dawkins replies to an earlier claim of Rose's that his work fosters racism and neo-Nazi ideals in

Dawkins, R. "Selfish Genes in Race or Politics." *Nature,* 289 (1981), 528.

See also his extended response to the hostility of Midgley's claims in

Dawkins, R. "In Defence of Selfish Genes." *Philosophy,* 56 (1981), 556–573.

For a discussion of the political views of Lewontin and their intertwining with his biological work, see

Lumsden, C., and E. O. Wilson. "Genes, Mind, and Ideology." *The Sciences,* 21 (November 1981), 6–8.

Lewontin's remark about taking his job as a political activity can be found in the *Chronicle of Higher Education,* October 23, 1973.

A full-scale attack on sociobiology from a biological as well as political standpoint is contained in
    Biology as a Social Weapon, Science for the People Collective, eds. Minneapolis, MN: Burgess, 1977.
    Lewontin, R., S. Rose, and L. Kamin. *Not in Our Genes.* New York: Pantheon, 1984.

Interestingly enough, prior to the offensive launched by the Boston Group, the professional reviews of Wilson's book *Sociobiology* had been quite favorable. Good examples are found in the following collection:
    "Multiple Reviews of Wilson's *Sociobiology.*" *Animal Behavior,* 24 (1976), 698–718.

In fact, of the fourteen reviewers contributing to the above collection, only one definitely comes down on the negative side of the ledger. While speaking of reviews, of special interest is the review by Elliott White of Kitcher's inflammatory book cited under General References above. In this discussion, White argues convincingly against the egalitarian basis of many of the Boston Group's most bitter complaints against Wilson. For the full review, see
    White, E. "Review of Kitcher, P., *Vaulting Ambition: Sociobiology and the Quest for Human Nature.*" *Journal of Social and Biological Structures,* 11 (1988), 283–286.

SO-SO BIOLOGY

Sahlins's criticism of the practical aspects of kin selection are contained in his scathing critique of sociobiology:
    Sahlins, M. *The Use and Abuse of Biology: An Anthropological Critique of Sociobiology.* Ann Arbor, MI: University of Michigan Press, 1976.

Another critique of the idea of biological determinism worth noting is
    Thompson, J. "Human Nature and Social Explanation," in *Against Biological Determinism,* S. Rose, ed., pp. 30–49. London: Allison and Busby, 1982.

Dawkins's idea of a hereditary agent playing the role in culture that genes do in biology has been put forward a number of times. In addition to the meme presented in Dawkins's book *The Selfish Gene,* and the culturgen of Lumsden and Wilson, the same concept has been discussed under the rubric of a *sociogene* in
    Swanson, C. *Ever-Expanding Horizons.* Amherst, MA: University Massachusetts Press, 1983.

The idea of using biological evolutionary concepts to try to model the process of cultural change has also been pursued by many authors, some with a vengeance. Two relatively recent treatments showing the kind of mathematical level to which the idea has risen are

Boyd, R., and P. Richerson. *Culture and the Evolutionary Process.* Chicago: University of Chicago Press, 1985.

Cavalli-Sforza, L., and M. Feldman. *Cultural Transmission and Evolution: A Quantitative Approach.* Princeton, NJ: Princeton University Press, 1981.

CONFLICTING RATIONALITIES AND THE DILEMMA OF COOPERATION

By far the best non-mathematical, introductory account I know of treating game theory for the social scientist is

Colman, A. *Game Theory and Experimental Games.* Oxford: Pergamon Press, 1982.

This book is filled with interesting examples of different types of games modeling every kind of human strategic interaction from arms races to confrontations over moral philosophy. But if you want the mathematics behind the results discussed in Colman, you'll have to go elsewhere. One good place is

Jones, A. J. *Game Theory: Mathematical Models of Conflict.* Chichester, UK: Ellis Horwood, 1980.

The Prisoner's Dilemma has by now been the subject of well over one thousand research articles and numerous book-length accounts. Still one of the best is

Rapoport, A., and A. Chammah. *Prisoner's Dilemma: A Study in Conflict and Cooperation.* Ann Arbor, MI: University of Michigan Press, 1965.

The fascinating computer tournaments of Axelrod are described in

Axelrod, R. *The Evolution of Cooperation.* New York: Basic, 1984.

See also the easily accessible popular discussion in

Hofstadter, D. "Computer Tournaments of the Prisoner's Dilemma," in *Metamagical Themas,* pp. 715–734. New York: Basic, 1985.

In a related work, Peter Corning argues for the idea of egoistic cooperation as a theory of progressive evolution. Corning notes that in a world of 2 million living species, only about ten thousand can be said to be eusocial. He asks how such islands of cooperation can emerge in a sea of conflict. For his answer see

Corning, P. *The Synergism Hypothesis.* New York: McGraw-Hill, 1983.

An introductory account of Axelrod's more recent work on the Norms Game is found in

Axelrod, R. "Laws of Life." *The Sciences,* 27 (1987), No. 2, 44–51.

BRINGING IN THE VERDICT

During the course of reviewing Melvin Konner's book *The Tangled Wing,* an extended meditation on the biology of human emotions, the noted science journalist Horace Freeland Judson reviews many of the attacks on sociobiology, concluding that it has by no means lost the war. His arguments are given in

Judson, H. F. "An Imperial Presence." *The Sciences,* 23 (1983), 20–23.

CHAPTER FOUR

GENERAL REFERENCES

An encyclopedic source (literally) for information about all aspects of language is

> *The Cambridge Encyclopedia of Language,* D. Crystal, ed. Cambridge: Cambridge University Press, 1987.

An exploration of the thesis that language is really the interplay between systems of grammar and human behavior is carried out in the following very readable, almost popular, volume:

> Farb, P. *Word Play: What Happens When People Talk.* New York: Knopf, 1974.

The interconnections of information theory, languages, and codes like DNA are presented in a form suitable for popular consumption in

> Campbell, J. *Grammatical Man.* New York: Simon and Schuster, 1982.

A standard textbook account of language in its many manifestations is

> Fromkin, V., and R. Rodman. *An Introduction to Language,* 3rd Edition. New York: Holt, Rinehart and Winston, 1983.

For a Trivial Pursuit–type miscellany of fascinating facts about the peculiarities of the world's languages, such as the fact that German was almost adopted as the official language of the United States, or that the complete form of the Spanish insult *itú madre!* consists of five syllables that are often just whistled or beeped out on the horn of a car, see

> Berlitz, C. *Native Tongues.* New York: Grosset and Dunlap, 1982.

A detailed account of all the major schools of linguistic thought from de Saussure to the modern London school is provided in

> Sampson, G. *Schools of Linguistics.* Stanford, CA: Stanford University Press, 1980.

An ever-increasing amount of evidence is coming to light suggesting that human language origins are biologically based in evolutionary changes in our vocal mechanisms, along with corresponding changes in neural control circuits in the brain. One of the prime exponents of this view is Philip Lieberman of Brown University, who gives a nontechnical introduction to his ideas in

> Lieberman, P. "Voice in the Wilderness." *The Sciences,* 28, No. 4 (1988), 23–29.

For more technical accounts of the same type, but applied to the even broader issues of general intelligence, see

> *Intelligence and Evolutionary Biology,* H. and I. Jerison, eds. Berlin: Springer, 1988.

DUMB DOGS AND CLEVER HANS

Interspecies communication seems to hold a continuing fascination for humans of all ages, an instinctual urge that can be seen by our predilection for keeping house pets. Several works detailing the current state of this ongoing effort to talk with the animals are

Animal Intelligence: Insights into the Animal Mind, R. Hoage and L. Goldman, eds. Washington, D.C.: Smithsonian Press, 1986.

The Clever Hans Phenomenon: Communication with Horses, Whales, Apes, and People, T. Sebeok and R. Rosenthal, eds., Annals of the New York Academy of Sciences, Vol. 364. New York: New York Academy of Sciences, 1981.

Crail, T. *Apetalk & Whalespeak: The Quest for Interspecies Communication.* Chicago: Contemporary Books, 1983.

Griffin, D. *Animal Thinking.* Cambridge, MA: Harvard University Press, 1984.

Griffin, D. *The Question of Animal Awareness,* Revised Edition. Los Altos, CA: Kaufman, 1981.

Wade, N. "Does Man Alone Have Language? Apes Reply in Riddles, and a Horse Says Neigh." *Science,* 208 (June 20, 1980), 1349–1351.

The Hoage and Goldman volume contains the papers presented at a 1983 symposium addressing the issues of animal cognition. It represents an excellent survey of the entire field by the practitioners themselves. The *Clever Hans* book and the Wade article zero in not only on the question of animal communication, but also on the equally important problem of investigator deception. How can we really separate the effects of real animal communication from the kinds of cues given by their masters, wittingly or not? Crail's book is a popular introduction to the entire program of research on animal communication, ranging from the Gardners' work with chimpanzees to Lilly's efforts to communicate with the dolphins. The two books by Griffin discuss his lifelong efforts to try to understand the cognitive processes of animals and the question of whether or not it makes sense to speak of animal consciousness. Taken together, these items cover just about everything that an interested reader would need to know to get to the forefront of current research on this eternally tantalizing topic.

VERBAL BOTANY AND UNIVERSAL GRAMMAR

A quick overview of the development of linguistics as a science is given for the general reader in

Gardner, H. *The Mind's New Science.* New York: Basic, 1985.

This volume also serves as the best possible nontechnical introduction to the entire area now covered by the umbrella term *cognitive science.* For a more detailed look at linguistics per se, see the Sampson book cited earlier.

A good discussion of the entire problem of language acquisition, albeit from a decidedly Chomskian point of view, can be found in

Lightfoot, D. *The Language Lottery.* Cambridge, MA: MIT Press, 1982.

According to linguistic folklore, Chomsky's original manuscript, "The Logical Principles of Linguistic Theory," was prepared during his tenure as a junior fellow at Harvard. The MIT Press declined to issue the work and, as the story goes, a representative of the Dutch house Mouton picked up vibrations

about the book from one of its representatives who was curious about the excerpted version that Chomsky was then using for his classes at MIT. The rest
is history. The actual citation for this pathbreaking work is

    Chomsky, N. *Syntactic Structures.* The Hague: Mouton, 1957.

THE NOAM OF CAMBRIDGE

By now, Chomsky's ideas about language, mind, politics, and life have been
chronicled in so many places and in so many different ways that there's literally an account for every intellectual taste and purse. Probably much of the
reason for this widespread interest in his ideas is due to his role as one of the
most vocal opponents of the U.S. policy in Vietnam. In fact, one reporter for
The New York Times was surprised to find that Chomsky was a famous linguist
and that his linguistics had something to do with his public role as a political
figure. Since I personally don't find that much connection between his linguistics and his politics, I haven't dwelt upon the latter in this chapter. However,
for those readers wanting more details in this direction, as well as full accounts
of the Chomskian revolution, linguistically speaking, two of the best sources
are the biographies

    Leiber, J. *Noam Chomsky.* Boston: G. K. Hall, 1975.
    Lyons, J. *Noam Chomsky,* Revised Edition. London: Penguin, 1977.

The Lyons book is readily available worldwide in paperback and gives an excellent introductory account of Chomsky's life and thoughts. However, for those
wanting more than just a surface account of the ideas, but without the background or interest for attacking a full-scale technical treatment, the book by
Leiber is hard to beat. Too bad it's so difficult to find. But the search is definitely worth the effort. For a verbatim account of Chomsky's views on linguistics, psychology, sociobiology, Piaget, Skinner and much more, see

    Gliedman, J. "Interview with Noam Chomsky." *Omni,* 1979.
    "The Ideas of Chomsky," in *Men of Ideas,* B. Magee, ed. Oxford: Oxford
    University Press, 1978.

Relatively accessible technical accounts of transformational grammars are
given in the Lightfoot book noted above, as well as in

    Smith, N., and D. Wilson. *Modern Linguistics: The Results of the Chomsky's
    Revolution.* Bloomington, IN: Indiana University Press, 1979.

Perhaps the most readable discussions by Chomsky himself on these matters
are contained in his general lectures:

    Chomsky, N. *Language and Mind,* Enlarged Edition. New York: Harcourt,
    Brace, Jovanovich, 1972.
    Chomsky, N. *Reflections on Language.* New York: Pantheon, 1975.

A critical assessment of Chomsky's theories on linguistics as they stood at
the end of the 1970s is found in the collection

    On Noam Chomsky: Critical Essays, 2nd Edition, G. Harman, ed. Amherst,
    MA: University of Massachusetts Press, 1982.

Of special interest in this book are the articles by John Searle and Robert
Lees, the first a reprint of Searle's well-known 1972 article in *The New York*

Review of Books, which serves as an eminently readable introduction to the whole corpus of Chomsky's thoughts in linguistics. The Lees contribution is the review of *Syntactic Structures* in the journal *Language* that sparked off the Chomskian revolution. It's perhaps not without interest to note that Lees, as well as being a linguist, is also a chemical engineer and was working at the MIT Research Lab of Electronics at the time of writing this review. Thus, he was uniquely prepared to understand and appreciate what was at the time the novel, almost engineering-oriented nature of Chomsky's approach. As a further offering from this very informative volume, let me quote the poem by John Hollander showing that Chomsky's famous "colorless green ideas sleep furiously" may have semantic content, or at least utility, after all:

COILED ALIZARINE
for Noam Chomsky

Curiously deep, the slumber of crimson thoughts:
    While breathless, in stodgy viridian,
Colorless green ideas sleep furiously

POSITIVELY REINFORCING

Skinner's ideas on behavior and mind have by now entered into what one could almost term the folk wisdom of American popular psychology, having been explicated in innumerable books and articles. A worthwhile recent account putting Skinner's behaviorist notions into the context of modern ideas on thought and mind is given in

Flanagan, O. *The Science of the Mind.* Cambridge, MA: MIT Press, 1984.

This book, incidentally, is also an excellent reference for the ideas of Piaget and their relationship to the mainstream of current thinking on minds and machines.

Chomsky's notorious review of Skinner's *Verbal Behavior* was originally published in the widely circulated periodical *Language,* and served as one of the major stepping-stones for the ascendancy of Chomskian ideas into the dominant position not only in modern linguistics, but in psychology as well. The original review is

Chomsky, N. "Review of Skinner's *Verbal Behavior.*" *Language,* 35 (1959), 26–58.

Undaunted by the decline of behaviorism as a significant line of thought in modern psychology, even in retirement Skinner continues to not only preach the behaviorist gospel, but also to practice what he preaches by living his daily life in a modern version of his Skinner box. For a journalistic account of Skinner at age eighty-three, see

Goleman, D. "The Behaviorist Box of B. F. Skinner." *International Herald Tribune,* August 28, 1987.

OUT OF THE MOUTHS OF BABES

Piaget is usually counted as one of the founders of the so-called structuralist school of thinkers, another being the famed French anthropologist Claude

Lévi-Strauss. Interestingly, these two pioneers took diametrically opposed positions on the role of language in shaping thought processes. Piaget, as we know, felt that language makes only a small contribution to thought, while Lévi-Strauss was of the opinion that one starts with language, which then plays a determining role in thought. For an account of both men and their lives, work, and roles in the development of the structuralist movement, the following book is hard to beat:

Gardner, H. *The Quest for Mind: Piaget, Lévi-Strauss, and the Structuralist Movement,* 2nd Edition. Chicago: University of Chicago Press, 1981.

For another good source of critical analysis of Piaget's role in establishing the cognitive thrust of modern psychology and its consequent influence on theories of the mind, see the Flanagan book cited in the preceding section. This book also provides a good account of the developmental theories of the psychologist Lawrence Kohlberg regarding the stages of evolution of morals. In Kohlberg's view, there is an objective moral "good," which becomes apparent in a half-dozen or so stages of development. Basing his theory of moral development upon Piaget's stages of cognitive development, Kohlberg claims to be able to resolve the debate between the Kantians, who cling to an absolute categorical imperative, and the followers of Mill, who argue for a kind of pleasure-maximizing utilitarianism. According to Kohlberg's extension of Piaget's stages, the hands-down winner of this particular fight is Kant.

IT'S ALL A QUESTION OF SEMANTICS

The work of Sapir and Whorf contending that one's view of the world is not only influenced but actually determined by one's language is outlined in the Sampson book noted earlier under General References. For Whorf's own account, see the following collection of reprints of his articles:

Language, Thought, and Reality: Selected Writings of Benjamin Whorf, J. B. Carroll, ed. Cambridge, MA: MIT Press, 1956.

A penetrating discussion of the relevance of Chomsky's ideas vis-à-vis those of relativists like Sapir and Whorf within the context of literary analysis is given in

Steiner, G. "Whorf, Chomsky, and the Student of Literature," in *On Difficulty: Selected Essays*. Oxford: Oxford University Press, 1978.

An introduction to the work of Sampson on an evolutionary approach to linguistics is given in

von Schilcher, F., and N. Tennant. *Philosophy, Evolution and Human Nature.* London: Routledge and Kegan Paul, 1984.

For a more extensive discussion, see

Sampson, G. "Linguistic Universals as Evidence for Empiricism." *Journal of Linguistics,* 14 (1978), 129–375.

Sampson, G. *Making Sense*. Oxford: Oxford University Press, 1980.

For an assessment of some of Sampson's views in the above volume, see the following review, which questions Sampson's ability to deal with the "poverty of the stimulus" problem:

Lightfoot, D. "Review of *Making Sense*," in *Journal of Linguistics,* 18 (1982), 426–431.

The Watchmaker Parable underlying Sampson's evolutionary approach to the building-up of a hierarchical language structure is presented in
Simon, H. "The Architecture of Complexity," in *The Sciences of the Artificial,* 2nd Edition, Cambridge, MA: MIT Press, 1981.

From a computational point of view, the result of Peters and Ritchie shows that a Chomskian transformational grammar is capable of computing (in the formal sense made specific in the artificial intelligence chapter) anything that can be computed. A strong argument can also be made that to survive, humans must also be able to compute in some abstract sense. The question is then how much computing power we really need in order to survive. Since presumably evolution has endowed us with computing power "from below," we embody an amount of computing capability that's sufficient for our needs, but little more. Some have claimed, therefore, that it's unreasonable to suppose that our brains must necessarily be modeled by the most powerful type of computing machine that's theoretically possible. It's at this point that the Montague grammars, with their computational limitations to characterizing only context-sensitive languages, begin to look interesting. For further technical details on the structure of such grammars, see
Montague, R. *Formal Philosophy.* New Haven, CT: Yale University Press, 1974.

For a summary of recent work building upon the foundations laid by Montague, see
Gazdar, G. "Generative Grammar," in *New Horizons in Linguistics,* Vol. 2, J. Lyons et al, eds., pp. 122–151. London: Penguin, 1987.

SHOOT-OUT AT THE ROYAUMONT CORRAL

The definitive account of the goings-on at Royaumont is given in
Language and Learning: The Debate Between Jean Piaget and Noam Chomsky, M. Piattelli-Palmarini, ed. Cambridge, MA: Harvard University Press, 1980.

This volume presents not only the salvos fired by both of the principals, but also extensive rumblings from the "chorus," as well as detailed postmortems by other commentators on the cognitive science scene. It's interesting to note that the debate resulted in a living case study of Piaget's ideas of accommodation and assimilation, since Chomsky seemed to insist that others accommodate their own views to his own, while Piaget held open the possibility of widening his own views to assimilate the Chomskian criticisms into his system. Probably the best that can be said about the outcome is that the two views are complementary in much the same way that waves and particles are complementary in quantum theory. A popularized account of the debate can be found in
Gardner, H. "Encounter at Royaumont." *Psychology Today,* July 1979, pp. 14–16.

Illuminating and compact introductions to Chomsky's current thinking on the mind can be found in his recent semipopular accounts based on lecture series in San Diego and Managua:

Chomsky, N. *Language and Problems of Knowledge: The Managua Lectures.* Cambridge, MA: MIT Press, 1988.

Chomsky, N. *Modular Approaches to the Study of the Mind.* San Diego, CA: San Diego State University Press, 1984.

In the San Diego lectures, Chomsky gives a particularly concise summary of the problems surrounding mental representations, assuming they exist. According to his account, they can be divided into three categories:

- *The Syntax Problem:* Of what kinds of elements are the representations composed and how are they put together?
- *The System Problem:* How are the various mental modules organized and interconnected?
- *The Rule Problem:* Can we characterize mental representations in terms of a system of rules that determines their properties?

RULES AND REPRESENTATIONS

The claim that human cognitive faculties can be described by rules acting on mental representations is the very essence of the machine metaphor that underpins the hopes of the artificial intelligentsia in particular, and the cognitive scientists in general. For a nice textbook introduction to cognitive science, see

Stillings, N., et al. *Cognitive Science: An Introduction.* Cambridge, MA: MIT Press, 1987.

An extensive account of Chomsky's ideas on the question of rules and mental representations is presented in his book *Rules and Representations* (New York: Columbia University Press, 1980). His major points are excerpted, together with extensive peer commentary, in

Chomsky, N. "Rules and Representations." *Behavioral and Brain Sciences,* 3 (1980), 1–61.

The system-theoretic perspective showing the essential equivalence of external and internal rules, at least from a mathematical point of view, is developed in detail in

Casti, J. "Behaviorism to Cognition: A System-Theoretic Inquiry into Brains, Minds, and Mechanisms," in *Real Brains, Artificial Minds,* J. Casti and A. Karlqvist, eds., pp. 47–75. New York: Elsevier, 1987.

CHAPTER FIVE

GENERAL REFERENCES

For a general overview of current AI principles and practice, the following books are particularly good, giving an easily accessible account of many of the ideas and actors on today's AI scene:

Johnson, G. *Machinery of the Mind.* New York: Times Books, 1986.

Waldrop, M. *Man-Made Minds.* New York: Walker, 1987.

For an easily understood introduction to some of the technical ideas that I've only touched upon, see

Aleksander, I., and P. Burnett. *Thinking Machines.* Oxford: Oxford University Press, 1987.

Haugeland, J. *Artificial Intelligence: The Very Idea.* Cambridge, MA: MIT Press, 1985.

Much of the early history of AI up to the mid-seventies, as well as in-depth interviews and portraits of many of the players in our game, such as Simon, Newell, Dreyfus, and Feigenbaum, is given in the work

McCorduck, P. *Machines Who Think.* San Francisco: Freeman, 1979.

In 1983 the New York Academy of Sciences sponsored a meeting devoted to all aspects of the scientific, intellectual, and social impact of the computer. Part of that workshop was a round-table discussion on the question of what we have termed strong AI, human. The transcript of that discussion provides a good background to the entire spectrum of matters considered here. It can be found in the volume

Computer Culture, H. Pagels, ed., Annals of the New York Academy of Sciences, Vol. 426. New York: New York Academy of Sciences, 1984.

The systems interface of AI, neuroscience, and cognitive psychology, together with an exposition of some of the top-down and bottom-up conflicts, is explored in

Boden, M. *Computer Models of the Mind: Computational Approaches in Theoretical Psychology.* Cambridge: Cambridge University Press, 1988.

Mindwaves, C. Blakemore and S. Greenfield, eds. Oxford: Blackwell, 1987.

Real Brains, Artificial Minds, J. Casti and A. Karlqvist, eds. New York: Elsevier, 1987.

The theme of thinking machines and their possible technological, social and psychological implications for man has long been a staple of the science fiction community. Some of my favorites in this line are

Hogan, J. P. *Two Faces of Tomorrow.* New York: Ballantine, 1979.

Jones, D. F. *Colossus.* New York: Berkeley, 1976.

Ryan, T. J. *The Adolescence of P1.* New York: Macmillan, 1977.

Each of these books deals with the general theme of a cognitive computer run amok, threatening human supremacy, and finally yielding its usurped control back to its human masters. It's probably stories like these that give the Weizenbaums of the world nightmares, but for the rest of us they offer a sugar-coated lesson in how thinking machines might actually come about, and the kinds of behavior they might display.

THE TURING TEST AND THE CHINESE ROOM

The Imitation Game was first suggested by Alan Turing in the fundamental paper

Turing, A. "Computing Machinery and Intelligence." *Mind,* 59 (1950).

This paper has since been reprinted in many places, perhaps the most easily accessible being

Hofstadter, D., and D. Dennett. *The Mind's I.* New York: Basic, 1981.

This volume is highly recommended as a treasure trove of additional original material, together with extensive editorial commentary, on the entire spectrum of issues pertaining to minds, brains, machines, souls, and self.

Searle's original paper in which he presents the Chinese Room thought experiment is
Searle, J. "Minds, Brains, and Programs." *Behavioral and Brain Science,* 3 (1980), 417–424.

This already classic paper has also been reprinted a number of times, including an appearance in the Hofstadter and Dennett volume just cited. However, I strongly recommend the original reference as it also contains extensive peer commentary by twenty-seven of the most prominent workers in the field, as well as Searle's rejoinder to their remarks.

Alan Turing was truly one of the unsung heros of the Second World War, his breaking of the German command code ranking with the development of the atomic bomb as a pivotal factor in the war's outcome. However, in contrast to von Neumann, Oppenheimer, Teller, & Co., Turing and his work both faded into a totally undeserved obscurity following the war, with even his position in academic circles being relatively anonymous. It is only in the last decade or so that Turing's real genius has been given public recognition, much of it attributable to the outstanding biography:
Hodges, A. *Alan Turing: The Enigma.* New York: Simon and Schuster, 1983.

This work tracing Turing's life and career, together with his tragic suicide, has recently been produced as the play *Breaking the Code,* which has had a successful run on the London and New York stages, further exposing to the general public Turing's fundamental contributions both to science and to his country.

The work begun in the late 1940s examining the interface between brains and machines represents the germ of the idea that is now flourishing under the rubric "cognitive science." An excellent account for the general reader of the history, objectives, and current programs in this field is
Gardner, H. *The Mind's New Science: A History of the Cognitive Revolution.* New York: Basic, 1985.

FORMAL SYSTEMS, MACHINES, AND TRUTHS

A wonderful introduction to the charms and wiles of formal systems as well as much, much more is the *tour de force*
Hofstadter, D. *Gödel, Escher, Bach: An Eternal Golden Braid.* New York: Basic, 1979.

In this Pulitzer Prize–winning masterpiece, Hofstadter introduces what amounts to the manifesto of the bottom-up school of AI by means of a series of Lewis Carroll–like dialogues, thought experiments, and philosophical speculations elucidating the intricacies of formal systems, Turing machines, Gödel's theorems, Zeno's paradoxes, the Turing-Church Thesis, the theory of evolution, self-reference, and a whole lot more. Hofstadter probably did irreparable harm to his standing in the mainstream AI community by having the temerity to write such a call to arms, especially one committing what in academia are the cardinal sins of being both intelligible and popular with the public. But he did the rest of us an invaluable service by wrapping up this circle of ideas in such an entertaining, informative, and easily digestible package. Highly recommended.

A good introduction to the idea of a Turing machine is given in the Haugeland book cited earlier. See also
  Rucker, R. *Infinity and the Mind.* Boston: Birkhäuser, 1982.

Hilbert's formalist program for mathematics is based upon the idea that mathematics can be viewed as an activity in which we derive certain strings of symbols from certain other strings of symbols according to a set of rules. To avoid infinities, Hilbert required that only finitistic methods be used, where a method is finitistic if it involves no infinite searches and can be specified in a finite number of steps. It was Hilbert's view that one could find a finitist proof of the consistency of mathematics. As noted in the text, Gödel's Incompleteness Theorem shattered this illusion once and for all by showing that not only is any given formal system incomplete, but that there is no finitely given formal system that can prove all true statements about the arithmetic of real numbers. A good, but slightly technical, reference on these matters is
  Webb, J. *Mechanism, Mentalism, and Metamathematics.* Dordrecht, Netherlands: Reidel, 1980.

Somewhat less technical introductions to Hilbert's program, as well as to Gödel's results, are the Hofstadter and Rucker books discussed above, as well as
  Wang, H. *From Mathematics to Philosophy.* New York: Humanities Press, 1974,

from which the Gödel quote regarding the possibility for thinking machines was taken. For an account of Chaitin's work on information-theoretic versions of Gödel's Theorem as well as much more on the relationships of machines, formal systems, computability, and biology, see the collection
  Chaitin, G. *Information, Randomness and Incompleteness.* Singapore: World Scientific, 1987.

Also of interest in this same connection is the more popular treatment in
  Rucker, R. *Mind Tools.* Boston: Houghton-Mifflin, 1987.

"STRONG" VS. "WEAK" AI, BRAINS, AND MINDS

An excellent discussion of the origins and goings-on at the pioneering Dartmouth summer gathering is provided by McCorduck in the volume cited under General References. It makes particularly interesting reading to look at the interviews with McCarthy, Minsky, Simon, and others present at the meeting, comparing their feelings at the time about the future course of AI with the way things have actually worked out.

The various categories of "strong" and "weak" AI are discussed in considerably more detail in

Gunderson, K. *Mentality and Machines,* 2nd Edition. Minneapolis: University of Minnesota Press, 1985.

Gunderson provides not only a useful categorization for sharpening the "Can machines think?" question, but also offers an extremely thought-provoking critique of Turing's Imitation Game. On the fundamental question, Gunderson concludes that without addressing the mind-body relationship, no progress is possible on strong AI. But to do this, we need a first-person perspective to be somehow encoded into an essentially third-person set of descriptions.

As part of his work on the theoretical foundations of computing and machines, von Neumann discovered that there was no theoretical barrier to the idea of a self-reproducing machine. Further, he showed that such a machine would necessarily have to contain an encoded description of itself, i.e., be capable of self-reference in some definite sense. Thus, it's doubly odd that he seemed so pessimistic about the idea of a computer duplicating the cognitive powers of the human brain. Von Neumann's final (unfinished) work, in which he lays out some of his thoughts on the matter, is the text of his Silliman Lectures:

von Neumann, J. *The Computer and the Brain.* New Haven, CT: Yale University Press, 1958.

Another excellent volume exploring the brain-mind-machine connection is

Arbib, M. *Brains, Machines, and Mathematics,* 2nd Edition. New York: Springer, 1987.

TOP-DOWN SYMBOL CRUNCHING

The treatment given here of the underlying principles of the Simon and Newell top-down programs really doesn't do justice to the ideas employed. The Haugeland and McCorduck books cited above provide a balanced historical and semitechnical view. But as always in matters of this sort, it's preferable to hear from the protagonists themselves. For this, the best nontechnical introduction is the classic book

Simon, H. *The Sciences of the Artificial,* 2nd Edition. Cambridge, MA: MIT Press, 1981.

For an introductory but illuminating account of SHRDLU, see Hofstadter's magnum opus. The quote by Winograd can be found in

Waldrop, M. "Machinations of Thought." *Science '85,* March 1985, p. 44.

Schank gives a popular account of his work on scripts in

Schank, R., and P. Childers. *The Cognitive Computer: On Language, Learning, and Artificial Intelligence.* Reading, MA: Addison-Wesley, 1984.

For a detailed blow-by-blow record of the development of a script-following program in Wilensky's lab at Berkeley, as well as a firsthand account of the battles between the AIers and Dreyfus-Searle, see

Rose, F. *Into the Heart of the Mind: An American Quest for Artificial Intelligence.* New York: Harper and Row, 1984.

An illuminating account of the problems of getting computers to "understand" is given in

Winograd, T., and F. Flores. *Understanding Computers and Cognition.* Reading, MA: Addison-Wesley, 1986.

BOTTOM-UP EMERGENCE

The first salvo fired in Hofstadter's bottom-Up AI program was his *Gödel, Escher, Bach* book cited earlier. Subsequently, he put down his philosophy on the dream underlying mainline, top-down AI and his objections to it in the paper

Hofstadter, D. "Waking Up from the Boolean Dream, or, Subcognition as Computation," in *Metamagical Themas,* pp. 631–665. New York: Basic, 1984.

For details about the principles on which *Jumbo* is based, along with additional information about its inner workings, see

Hofstadter, D. "The Architecture of *Jumbo.*" *Proceedings of the 2nd Machine Learning Workshop,* 1983, pp. 161–170.

The problems of identifying letterforms, as well as those involved in trying to get a program to do analogies, are discussed in more detail in the *Metamagical Themas* volume. A popular account of the Hofstadter position vis-à-vis "classical" AI, together with a consideration of the competing personalities as well as their programs, is found in

Gleick, J. "Exploring the Labyrinths of the Mind." *The New York Times Magazine,* August 21, 1983, p. 23.

The flavor of the guerrilla warfare being waged between the competing AI schools is captured in the acerbic commentary by Newell on Hofstadter's views reported in

The Study of Information: Interdisciplinary Messages, F. Machlup and U. Mansfield, eds. New York: Wiley, 1983.

Hofstadter's rejoinder can be found in the postscriptum to his "Boolean Dream" article referenced above.

For a detailed view of Marvin Minsky's thinking on mental *kollectivs,* see his book

Minsky, M. *The Society of Mind.* New York: Simon and Schuster, 1987.

Also of interest is the Minsky and Papert treatment of perceptrons, which is reported in the following new edition written to take account of the revival of the perceptron idea in the new connectionism:

Minsky, M., and S. Papert. *Perceptrons,* Enlarged Edition. Cambridge, MA: MIT Press, 1988.

Lenat's evolutionary approach to bottom-up cognition is discussed in the Waldrop work cited earlier.

A popular introduction to the general philosophy and program of the new connectionists is

"Seeking the Mind in Pathways of the Machine." *The Economist,* June 29, 1985, p. 83.

A vastly more detailed, technical account of the entire effort is given in
Parallel Distributed Processing, Vol. 1: *Foundations,* Vol. 2: *Psychological and
Biological Models,* J. McClelland and D. Rumelhart, eds. Cambridge, MA:
MIT Press, 1986.

PHILOSOPHERS AGAINST: THEY'LL NEVER THINK!

An excellent paper summarizing all the arguments of the "Computers can't
think" crowd is
Grabiner, J. "Artificial Intelligence: Debates About Its Uses and Abuses."
Historica Mathematica, 11 (1984), 471–480.

Details of the Dreyfus arguments against the AI community are given in the
volumes
Dreyfus, H. *What Computers Can't Do: The Limits of Artificial Intelligence,*
Revised Edition. New York: Harper Colophon, 1979.
Dreyfus, H., and S. Dreyfus. *Mind over Machine.* New York: Free Press,
1986.

Many more details of the historical development of the Dreyfus case, along
with extensive commentary and interviews with his opponents, are found
in McCorduck's book cited above. For a more specific, technically based
attack, see
Wilks, Y. "Dreyfus' Disproofs." *British Journal for the Philosophy of Science,*
27 (1976), 177–185.

The original reference for Lucas's argument from Gödel is
Lucas, J. "Minds, Machines, and Gödel." *Philosophy,* 36 (1961), reprinted in
Minds and Machines, A. Anderson, ed. Englewood Cliffs, NJ: Prentice-Hall,
1964.

Objections to Lucas are put forward in Hofstadter's *Gödel, Escher, Bach.* More
technical arguments are given in
Benacerraf, P. "God, the Devil, and Gödel." *The Monist,* 51 (1967), 9–32.

Searle's Chinese Room–style arguments against strong AI are amplified in
his Reith Lectures given on the BBC. These lectures have been published as
the book
Searle, J. *Minds, Brains, and Science.* Cambridge, MA: Harvard University
Press, 1984.

The extreme generality of the Turing Test, together with a spectrum of entertaining arguments supporting its claim as an indicator of intelligence, is
explored by one of the leading philosophers supporting AI in
Dennett, D. "Can Machines Think?" in *How We Know,* M. Shafto, ed. New
York: Harper and Row, 1985.

THE MORALIST AND THE MYSTIC

A detailed summary of Weizenbaum's arguments involving the humanity-ver-
sus-machine issue is found in
Weizenbaum, J. *Computer Power and Human Reason.* San Francisco: Freeman, 1976.

As far as I can tell, no one has been convinced of the case Weizenbaum tries to make and his arguments are seldom heard any longer. However, when the book was first published there were many impassioned discussions about the points raised. For an account of these, see the McCorduck book already cited.

Rucker's mystical views are outlined in his book noted above, as well as in the paper

Rucker, R. "Towards Robot Consciousness." *Speculations in Science and Technology,* 3 (1980), 205–217.

BRINGING IN THE VERDICT

For a somewhat more detailed, technical account of my views expressed here on matters of self-reference as it pertains to a system's ability to contain model of itself, as well as my contentions about the distinction between a model and a simulation, the interested reader should consult Chapter Seven of

Casti, J. *Alternate Realities: Mathematical Models of Nature and Man.* New York: Wiley, 1989.

CHAPTER SIX

GENERAL REFERENCES

Numerous popular and semipopular treatments of the ETI question have been published in recent years, examining the topic from various points of view. Here is one of the best:

Shklovskii, J., and C. Sagan. *Intelligent Life in the Universe.* San Francisco: Holden-Day, 1966.

This volume really kicked off the SETI era in several ways. First of all, it is a thorough, scientifically documented, and literate account of all aspects of the SETI question, circa the mid-sixties. Furthermore, the book represents a unique kind of collaboration between the Russian Shklovskii and the American Sagan, which originally began as just a translation of a similar book in Russian by Shklovskii, but turned into a major collaborative venture on what amounts to a different book. With the exception of some of the experimental work, most of the material covered is still relevant today and can be read with profit. Highly recommended.

A more recent account of theoretical and experimental ETI is given in semipopular form by

Baugher, J. *On Civilized Stars: The Search for Intelligent Life in Outer Space.* Englewood Cliffs, NJ: Prentice-Hall, 1985.

James Trefil is a physicist at George Mason University in Virginia, and a man well known for his popular books on the wonders of physics and Nature. His colleague Robert Rood is an astronomer interested in ETI. Over a few beers at a local pub they put their heads together and started speculating about the ETI question, trying to address the issues from as unbiased a viewpoint as possible within the constraints of human prejudice. Their conclusions (which are not the same for each author) are presented in the popular account

Rood, R., and J. Trefil. *Are We Alone? The Possibility of Extraterrestrial Civilizations.* New York: Scribner's, 1981.

An excellent source for many of the pioneering ETI papers, as well as a representative selection of readings outlining various aspects of the ETI question, is the collection

The Quest for Extraterrestrial Life: A Book of Readings, D. Goldsmith, ed. Mill Valley, CA: University Science Books, 1980.

It has been argued that the resolutely anthropomorphic bias of most SETI work may blind us to how aliens might communicate and think. For a fascinating account of this aspect of SETI from the point of view of a psychologist, see

Baird, J. *The Inner Limits of Outer Space.* Hanover, NH: University Press of New England, 1987.

No topic has provided more ammunition for the science fiction writers than contact with ETI in all its possible forms. To my eye, some of the best accounts focusing on radio contact are

Gunn, J. *The Listeners.* New York: Scribner's, 1972.

Lem, S. *His Master's Voice.* New York: Harcourt Brace Jovanovich, 1983.

McDevitt, J. *The Hercules Text.* New York: Berkley, 1986.

Sagan, C. *Contact.* New York: Simon and Schuster, 1985.

These volumes all have the same basic theme: the receipt, translation, and interpretation of a signal, and the way in which human hopes, fears, and interactions are affected by the knowledge that ETI exists. Each of these volumes has its own answer to the question "What does communication with ETI mean for mankind?," my own favorite being the less than gushing account given by Lem.

The literature on direct contact is so enormous that it's impossible to give even a representative sampling of the many themes that have been explored. Instead let me list just a smattering of my personal favorites:

Bova, B. *Voyagers.* New York: Doubleday, 1981.

Crichton, M. *Sphere.* New York: Knopf, 1987.

Forward, R. *The Dragon's Egg.* New York: Ballantine, 1980.

Lem, S. *Solaris.* London: Faber and Faber, 1971.

McCollum, M. *Life Probe.* New York: Ballantine, 1983.

Moffitt, D. *The Jupiter Theft.* New York: Ballantine, 1977.

The usual Hollywood vision of how we would react to the landing of an alien vessel is something along the lines depicted in *Close Encounters,* where everyone is calm, peaceful, and full of cosmic harmony and goodwill. Some observers, myself included, feel considerably less sanguine about the possibility. The results of Orson Welles's Mercury Theatre radio broadcast of *The War of the Worlds* on Halloween 1938 suggest that the most likely outcome of such direct contact will be nothing less than sheer terror. This aspect of SETI appears to await an enterprising investigator from the psychological community.

THE FERMI PARADOX AND PROJECT OZMA

While putting this section together, I wanted to track down as precisely as possible the moment that SETI entered the experimental phase with the beginning of Project Ozma. It seemed to me that such a significant beacon on the SETI landscape would be well chronicled, especially since it happened only a couple of decades ago. As an indication of why scientists make poor historians, I found the following dates in 1960 given: April 8 (Baugher, 1985), April 11 (Papagiannis, 1985), autumn (Shklovskii and Sagan, 1966), May-June-July (McGowan and Ordway, 1966), early 1960 (Sagan, in a 1974 article), spring (Rood and Trefil 1981), and, worst of all, no date at all offered by Frank Drake himself, in an account of the experiment published only *one year* after it had been completed. What a mess! The date of April 11 quoted in the text is taken from personal accounts of the Ozma search given at a twenty-fifth-anniversary *Fest* held at the National Radio Astronomy Observatory, the proceedings of which are reported in

The Search for Extraterrestrial Intelligence, K. Kellerman and G. Seielstad, eds. Green Bank, WV: NRAO, 1986.

Living in the modern age of risk-averse science, NSF peer review, and unimaginative scientific apple polishing, I find it refreshing to read Drake's account of how there was no proposal, no committee, no referees, no studies, just an OK from NRAO Director Otto Struve. In short, science as it should be—done by scientists and not by congressmen, NSF program managers, university veeps, or political action groups.

The classic paper advocating the 1420-MHz "waterhole" frequency as the natural place to look for ETI is

Cocconi, G., and P. Morrison. "Searching for Interstellar Communications." *Nature,* 184 (1959), 844.

Plans for Project Ozma and the publication of the Cocconi and Morrison paper were progressing along totally independent lines. So when the paper appeared, NRAO Director Otto Struve was apparently quite agitated, wanting to ensure that appropriate credit for the idea of a search would go not to Cocconi and Morrison, but rather to the newly founded NRAO. As a preemptive strike, Struve totally changed a talk scheduled the following week at MIT to emphasize the Ozma project, thereby putting the NRAO on record with the idea. Struve was clearly a man with a deep understanding of the ways of credit in science, not to mention the bureaucratic one-upsmanship needed to keep a fledgling organization visible where it counted—with the funding agencies.

THEORETICAL ETI: THE DRAKE EQUATION

The Drake equation was first formulated at a meeting in November 1961 at the National Radio Astronomy Observatory, only a year after the Ozma search. Since then many alternate formulations have been offered, although the key astrophysical, biological, and sociocultural components have remain unchanged.

One of the major objections to the use of the Drake equation in ETI studies is that each of its components can itself be decomposed into an almost endless list of "sub-Drake" equations. For example, the term f_l, involving the likelihood of the appearance of life, lumps into one number a large collection of separate steps, each of which has its own probability of occurrence. Carrying out this kind of microanalysis on each of the terms leads to a "super" Drake equation containing about as many terms as you wish. If each such term has a likelihood of less than one, multiplying them together creates an estimate for N that is as low as your prejudices require. The counterargument to the resulting claim that the Drake equation is useless is to claim that the arbitrarily small estimates of N arise from the *assumed* independence of the individual terms. If some are dependent, then all bets are off and the equation can again be employed. More details on all these matters are found in the NRAO volume cited above.

SLICES OF THE ETI PIE
Far more detailed accounts of the various slices of the ETI pie are found in the Sagan and Shklovskii, Rood and Trefil, and Baugher volumes noted above.

The simulations of possible planetary systems are taken from
Dole, S. "Computer Simulation of the Formation of Planetary Systems." *Icarus,* 13 (1970), 494–508.

Hart's calculations showing the narrow path that the Earth had to tread in order to avoid becoming either a frozen wasteland or a Turkish bath are given in
Hart, M. "Habitable Zones About Main Sequence Stars." *Icarus,* 37 (1979), 351–357.

More recent calculations show that perhaps the CHZ is not as small as Hart imagined. These models, based upon the idea that concentrations of carbon dioxide in the atmosphere would be enough to prevent water from freezing, even on planets far from their parent star, push the CHZ for Earth-like planets from Hart's estimate of 0.95–1.05 AU to 0.95–1.5 AU, an increase of almost 50 percent. For details, see the account
"Model Atmospheres Show Signs of Life." *New Scientist.* January 7, 1988, p. 41.

Discussions of Miller's classic experiment are found in almost every book on ETI, this one not excepted. The current guru of this type of investigation aimed at showing how life could (must?) have arisen on Earth is Cyril Ponnamperuma of the University of Maryland. A good account of the present state of this arcane chemical art is
Ponnamperuma, C. "Primoridial Organic Chemistry," in *Extraterrestrials: Where Are They?*, M. Hart and B. Zuckerman, eds. New York: Pergamon, 1982.

ANTHROPOMORPHISMS, CHAUVINISMS, AND ETI NUMEROLOGY
In Table 6.1, the Hart estimates for the value of N are found in
Hart, M. "*N* Is Very Small," in *Strategies for the Search for Life in the Universe,* M. Papagiannis, ed., pp. 19–25. Dordrecht, Netherlands: Reidel, 1980.

When interpreting the statistical estimates given by Sturrock, one should be extremely careful to note that the legitimacy of the conclusions regarding the confidence levels for N are entirely dependent upon the accuracy of the various estimates that went into the statistical analyses. Thus, while Sturrock's statistical wizardry may be beyond reproach, if the raw data regarding N that formed the basis for his calculations is hopelessly adrift, then so is the credibility of the final conclusions. Readers interested in a complete account of Sturrock's analysis can find it in

Sturrock, P. "Uncertainty in Estimates of the Number of Extraterrestrial Civilizations," in *Strategies for the Search for Life in the Universe,* M. Papagiannis, ed., pp. 59–72. Dordrecht, Netherlands: Reidel, 1980.

The complete transcript of the historic Byurakan meeting, along with a number of supplementary documents including a discussion of the notion of subjective probability, is found in the following volume which is must reading for anyone curious about ETI:

Communication with Extraterrestrial Intelligence, C. Sagan, ed. Cambridge, MA: MIT Press, 1973.

Informal, personal accounts of the goings-on at this Armenian gathering by two of the participants are given in

McNeill, W. "Journey from Common Sense." *University of Chicago Magazine,* 64 (May–June 1972), 2–14.

Dyson, F. "Letter from Armenia." *The New Yorker,* November 6, 1971, p. 126.

These accounts illustrate that even such cerebral gatherings are not without their lighter side, as evidenced at Byurakan by Joseph Shklovskii's response to someone's slightly harebrained idea that there was a strong correlation between sunspot maximums and the appearance of notable examples of human creativity. Shklovskii observed that "this theory was obviously concocted during a period of a deep sunspot minimum!"

Dyson is one of America's most publicly visible physicists, having been involved not only in pioneering research in quantum theory, but also serving as a member of the Orion Project devoted to the design of a low-cost vehicle for human space travel. In addition, he has been a tireless crusader for a more sane view of the dangers of uncontrolled nuclear weaponry. His autobiography, detailing his feelings on these issues for a general audience, is

Dyson, F. *Disturbing the Universe.* New York: Harper and Row, 1979.

A far different side of Dyson's life, one showing that even eminent theoretical physicists are not immune to the kinds of generational parent-child conflicts that plague the rest of us, is provided by the candid profile of Dyson and his son, George, given in

Bower, K. *The Starship and the Canoe.* New York: Holt, Rinehart and Winston, 1978.

EXPERIMENTAL SETI: HOW SHOULD WE LISTEN?

Up-to-date accounts of the various types of radio searches for ETI are given in these survey articles:

Papagiannis, M. "Recent Progress and Future Plans on the Search for Extraterrestrial Intelligence." *Nature,* 318 (1985), 135–140.

Tarter, J. "SETI Observations Worldwide," in *The Search for Extraterrestrial Intelligence,* K. Kellermann and G. Seielstad, eds., pp. 79–98. Green Bank, WV: NRAO, 1986.

The NRAO volume also contains a number of other papers outlining the specific details of a variety of SETI radio searches underway or planned, including details of the NASA program.

Dyson's idea of dismantling one's home solar system to create a material sphere surrounding its sun was initially proposed in the one-page note

Dyson, F. "Search for Artificial Stellar Sources of Infrared Radiation." *Science,* 131 (1960), 1967.

For some strange, inexplicable reason, this idea appears to have captured the fancy of the Russians, and several Soviet searches have been conducted looking for such "hot" sources of infrared radiation. Somehow the idea has never seemed as appealing to American astronomers and, as far as I can tell, it's currently on the back burner of U.S. SETI activity.

For an account of Michael Papagiannis's arguments for why the asteroid belt might be a good place to look for ETI, see

Papagiannis, M. "Colonies in the Asteroid Belt, or a Missing Term in the Drake Equation," in *Extraterrestrials: Where Are They?,* M. Hart and B. Zuckerman, eds., pp. 77–86. New York: Pergamon, 1982.

WHAT ARE WE LISTENING FOR?—THE SYNTAX AND SEMANTICS OF SETI

A very readable popular account of the entire SETI issue, including a number of interesting graphics illustrating the communication problem, is given in

McDonough, T. *The Search for Extraterrestrial Intelligence: Listening for Life in the Cosmos.* New York: Wiley, 1987.

Pictorial radio messages are not the only type of language that's been suggested for communicating with ETI. Some years ago, Dutch mathematician Hans Freudenthal developed a purely logical, nonverbal, semantic-based language called LINCOS (for Lingua Cosmica) for such messages to the stars. While the study of terrestrial languages includes grammar, syntax, and phonemics, LINCOS is designed entirely in terms of semantics. It consists of a coded system of units that are clearly enumerated as chapters and paragraphs. This structure facilitates the interpretation of the message, as the semantic content can be derived from logic external to the linguistic system itself. A LINCOS transmission begins with the most elementary concepts of mathematics and logic, since the language must describe itself before it can be used as a communications medium. After this "self-definition" phase, the language goes on to logically develop more complicated concepts of the natural, social, and behavioral sciences. For a detailed description of the language and its use, see

Freudenthal, H. *LINCOS: Design of a Language for Cosmic Intercourse.* Amsterdam: North-Holland, 1960.

The initial idea for the plaque on *Pioneer 10* seems to have come from the science writer Eric Burgess, who realized that the probe would become the first human artifact ever to leave the solar system. He got in touch with a colleague, writer Richard Hoagland, who in turn contacted Carl Sagan, and the rest was history. As an irrelevant footnote to the whole episode, Sagan's former wife Linda Saltzman was responsible for the drawings of the naked male and female figures that caused all the ruckus about space pornography.

An excellent account of the development of the far more ambitious project to put a record of Earth on the Voyager probes is given in the McDonough book cited earlier, as well as the following volume produced by the project team itself:

Sagan, C., F. Drake, A. Druyan, T. Ferris, J. Lomberg, and L. Saltzman Sagan. *Murmurs of Earth.* New York: Ballantine, 1978.

Here is the solution to Frank Drake's hypothetical message from ETI:

Decoded message from ETI

The picture shows a figure of a humanoid whose home star is given along the left border surrounded by the nine planets of its solar system. The figure's upper-right corner shows diagrams of carbon and oxygen, suggesting that the ETI's body chemistry is similar to our own. Immediately to the right of the first five planets are shown the first five positive integers, written in binary fashion with a parity bit, i.e., $1 = 10$, $2 = 100$, $3 = 111$, $4 = 1000$, $5 = 1011$. Note that the parity bit causes each number to have an odd number of 1's. To the right of the alien, and connected by a diagonal line, is a cartoonist's balloon inside of which are three numbers (you can tell they're numbers because they each have an odd number of 1's). There is the number 5 next to planet 2; 868 by planet 3; and about 4 billion next to planet 4. Presumably these numbers reflect the populations of aliens on these planets, indicating an exploratory expedition on the second planet and a colony on the third, while planet 4 is the home planet. To the right of the alien is its height of 31 "units," which it is logical to assume are the natural units of the message itself, the wavelength of the transmitting signal. The line of four blocks underneath the alien itself might be interpreted as the alien's code for itself, since it can't be a number (because it has an even number of 1's).

Further amplification and elaboration of Ball's list of possibilities for ETI is given in his original article:

Ball, J., "Extraterrestrial Intelligence: Where Is Everybody?" in *The Search for Extraterrestrial Life: Recent Developments,* M. Papagiannis, ed., pp. 483–486. Dordrecht, Netherlands: Reidel, 1985.

N > 1: ETI EXISTS!

The argument given here about the outrageous costs of mounting a manned exploration of even a nearby star system is given in detail in

Drake, F. *"N Is Neither Very Small Nor Very Large,"* in *Strategies for the Search for Life in the Universe,* M. Papagiannis, ed., pp. 27–34. Dordrecht, Netherlands: Reidel, 1980.

Other arguments conclude just the opposite, saying that star travel is well within our projected pocketbook. For an account of these claims, see

Interstellar Migration and the Human Experience, B. Finney and E. Jones, eds. Berkeley, CA: University of California Press, 1985.

Sagan, C. "Direct Contact Among Galactic Civilizations by Relativistic Interstellar Spaceflight," *Planetary and Space Science,* 11 (1963), 485.

Singer, C. "Settlements in Space, and Interstellar Travel," in *Extraterrestrials: Where Are They?,* M. Hart and B. Zuckerman, eds., pp. 46–61. New York: Pergamon, 1982.

THE SHAPE OF ETIS TO COME

The possibilities for alien anatomy are virtually endless, with the science fiction literature having at one time or another explored most of them. For those like myself having a congenital weakness for such kinds of speculation, the following volume of artistic interpretations of ETI is absolutely must reading:

Barlowe, W., and I. Summers. *Barlowe's Guide to Extraterrestrials: Great Aliens from Science Fiction Literature.* Leicester, UK: Windward, 1979.

Speculations not only on the nature of ETI's anatomy, but also on social structures and life-styles, can be found in

Cultures Beyond Earth, M. Maruyama and A. Harkins, eds. New York: Vintage, 1975.

Forward, R. "When You Live Upon a Star . . . ," *New Scientist,* December 24, 1987, pp. 36–38.

Jonas, D. and D. *Other Senses, Other Worlds.* New York: Stein and Day, 1976.

ETI? THERE'S NO SUCH THING: $N = 1$

For light comic relief, the article below by Adler is tough to beat. For some low-level laughs, have a look at

Adler, A. "Behold the Stars." *Atlantic Monthly,* 234 (1974), 109.

Michael Hart's pathbreaking ETI paper, trying to show that the emperor has no clothes, is

Hart, M. "An Explanation for the Absence of Extraterrestrials on Earth." *Quarterly Journal of the Royal Astronomical Society,* 16 (1975), 128–135.

The countervailing claim that "absence of evidence is not evidence of absence" can be found in

Cox, L. "An Explanation for the Absence of Extraterrestrials on Earth." *Quarterly Journal of the Royal Astronomical Society,* 17 (1976), 201.

Tipler's classic contribution to the SETI debate was first published in

Tipler, F. "Extraterrestrial Intelligent Beings Do Not Exist." *Quarterly Journal of the Royal Astronomical Society,* 21 (1980), 267–281.

For a personal account of the behind-the-scenes machinations surrounding publication of the above paper, as well as additional commentary, the reader should consult Tipler's contribution to the volume

Rothman, T., et al. *Frontiers of Modern Physics.* New York: Dover, 1985.

Not to be upstaged by Tipler's arguments, Carl Sagan, the paper's original referee, had lots of time to muster his ammunition against Tipler's claim, which he termed the solipsist approach to ETI. See

Sagan, C., and W. Newman. "The Solipsist Approach to Extraterrestrial Intelligence." *Quarterly Journal of the Royal Astronomical Society,* 24 (1983), 113–121.

The anthropic argument given by Carter for the nonexistence of ETI is most easily accessed in

Barrow, J., and F. Tipler. *The Anthropic Cosmological Principle.* Oxford: Oxford University Press, 1986.

Accounts of Simpson's biological objections to ETI can be found reprinted in the Goldsmith volume cited above under General References, while Mayr's arguments are given in the volume

Extraterrestrials: Science and Alien Intelligence, E. Regis, ed. Cambridge: Cambridge University Press, 1985.

This volume, incidentally, contains a wealth of additional material on all sides of the SETI question and is highly recommended as a general reference.

Nicholas Rescher's arguments for the likelihood of ETI's science being "weird" by our standards can be found in "Extraterrestrial Science," pp. 83–116 in the Regis book cited above. In this same connection, see Regis's article "SETI Debunked," pp. 231–244 in the same volume.

For the arguments by Eigen, Schuster, and Dawkins regarding the ability of complex systems to form "randomly" using the ratcheting principle, see their popular works:

Eigen, M., and P. Schuster. *The Hypercycle: A Principle of Self-Organization.* Berlin: Springer, 1979.

Dawkins, R. *The Blind Watchmaker.* London: Longman, 1986.

SUMMARY ARGUMENTS

In Table 6.4, I have noted Freeman Dyson's argument for comets as a likely home for ETI. While this doesn't exactly constitute a claim that ETI exists, it's an intriguing idea for getting around in the universe: Just hitch a ride on a comet and your energy problems are over since you can let Nature can pay the bill. Dyson's principal claim is that since there are a lot of comets around, each of which contains an abundance of free raw material, this would be a likely way for a cost-conscious ETI to go if it wanted to look over the galaxy—provided it wasn't in a big hurry!

CHAPTER SEVEN

GENERAL REFERENCES

Bookshops are literally overflowing with volumes at all degrees of technical sophistication offering to "explain" the paradoxes of the quantum world to the uninitiated. Many of them do a pretty good job; some are misleading; others are just artless junk. Among the works in the first category, one stands out in my mind as being the hands-down winner when it comes to a thoroughly readable, highly enlightening, vastly entertaining, well-illustrated nontechnical treatment of quantum mischief. That volume, upon which I have shamelessly modeled some of the earlier sections of this chapter, is

Herbert, N. *Quantum Reality: Beyond the New Physics.* New York: Doubleday, 1985.

There seems to be something about quantum theory that brings out the poet in writers who attempt to convey the ideas to the general reader. In addition to the Herbert book above, three other accounts highly recommended for the not especially technically inclined are

Pagels, H. *The Cosmic Code.* New York: Simon and Schuster, 1982.

Rae, A. *Quantum Physics: Illusion or Reality?* Cambridge: Cambridge University Press, 1986.

Squires, E. *The Mystery of the Quantum World.* Bristol, UK: Hilger, 1986.

In the last decade or so, it has become a bit of a fad to try to link quantum reality as discussed here with all sorts of mystical ideas having their roots in various Eastern religions. While I hold no particular brief for these efforts, I feel that, like anyone attempting to capture the uncapturable, some of the authors do a better job of hitting the target than others. A volume worthy of honorable mention in this connection is

Zukav, G. *The Dancing Wu Li Masters.* New York: Morrow, 1979.

Two volumes of the "quantum theory ⟷ mystical world" type that in my opinion a discriminating reader can safely miss are

Toben, B., and F. Wolf. *Space-Time and Beyond.* New York: Dutton, 1975.

Wolf, F. *Star Wave.* New York: Macmillan, 1984.

Both of these volumes (especially the second) are of the sort that contribute to the enthusiasm with which most physicists regard quantum reality research as being a highly suspect activity, if not downright unscientific or even unprofessional. Strangely enough, author Fred Wolf is a trained physicist whose earlier book *Taking the Quantum Leap* won the American Book Award for science exposition. That effort was, in my view, a successful attempt to explain the concepts and principles of the quantum world to a general audience. However, with *Star Wave,* a fairly evident attempt to reach an even wider audience, the author flies off the track with a host of outrageous speculations about quantum theory and its relevance to new laws of psychology, love, hate, sanity, mind control, death, reincarnation, and a whole lot more. While this sort of thing probably does sell books, it doesn't do much to further the understanding of the limitations of quantum theory for curing the worlds ills. While I'd never endorse any kind of "ban the book" initiative, I would feel more comfortable if books like this were not around.

In a more positive vein, the history of both the ideas and the people of quantum mechanics is vividly portrayed in the following works:

Cline, B. *Men Who Made a New Physics.* Chicago: University of Chicago Press, 1987.

Jammer, M. *The Philosophy of Quantum Mechanics.* New York: Wiley, 1974.

The Jammer volume is fairly technical in parts, but gives an insider's account of the back room discussions, as well as the personality factors, that underlie how the Copenhagen Interpretation came to quantum ascendancy. The Cline book is a purely nontechnical version of the same people and events, written in a clear, informative fashion by a science writer. Both books are to be highly praised for the light they shed on the human factor in the creation of a scientific revolution.

BUILDING THE STAGE

A more detailed account of Wheeler's "contextual" twenty-questions game, as well as a nice, compact introduction to the basic ideas of quantum theory, is found in the first part of

The Ghost in the Atom, P. Davies and J. Brown, eds. Cambridge: Cambridge University Press, 1986.

This is an extremely interesting little book, the bulk of which is a collection of transcripts of interviews originally broadcast on BBC radio with many of the main actors in the modern quantum-reality game, including Wheeler, Bell, and Bohm.

GHOSTS IN THE ATOM

For a lavishly illustrated and somewhat more detailed discussion of the double-slit experiment, the following popular treatment is hard to beat:

Hey, T., and P. Walters. *The Quantum Universe*. Cambridge: Cambridge University Press, 1987.

For the overall idea using waveform families, prisms, and the like to explain the Schrödinger solution to the Description Problem, I am indebted to the outstanding treatment provided in Herbert's book cited above. The reader is strongly encouraged to consult Herbert for a more leisurely account. Incidentally, to be perfectly accurate, the quantities termed c_i in the text are related to the actual values of the quantum wave function $W(x, t)$, which is complex-valued. This is necessary for $W(x, t)$ to display the needed wavelike behavior. Thus the elements c_i are not real numbers but complex quantities, implying that when we compute the probabilities of various experimental outcomes, we should really use $c_i c_i^* = |c_i|^2$, where $|\cdot|$ is the complex modulus, and not the simpler c_i^2 of the text. A good source for a proper discussion of these matters is the well-known textbook

Feynman, R., R. Leighton, and M. Sands. *The Feynman Lectures on Physics*, Vol. III. Reading, MA: Addison-Wesley, 1965.

MEASUREMENT TO MEANING

The quote in the text illustrating the kind of misinformation in circulation regarding the Heisenberg Uncertainty Principle was taken from *An Incomplete Education* by J. Jones and W. Wilson (New York: Ballantine, 1987), p. 489. Here's another from a different source:

It seems to me that we can apply the Heisenberg uncertainty principle to the problem of the meaning of words. Writers, poets, etc. use words in a very large, general sense, but for them they have a very special meaning. The single word has a very special function in their description. In contrast, in the sciences words are very sharply defined and have a very short-range validity. But this fact made it possible that this word is understood universally. Restricting the domain of validity of the word produces, on the other hand, a gain in universality.

This statement, made by a physicist at an interdisciplinary meeting aimed at bringing scientists, writers, musicians, and others together, might (by a charitable interpretation) be thought of as an appeal to the Heisenberg Uncertainty Principle as a metaphor. But surely the author cannot be claiming that a word used in a specialized sense is in any meaningful way "conjugate" to that same word used in an everyday manner. To my mind, it's an open question whether or not use of Heisenberg's principle in this kind of metaphorical sense helps or hinders the process of bringing science back into contact with the mainstream of intellectual life.

THE ROMANTIC REALITIES

To cover the evolution of thought from Copenhagen to Austin and way stations in between, the following collection of reprints and commentary is must reading:

Quantum Theory and Measurement, J. A. Wheeler and W. Zurek, eds. Princeton, NJ: Princeton University Press, 1983.

Of special interest in this volume is a series of papers and lectures detailing the ongoing battle between Einstein and Bohr over the adequacy of the Copenhagen Interpretation.

While von Neumann's "Cut Theorem" appeared to have pulled the rug out from under the naïve realists, it should be kept in mind that von Neumann was a mathematician, not a physicist. As a result, the assumptions he made in his quantum bible were the kind that led to a mathematically elegant theory, but not ones that in retrospect necessarily appear to be physically appropriate. In fact, John Bell in a recent interview went so far as to call von Neumann's proof "silly." But to illustrate the Great Man Theory of science, von Neumann's immense prestige as a mathematician convinced the physicists that it must be so if von Neumann said it, thus setting quantum reality research back at least thirty years. For those eager to see what the nature of these dubious assumptions are, the English version of the bible should be consulted:

von Neumann, J. Mathematical Foundations of Quantum Mechanics, R. Beyer, trans. Princeton, NJ: Princeton University Press, 1955.

For a wide range of ideas about the ways in which science might shed some light on the problem of consciousness, see the following volume, which reports the proceedings of a meeting on the topic involving such luminaries as Bohm, Fritjof Capra, and Nobel laureate Brian Josephson, meeting with a group of French and Spanish thinkers on the matter:

Science and Consciousness: Two Views of the Universe, M. Cazenave, ed. Oxford: Pergamon, 1984.

Schrödinger originally put forth his cat paradox in 1935 in the German journal Naturwissenschaften. An English translation appeared in 1955, coincidentally with the English version of von Neumann's book. An easily accessible source for the Schrödinger paper is the Wheeler and Zurek compendium already cited. For a full treatment of Wigner's views on the quantum reality issue, as well as his always insightful reflections on mathematics, physics, and their mutual dependence, see the collection of papers and essays

Wigner, E. Symmetries and Reflections. Bloomington, IN: Indiana University Press, 1967.

Wheeler has been a tireless campaigner for the measurement option view of quantum reality, having written numerous articles and books all hammering home the idea that observers have a choice in creating the kind of reality they see. Two good summaries of his ideas are given in

Wheeler, J. A. "Beyond the Black Hole," in Some Strangeness in the Proportion, H. Woolf, ed., pp. 341–375. Reading, MA: Addison-Wesley, 1980.

Wheeler, J. A. "How Come the Quantum?," Annals of the New York Academy of Sciences, Vol. 480, pp. 304–316. New York: New York Academy of Sciences, 1986.

A good account of an Earth-based Delayed-Choice Experiment using mirrors and light beams is found in the Squires book cited under General References.

Like Schrödinger, Heisenberg was much concerned with the philosophical implications of quantum theory, including his own Duplex Interpretation. In his later years, Heisenberg published a number of volumes of essays outlining his reflections on these and other matters. One of the best is

Heisenberg, W. *Physics and Beyond.* New York: Harper and Row, 1971.

Borges is far from the only writer who has mined the seemingly inexhaustible vein of "alternative realities" for material to entertain his readers. In my view, two of the best efforts in this direction by the sci-fi fraternity are

Hogan, J. *The Proteus Operation.* New York: Bantam, 1985.
Moore, W. *Bring the Jubilee.* New York: Fantasy House, 1952.

Both of these books deal with "what ifs," involving branches of the universe in which the Confederacy (for Moore) or the Nazis (for Hogan) won their respective wars. I won't spoil the reader's enjoyment by giving away the plots, other than to say that they both involve the usual twists of time travel to set things "straight" somehow. All in all, good fun.

On the side of sober science, the best source for Everett's work is the volume *The Many-Worlds Interpretation of Quantum Mechanics,* B. de Witt and N. Graham, eds. Princeton, NJ: Princeton University Press, 1973.

In addition to reprints of Everett's key papers, this volume also includes an assessment of the idea by Wheeler as well as a more introductory account by de Witt. For an account of Deutsch's version of the MWI, together with a discussion of an experiment that at least in principle would allow us to make contact with such worlds, see the BBC interview volume edited by Davies and Brown cited above.

THE DOGWORK REALITIES

Einstein's objections to the romantic realists have been chronicled in virtually every one of the thousands of accounts of his life and times. Generally these accounts introduce Einstein's naïve realist views by quoting his famous remark, "God does not play dice with the universe," or words to that effect. In my opinion the best statement of Einstein's thoughts is, of course, from Einstein himself. It is reported in his autobiography, which forms the first part of the volume

Albert Einstein: Philosopher-Scientist, Vol. 1, P. A. Schilpp, ed. Lasalle, IL: Open Court, 1949.

The quantum-logical explanation for the Three-Polarizer Paradox is well explained in the Herbert volume noted under General References. For a fine discussion of the entire quantum-logic idea using only a small amount of undergraduate mathematics, the reader is urged to consult

Gibbins, P. *Particles and Paradoxes: The Limits of Quantum Logic*. Cambridge: Cambridge University Press, 1987.

A completely nontechnical overview of the Quantum Potential Interpretation is provided by the editors in the introduction to the following volume of essays in honor of David Bohm upon his retirement. The introduction traces the development of Bohm's thinking on the matter from his first days in Princeton to his current ideas on the holographic universe. This account is followed by a slightly more technical discussion by Bohm himself, as well as a number of papers of varying degrees of difficulty by other heavies such as Bell, Feynman, and Finkelstein. All in all, a volume to be highly prized, praised, and perused:

Quantum Implications: Essays in Honor of David Bohm, B. Hiley and F. David Peat, eds. London: Routledge and Kegan Paul, 1987.

Many of Bohm's philosophical ideas underpinning the Quantum Potential Interpretation are covered in the book

Bohm, D. *Causality and Chance in Modern Physics*. Philadelphia: University of Pennsylvania Press, 1957.

For those interested in Bohm's current thinking about the "holographic universe," the following collections of interviews should prove illuminating:

Dialogues with Scientists and Sages, R. Weber, ed. London: Routledge and Kegan Paul, 1986.

The Holographic Paradigm, K. Wilber, ed. Boulder, CO: Shambhala, 1982.

A historical account of the origin of the quantum potential idea is given by de Broglie in

Broglie, L. de. "Interpretation of Quantum Mechanics by the Double Solution Theory." *Annales de la fondation Louis de Broglie,* 12 (1987), 399–421.

The original sources for the Absorber Theory are two papers by Wheeler and Feynman in *Reviews of Modern Physics* in 1945 and 1949. The modern incarnation of the theory according to Cramer is briefly described in the popular article

Cramer, J. "The Alternate View: The Quantum Handshake." *Analog Science Fact/Fiction,* November 1986.

More technical treatments are given in

Cramer, J. "An Overview of the Transactional Interpretation of Quantum Mechanics." *International Journal of Theoretical Physics,* 27 (1988), 227–236.

Cramer, J. "The Transactional Interpretation of Quantum Mechanics." *Reviews Modern Physics,* 58 (1986), 647–687.

THE BELL TOLLS FOR LOCALITY

I am indebted to Euan Squires's treatment in his book cited under General References for the idea of the Alexander and Anastasia telepathy experiment to illustrate the concepts behind Bell's Theorem. For another kind of story illustrating the same principles using flashing colored lights, see

Mermin, D. "Is the Moon Really There When Nobody Looks? Reality and the Quantum Theory." *Physics Today,* April 1985, pp. 38–47.

The famous EPR Paradox is described in virtually every introductory treatment of the quantum reality question, including all the volumes listed under General References. The original paper can be found in the Wheeler and Zurek collection noted above.

An especially good elementary account of the derivation of the Bell inequality using the colorful image of a nail gun instead of our electron-pair generator is given by Pagels in his book noted under General References. For a treatment by the master himself, see the original papers, which are reprinted both in the Wheeler and Zurek volume and in
Bell, J. S. *Speakable and Unspeakable in Quantum Mechanics.* Cambridge: Cambridge University Press, 1988.

Bell's recollections about the origin of his theorem, as well as his thoughts on Eastern religions, von Neumann's proof, and current trends in quantum reality, are all reported in
"Interview with John Bell," *Omni,* May 1988, 85ff.

A slightly technical but still eminently readable discussion of the Aspect experiments, Bell's Theorem, and the inviability of any local hidden-variable kind of reality is
Rohrlich, F. "Facing Quantum Mechanical Reality." *Science,* 221 (September 23, 1983), 1251–1255.

A popular-science account of the Bell result is given in
d'Espagnat, B. "The Quantum Theory and Reality." *Scientific American,* 241 (November 1979), 128–140.

Finally, a fairly technical reference addressing hidden variables and all the problems of quantum realism is
Redhead, M. *Incompleteness, Nonlocality, and Realism.* Oxford: Oxford University Press, 1987.

One special point worthy of note about the foregoing volume is its treatment of the so-called Kochen-Specher Paradox. The essence of this additional quantum paradox is that on the one hand, common sense (again!) would lead us to expect that the algebraic structure of the operators representing attributes should be mirrored in the algebraic structure of the set of attribute values themselves. But if this kind of "mirroring" holds, then, Kochen and Specher show, it is impossible to assign values to all attributes in all quantum states.

IN THE BEGINNING, THE VERY BEGINNING
An entertaining and informative account of the Wilson-Penzias discovery of the "whispers of the the cosmos" is given in the following treatment of the men and the science at Bell Labs:
Bernstein, J. *Three Degrees Above Zero.* New York: Scribners, 1984.

For an introductory treatment of those first few moments of the universe *after* the mysterious origin, there's no better source than
Weinberg, S. *The First Three Minutes.* New York: Basic, 1977.

Two very readable discussions of the Eddington-Dirac "numerology" discoveries are
Carr, B., and T. Rothman. "Coincidences in Nature and the Hunt for the Anthropic Principle," in *Frontiers of Modern Physics,* T. Rothman et al., eds., pp. 108–130. New York: Dover, 1985.
Rothman, T. "A 'What You See Is What You Beget' Theory." *Discover,* May 1981, pp. 90–99.

The definitive treatment of all aspects of the anthropic principles is
Barrow, J., and F. Tipler. *The Anthropic Cosmological Principle.* Oxford: Oxford University Press, 1986.

Many of the topics that have occupied us in the preceding chapters, including the origin of life, quantum reality, the existence of ETI, and much, much more, are examined in detail from the anthropic perspective in this seven hundred-page treatise. While the discussion may be a bit too technical for the general reader in places, there's so much material in this encyclopedic volume that everyone will find something to justify the cost of the book. It's truly a "don't miss it" kind of volume. Less breathtaking, but still excellent, accounts of anthropic ideas for the general reader are given in
Boslough, J. *Stephen Hawking's Universe.* New York: Morrow, 1985.
Gale, G. "The Anthropic Principle." *Scientific American,* December 1981, pp. 114–122.
Greenstein, G. *The Symbiotic Universe.* New York: Morrow, 1988.
Leslie, J. "Anthropic Principle, World Ensemble, Design." *American Philosophical Quarterly,* 19 (1982), 141–151.
Rees, M. "The Anthropic Universe," *New Scientist,* August 6, 1987, pp. 44–47.

Critics of anthropic reasoning have put forward a spectrum of reasons why such ideas have no place in *real* physics. Steven Weinberg, for example, says that "I certainly wouldn't give up attempts to make the anthropic principle unnecessary by finding a theoretical basis for the value of all the constants. It's worth trying, and we have to assume that we shall succeed, otherwise we surely shall fail." A somewhat less delicate critique is
Pagels, H. "A Cozy Cosmology." *The Sciences,* 25, No. 2 (1985), pp. 34–38.

In this article, Pagels notes that Dicke himself now thinks that the anthropic principles are worthless unless there was an element of arbitrariness in the origin of the universe. The argument is simple: If the values of the fundamental constants were fixed by the laws prevailing at the beginning, then the question of the origin of life was settled at the outset and the anthropic principles are unnecessary. But if there is some randomness in the way the constants are set, then Dicke thinks the anthropic-style reasoning may have some utility after all.

A thorough discussion of the quantum cosmology question is given in the Barrow and Tipler book noted above. For introductory accounts of how the universe could have arisen out of nothing more than a quantum fluctuation in the vacuum, see

Tryon, E. "Is the Universe a Vacuum Fluctuation?" *Nature,* 246 (1973), 396.

Vilenkin, A. "Creation of the Universe from Nothing." *Physics Letters,* B117 (1982), 25.

Padmanabhan, T. "Quantum Cosmology—Science of Genesis?" *New Scientist,* September 24, 1987, pp. 60–63.

Speculations from a scientific standpoint on the final state of the universe appear to be of rather recent vintage, one of the original papers being

Dyson, F. "Time Without End: Physics and Biology in an Open Universe." *Reviews of Modern Physics,* 51 (1979), 447.

A rather thorough discussion of this fascinating topic, emphasizing, of course, the anthropic perspective, is given in Barrow and Tipler.

INDEX

—

accommodation, 240
adaptation, 149, 154, 175, 177
adaptive traits, 149
addition, 271–272
Adler, Alfred, 397–398
advanced waves, 465
AI, 286–288, 495–496
  and antibehaviorism, 322–324
  bottom-up school of, 299–309
  connectionist school of, 309–314
  morality of, 325–328
  and phenomenology, 315–320
  strong, 286–288

top-down school of, 290–299
  weak, 286
Aleksander, Igor, 310
alien life forms, 394–397
alphabet, 268, 275
Altmann, Sydney, 90
altruism, 171–172, 195
amino acids, 75, 77, 84–85, 98, 356
  left-handed, 85
  right-handed, 85
analogy, 176
animal aggression, 156–158
anthropic principles, 479–484

anthropic principles (*cont.*)
  arguments against, 482–484
  final, 482, 487
  participatory, 482, 487
  strong, 481
  weak, 404, 479–480, 485
anticodon, 78
Arecibo message, 380–382
Aristotle, 16
  and logical deduction, 16, 18
  and nature of reality, 18, 39
  theory of causation, 130
  will of, 16
Armer, Paul, 316, 319
Arrhenius, Svante, 116
artificial intelligence, *see* AI
artificial plants, 515
ASCII code, 277
Aspect, Alain, 474
assimilation, 240
attributes, 417
  contextual, 442
  dynamic, 417, 440
  static, 417, 440
Austin Interpretation, 446–450
Automated Mathematician (program), 306–307
Axelrod, Robert, 200–203

Bacon, Francis, 19
  and principle of induction, 19
Ball, John, 386
  response to Fermi Paradox, 386
Barrow, John, 487
Bauer, Henry, 67
  critique of Velikovsky, 501
behavioral traits, 153
  genetically altruistic, 154
  genetically selfish, 153
  phenotypically altruistic, 153
  phenotypically selfish, 153
  and sexual selection, 163–165
behaviorism, 233–234, 237
  radical, 234, 236
belief systems, 62, 66–67, 125, 506–507
Bell, Jocelyn, 2–5
  and Nobel Prize, 5–6

Bell, John, 471, 476
Bell's Theorem, 471–474
Benacerraf, Paul, 321
Big Bang Theory, 476–477, 483
Bigelow, Julian, 39
Black, David, 348
Blinker, 133
Block, 133, 136
Bloom, Allan, 498
Bloomfield, Leonard, 214
Boas, Franz, 214
Bohm, David, 461–462, 464
Bohr, Niels, 441–443, 456–457
Boltzmann, Ludwig, 48–50
  suicide, 49–50
  and theory of heat, 49
Boltzmann machine, 310–313
Borges, Jorge Luis, 453–454
Boston Group, *see* Science for the People Sociobiology Study Group
Boyle's Law, 23–24
Bracewell, Ronald, 401
Broglie, Louis de, 462
Byurakan SETI meeting, 367, 387–388
  final resolution, 388

Calvin, Melvin, 73
carbon chauvinism, 363
Carter, Brandon, 403, 483–484
  arguments against ETI, 403–404
categorical imperative, 188
Cech, Thomas, 90
cell, 76
  cytoplasm, 76
  eukaryotic, 76, 101–102
    diagram of, 77
  nucleus, 76
  prokaryotic, 76
    fossil evidence for, 102
  reproduction process, 80–82
cellular parasitism, 101
Central Dogma of Molecular Biology, 82, 108, 147
Central Dogma of Social and Behavioral Biology, 147, 188, 518

Central Problem of Modern
    Linguistics, 216
Central Tenet of Human
    Sociobiology, 152
Chaitin, Gregory, 279–280
Chaitin's Theorem, 280
chess, 269
Chinese Room test, 265–267
chirality, 85
chloroplasts, 101
Chomskian theory of language,
    219–229
Chomsky, Noam, 218–219, 222
    debate with Piaget, 250–253
    and linguistics, 230
    and psychology, 230
    and sociobiology, 252–253
    views of Skinner's radical
        behaviorism, 236–237
chromosome, 76
Clauser, John, 474
Clever Hans, 210–211
coacervates, 96–98, 355
Cocconi, Giuseppe, 342
codon, 77, 102
coevolutionary circuit, 178–182
    diagram for, 180–181
Colby, Kenneth, 325, 327
Complementarity Principle, 442
complexity, 279–280
computer, 275–278
    as formal system, 278
    logical unit, 276
    memory unit, 276
    output unit, 276
    program, 276
    and souls, 329–330
    universal, *see* Turing machine
    viruses, 138–139
Comte, August, 31
    and evolution of knowledge, 31
conceptual dependency graph, 291
consciousness, 329, 444–445
continuously habitable zone, 351
controlled experiment, 20
Conway, J. H., 132
cooperative behavior, 199
Copenhagen Interpretation, 441–443
Copernican Principle, 493

Cox, Laurence, 401
Cramer, John, 466
creationism, 122–123
    and Arkansas Act 590, 123–124
Creation Research Society, 122, 124
Crick, Francis, 82, 115, 117
Cryer, 395–397
crystal growth, 111–113
crystallizer experiment, 115
culture, 359
    emergence of, 359–360
culturgen, 178, 195
Cut Theorem, 445, 549
Cygnans, 394–395

Dartmouth conference on AI,
    285–286
Darwin's Formula, 148
Dawkins, Richard, 151, 175, 195,
    204, 409
deep structure, 229
Delayed-Choice Experiment, 447–449
    diagram of, 448
Dennett, Daniel, 266, 324
Deutsch, David, 455
Dicke, Robert, 479–480
Dictionary Correspondence Theorem,
    434
Dirac, Paul, 478–479
dissipation, 486
distributive law of logic, 459
Dixon, Robert, 375
DNA, 76–77, 132, 406–407
    double-helix structure, 78
Dole, Stephen, 348
Domestic Bliss vs. He-man game,
    164–166, 183
double-slit experiment, 420–423
    with bullets, 420
    with electrons, 422–423
    with water waves, 420–421
Drake, Frank, 341–342, 375, 382, 387
Drake equation, 343–345
    estimates for N, 365
    statistical analysis of, 365–367
Dreyfus, Hubert, 315–317, 320, 333
    and RAND Corporation, 316–317
Dreyfus, Stuart, 316–317

Dual-Origin Theory (Double-Origin
    Hypothesis), 99–100, 102
Duplex Interpretation, 450–453
Dyson, Freeman, 47, 367–368, 372
    on philosophy of science, 47
Dyson sphere, 372–373

Eater, 133, 136
Eddington, Sir Arthur, 478
Eigen, Manfred, 409
Eigen experiment, 89, 104
Einstein, Albert, 415, 419, 456–
    457
electron spin, 431–432
ELIZA (program), 325–326
empirical laws, 22–23
entropy, 49, 304
environment, 148, 154
epigenetic rules, 179–181
EPR Paradox, 470
    Bohm version with electrons,
        470–471
equilibration, 240
error catastrophe, 94
ETI, 496–497
    direct contact via space travel,
        389–391
    direct visitation by, 391–397
    factorization arguments against,
        398, 405–409
    observation arguments against,
        399–404
Eurisko (program), 307
Everett, Hugh, III, 454
evolution, 92
    biological, 92, 353
    chemical, 92, 353
    convergent, 392–393
evolutionary game theory, 158–162
evolutionary stable strategy (ESS),
    160, 200–201
Extended General Theory, 244–245
extraterrestrial intelligence, *see* ETI

falsification (refutability), 33
Feigenbaum, Edward, 290, 319

Fermi Paradox, 340
Feyerabend, Paul, 37–38
    and scientific method, 37–38
    student experiences of, 504
Feynman, Richard P., 430
finite-state Markov process, 224
Finkelstein, David, 461
fitness, 149–150
    genetic, 150, 154
    inclusive, 167, 193
        in order Hymenoptera, 167–170,
            172
    maximization, 175
    phenotypic, 150, 154
Flanagan, Owen, 185
formal cause, 130–131
formalist program for mathematics,
    279, 533
Formal Mode, 130–131
Fox, Sidney, 97–98
frames, 297–299
Fundamental Question of the
    Philosophy of Science, 26

Gardner, Martin, 482
Gell-Mann, Murray, 48
gene, 76
    regulatory, 76
    structural, 76
gene inflation, 175
Gene-Protein Linkup Problem, 84,
    89
generative semantics, 245
genetic code, 77, 84
    second, 512
    table for, 79
    translation, 77–79
        diagram of, 81
genetic deterioration, 360–361
genetic determinism, 151
genome, 93
genotype, 147–148, 150, 154
Gish, Duane, 124
Glider Gun, 133, 136
    diagram of, 135
God, 487
    relationship with man, 507

Gödel, Kurt, 279, 281
  on thinking machines, 284
Gödel sentence, 281
Gödel's Theorems, 279–282, 285
  as arguments against AI, 320–322
Gold, Thomas, 387
  and pulsars, 500
Gould, Stephen Jay, 182, 187, 192, 206
grammar, 213, 253
  decidable, 249
  finite-state, 223–225
  of formal system, 269, 275
  generative, 215, 223
  Montague, 246–248
  phrase-structure, 223, 225–228
  transformational, 223, 228–229, 245
  universal, 215, 219–221, 223
Granger, Richard, 305–306
grassland spiders and ESS, 162–163
group selection, 156–157, 171
Guth, Alan, 486

Haldane, J.B.S., 70, 96, 166
Hamilton, William, 166
Harris, Zellig, 218
Hart, Michael, 351–352, 399–402, 406
  arguments against ETI, 400–402, 406–408
Hawk-Dove game, 158–160
Hawk-Dove-Indecisive game, 161
Hawking, Stephen, 446
Heisenberg, Werner, 450–453
Heisenberg Uncertainty Principle, 434–438, 440, 442, 548
heredity, 149
Hewish, Anthony, 3–5
hidden variables, 439–440, 457–458, 470–471, 474
Hilbert, David, 279–280
Hinton, Geoffrey, 310
Hofstadter, Douglas, 266, 300–301, 306
Horowitz, Paul, 376–377
Hoyle, Sir Fred, 5, 117–118
humanism, 329
hydrated electron, 510–511

hypercycle, 92–95
hypothetical ETI message, 543–544

Imitation Game, 261–265
incest avoidance, 176
incompleteness theorem, 279–280, 321
inconsistency theorem, 279
induction, 19
  problem of, 20, 30–33
inflationary universe theory, 486–487
information-processing machine, 254–255
Initial State Paradox, 485–487
instrumentalism, 25, 46
intelligence, 357
  probability of emergence of, 357–359
internal dynamics, 256
interpretation, 272
interpretive semantics, 245
investigator interference, 128
irrationalist, 46

Jumbo (program), 301–304
Jung, Carl, 411
  on alchemy, 411
"junk" DNA, 85–86, 91, 509
Just So stories, 177, 194, 196

Kalman, Rudolf, 25
Kammerer, Paul, 50
  and midwife toad, 50–51
  suicide, 51
Kardashev, Nikolai, 372
Kinetic Theory of Gases, 23–24
kin selection, 166–167, 171–172, 184, 195
  coefficient of, 167
Kitcher, Philip, 175
knowledge of language, 231
Kochen-Specher Paradox, 552
Kohlberg's theory of morals, 528
Kolmogorov, Andrei, 280
Kraus, John, 375

Kuhn, Thomas, 39
  and scientific paradigms, 39–40
  and Fivefold Way, 44–45
  compared with Popper and
    Lakatos, 45

Lakatos, Imré, 35
  and scientific research programs,
    35
Lamarckian inheritance, 82
Langton, Christopher, 137
language, 211–217
  common characteristics, 213
  context-free, 248
  context-sensitive, 248
  hierarchical structure, 247–248
  origins, 211–212
  Rule Problem, 530
  Syntax Problem, 530
  System Problem, 530
language acquisition, 216–217, 495
language competence, 231
law of effect, 236
Lenat, Douglas, 306
Levins, Richard, 187, 192
Lewontin, Richard, 187, 190–192
Liar's Paradox, 281
life, 74
  functional activities, 74, 137–138
  probability of, 353–357
Life game, 132–136
lifetime of civilization, 360–362
LINCOS (Lingua Cosmica), 542
linguistic determinism, 242
linguistic relativism, 242
linguistic research, 212–214
  empiricists (localists), 212, 214,
    231, 243
  rationalists (globalists), 212, 217,
    231
locality, 474
logical positivism, 27, 32
logical positivists, 31
Logic Theorist (program), 292–293
Lorenz, Konrad, 155–156
  theory of animal aggression, 156,
    158

Lucas, John, 320, 333
Lumsden, Charles J., 178
Lumsden-Wilson Thesis, 152–153
Lysenko, T. D., 70

McCarthy, John, 285, 327
McCarthyism, 461–462
McCracken, Daniel, 328
machine, *see* computer
Mahfouz, Naguib, 213
Many-Universes Theory, 485
Many-Worlds Interpretation,
    453–456
Margulis, Lynn, 101
material cause, 130
Material Mode, 130
mathematical system theory, 255–256
Mayr, Ernst, 407
Meaning Circuit, 450–451
means-end analysis, 292
mechanism, 329
meme, 195
mentalism, 256
mental modules, 251
mental representations, 254–256
Merton, Robert K., 51
  and norms of science, 51–52, 55
metabolism, 74, 96
microworlds, 295–296
Milgram, Stanley, 143–145
  teaching experiment, 143–145
Miller, Stanley, 70–71, 95, 99
Miller experiment, 71–73, 96–97, 100,
    103, 354
  diagram of, 72
Minsky, Marvin, 307–308
Mirror Hypothesis, 386
mitochondria, 101
model, 21–22, 64, 272, 338–339
  of cognitive processes, 268
  mathematical, 22, 24
modeling relationship, 338
Monod, Jacques, 250
Moore neighborhood, 132
morphemes, 213
Morris, Henry, 124
Morrison, Philip, 342, 387, 402

multiplier effect, 197
Mystery of the Quantum World, 423
mysticism, 329
myth, 16–17

naïve realism, 415, 457–458
  basic tenets of, 415
naked genies, 89
NASA SETI program, 375–377
  all-sky survey, 375–376
  targeted search, 376
neutron star, 2, 5–6
Newell, Allen, 290, 305
Newton, Isaac, 21
  and idea of mathematical model, 21
Newton's Second Law, 417–418
Niessert, U., 94–95
nondistributive lattices, 458
normal science, 42
Norms game, 202–203
nucleic acids, 75–76, 356
nucleotides, 75, 84
  bases, 75
    pairing rules, 76–77, 80
  nucleotide synthesis, 100–101

objectivity, 474
Ohio State SETI project, 378
Oparin, Alexander, 69–70, 96
Oparin-Haldane Hypothesis, 69. See
    also Primordial Soup Theory
operant behavior, 234
operations science, 129
organelles, 101
Orgel, Leslie, 90, 95, 99, 108
origin of life, 90, 493–494
  Cairns-Smith Clay Theory,
    109–114
  creationist view, 122–124
  Directed Panspermia Theory,
    116–117
  Dyson theory 105–107
  Eigen scenario, 91–92
    problems with, 93
  Fox's scenario, 98
    problems with, 99

Gilbert scenario, 90–91
  problems with, 93
Hoyle and Wickramasinghe
    Disease Theory, 119–120
Hoyle and Wickramasinghe
    Lifecloud Theory, 118–119
Oparin's scenario, 97
  problems with, 96–97, 99
Shapiro-Dyson scenario, 108
Shapiro theory, 102–104
origins science, 129
overstabilization, 361

Pagels, Heinz, 482
Pais, Abraham, 456
Panspermia Theory, 116
Papagiannis, Michael, 372
Papert, Seymour, 319
paradigm, 39–43, 64
paradigm shift, 42–44
parental investment, 164
parental manipulation, 172
Pavlov, Ivan, 233
Penzias, Arno, 476
peptide, 98
perceptron, 308–309
Petri Dish Hypothesis, 386
phenotype, 148, 150, 154
philosophy of science, 47
  comparison table, 47
phonemes, 213
phrase marker, 226
phrase structure rules, 225
physicalism, 256
Piaget, Jean, 237–239
  comparison with Skinner and
    Chomsky, 242
  and language acquisition, 241
Piagetian stages of mental
    development, 239–240
pilot wave, 462
Pioneer 10 plaque, 382
Planck, Max, 419
planetary bias, 363–364
Planetary Society, 377–378
planetary systems, 346–351
  double, 352

planetary systems (*cont.*)
  isolated, 350
  suitable for life, 351–353
polarization, 459
polymer chains, 354–355
Ponnamperuma, Cyril, 73
Popper, Sir Karl, 32–33
  vs. logical positivism, 34
population collapse catastrophe,
    94–95
potentia, 451–453
poverty of the stimulus, 216
primordial soup, 92
Primordial Soup Theory, 69, 100,
    127
  difficulties with, 127–128
Principle of Continuity, 492–493
Principle of Mediocrity, 341, 347,
    352, 366, 371, 403, 407, 411,
    483
Principle of Plentitude, 343
Prisoner's Dilemma game, 198–202
  computer tournament, 200–202
private events and language, 236
probability, 365
  subjective, 366
Problem of Auxiliary Hypotheses,
    33
Problem of Genetic Constraints,
    193–194, 196
Project Ozma, 341–342, 371
  dates for, 539
Project Sentinel, 377
proof sequence, 273, 275
proteinoid, 97–98
proteins, 75–76
protein structure, 84
Proxmire, William, 378–379
pseudoscience, 57–62
  hallmarks of, 57–59
pulsar, 5–6

quantum cosmology, 484–488
Quantum Description Problem,
    424–428, 430
Quantum Interpretation Problem,
    438–440

orthodox view, 438, 442
  reactionary view, 438–439
quantum logic, 458–461
Quantum Measurement Problem,
    432–434, 440, 443, 445, 453
quantum object, 440, 443
quantum potential, 462–464
Quantum Potential Interpretation,
    461–465
quantum reality, 497
quantum wave function, 425–428,
    440, 462
  collapse of, 431, 454–455, 464,
    466
quasi-species, 92

radio noise on Earth, 370
radiotelescope, 368–371
  Arecibo, 379
  frequency range, 369–371
  search direction, 371–373
  sensitivity, 371
Rapoport, Anatol, 201
rationalist, 46
rationality, 199
  collective, 199
  individual, 199
realism, 24, 46
reality, 417–419
  consciousness-created, 449
  contextual, 417
  Newtonian, 417–418, 429
  objective, 417
  observer-created, 449
reciprocal altruism, 173
recognition physics, 419
reducing mixture, 69, 71
reification, 194
relativism, 25, 46
replication, 74
  protein, 103
replicator, 148, 154
Rescher, Nicholas, 405
  argument against ETI, 405–406
retarded waves, 465
Rhine, Joseph B., 467
ribosome, 77

RNA, 76–77
  exon, 91
  messenger (mRNA), 76
  replication, 89–90
  self-catalytic, 90–91
  transfer (tRNA), 78
Rothman, Tony, 482–483
Rucker, Rudy, 328–329
rules of inference, 269, 275
Rumelhart, David, 310

Sagan, Carl, 377–378, 382, 387, 398
  and SETI program, 384
Sahlins, M., 195
Sampson, Geoffrey, 246
Sapir, Edward, 242
Sapir-Whorf Hypothesis, 242–243
Saussure, Ferdinand de, 214
Schank, Roger, 297
Schrödinger, Erwin, 424, 445
  quantum description, 424–429
Schrödinger's Cat, 445–446
Schuster, Peter, 94, 409
Schwartz, Barry, 193, 207
science, 11
  ideology of, 13–14, 56
  logical structure of, 13
  public conceptions of, 11
  as social activity, 52–56
science and religion, 62–66
  comparison table, 65
  differences between, 64, 124–125
  possible reconciliations, 65–66
Science for the People Sociobiology
    Study Group, 187–190, 197,
    204, 206
scientific method, 13, 46
scientific research programs, 35–36
  hard core of, 35
  negative heuristic of, 35
  positive heuristic of, 35
  protective belt of, 35
scientific theory, 23–24
  criteria for, 129
scientism, 67
Scopes trial, 121–122
Scrabble, 269–270

scripts, *see* frames
Searle, John, 288, 322, 334–335
selection, 149, 176
  natural, 150, 154
selfish gene, 175–176
selfish RNA, 94–95
self-reference, 335
self-repair, 74
self-reproducing automaton, 131
  requirements for, 131–132
semantic network, 291
semantics, 213–214
Shapere, Dudley, 44
Shapiro, Robert, 102
Shklovskii, I. S., 373, 376
short-circuit catastrophe, 94–95
SHRDLU (program), 295–296
Simon, Herbert, 246, 290, 299
Simpson, George Gaylord, 407
simulation, 338–339
  of cognitive processes, 267
Skinner, B. F., 232–236
  and language acquisition, 236
  and pigeon guidance system, 235
Skinner box, 235
Smith, John Maynard, 158, 201
social behavior, 146
  human, 146
social Darwinism, 187
Society of Mind, 307–308
sociobiology, 494–495
  and animal behavior, 170
  and falsification, 196
  and morals, 185–186
  and Prisoner's Dilemma game,
    202–203
  political objections to, 186–192
  and religion, 184–185
  scientific objections to, 177–178,
    192–198
  and sexism, 182–184
Solzhenitsyn, Alexander, 206
Spectral Area Theorem, 435, 437
Spencer, Herbert, 187
Spiegelman experiment, 88–89
Spiegelman monster, 88
Spielberg, Steven, 377
spontaneous generation, 508

stars, 345
  binary system, 346
  G-type, 346, 363, 371, 373
  rate of formation of, 345–346
star-type chauvinism, 363
state, 283–284
  brain, 283–284, 331
  machine, 283–284, 287, 331
  mental (cognitive), 283–284, 287, 331
Steady-State Theory, 483
Steiner, George, 243
stimulus-response behavior, 232
stimulus-response pattern, 254, 256
strategy, 160
  evolutionary stable (ESS), 160
  uninvadable, 160
Strategy of Sociobiology, 154
string, 268, 275
  admissible, 269
  provable, 273–274
structural linguistics, 218
Sturrock, Peter, 366–367
Summerlin, William T., 53
  and patchwork mouse, 53–54
superluminal signaling, 465, 471
  arguments against, 475–476
surface structure, 229
Sutherland, N. S., 328
syntax, 214, 222–223
synthetases, 88, 93, 102
system, 268
  complete, 274–275, 279
  consistent, 274–275, 279
  external description, 254–255
  formal, 268–273, 279–281, 288
  internal description, 254–255
  states, 256

Tarter, Jill, 375
Teilhard de Chardin, Pierre, 487
theorem, 271, 273, 275
Theory of Relativity, 418–419
  General, 419
  Special, 418, 463
Three-Coin Problem, 292–293
Three-Polarizer Paradox, 459–460
Tinbergen, Niko, 207

Tipler, Frank, 401–403, 487
  dispute with Sagan, 402
TIT FOR TAT, 201–202
toolmaking, 357–358
Transactional Interpretation, 465–467
transcription, 77
transformational rules, 228
Trivers, Robert, 173
truth, 274
  formalizable, 278
  logical, 274
Tryon, Ed, 486
Turing, Alan, 264–265, 288
Turing-Church Thesis, 278, 285
Turing machine, 278–279, 285, 288
Turing Test, see Imitation Game
twenty-questions game, 416
Types I, II, and III civilizations, 372–373

UFO Hypothesis, 386
Urey, Harold, 71, 99

variation, 149
Velikovsky, Immanuel, 7–9, 59, 61
  comparison with Bell and Hewish, 9
  and Worlds in Collision
    controversy, 7–9
Verification Principle, 27, 32
Verschuur, Gerritt, 374
Vienna Circle, 27, 29
von Neumann, John, 39, 131, 288–290, 442–445
  on thinking machines, 289–290
von Neumann probe, 401
Voyager probes videodisk, 382–385

Watchmaker Parable, 246–247, 409
waterhole frequency, 370, 376
Watson, John B., 232–233
waveform family, 425–428, 434, 437, 440
  conjugate, 434
Weizenbaum, Joseph, 325–328, 332

Wheeler, John A., 416, 419, 447, 482

Wheeler-Feynman absorber theory, 465

Whitley, C. H., 321

Whole Environment Evolution Synthesizer (WEES), 86–87

Whorf, Benjamin, 241–243

Wickramasinghe, Chandra, 126

Wigner, Eugene, 445

Wigner's Friend, 445–446

Wilensky, Robert, 319

Wilson, Edward O., 174, 187–190

Wilson, Robert, 476

Wilson's Ladder, 175

Winograd, Terry, 295–296

Wittgenstein, Ludwig, 27–30
    and logical structure of language, 27–29
    and picture theory of language, 31

words, 213

WOW signal, 375

Zoo Hypothesis, 386

Grateful acknowledgment is made to the following individuals and publishers for permission to reproduce material used in creating the figures in this book. Every effort has been made to locate the copyright holders of material used here. Omissions brought to our attention will be corrected in future editions.

Cambridge University Press for Figures 1.1, 7.1, 7.2, and 7.4, which are reproduced from T. Hey and P. Walters, *The Quantum Universe;* Figure 2.7, which is reproduced from F. Dyson, *Origins of Life;* and Figure 2.8, which is reproduced from A. Cairns-Smith, *Seven Clues to the Origin of Life.*

Harper & Row for Figure 1.3, which is reproduced from I. Barbour, *Myths, Models, and Paradigms.*

Transworld Publishers for Figure 2.3, which is reproduced from J. Gribbin, *In Search of the Double Helix.*

Basil Blackwell, Ltd., for Figures 2.2, 2.5, and 2.9, which are reproduced from A. Scott, *The Creation of Life,* and for Figure 5.1, which is reproduced from *Mindwaves,* C. Blakemore and S. Greenfield, eds.

Basic Books for Figure 2.4, which is reproduced from D. Hofstadter, *Metamagical Themas: Questing for the Essence of Mind and Pattern.*

Reidel Publishing Co. for Figure 2.6, which is reproduced from N. Lahav, "The Synthesis of Primitive 'Living Forms': Definitions, Goals, Strategies and Evolution Synthesizers," *Origins of Life,* 16 (1985–86), 129–149.

Elsevier Science Publishing Co. for Figure 3.2, which is reproduced from D. Barash, *Sociobiology and Behavior.*

W. H. Freeman and Company for Figure 3.3, which is reproduced from J. Maynard Smith, "The Evolution of Behavior," *Scientific American,* September 1978.

Harvard University Press for figure of coevolutionary circuit in the "To Dig Deeper" section for Chapter Three, which is reproduced from C. Lumsden and E. O. Wilson, *Genes, Minds, and Culture.*

MIT Press for Figure 4.4, which is reproduced from B. Whorf, *Language, Thought, and Reality;* Figure 4.5, which is reproduced from D. Lightfoot, *The Language Lottery;* and Figures 6.4 and 6.5, which are reproduced from *Communication with Extraterrestrial Intelligence,* C. Sagan, ed.

Routledge and Kegan Paul, Limited, for Figure 4.6, which is reproduced from F. von Schilcher and N. Tennant, *Philosophy, Evolution and Human Nature.*

Atheneum for the poem "Coiled Alitarine" from J. Hollander, *The Night Mirror,* 1971.

Houghton-Mifflin Co. for Figure 5.2, which is reproduced from R. Rucker, *Mind Tools,* illustration by the Design Group, Nancy Blackwell, Susan Micklem and Sarah Micklem.

Petrocelli Books, Inc., for Figure 5.3, which is reproduced from P. Jackson, *Introduction to Artificial Intelligence.*

Academic Press, Inc., for Figure 5.4, which is reproduced from T. Winograd, *Understanding Natural Language;* Figures 6.1 and 6.2, which are reproduced from S. Dole, *Icarus,* 13 (1970), 500–504; and Figure 6.6, which is reproduced from G. Verschurr, *Icarus,* 19 (1973), 329.

Michael Arbib for Figure 5.6, which is reproduced from M. Arbib, *Brains, Machines, and Mathematics,* McGraw-Hill.

National Radio Astronomy Observatory for Figures 6.3 and 6.8, which are reproduced from *The Search for Extraterrestrial Intelligence,* K. Kellerman and G. Seielstad, eds.

The Ohio State University Radio Observatory for Figure 6.7.

Prentice-Hall, Inc., for Figure 6.12, which is reproduced from J. Baugher, *On Civilized Stars.*

Windward Press, Ltd., for Figures 6.12 and 6.13, which are reproduced from *Barlowe's Guide to Extraterrestrials,* W. Barlowe and I. Summers, eds.

Professor Frank Drake for the figure depicting the solution to the alien message shown in the "To Dig Deeper" section for Chapter Six.

Doubleday & Co. for Figures 7.5, 7.7, and 7.9, which are reproduced from N. Herbert, *Quantum Reality,* 1985.

John A. Wheeler for Figure 7.8, which is reproduced from J. A. Wheeler, "How Come the Quantum?," *Annals of the NY Academy of Sciences,* Vol. 480, 1986.

Justin Leiber for Figure 4.3, which is reproduced from his book *Noam Chomsky,* New York: St. Martin's Press, 1975.

THE AUTHOR

John L. Casti completed a Ph.D. in Mathematics from the University of Southern California in 1970. Following tours of duty at The RAND Corporation and the University of Arizona, he left the USA in 1974 to take up a post as one of the first research staff members of the International Insitute for Applied Systems Analysis (IIASA) in Vienna, Austria, where he worked on problems of system modeling and applied systems analysis. In the autumn of 1986 he joined the faculty of the Technical University of Vienna.

His current research interests center about the development of a coherent theoretical framework for naturally incorporating the characteristic features of living systems, self-repair and replication, into the standard Newtonian framework generally used to model natural phenomena. He currently divides his time between the United States and Europe, where he is engaged in preparing a book on the circle of questions surrounding problems of uncertainty, randomness, prediction and explanation in modern science.